Handbook of Experimental Pharmacology

Volume 124/II

Springer
Berlin
Heidelberg
New York
Barcelona
Budapest
Hong Kong
London
Milan
Paris
Santa Clara
Singapore
Tokyo

Drug Toxicity in Embryonic Development II

Advances in Understanding Mechanisms of Birth Defects: Mechanistic Understanding of Human Developmental Toxicants

Contributors

B.D. Abbott, M. Barr, B. Bielec, B.R. Danielsson, G.P. Daston
L. Earl Gray, Jr., R.H. Finnell, M. Fujinaga, W.R. Kelce, S. Klug
D.M. Kochhar, H. Nau, D. Neubert, R. Neubert, R.M. Pauli
J.M. Rogers, R. Stahlmann, W.S. Webster

Editors

R.J. Kavlock and G.P. Daston

Springer

ROBERT J. KAVLOCK, Ph.D.
Director, Reproductive Toxicology Division (MD-71)
National Health and Environmental
Effects Research Laboratory,
US Environmental Protection Agency
Research Triangle Park, NC 27711
USA

GEORGE P. DASTON, Ph.D.
The Procter and Gamble Company
Miami Valley Laboratories
P.O. Box 538707
Cincinnati, OH 53239
USA

With 58 Figures and 59 Tables

ISBN-13:978-3-642-64409-2 Springer-Verlag Berlin Heidelberg New York

Library of Congress Cataloging-in-Publication Data. Drug toxicity in embryonic development I: advances in understanding mechanisms of birth defects/editors, R.J. Kavlock and G.P. Daston. p. cm. – (Handbook of experimental pharmacology: vol. 124/I-II) Includes bibliographical references and index. Contents: V. I. Morphogenesis and processes at risk – v. II. Mechanistic understanding of human developmental toxicants.

ISBN-13:978-3-642-64409-2 e-ISBN-13:978-3-642-60447-8
DOI: 10.1007/978-3-642-60447-8

2. Fetus–Effect of drugs on. 3. Abnormalities. Human–Etiology. I. Kavlock, Robert J. II. Daston, George P. III. Series. QP905.H3 vol. 124/I-II [RA1238] 615.1 s–dc20 [616.043] 96-24471

Softcover reprint of the hardcover 1st edition 1997

Cover design: Design & Production GmbH, Heidelberg

Typesetting: Scientific Publishing Services (P) Ltd, Madras

SPIN: 10538704 27/3136/SPS – 5 4 3 2 1 0 – Printed on acid-free paper

Dedicated to
Dr. Casimer T. Grabowski, our mentor, who encouraged us to pursue careers in understanding the mechanisms by which environmental agents might cause birth defects and who gave us the foundation to be successful scientists

Preface

Having received the invitation from Springer-Verlag to produce a volume on drug-induced birth defects for the *Handbook of Experimental Pharmacology*, we asked ourselves what new approach could we offer that would capture the state of the science and bring a new synthesis of the information on this topic to the world's literature. We chose a three-pronged approach, centered around those particular drugs for which we have a relatively well established basis for understanding how they exert their unwanted effects on the human embryo. We then supplemented this information with a series of reviews of critical biological processes involved in the established normal developmental patterns, with emphasis on what happens to the embryo when the processes are perturbed by experimental means. Knowing that the search for mechanisms in teratology has often been inhibited by the lack of understanding of how normal development proceeds, we also included chapters describing the amazing new discoveries related to the molecular control of normal morphogenesis for several organ systems in the hope that experimental toxicologists and molecular biologists will begin to better appreciate each others questions and progress.

Several times during the last two years of developing outlines, issuing invitations, reviewing chapters, and cajoling belated contributors, we have wondered whether we made the correct decision to undertake this effort. However, now that we can look back on the finished product, we are confident that our initial goal was achieved and that this volume of the *Handbook* fills a void in the literature. In particular, we feel the volume complements the excellent descriptive compendia of the development effects of chemicals currently available by its focus on assessing mechanisms and modes of action. We are very pleased to have had the pleasure to have worked with the contributing authors, and hope that they forgive us for the "can you embellish on this....can you illustrate the process....can you condense this section....have you thought about this issue..." questions that we asked of them in the review process. Already we know we have some gaps in the coverage, but believe we have accomplished a synergism in combining the three sections as we did. Hopefully the work will stand the test of time similar to Jim Wilson's

Environment and Birth Defects, which although published in 1973, is still given to new students and trainees as a starting point for how to begin assessing the potential hazards and risks of chemicals to the developing embryo.

Research Triangle Park, NC, USA
Cincinnati, OH, USA
March 1996

ROBERT J. KAVLOCK
GEORGE P. DASTON

List of Contributors

ABBOTT, B.D., Reproductive Toxicology Division (MD-67), National Health Effects and Environmental Research Laboratory, U.S. Environmental Protection Agency, Research Triangle Park, NC 27711, USA

BARR, M., Jr., Pediatric Genetics, Teratology Unit, D1109 MPB 0718, University of Michigan Medical Center, Ann Arbor, MI 48109–0718, USA

BIELEC, B., Department of Veterinary Anatomy and Public Health, College of Veterinary Medicine, Texas A&M University, College Station, TX 77843–4458, USA

DANIELSSON, B.R., Astra Safety Assessment, 151 85 Södertälje, Sweden, and Department of Toxicology, Uppsala University, 75124 Uppsala, Sweden

DASTON, G.P., The Procter and Gamble Company, Miami Valley Laboratories, P. O. Box 538707, Cincinnati, OH 53239, USA

EARL GRAY, L., Jr., Endocrinology Branch (MD-72), Reproductive Toxicology Division, National Health and Environmental Effects Research Laboratory, U.S. Environmental Protection Agency, Research Triangle Park, NC 27711, USA

FINNELL, R.H., Department of Veterinary Anatomy and Public Health, College of Veterinary Medicine, Texas A&M University, College Station, TX 77843–4458, USA

FUJINAGA, M., Department of Anesthesia, Stanford University School of Medicine, Stanford, CA 94304, and Anesthesiology Service, Palo Alto V.A. Medical Center, Palo Alto, CA 94304
Mailing address: 3801 Miranda Avenue, 112A, Palo Alto, CA 94304, USA

KELCE, W.R., Endocrinology Branch (MD-72), Reproductive Toxicology Division, National Health and Environmental Effects Research Laboratory, U.S. Environmental Protection Agency, Research Triangle Park, NC 27711, USA

Klug, S., Free University of Berlin, University Medical Center Benjamin Franklin, Fachbereich Humanmedizin, Garystr. 5, 14195 Berlin, Germany

Kochhar, D.M., Department of Pathology, Anatomy and Cell Biology, Thomas Jefferson University, 1020 Locust Street, Philadelphia, PA 19107, USA

Nau, H., Free University of Berlin, University Medical Center Rudolf Virchow, Standort Charlottenburg, Institute of Toxicology and Embryopharmacology (WE18) Garystr. 5, 14195 Berlin, Germany

Neubert, D., Free University of Berlin, University Medical Center Benjamin Franklin, Institute of Toxicology and Embryopharmacology, Garystr. 5, 14195 Berlin, Germany

Neubert, R., Children's Hospital, Kaiserin Auguste-Victoria House, University Medical Center Benjamin Franklin, Free University of Berlin, Garystr. 5, 14195 Berlin, Germany

Pauli, R.M., Departments of Pediatrics and Medical Genetics, University of Wisconsin-Madison, Clinical Genetics Center #353, 1500 Highland Avenue, Madison, WI 53705, USA

Rogers, J.M., Perinatal Toxicology Branch (MD-67), Developmental Toxicology Division, Health Effects Research Laboratory, US Environmental Protection Agency, Research Triangle Park, NC 27711, USA

Stahlmann, R., Free University of Berlin, University Medical Center Benjamin Franklin, Fachbereich Humanmedizin, Garystr. 5, 14195 Berlin, Germany

Webster, W.S., Department of Anatomy and Histology, University of Sydney, Sydney, N.S.W. 2006, Australia

Contents

Section III: Pathogenesis and Mechanisms of Drug Toxicity in Development

CHAPTER 23

Anticonvulsant Drugs: Mechanisms and Pathogenesis of Teratogenicity
R.H. Finnell, B. Bielec, and H. Nau 121

CHAPTER 24

Cardiovascular Active Drugs

CHAPTER 25

Anticoagulants

CHAPTER 26

Antiviral Agents

Contents of Companion Volume 124/I

Section III
Pathogenesis and Mechanisms of Drug Toxicity in Development

CHAPTER 21

Retinoids

D.M. KOCHHAR

A. Introduction

Natural and synthetic compounds possessing a chemical structure and functional properties similar to vitamin A are called retinoids. Vitamin A is an essential micronutrient for humans because it is not synthesized de novo in the body. The normal diet provides it either as provitamin A (or carotenoids), found in plants and vegetables, or as preformed vitamin A, found in animal products (BENDICH and LANGSETH 1989). Vitamin A was discovered almost 75 years ago by MCCOLLUM and DAVIS (1913) as a lipid-soluble growth factor present in certain foods. Further studies showed that vitamin A not only maintained growth, but also prevented xerophthalmia and night blindness. Vitamin A-deficient children are anemic and grow poorly, have more infections, and are more likely to die than their peers (SOMMER 1989). There is overwhelming evidence that vitamin A, presumably through its metabolite retinoic acid (RA), participates in organogenesis at several stages and sites during normal embryonic development. The offspring of female rats fed vitamin A-deficient diets prior to and during gestation show numerous malformations in the eyes, the genitourinary system, the diaphragm, and, less frequently, the heart (WILSON and WARKANY 1950).

Examples of natural (retinol, all-*trans* RA, 13-*cis* RA, 4-oxo-all-*trans* RA, 9-cis RA) and several typical synthetic retinoids are listed with their chemical structures in Table 1. Besides the important role of retinol and RA as micronutrients in growth and development, the retinoids are now viewed as drugs for treatment of oncologic and dermatologic diseases. History does not record whether COHLAN (1953) was surprised at his findings that hypervitaminosis A was teratogenic in rats. To date, while there is only indirect evidence that vitamin A deficiency compromises the ability of a pregnant female in maintaining and delivering a normal baby, there is indisputable evidence that ingestion of vitamin A-related therapeutics during pregnancy results in fetal malformations (ROSA et al. 1986; ANONYMOUS, 1987). A very high percentage of human babies exposed in utero to 13-*cis* RA show a spectrum of anomalies which only partially overlap with the vitamin A deficiency syndrome (VADS). Malformations of the face and other craniofacial derivatives, hindbrain derivatives, and heart are typical components of human retinoid embryopathy (LAMMER et al. 1985). The circumstance that both deficiency and an excess of

Table 1. Structures and names or identifying code numbers of natural and representative synthetic retinoids

Serial number	Structure	Name or code number
1	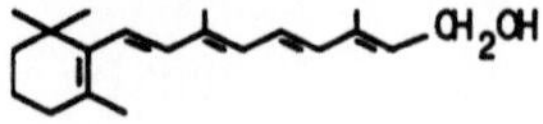	Retinol (vitamin A)
2	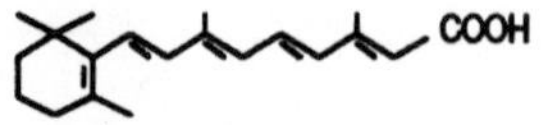	All-*trans* RA (tretinoin)
3	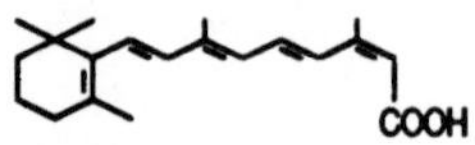	13-*cis* RA (isotretinoin)
4	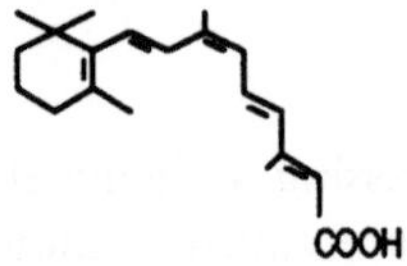	9-*cis* RA
5	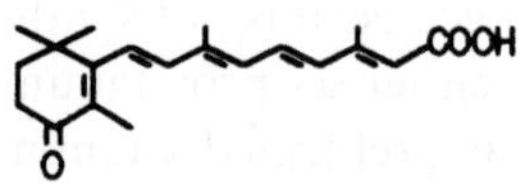	4-oxo-all-trans RA
6	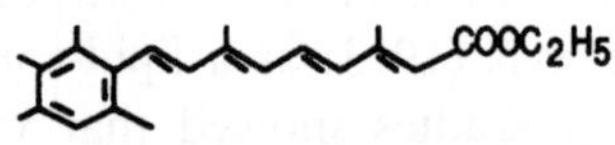	Etretinate
7	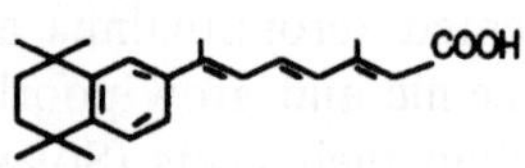	Ro13-6307
8	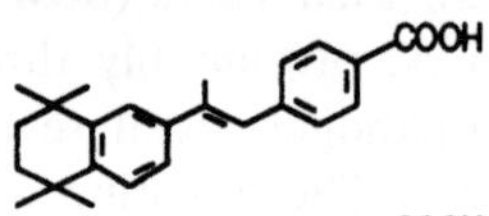	TTNPB
9	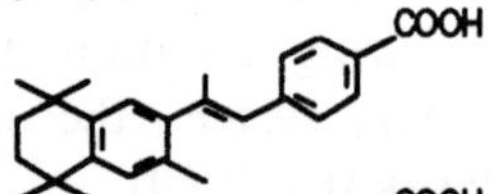	3-methyl TTNPB
10	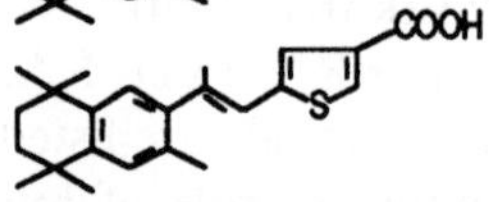	AGN 191701
11	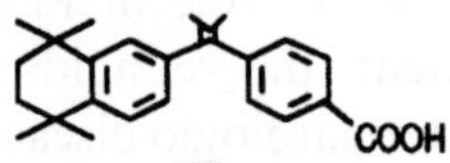	SR 11217
12	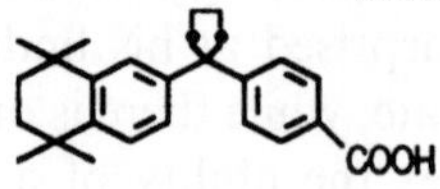	SR11237
13	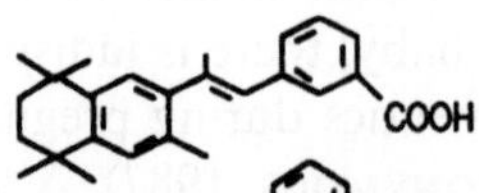	AGN 192240
14	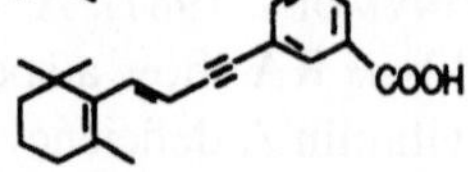	AGN 190727

RA, retinoic acid; TTNPB, 4-[2-(5,6,7,8-tetramethyl-2-naphthalenyl) propen-1-ly] benzoic acid;

vitamin A are teratogenic has stimulated research not only in understanding the mode of teratogenic action, but also in unraveling the role natural retinoids may play in directing the developmental events of the normal embryo (HOFMANN and EICHELE 1994; NAU et al. 1994). Attention has recently focused on a series of endogenous proteins which serve as nuclear receptors for natural retinoids as a means to discover how retinoids intervene in diverse cellular functions and which of their cellular and molecular targets are crucial to the developing embryo (MANGELSDORF et al. 1994; CHAMBON, 1994; LINNEY and LAMANTIA 1994).

In this chapter we summarize information on the types of congenital malformations associated with vitamin A deficiency or excess in laboratory animals and in humans. We also discuss various lines of evidence for and against the involvement of retinoid receptors as mediators of retinoid-induced teratogenesis. The major objective of the whole exercise is to deduce pathogenetic mechanisms initiated by retinoids in the embryo which result in the cascade of events leading to fetal malformation.

B. Vitamin A Deficiency in Development

The reports by Warkany and colleagues in the 1940s showed convincingly that dietary deficiencies in pregnant animals could adversely affect their developing embryos and result in malformations (see WILSON 1977). Earlier, HALE (1933) had observed the birth of piglets without eyeballs in a vitamin A-deficient sow. Although severely deficient animals are infertile, less severe deficiency in various species (rat, pig, rabbit, monkey) produces a spectrum of fetal malformations affecting the eyes, brain, heart, blood vessels, and urogenital system (WALLINGFORD and UNDERWOOD 1986; PALLUDAN 1976). In view of the paucity of definitive human studies, the results of a study on pregnant rhesus monkeys should be noted (O'TOOLE et al. 1974). Only four out of ten vitamin A-deficient females who mated successfully carried their pregnancies to term, and these four infants had various eye anomalies, including congenital or delayed xerophthalmia, cataract, and ciliary body hyperplasia. Two of these four infants also had metaplasia of tracheal/esophageal epithelia, and one had mild hydronephrosis. It is noteworthy that other gross anatomic defects seen in fetuses of deficient rats and pigs were not encountered in the monkey infants.

There is insufficient evidence to show that eye and other anomalies in human babies can occur in maternal vitamin A deficiency (WALLINGFORD and UNDERWOOD 1986). Xerophthalmic women and those suffering from night blindness, apparently due to vitamin A deficiency, have reportedly had children afflicted with congenital xerophthalmia, anophthalmia, microphthalmia, and other eye defects. Evidence that the deficiency was the direct and the only cause of these anomalies remains to be established.

Evidence for a direct role of vitamin A in embryogenesis came from the studies by WILSON et al. (1953) in which they were able to ameliorate maldevelopment in the rat pups by supplementation of the maternal diet with vitamin A. Wilson et al. (1953) discovered that the most vulnerable period of rat gestation to vitamin A-deficiency coincided with active stages of organogenesis; administration of vitamin A to the deficient rats at progressively earlier times during pregnancy resulted in a progressive reduction in offspring affected by malformation. Although these experiments were well conducted, nutritional studies alone did not resolve the question of whether or not some component malformations of the VADS were due to some maternal dysfunction secondary to the state of vitamin A deficiency. Attempts have recently been made to address this question in transgenic mice by observing developmental outcome after blocking vitamin A function through molecular inactivation of retinoid receptors.

C. Retinoid Receptors

Within the last few years tremendous advances have been made in the understanding of the mechanism of action of RA at the molecular level. Recently, several nuclear receptors which bind retinoids have been cloned (Mangelsdorf et al. 1994). There are two classes of receptors, termed RA receptors (RAR) and retinoid X receptors (RXR). Each class is represented by three genes, termed α, β, and γ. The sequences of RAR α, -β, and -γ are more homologous to each other than to the RXR genes. Further diversity in each receptor family occurs through the use of multiple promoters and alternative splicing to generate receptor isoforms that differ at the N terminus. Thus each of the RAR has at least two major isoforms (Mangelsdorf et al. 1994; Ruberte et al. 1991b). Analysis of the deduced amino acid sequence has demonstrated that these receptors are transcription factors and are quite homologous to the steroid/thyroid hormone receptors (Fig. 1). Each contains a highly conserved DNA-binding domain (domain C), a well-conserved ligand-binding domain (domain E), and four additional domains which are not as well conserved (domains A, B, D, and F). The isoforms of each receptor type differ in the amino terminal domains A and B, but domains C-F are identical. The RAR have a high-affinity binding site for both all-*trans* RA, and 9-*cis* RA, while RXR bind only 9-*cis* RA (Levin et al. 1992; Heyman et al. 1992). In addition, transactivation assays and retinoid-binding assays with these receptors have shown that each displays a different affinity for RA and a number of RA analogues (Crettaz et al. 1990; J.M. Lehmann et al. 1992; Jong et al. 1993; Allegretto et al. 1993). Finally, a unique pattern of expression of each RAR and RXR along with several of the isoforms has been reported in the developing embryo and the adult (Dolle et al. 1989b, 1990; Mendelsohn et al. 1991, 1992; Ruberte et al. 1993). Taken together, these data suggest that each of the RAR and RXR may have unique functions within specific tissue during development and in the adult.

The RAR and RXR have each been demonstrated to bind RA and to alter the rate of transcription of specific genes via a RA response element (RARE)

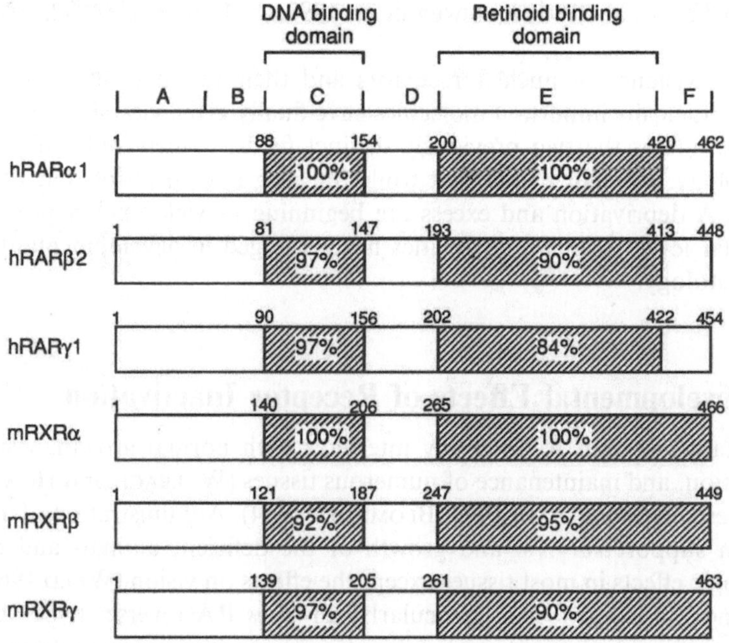

Fig. 1. Amino acid sequences of the three retinoic acid receptors (*RAR*; from human) and the three retinoid X receptors (*RXR* from mouse). The number of amino acids in each receptor and the positions of amino acids corresponding to the boundaries between different regions (*A–F*) are shown. The percentage identity in the DNA-binding domains and in the ligand-binding domains within the RAR subfamily (relative to RARα1) and within RXR subfamily (relative to RXRα) are indicated. (Adapted from CHAMBON 1994)

or retinoid X response element (RXRE) within the promoter of target genes (MANGELSDORF et al. 1994). Two distinct pathways for transcription activation of responsive genes by retinoid have been identified. In the first pathway, one of the RAR and one of the RXR bind as heterodimers to specific half-sites in the promoters of the target genes in such a way that the RAR component is located downstream at the binding site. All-*trans* RA and certain synthetic retinoids, e.g., 4-[2-(5,6,7,8-tetramethyl-2-naphthalenyl) propen-1-ly] benzoic acid (TTNPB), require such heterodimers (RXR/RAR) to activate transcription. The other pathway utilizes a different type of binding site which permits RXR to bind as homodimers (RXR/RXR); certain synthetic retinoids preferentially activate transcription through such a mechanism. Interestingly, 9-*cis* RA and certain other synthetic retinoids are able to bind and activate both receptor pathways (KUROKAWA et al. 1994). In a later section, we describe differences in teratogenic potency that we have observed among a number of receptor-selective synthetic retinoids. Furthermore, RXR are capable of forming heterodimers with other nuclear receptors such as thyroid hormone receptor, vitamin D receptor and peroxisome proliferator activator receptor (YU et al. 1991; KLIEWER et al. 1992a,b; LEID et al. 1992; ZHANG et al. 1992a,b).

The existence of nuclear receptors and their putative interactions with other biologically important molecules have finally given vitamin A an identity which straddles the two previously distinct fields, namely nutrition and endocrinology. It is fair to say that while the long known pleiotropic effects of vitamin A deprivation and excess are beginning to yield explanations at the molecular level, further complexities have emerged in developmental biology and teratology.

D. Developmental Effects of Receptor Inactivation

Postnatally, vitamin A deficiency interferes with normal growth, vision, reproduction, and maintenance of numerous tissues (WOLBACH and HOWE 1925; see reviews in SPORN et al. 1994; BLOMHOFF 1994). Administration of all-*trans* RA can support survival and growth of the deficient animals and prevent pathologic effects in most tissues except the effects on vision (WALD 1968; VAN PELT and DE ROOIJ 1991). Particularly, all-*trans* RA reverses hyperkeratosis of the skin and the widespread keratinzing squamous metaplasia of the visceral epithelia which are classic symptoms of vitamin A deficiency. For these reasons and because all-*trans* RA and 9-*cis* RA are the only known endogenous ligands recognized by the nuclear retinoid receptors, current opinion holds that these RA isomers perform most of the natural functions of vitamin A.

It is uncertain at present whether the RA isomers alone can ameliorate all of the detrimental effects of vitamin A deficiency in embryonic development. Pregnancy in severely vitamin A-deficient rats can be sustained by supplementation with RA, but animals maintained on RA alone usually resorb their conceptuses during late gestation, and this is preceded by histological abnormalities in the placenta (THOMPSON et al. 1964; NOBACK and TAKAHASHI 1978). This indicates that retinol itself or one of its metabolites, other than RA, is indispensable for some functions in certain tissues.

To gain further information on the role of retinoids in development, investigators have now turned to another strategy in lieu of the older approach of nutritional deprivation. This approach has involved the use of transgenic mouse embryos which are genetically deficient in one or more of the retinoid receptors. Since a complete absence of all retinoid receptors in the same animal would be expected to be embryolethal, investigators have used two main approaches to disrupt receptor function in mice.

In one approach exemplified by SAITOU et al. (1995) and IMAKADO et al. (1995), functions mediated by RAR in transgenic mice are interrupted through an expression of dominant negative mutant proteins which dimerize with RXR but are transcriptionally inactive. To avoid early embryolethality, the authors used keratin gene promoters to target expression in the developing epidermis in a stage-specific manner. Dramatic effects were found in the skin of newborn mice, which was thin, fragile, dry, and shiny with delayed hair formation. These mutant animals died within 24 h after birth, presumably due to an aberrant barrier function of the affected skin. Histological and immunolocalization studies showed that keratinocyte differentiation and maturation were profoundly suppressed. Gross examination revealed no other phenotypic changes in organogenesis, which is not surprising since the functional disruption of the receptors was targeted to the epidermis. This strategy should be of great value in examining the developmental role of retinoids in other systems (ANDERSEN and ROSENFELD 1995).

In the second approach, reviewed by CHAMBON (1994), investigators have used gene "knockout" approach through directed mutagenesis via homologous recombination to create null mutant mice for individual RAR and RXRα. Surprisingly, newborn mice homozygous null for RARα1 isoform, RARβ2/β4 isoforms, or RARγ2 isoform did not exhibit any obvious congenital anomalies, and the mice were healthy and fertile (LOHNES et al. 1993; LUFKIN et al. 1993; MENDELSOHN et al. 1994b). The lack of any defective development in such null mutants raised the likelihood that perhaps gene disruptions abolishing the expression of all isoforms of the receptor were necessary for obtaining altered phenotypes. This has turned out to be true, but only partially.

It has been observed that RARα gene is ubiquitously expressed in the mouse embryo. It was therefore unexpected that RARα null homozygotes underwent virtually uninterrupted organogenesis until birth; these mutants were, however, seriously impaired, since they suffered early postnatal mortality and the males showed patchy degeneration of the seminiferous tubules in the testes (LUFKIN et al. 1993). In contrast to the ubiquitous distribution of RARα, RARγ transcripts show a more restricted distribution in the embryo, being localized in squamous epithelia and in precartilagenous mesenchymal condensations of the limbs and other sites (DOLLE et al. 1989b, 1990; RUBERTE et al. 1990). Again, RARγ null mutants showed early postnatal mortality and male sterility, but harbored only a few minor congenital defects, which were much milder than what would be expected from the spatiotemporal distribution of this receptor in the embryo (LOHNES et al. 1993). For instance, no limb malformations occurred even though RARγ is known to be uniformly expressed during early morphogenesis and thereafter becoming restricted to precartilage condensations (DOLLE et al. 1989b; RUBERTE et al. 1990). In spite of the disappointingly modest phenotypic changes associated with RARα and RARγ null genotypes, which mimicked at best only a small spectrum of the congenital anomalies in vitamin A-deficient animals, one aspect that provided an encouraging basis for further work was early death in these animals. Newborn mice of either genotype were often cannibalized by their dams; survivors exhibited a slower growth rate, became emaciated, and eventually suffered premature death. This sequence of events is reminiscent of the animals raised on a vitamin A-deficient diet, suggesting that the organism requires a functional vitamin A signaling system, including receptors, to maintain normal homeostasis and survival.

Although the overall spatiotemporal pattern of expression in the developing embryo is unique to each retinoid receptor, some tissues express more than one receptor during organogenesis (RUBERTE et al. 1991a). One can envisage a functional overlap between receptors which permits one receptor type to maintain normal development in the absence of expression of the knocked-out receptor in mutant embryos. The validity of this suggestion of redundancy of multiple retinoid receptors is borne out in further studies on double null mutants where disruptions of receptor genes in different combinations eventually recapitulated almost all of the congenital anomalies of the fetal VADS, including defects in the eye and the cardiovascular, urogenital, and the respiratory systems (CHAMBON 1994; LOHNES et al. 1994; MENDELSOHN et al. 1994a).

Even if these null mutants only partially answered the question of how vitamin A participates in normal development, the studies yielded a few important lessons. First, it could now be argued convincingly that the biological functions of vitamin A are indeed mediated by the retinoid receptors. Second, it could be suggested that the mere presence of a transcript in a cell or tissue was not sufficient proof that the product was actually needed at that site (CHAMBON 1994). For instance, no limb malformations occurred in any of the single or double null mutants even when each of the RARα, -β, and -γ are expressed in that tissue. Third, the most intriguing discovery recently reported concerned the developmental function of RXRα; RXRα null mutant embryos died in utero between 13–16 days of gestation and showed heart and eye malformations (SUCOV et al. 1994; KASTNER et al. 1994). These observations strongly suggested that RXRα/RAR heterodimers mediate retinoid signaling in vivo.

In addition to nuclear receptors, two cytoplasmic RA-binding proteins (CRABP I and CRABP II) had been identified in all vertebrates even before the existence of the nuclear receptors were discovered; they have long been implicated in the control of RA function at its sites of action within the target cells (ONG and CHYTIL 1975; NAPOLI 1994; FIORELLA and NAPOLI 1994; MADEN 1993; DENCKER et al. 1990; BOYLAN and GUDAS 1991). The results of gene disruptions of each of these proteins in single or even double mouse mutants were again unexpected; both CRABP I and CRABP II knockout mice were viable, healthy, and normal in every way except that a number of fetuses showed an extra digital segment in their autopods (GORRY et al. 1994; LAMPRON et al. 1995; FAWCETT et al. 1995). Thus both binding proteins appeared to be dispensable for mouse development and adult life.

E. Retinoid-Induced Teratogenesis

I. Hypervitaminosis A

Maternal hypervitaminosis A during pregnancy is teratogenic in several animal species, and evidence from several sources indicates that fetal malformations are due to the direct action of vitamin A or its derivatives on the embryo rather than the result of any of the toxicologic effects on the mother.

COHLAN (1953, 1954) reported teratogenic effects of maternal hypervitaminosis A in pregnant rats. Healthy rats on a normal diet were given orally a preparation of "natural vitamin A," presumably vitamin A esters, beginning 2–4 days after mating and ending at gestational day (GD) 16 (duration of gestation in the rat is 22 days). A number of exposed neonates presented such malformations as exencephaly, cleft lip and/or cleft palate, brachygnathia, and various eye defects. GIROUD and MARTINET (1959) extended these findings by showing that restricting dosing of rats only to a few days of gestation altered the pattern of anomalies which were stage dependent. In addition to anencephaly, facial dysmorphia, spina bifida, syndactyly, and encephalocele, these authors also described histological abnormalities in the teeth, kidneys, and ureters of rat pups. To further refine stage dependency of malformation, KALTER (1960) gave pregnant inbred mice a single oral dose of vitamin A palmitate (10 000 IU per dose) between GD 7 and 12 (birth occurs on the 19th day). A large number of embryos resorbed after treatment on GD 7–9. The survivors showed a specific cluster of malformations depending on the day of exposure (KALTER and WARKANY 1961). Anomalies produced affected the skin, ears, eyes, face, mouth, teeth, tongue, palate, thymus, ribs, brain, spinal cord, heart, great vessels, kidney, ureter, bladder, genital ducts, rectum, anus, tail, and limbs.

Subsequent studies have shown that embryos of nearly all species of experimental animals are susceptible to maternal hypervitaminosis A (GEELEN 1978). It was later shown that all-*trans* RA was a much more potent teratogen than retinol and that the whole spectrum of congenital malformations due to maternal hypervitaminosis A was mimicked when all-*trans* RA was used alone in dosing the pregnant animals (KOCHHAR 1967; SHENEFELT 1972). Since retinyl esters and retinol are capable of conversion to all-*trans* RA, the latter has been presumed to be the derivative most responsible for vitamin A-induced teratogenesis.

Unequivocal evidence that vitamin A is teratogenic in humans comes only from the experience with 13-*cis* RA, to which a large number of women became inadvertently exposed during pregnancy, resulting in a fetal malformation syndrome now known as retinoid embryopathy. This rather definitive evidence has made it possible to sort out reports from among a group of suspected case histories numbering less than two dozen where the involvement of hypervitaminosis A in fetal malformations could be ascertained with some degree of certainty.

Epidemiological studies undertaken to estimate the risk of birth defects due to hypervitaminosis A have so far been inconclusive, mainly because a large sample size is required in view of the infrequent occurrence of vitamin A overdose (VALLET et al. 1985; WERLER et al. 1989). Using data from the Spanish hospital-based, case control registry, MARTINEZ-FRIAS and SALVADOR (1990) found no statistically significant overall association between exposure to vitamin A and birth defects. The authors cautioned, however, that the data tended to suggest that some risk of teratogenic effect may exist for maternal exposure to 40 000 IU or higher.

II. Retinoic Acid and Retinoic Acid Isomers

1. Laboratory Animals

All-*trans* RA is consistently referred to in the literature as a "potent" teratogen. The reason for this is that it is capable of eliciting a teratogenic response in all animal species tested so far, and this occurs at a dose which rarely produces ill effects in the dam. Even a brief exposure during the period of organogenesis is sufficient to cause terata in virtually 100% of the fetuses of mice, rats, rabbits, and hamsters (Geelen 1978; Willhite 1990; Agnish and Kochhar 1993; Kochhar and Satre 1993).

Two generalizations regarding teratogenicity have emerged from a large number of experimental studies (Agnish and Kochhar 1993; Shenefelt 1972; Kochhar 1973; Kistler 1981; Satre and Kochhar 1989):

1. The teratogenic response is dependent on the developmental stage. Exposure of the early postimplantation embryo (GD 8–10 in the rat) frequently results in craniofacial and overt central nervous system (CNS) defects. Exposure occurring on GD 12–14 often is associated with limb and genitourinary defects.
2. The teratogenic response is dose dependent. Higher doses not only increase the frequency and severity of defects, they also cause embryolethality. One needs relatively low doses during early phases and higher doses during later phases of organogenesis for the teratogenic response. During the critical developmental stage of a given organ, the teratogenic response is proportional to the amount of all-*trans* RA localized in the embryo.

All-*trans* RA yields several metabolites in the body through isomerization and oxidation, e.g., 13-*cis* RA, 9-*cis* RA, and 4-oxo RA; many of these are also teratogenic (Fig. 2; Satre et al. 1989; Willhite 1990; Kochhar et al. 1984, 1995). Since teratogenic features of 13-*cis* RA are important for comparison with its effects on the human embryo, this isomer of all-*trans* RA deserves special mention.

All therapeutic agents meant for human use are first tested for teratogenic activity in laboratory animals. In the established protocol, the dams are dosed daily during the postimplantation period so that not only the susceptibility of all aspects of organogenesis are included in the screening, but cumulative effects of the agent are also revealed. In contrast to the multiple-dose protocol, studies that aim to uncover the cellular and molecular basis of action of the teratogen prefer to employ single-dose regimens. Both types of studies have shown that 13-*cis* RA produces developmental effects in embryos similar to those from all-*trans* RA, but its potency in certain species, e.g., rats and mice,

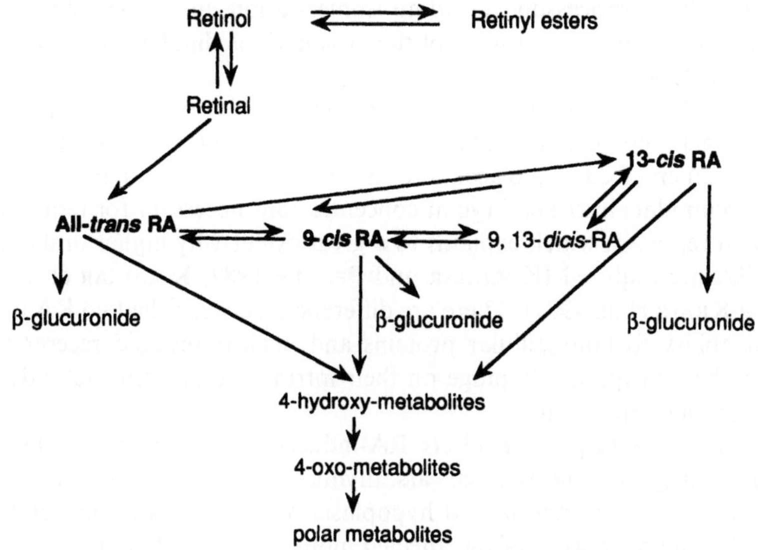

Fig. 2. Cellular pathways for the generation and metabolism of retinoic acid (*RA*) isomers

is lower than all-*trans* RA. In pregnant rats, while a daily dose of up 50 mg/kg given on GD 7–15 produced no fetal effects, doses higher than 100 mg/kg spared none of the exposed fetuses (AGNISH and KOCHHAR 1993). Cleft palate, pinna defects, exencephaly, microcephaly, spina bifida, and open eyes were the most frequently observed anomalies. Using the same protocol, a daily dose of 15 mg all-*trans* RA/kg affected all litters, and in most of the litters 65%–100% of the fetuses were malformed (AGNISH and KOCHHAR 1993). When lowest teratogenic doses of 13-*cis* RA and all-*trans* RA under the multiple-dose regiments are compared, it becomes clear that rats and mice require much higher doses of 13-*cis* RA than all-*trans* RA for equivalent teratogenic effects (AGNISH and KOCHHAR, 1993; KISTLER 1987; HUMMLER et al. 1990; HENDRICKX and HUMMLER 1992).

In monkeys, however, 13-*cis* RA appears to be slightly more active as a teratogen than all-*trans* RA. HUMMLER et al. (1990) administered 2.5 mg 13-*cis* RA/kg daily on GD 10–25 (followed by the same dose given twice daily on GD 26 and 27) to pregnant cynomolgus monkeys (*Macaca fascicularis*) and found that all fetuses ($n = 7$) were malformed. The lowest teratogenic dose of all-*trans* RA in the cynomolgus monkey turned out to be in the range of 5–10 mg/kg, although the gestational days of exposure were not entirely the same as in the 13-*cis* RA study (HENDRICKX and HUMMLER 1992).

Because the maternal doses of 13-*cis* RA at which the human embryo has responded with malformations range from 0.5 to 1.5 mg/kg, it follows that humans are more sensitive to this retinoid than are cynomolgus monkeys (LAMMER et al. 1985). Differences between species in bioavailability after oral doses, metabolic processing including isomerization, and other pharmacokinetic parameters are among some of the explanations that have been advanced (NAU et al. 1994).

Some effort has been made in previous studies to address the equally puzzling phenomenon in rodents of a large disparity in teratogenic doses of the two RA isomers. Reduced placental transfer in mice and the inability of 13-*cis* RA to accumulate in the embryo at concentrations necessary for teratogenesis have been reported, which explains to a great extent why higher oral doses of 13-*cis* RA are required (KOCHHAR and PENNER 1987; KOCHHAR et al. 1987; CREECH-KRAFT et al. 1987). There are differences between the two RA isomers in their ability to bind cellular proteins and nuclear retinoid receptors, but whether these properties impinge on their intrinsic teratogenic activities is at present an open question.

The primary targets of 13-*cis* RA-induced embryopathy in the cynomolgus monkey were the external ears, thymus, heart, and brain (HUMMLER et al. 1990). Microtia or anotia and hypoplasia of the vermis in the cerebellum constituted the phenotype of the affected monkey fetuses. Like the 13-*cis* RA-treated fetuses, the monkey fetuses exposed to all-*trans* RA had external ear and temporal bones defects, but they showed no thymus or heart anomalies (HENDRICKX and HUMMLER 1992). All-*trans* RA-treated fetuses also had other craniofacial anomalies not seen in 13-*cis* RA-treated fetuses, such as mandibular hypoplasia and cleft palate, and showed no thymus or heart anomalies (HENDRICKX and HUMMLER 1992). These differences between phenotypes may be ascribed to the small sample size inevitable in primate studies as well as to differences in the protocols followed in the two studies (HUMMLER et al. 1990; HENDRICKX and HUMMLER 1992). However, the concordance seen between the phenotypes of human infants exposed prenatally to 13-*cis* RA and of the cynomolgus monkeys in these and other studies is striking (WILSON 1974; FANTEL et al. 1977; HENDRICKX et al. 1980). The high incidence of embryolethality in monkeys is also similar to results in humans; LAMMER et al. (1985) reported a 20% abortion rate in women who took 13-*cis* RA within the first 10 weeks of pregnancy.

2. Humans

13-*cis* RA (isotretinoin) was marketed under the trade name Accutane in 1982 for treatment of severe recalcitrant cystic acne, and in 1983 the first cases of birth defects and instances of spontaneous abortions among drug recipients came to light (ROSA 1993).

By April 1990, more than 40 spontaneous abortions and 87 instances of birth defect outcomes were attributable to 13-*cis* RA. A majority of these occurrences were in the United States and very few were reported in other countries. The pattern of dysmorphogenesis included one or more components of malformations in the craniofacial complex and cardiovascular systems and the CNS, and these occurred at dosages which were within the therapeutic range of 0.4–1.5 mg/kg daily (Rosa et al. 1986; Rosa 1993; Lammer et al. 1985; Braun et al. 1984; Benke 1984; Lott 1983; Cruz 1984; Hill 1984; Fernhoff and Lammer 1984; Lammer and Opitz 1986; Anonymous 1991; Rizzo et al. 1991).

The most striking feature of the syndrome is the absence or reduced size of the external ears and canals, which occurred in 62 out of 87 cases. Other craniofacial anomalies were facial asymmetry and facial palsy, micrognathia with cleft palate, and maldevelopment of the cranial bones. Thymus was reduced or absent in seven cases.

The most frequent occurrence of defects was in the CNS, including hydrocephalus, cerebellar hypoplasia, absence of vermis, and structural malformation of the cerebral cortex. Lammer and Armstrong (1991) concluded that various CNS defects arise from maldevelopment of derivatives of the rhombencephalic alar plate. Preliminary findings indicate that 52% of 5-year-olds exposed during the first trimester show intellectual deficits, and these children included many (37.5%) who showed no structural CNS malformations at birth (Adams and Lammer 1991, 1993). It is not known at present but it is probable that some neurobehavioral deficits may be encountered in children exposed to 13-*cis* RA during the fetal period, since the period of CNS development and maturation extends beyond the first trimester.

Cardiovascular defects occurred less frequently (38 out of 87 cases) and were a less characteristic feature of the retinoid syndrome than the other defects described above. These defects could be classified as conotruncal or branchial arch tissue defects (Lammer and Opitz 1986). Frequently observed anomalies were transposition of the great vessels, tetralogy of Fallot, truncus arteriosus communis, ventricular septal defects, interrupted aortic arch, retroesophageal right subclavian artery, and hypoplasia of the aorta (Lammer and Opitz 1986).

Although limb defects were not encountered among the cases documented earlier, Rizzo et al. (1991) recently described a child and a fetus who showed unusual limb reduction defects in addition to other components of the retinoid syndrome.

Lammer et al. (1988) have estimated the risk for fetal malformations following maternal exposure to 13-*cis* RA during the first trimester. In this prospective study of 59 pregnancies, nine resulted in spontaneous abortions, one in a malformed live birth, and 37 in live birth without apparent evidence of major malformations; this represented a risk of 23% for fetuses that reach 20 weeks of gestation (Lammer et al. 1988).

It may be instructive to consider why so many pregnant women were exposed, despite warnings of the teratogenic hazard of 13-*cis* RA. Cosmetic considerations allowed a more widespread prescription of a very effective drug intended only for recalcitrant acne. The failure of contraceptive devices in use during treatment with 13-*cis* RA accounted for a full one third of exposures (Lammer et al. 1985). Some other instances of exposure could have been prevented by various means, e.g., better communication between the dermatologist and the obstetrician and between the physician and the patient, and knowledge of contraceptive methods (ROSA 1993).

All-*trans* RA has also been used for a number of years under the trade name Retin-A for the treatment of acne as a cream for skin application only. Its use has become more widespread since 1988 following reports suggesting its effectiveness in preventing wrinkles and ameliorating photodamage. Sporadic reports have appeared in the literature of a teratogenic outcome associated with the topical use of all-*trans* RA, but chance occurrences rather than causal relationships are usually indicated (ROSA 1993; AGNISH and KOCHHAR 1993; CAMERA and PREGLIASCO 1992; JICK et al. 1993). Experimental studies suggest a relatively low teratogenic potential of topically applied all-*trans* RA, mainly because only limited concentrations of the drug can be externally applied, and further attenuation in the maternal circulation restricts embryonic exposure (ZBINDEN 1975; WILLHITE et al. 1990; NAU 1993). However, an adverse outcome in human embryos resulting from topical synthetic retinoids remains a possible concern.

III. Synthetic Retinoids

1. Laboratory Animals

Innumerable chemical modifications have been introduced into either end of the retinol molecule and into the side chain to generate synthetic retinoids so as to assess and compare their biological and pharmacologic properties (Table 1). Since teratogenic effects are one of the most serious side effects, a number of the synthetic retinoids have been tested in pregnant animals (WILLHITE 1990; KOCHHAR et al. 1987). In the teratology protocol used by us, pregnant mice are given a single oral dose on GD 11, followed by morphologic evaluation of near-term fetuses; incidences of resorption, growth retardation, palatal cleft, and limb reduction defects are consistent and reliable end points with which to compare relative teratogenic potencies of synthetic retinoids (KOCHHAR 1973; KOCHHAR et al. 1984; 1987). In this bioassay, 13-*cis* RA is about four fold less active as a teratogen than all-*trans* RA, but is equal in potency to retinol. Etretinate, an aromatic analogue which has been marketed under the trade name Tigason for the treatment of psoriasis, is twice as potent as all-*trans* RA. Interestingly, certain chemical modification of the molecule makes the resultant compound a supremely potent teratogen; Ro 13–6307, which differs from all-*trans* RA in having an aromatic ring inserted in its side chain next to the modified original ring, is 40-fold more active than all-*trans* RA as a teratogen (Table 1; KOCHHAR and PENNER 1992). By far the most active teratogens are benzoic acid derivatives of all-*trans* RA, termed arotinoids, e.g., TTNPB, which is 700-fold more potent that all-*trans* RA (FLANAGAN et al. 1987; KISTLER et al. 1990; KOCHHAR et al. 1987; ZIMMERMANN et al. 1985). The probable reasons for such vast differences in teratogenic potencies are discussed below.

It has emerged from structure–activity correlations that an acidic end group and a side chain of sufficient length are necessary for a retinoid to act as a teratogen; while a ring is required at one end of the molecule for activity, the retinoid can still be an effective teratogen with changes in the ring structure that alter its lipid solubility (WILLHITE 1990).

2. Humans

Etretinate, an aromatic retinoid with an ethyl ester end group, marketed under the trade name Tigason (Tegison outside the United States, marketed by Sautier Laboratories, Geneva), was reported to physicians by the manufacturer in 1983 to have caused four malformed cases among patients under treatment for psoriasis. Three other cases were reported in subsequent publications (HAPPLE et al. 1984; ROSA 1993). ROSA (1993) concluded from a summary of a total of 12 birth outcomes ascribed to etretinate that the anomalies showed dissimilarities to the 13-*cis* RA syndrome. Limb anomalies occurred more frequently (six cases), while ear, cardiac, and thymus defects were less common. In view of the small number of cases and the bias in reporting only the abnormal outcomes after etretinate therapy, risk estimates are not available.

LAMMER (1988) reported an unusual occurrence of a malformed baby who was conceived 11 months after discontinuation of etretinate therapy where the defects resembled the retinoid syndrome. Although others have questioned whether etretinate was really the causative agent in this case (BLAKE and WYSE 1988), it is known that daily dosing with etretinate results in its accumulation in the body, and detectable concentrations of etretinate and its metabolite acitretin have been detected in the circulation of patients even 1 year after cessation of treatment (PARAVICINI et al. 1981; MASSARELLA et al. 1985). There are other reports of fetal malformations where conception occurred several months after cessation of etretinate therapy (GROTE et al. 1985).

It is particularly disturbing that teratogenic effects occurred even when very low levels of the retinoid were detected in the circulation. If these reports are confirmed, etretinate would seem to be a very potent teratogen in humans. In view of its long half-life in the body, it has been recommended that women patients contemplating pregnancy avoid the use of this drug for at least 1, if not 2 years prior to conception (WARD et al. 1983; BOLLAG 1985). In the United States, the package insert for the drug warns that "the period of time during which pregnancy must be avoided after treatment is concluded has not been determined." Because of the shorter half-life, acitretin, an analogue of etretinate without the ethyl ester end group, is currently considered a more desirable drug than etretinate. Acitretin (under the trade name Soriatane) was recently approved by the Food and Drug Administration (FDA) for marketing in the United States. This drug is already available in several European countries.

F. Pathogenesis

Research into the cause of drug-induced congenital anomalies has identified the vulnerability of a number of developmental events that may be disrupted. Some of the events especially implicated in retinoid-induced teratogenesis are growth misregulation and ectopic cell death, disruption in tissue interactions and pattern formation, and interference in the cell differentiation processes. Many of these events are not mutually exclusive and may, in fact, be interdependent. As the major organ systems that are components of the retinoid syndrome, e.g., face, ear, brain, heart, aortic arches, and thymus, receive contributions from the neural crest cells, the latter may be a special target (Lammer et al. 1985; Anonymous 1991). Endogenous RA has been suggested to participate in organogenetic pattern formation in normal vertebrate embryos through transcriptional control of important developmental genes such as the homeobox (*Hox*) genes, and this action is most likely mediated by its nuclear receptors (for reviews, see Brockes 1989; Tabin 1991). This has strengthened the notion that an excess of RA may result in maldevelopment through an ectopic or chaotic expression of genes which disrupt pattern formation. If this turns out to be true, it is reasonable to suggest that most of the cellular changes that have been observed to date in retinoid-exposed embryos such as cell division, cell differentiation, cell interactions, migrations, and cell death could all emanate from a subtle disruption induced at the inception of organogenesis in the pattern of expression of developmental control genes (Langman and Welch 1967; Kochhar 1968, 1977).

I. Receptor Involvement in Teratogenesis

An examination of the teratogenic potencies of retinoids shows that some retinoids are much more potent than RA, while others are less active. As discussed above, the lower teratogenic potency of 13-*cis* RA relative to all-*trans* RA in rodents is known to be due to its metabolism and its pharmacokinetic behavior, which differ from those of all-*trans* RA (Creech-Kraft et al. 1987). Such metabolic/pharmacokinetic features vary among species, which is now known to be the reason why in primates 13-*cis* RA is equally, if not even more potent than all-*trans* RA (Nau et al. 1994). On the other hand, high teratogenic potency of some of the synthetic retinoids appears to be intrinsic rather than exclusively due to metabolic stability or to pharmacokinetic factors in the maternal/placental/embryonic compartments (Howard et al. 1989; Willhite 1990; Kochhar and Penner 1992). The likely intrinsic factors are cytoplasmic retinoid-binding proteins and, more importantly, the nuclear retinoid receptors. The precise function of binding proteins remains unresolved (Gorry et al. 1994; Lampron et al. 1995; Fawcett et al. 1995), but the receptors are indispensable in normal embryogenesis, as discussed above.

The assumption that the receptors, which function as transcription factors, are also involved in retinoid-induced teratogenesis implies that fetal anomalies are the result of alterations in the activity of genes responsive to the receptors. The role of retinoid receptors in teratogenesis needs to be explored further in order to understand the pathways of pathogenesis.

1. Retinoic Acid Receptors Versus Retinoid X Receptors

The identification of RAR and RAX as two distinct classes of nuclear receptors which mediate the biological/pharmacologic responses of retinoids has led to investigations into the role of the two response pathways in teratogenesis. Several recent studies have raised the likelihood that synthetic retinoids could be designed with limited or exclusive preference for binding and activation of RAR and/or RXR (BEARD et al. 1994; JONG et al. 1993; LEHMANN et al. 1992; BOEHM et al. 1994). The availability of a few such compounds, which are structural analogues of TTNPB, gave us the opportunity to seek correlations between their teratogenicity and their distinctive gene transactivational properties (JIANG et al. 1995).

The receptor activation assay which we employed involves cotransfection of HeLa cells with a chimeric receptor gene construct and an estrogen receptor-responsive reporter gene that cannot be activated by endogenous retinoid receptors usually present in mammalian cells. The molar concentration at which transactivation by each of the analogues was half-maximal (EC_{50}) is shown in Table 2. Several investigators have previously reported receptor activation activities of TTNPB and other natural and synthetic retinoids using other cell types as well as different gene constructs and reporter genes. Our results confirmed previous conclusions that TTNPB and all-*trans* RA bind and

Table 2. Comparison of molar concentrations of retinoids needed for receptor activation in transfected HeLa cells (JIANG et al. 1995)

Retinoid	Serial number[a]	Receptor activation EC_{50} (n*M*)			
		RAR			RXR
		α	β	γ	α
All-trans RA	2	5.0	1.5	0.5	NA
9-cis RA	4	102	3.3	6.0	13.0
TTNPB	8	21	4.0	2.4	NA
3-methyl TTNPB	9	4580	74	152	385
AGN 191701	10	> 1000	> 1000	> 1000	201
SR 11217	11	NA	NA	NA	11.8
SR 11237	12	NA	NA	NA	16.8
AGN 192240	13	NA	NA	NA	NA
AGN 190727	14	NA	NA	NA	NA

EC_{50}, 50% maximal activation; NA, not active (i.e., $EC_{50} > 10^4$ n*M*); RA, retinoic acid; RAR, RA receptor; RXR, retinoid X receptor; TTNPB, 4-[2-(5,6,7,8-tetramethyl-2-naphthalenyl) propen-1-ly] benzoic acid.
[a](see Table 1).

transactivate only the RAR subtypes, while 3-methyl TTNPB and 9-*cis* RA transactivate both the RAR and RXR subfamilies (LEHMANN et al. 1992; BOEHM et al. 1994; MANGELSDORF et al. 1990). While two of the five compounds in this initial study activated neither RAR nor RXRα, three compounds (serial numbers 10–12; see Table 1) were exclusive RXRα activators (Table 2). AGN 191701 (compound no.10) was unique among these RXR agonists since it did have some RAR-activating property at higher concentrations.

The in vivo teratogenicity of retinoid analogues was evaluated in pregnant mice which were given a single oral dose on GD 11 (KOCHHAR et al. 1984; JIANG et al. 1995). The results of these studies are summarized in Table 3. TTNPB was by far the most teratogenic of the compounds studied: all fetuses surviving after a maternal dose of 0.1 mg/kg were severely malformed (Table 3). These fetuses had severe craniofacial and limb defects of the type documented previously (Fig. 3) (KOCHHAR 1973, 1987). 3-Methyl TTNPB was over 100 times less teratogenic than TTNPB, consistent with differential effects of the two compounds in RAR transactivation. AGN 191701 produced no developmental effects at 1 or 10 mg/kg doses, and at 100 mg/kg it produced cleft palates in 35% of the exposed fetuses and mild limb defects in 53% of the fetuses. Thus AGN 191701 was over 1000 times less teratogenic than TTNPB, again reflecting their differential effects in RAR transactivation. As mentioned, AGN 191701 was a potent activator of RXRα, yet it was also a weak activator of RARβ and RARγ ; the latter property was considered by us to produce a teratogenic response, but only when administered to mice at high doses. The four compounds which were either RXR agonists (e.g., compounds

Table 3. Teratogenic Potencies of Retinoids in Mice, and in Limb Bud Cell Micromass cultures (JIANG et al. 1995)

Retinoid	Serial number[a]	Surviving fetuses malformed (%) at different doses (mg/kg)[b]					In vitro chondrogenic inhibition
		.01	0.1	1	10	100	IC_{50} (nM)
All-trans RA	2	–	–	0	23	100	31
9-cis RA	4	–	–	0	0	89	58
TTNPB	8	36	100	–	–	–	0.06
3-methyl TTNPB	9	–	0	22	89	100	8.0
AGN 191701	10	–	–	0	0	53	63
SR 11217	11	–	–	0	0	0	NA
SR 11237	12	–	–	0	0	0	NA
AGN 192240	13	–	–	0	0	0	NA
AGN 190727	14	–	–	0	0	0	NA

NA, not active (i.e., $IC_{50} > 10^4$ nM); RA, retinoic acid. TTNPB, 4-[2-(5,6,7,8-tetramethyl-2-naphthalenyl) propen-1-ly] benzoic acid.
[a]See Table 1.
[b]A single oral dose was given on day 11 of gestation; the fetuses were examined on day 17.

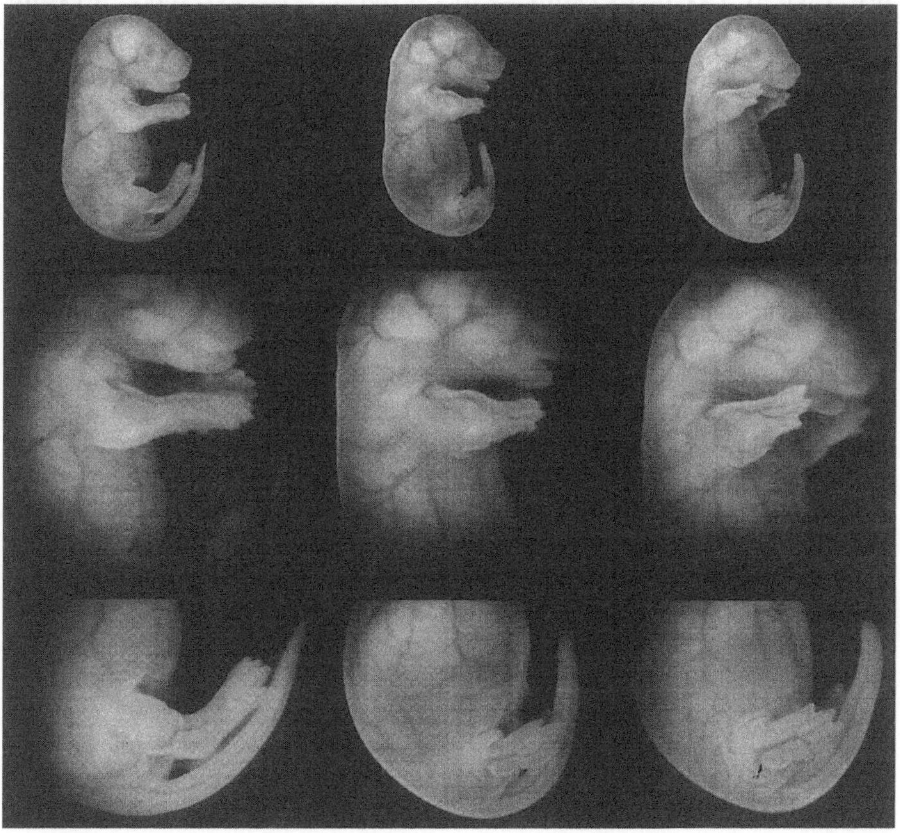

Fig. 3. Phenotypes of 4-[2-(5,6,7,8- tetramethyl-2-naphthalenyl) propen-1-ly] benzoic acid (TTNPB)-treated fetuses are compared with a normal gestational day (GD)-17 mouse fetus. *Left*, a live, normal fetus and its fore- and lindlimbs at higher magnifications. *Middle and right panels*, two fetuses from a dam treated on GD 11 with a single 0.1-mg/kg oral dose of TTNPB. All limbs of TTNPB-treated fetuses are phocomelic

no. 11 and 12) or completely inactive in receptor activation assay (e.g., nos. 13 and 14) also produced no teratogenic effects even at a dose of 100 mg/kg (Table 3).

To assess whether the differential teratogenicity of retinoids was intrinsic rather than due to maternal/placental factors, we used an in vitro assay comprising high-density micromass cultures of mouse embryo limb bud mesenchymal cells (AHRENS et al. 1977; KOCHHAR and PENNER 1992; KISTLER 1987; Table 3). KISTLER (1987) has reported that teratogenic activities of a number of structural analogues of TTNPB in rats and mice parallel their inhibitory activities in the limb bud cell cultures. As expected, TTNPB was highly active in the in vitro assay with a potency 500-fold greater than all-*trans*

RA (Table 3). 3-Methyl TTNPB was about 100 times less potent than TTNPB, a change which reflects the differential potencies of these compounds at the RAR. In addition, AGN 191701 was about 1000 times less potent than TTNPB in vitro, again reflecting their rank potencies in transactivation at the RAR. The two RXRα-specific agonists, compounds no. 11 and 12, which do not activate any of the RAR, were found to have no inhibitory activity in the chondrogenesis assay. Compounds no. 13 and 14 were similarly inactive.

The receptor activation profile of a given retinoid is by no means the only factor which influences its teratogenicity. As mentioned above, maternal/placental metabolism of natural retinoids and an access to the embryo of the active moieties play a definitive role (NAU et al. 1994). Arguably, the lack of teratogenicity of RXRα agonists, e.g., compounds no. 10–12, may simply be due to their inability to gain access to the embryo. The data from the in vitro chondrogenesis assay clearly indicated that the compounds no. 11 and 12 were intrinsically inactive. We have no information at present on the extent of placental transfer of compounds no. 11 and 12, but the results of a limited pharmacokinetic study on AGN 191701 (compound no. 10) showed that significant quantities of the intact compound were transferred to the embryo, where it could be detected for a period of 6 h after an oral dose of 10 mg/kg. Peak levels of AGN 191701 were 550 ng/ml in the maternal plasma and 284 ng/g wet weight in the embryo (JIANG et al. 1995; Fig. 4). The lack of a teratogenic response despite a significant level of placental transfer of AGN 191701 suggests that this compound has much diminished teratogenic potency compared to all-*trans* RA. A similar pharmacokinetic profile in the maternal plasma and embryo has been reported after a 10-mg/kg dose of all-*trans* RA, which, however, was associated with a fairly high teratogenic response (Fig. 4; CREECH-KRAFT et al. 1989; SATRE and KOCHHAR 1989).

We have now examined a total of 33 retinoids in the in vitro bioassay and have consistently found that, while potency at the RAR appears to be positively correlated with teratogenicity, RXR agonists show only diminished teratogenicity (Table 4). Most importantly, five of the tested retinoids activated neither RAR nor RXRα; all of these were nonteratogenic, supporting the suggestion that teratogenesis is mediated by receptors (Table 4). It would be important in further studies to examine the efficacy of these and other, similar compounds in therapeutic bioassays. It is interesting that, although all the RAR agonists in our study were active inhibitors in the chondrogenesis assay, their individual potencies varied over a wide range. No obvious correlation is apparent between the inhibitory activity of a retinoid and its transactivation activity at RARα, RARβ, or RARγ. Development of more RAR subtype-selective retinoids will be useful to gain further information.

In conclusion, preliminary correlations reported here lend support to the view that RXR agonists elicit a minimal or a much weaker teratogenic response than the RAR agonists. It has been established that RAR agonists preferentially act through RAR/RXR heterodimers (for reviews, see MANGELSDORF 1994; CHAMBON 1994; STUNNENBERG 1993). There is recent evidence

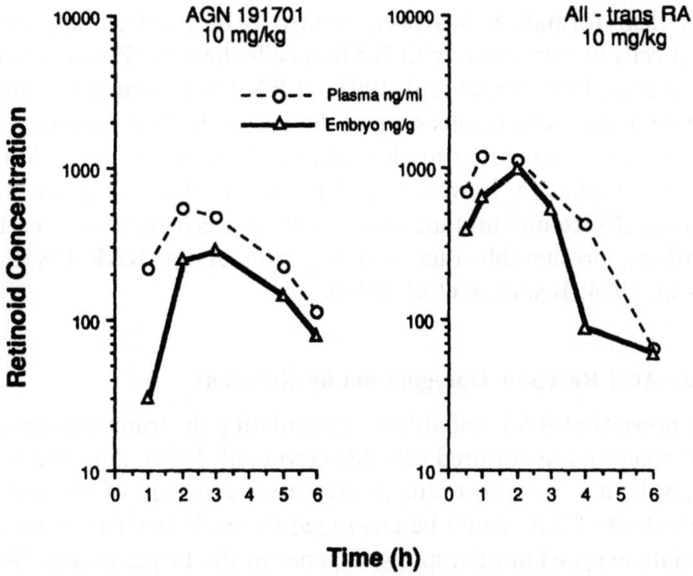

Fig. 4. Pharmacokinetic profile of AGN 191701 (Serial number 10; see Table 1) and all-*trans* retinoic acid (*RA*) in the maternal plasma and embryos. A single 10-mg/kg dose of AGN 191701 or all-*trans* RA was given by gavage to mice on day 11 of gestation, and whole embryos and venous plasma samples were collected hourly from different groups of mice for 6 h. Both AGN 191701 and all-*trans* RA undergo ready placental transfer to the mouse embryo, where their peak levels are maintained at 2–3 h after the dose. (Data from JIANG et al. 1995; SATRE and KOCHHAR 1989; CREECH-KRAFT et al. 1989)

Table 4. Teratogenic potencies of retinoic acid receptor (RAR)- and/or retinoid X receptor (RXR)-selective retinoids

Number of retinoids screened (total, $n = 30$)	Receptor preference	Relative teratogenic potency (RA, 1X)
10	RAR	45–1000X
4	RAR, RXR	3–25X
5	RXR, weak RAR	0.1–2X
6	RXR	0
5	Nonactivators	0

Gene transcriptional activation in HeLa cells and teratogenicity in limb bud cell cultures were determined by the methods described in the text.
RA, retinoic acid.

that the major role of RXR in transactivation is only to facilitate binding of RAR to the responsive promoters, and that it is RAR which is directly responsible for the transcriptional activity (STUNNENBERG 1993; VALCARCEL et al. 1994). Although it is not yet known whether RXR/RXR homodimers exist in nature, their formation has been shown to be preferentially induced in transfected cells by treatment with RXR-specific ligands (ZHANG et al. 1992b; KUROKAWA et al. 1994; YANG et al. 1993). If RXR homodimer formation were also to occur in vivo, our results suggest the possibility that response pathways mediated by such homodimers are less likely to lead to teratogenic lesions. It is important to emphasize, however, that RXR are essential in normal development, since RXRα null mutant embryos die in utero and show heart and eye malformations, presumably due to abrogation of the RAR/RXR function (SUCOV et al. 1994; KASTNER et al. 1994).

2. Retinoic Acid Receptor Upregulation by Retinoids

It is well known that RA is capable of upregulating the transcription of certain of its own receptors in cultured cells (ZELENT et al. 1991). An enhancement in receptor levels, if it were to occur in vivo after exposure of the embryo to a teratogenic dose of RA, could be envisaged to result in misregulation and/or inappropriate expression of responsive genes in the target tissues. Several investigators have, in fact, reported a higher expression than normal of the RARβ gene in embryonic tissues after in utero exposure to RA, which, in some cases, occurred ectopically (ROSSANT et al. 1991; ROWE et al. 1991; MENDELSOHN et al. 1991; ZIMMER and ZIMMER 1992; OSUMI-YAMASHITA et al. 1992; LYONS et al. 1992; HARNISH et al. 1990, 1992). Since many of these studies were conducted by the in situ hybridization method, which permitted only visual quantification, we briefly summarize results of our own studies, which utilized more quantitative method such as northern blots and RNase protection assay.

We examined the levels of mRNA of several isoforms of each RAR in normal and RA-treated embryos. Within 3–6 h after treatment of mice on GD 11 with all-*trans* RA (100 mg/kg), RARβ2 mRNA levels in the whole embryo increased seven fold, while both RARα2 and RARγ1 mRNA levels were elevated only two fold (HARNISH et al. 1992). Other isoforms were not affected. Since RA treatment of day-11 embryos produces especially limb defects in virtually every embryo, we then examined individual embryonic regions. Limb buds showed the highest elevations in RARβ2 mRNA levels (12-fold) compared to a moderate elevation in the head/craniofacial region (eight fold) and a small elevation in the remainder of the body (four fold) (HARNISH et al. 1992). In contrast, RARα2 and RARγ1 mRNA levels were elevated in all these tissues to a similar extent, which amounted to only about a two fold increase.

In order to seek further correlation between teratogenicity and RARβ2 mRNA levels in the retinoid-treated embryos, we initially compared the effects of two RA isomers in a time course study (JIANG et al. 1994). Since all-*trans* RA is considerably more potent than 13-*cis* RA in eliciting a teratogenic response in ICR mice, we compared the time course of induction of RARβ2 mRNA after a single dose of 100 mg all-*trans* RA/kg and three doses of 100 mg 13-*cis* RA/kg given 3 h apart. RARβ2 mRNA levels in the treated embryos were elevated by either isomer to manyfold higher than the levels in the tissues of the control, untreated embryos. In the limb buds, both isomers raised RARβ2 mRNA levels about 14-fold higher than the basal level. The RARβ2 mRNA levels were already elevated 3 h after treatment; thereafter, they plateaued until 9 h after treatment, when they underwent a decline to attain basal levels between 9 and 12 h after treatment. A similar time course in RARβ2 mRNA elevations was also observed in the head/craniofacial region and the remainder of the embryo; however, the extent of elevation of RARβ2 mRNA was lower (approximately eight fold and four fold, respectively). Treatment with lower doses of all-*trans* RA and 13-*cis* RA which have marginal or no effects resulted in an altered time course such that elevation in RARβ2 mRNA levels declined more rapidly, reaching a basal level at approximately 6 h after treatment.

Retinol, the precursor of RA in the embryo, was also capable of elevating RARβ2 mRNA levels in the limb bud, but the increase was delayed, apparently indicating that the metabolic conversion of retinol to RA preceded the effect on mRNA levels (Harnish et al. 1992). Finally, treatment of dams on GD 14, a time when embryos are relatively insensitive to RA, resulted in no elevation in RARα2 mRNA levels and greatly reduced elevation (two- to three-fold in all embryonic regions) in RARβ2 mRNA levels (Harnish et al. 1992). Recently, we have extended these findings to demonstrate that RARβ protein levels mirror the pattern described above for RARβ2 mRNA levels (Soprano et al. 1994). Therefore, the elevation in RARβ2 mRNA and protein correlates well with regions of the embryo, e.g., limb buds, which are specific targets for RA-induced teratogenesis on GD 11. These conclusions are supported by results of in situ hybridization studies by other investigators (Mendelsohn et al. 1991; Osumi-Yamashita et al. 1992; Rowe et al. 1991). It could, therefore, be concluded that some perturbation in the levels of RAR, particularly RARβ2, was undoubtedly correlated with the induction of teratogenesis by retinoids. Interestingly, this conclusion is challenged by the results of experiments where receptor null mutants were exposed to teratogenic doses of RA.

3. Retinoic Acid Effects in Receptor Null Mutants

Mouse embryos genetically deficient in certain retinoid receptors (e.g., RARα1 or all isoforms of RAR α, RARγ2 or all isoforms of RARγ, RARβ2, and RXRα) were individually examined to see whether their responses to RA were different in any way from those of the wild-type embryos. In general, all deficient embryos with the exception of RXRα null mutants were essentially as susceptible to RA-induced teratogenesis as the wild-type embryos (Lufkin et al. 1993; Lohnes et al. 1993; Mendelsohn et al. 1994b; H.M. Sucov et al., unpublished). Beside other malformations, typical limb reduction defects were just as frequent in RARβ2 null mutants as they were in the wild-type embryos (Mendelsohn et al. 1994b). It is too early to fully realize the significance of these experiments, but the implications from these studies are that none of the RAR alone is the sole mediator of any of the malformations. Further, it could be argued that the teratogenic effects of RA are really pharmacologic effects quite remote from the true physiological functions of RAR (CHAMBON 1994).

H.M. SUCOV et al. (unpublished) observed that RXRα null mutant embryos, which usually die by GD 16, were essentially resistant to RA-induced teratogenesis if treated at midgestation and examined prenatally. This indicated that the pharmacologic effects of RA occurred only if RXRα was available. This observation underscores the central role of RXR in the signaling pathways involving not only RAR and retinoids but also other hormones and their receptors. While embryos which lack one or the other RAR subtypes survive and struggle through critical developmental stages with the aid of the other RAR, they absolutely need at least a basal level of RXR to avoid developmental anomalies and death. Without RXR, the embryos are not only nonviable, but they are also inert to the presence of exogenous RA. This is perhaps the strongest argument in favor of the suggestion that the physiological and teratological effects of retinoids share at least one common mechanistic pathway.

II. Retinoic Acid-Regulated Molecules in Teratogenesis

Although no primary targets nor a comprehensive sequence of events for any of the defects have as yet been characterized, it is now known that RA modulates the expression of a number of genes of critical significance in the development of axial skeleton, face, brain, and limbs. Included in this group are many genes which have a prominent role in pattern formation (BAMFORTH 1994). Pattern formation is defined as a process of spatial organization of cells and tissues within a developing organ so that shape and function are coordinated in such a way that a whole, identifiable embryo emerges.

The first definitive evidence that the retinoids were involved in pattern formation came from an observation that endogenous RA extracted from chick limb buds and inserted back into the anterior margin of a normal chick wing bud induced a mirror-image duplication of the digits in the resultant limb (EICHELE 1989, for review). An analysis of the genesis of the new pattern suggested that RA had respecified information in the cells of the anterior margin so that they behaved as posterior margin cells, thus unfolding a new, yet precise mirror-image pattern (EICHELE 1989; BRICKELL and TICKLE 1989; TICKLE and BRICKELL 1991).

RA-induced transformation of cell fates in a developmental field from anterior to posterior position as observed in the limb bud is now interpreted to be due to the ectopic induction and/or repression of homeobox-containing *Hox* genes (IZPISUA-BELMONTE et al. 1991a; CHARITE et al. 1994; MORGAN and TABIN 1994; CONLON and ROSSANT 1992). Using a low oral dose of RA in pregnant mice, KESSEL and GRUSS (1991) described a number of subtle alterations in the axial skeleton in resultant fetuses, and these alterations followed the same time-dependent, craniocaudal sequence as changes in the expression of multiple, level-specific *Hox* genes. From these studies, (KESSEL 1992) proposed that normal development at each vertebral level is specified by a combination of functionally active *Hox* genes, which he called a "*Hox* code." According to this concept, exogenous RA interferes with the normal establishment of *Hox* codes and thus produces stage-dependent alterations in morphogenesis. It was recently reported that RA treatment of mouse embryos on GD 9 scrambled the spatio-temporal expression of two genes, *Hoxb1* and *Krox-20*, in the hindbrain and the associated neural crest cells; the treatment resulted in reduced size of the hindbrain and an absence of rhombomeric segmentation (MORRISS-KAY et al. 1991; MORRISS-KAY 1991). The anterior to posterior cell fate transformation may also be a logical explanation for malpositioning of the otic vesicle upon RA exposure in mouse, *Xenopus*, zebrafish, and human embryos (SULIK et al. 1988; ALTABA and JESSELL 1991; HOLDER and HILL 1991; LAMMER et al. 1985).

The mechanism for RA-induced limb defects in mice remains elusive. As already mentioned, RA administration to mice at midgestation produces skeletal dysplasia in fetuses which show long bone reduction and digital deformities (KOCHHAR 1973). As close scrutiny of the skeletal primordia of the limb bud has revealed, the effects of RA are likely to originate from a disruption of an early event in the process of cell differentiation (KOCHHAR 1977). Within 3–4 h of RA administration, the levels of RARβ2 transcripts in the apical ectodermal ridge (AER) and the central core mesenchyme are increased ten- to 20-fold compared to the pretreatment levels (MENDELSOHN et al. 1991; HARNISH et al. 1992). The AER is an inducer of distal limb outgrowth of mesenchyme underlying the AER (SAUNDERS and GASSELING 1968). Among other functions, the AER serves to suppress chondrogenesis in the subjacent mesenchyme in the so-called "progress zone," which is an important event in pattern formation (SUMMERBELL 1976; SOLURSH et al. 1981). RA-induced overexpression of RARβ2 in the AER may constitute one mechanism by which RA produces abnormalities in the skeletal patterning.

One could envisage that normal function of the retinoids in limb development would be to define regionally restricted cell differentiation domains. The presence of RARβ2-expressing cells lining the periphery of the digital cartilage segments and the absence of this receptor from the cartilage anlage of long bones which eventually forms in the central core of the mouse limb supports this suggestion (MENDELSOHN et al. 1991). Other evidence for a similar role of RA in development comes from hindbrain development in the chick and mouse embryos, where the expression of the *Hoxb1* gene is progressively restricted to only one segment, rhombomere 4 (MURPHY et al. 1989; WILKINSON et al. 1989; ZIMMER and ZIMMER 1992). It is now known that RA is responsible for this restriction by virtue of the presence of a conserved RARE in the negative regulatory region (repressor) in the *Hoxb1* gene, resulting in a sharpened segmentally restricted expression during rhombomere boundary formation (STUDER et al. 1994). The skeletal deformities in RA-treated fetuses may also be due, in part to, an elimination of sharp boundaries between cartilage-forming and non-cartilage-forming regions in response to an overexpression of RARβ2.

Recent studies have uncovered the involvement of a network of short- or long-range signaling molecules in epithelial–mesenchymal interactions essential for initial pattern formation in both chick and mouse limb buds (TABIN 1995). Fibroblast growth factor (FGF-4) and *sonic hedgehog* (*Shh*) are produced, respectively, by the AER and the ZPA (polarizing posterior mesenchyme) (NISWANDER et al. 1993; RIDDLE et al. 1993). These two molecules are reciprocally interlinked in determining the major patterning parameters along the proximodistal and anteroposterior axes of the early limb bud (NISWANDER et al. 1994; LAUFER et al. 1994; CHANG et al. 1994). Dorsalizing signal required for normal dorsoventral polarity is provided by *Wnt-7a,* which is synthesized by the dorsal ectoderm (PARR and MCMAHON 1995; YANG and NISWANDER 1995). Excess retinoids are likely to disrupt these interactions, based on the fact that RA can induce *Shh* mRNA levels ectopically in the chick wing bud and produce digital duplications (RIDDLE et al. 1993). Bone morphogenetic proteins (*BMP-2*) and a group of homeobox-containing genes (*Hoxd 9* to *-13*) are expressed in spatially distinct regions of the limb bud mesenchyme, indicative of different roles in pattern formation (DOLLE et al. 1989a; IZPISUA-BELMONTE et al. 1991b; KINGSLEY 1994). BMP-2 is a secreted protein of the transforming growth factor-β family, and its expression can be induced by *Shh*. BMP-2 and *Hoxd* are considered as downstream members of the patterning cascade (NISWANDER et al. 1994; ROSEN et al. 1994; LAUFER et al. 1994). These molecules are also likely targets, since RA can at least indirectly induce *Hoxd* genes and alter the expression pattern of other Hox genes in vitro and in vivo (KESSEL and GRUSS 1991; MAVILIO et al. 1988; KESSEL 1992). The influence of teratogenic doses of RA on any of these genes has yet to be investigated.

Although cell death is encountered in embryonic tissues after treatment with a number of unrelated teratogens, its occurrence in retinoid-treated embryos has elicited special attention (SCHWEICHEL and MERKER 1973; KOCHHAR 1977; SCOTT et al. 1980; KNUDSEN and KOCHHAR 1981; SULIK et al. 1988; ALLES and SULIK 1990; OSUMI-YAMASHITA et al. 1992; MOTOYAMA et al. 1994). Physiological cell death has long been recognized as an important component of pattern formation and organogenesis in normal embryos (GLUCKSMANN 1951; SAUNDERS 1966; FALLON and SAUNDERS 1968; LOCKSHIN and ZAKERI 1991). Of the two distinct processes by which cells die, necrosis and apoptosis, the cytological appearance of cells affected by RA favor apoptosis (LOCKSHIN and ZAKERI 1990; JIANG and KOCHHAR 1992). Necrosis denotes the process where acutely traumatized cells undergo rapid swelling of cytoplasmic and nuclear contents followed by lysis (FARBER 1985). Apoptosis defines a more active process where the affected cell controls, at least partially, its own demise; the cell shrinks and the nuclear contents condense (WYLLIE and MORRIS 1982; ARENDS and WYLLIE 1991).

From the foregoing discussion it appears unlikely that a paucity of cells in the developing tissue through cell death is the sole mechanism of RA-induced teratogenesis. The presence of apoptotic cells in the RA-treated embryos is more noticeable than some of the other molecular changes, since histological and histochemical methods have often been the ones most frequently applied. Moreover, apoptosis may be only one of the choices made by cells whose developmental program has been disrupted by exogenous retinoids. A number of molecular factors involved in apoptosis have been identified in various cell types (WILLIAMS 1991; WHITE et al. 1994). ZHANG et al. (1995) reported that one of these molecules, a peptide cross-linking enzyme termed transglutaminase II (TGase II), is induced in certain cell types by treatment with retinoids. It was shown that the enzyme induction involved the retinoid signaling pathway through RAR. Previous studies have suggested that the expression of TGase II is associated with the induction of apoptosis in several cell types, including the core mesenchymal cells in the mouse limb bud after RA treatment (PIACENTINI et al. 1991; CHIOCCA et al. 1988; JIANG and KOCHHAR 1992; FESUS et al. 1987). It remains to be seen whether TGase II or other molecular factors linked with apoptosis are also perturbed in other embryonic target organs and tissues.

Acknowledgments. I am indebted to my coworkers Dr. Dianne Soprano, Dr. Heng Jiang, and Dr. Rosh Chandraratna, whose unpublished work is cited. Thanks are also due to Dr. N.D. Agnish, who provided critical comments on the manuscript. The manuscript and illustrations were expertly prepared by Mr. John Penner. Funded by N.I.H. grant no. HD20925 and aided by funds from Allergan, Inc.

References

Adams J, Lammer EJ (1991) Relationship between dysmorphology and neuro-psychological function in children exposed to isotretinoin "in utero." In: Fujii T, Boer GJ (eds) Functional neuroteratology of short term exposure to drugs.Teikyo University Press, Tokyo, pp 159–169

Adams J, Lammer EJ (1993) Neurobehavioral teratology of isotretinoin. Reprod Toxicol 7: 175–177

Agnish ND, Kochhar DM (1993) Developmental toxicology of retinoids. In Koren G (ed) Retinoids: the risk-benefit ratio. Dekker, New York

Ahrens PB, Solursh M, Reiter RS (1977) Stage related capacity of limb chondrogenesis in cell culture. Dev Biol 60: 69–82

Allegretto EA, McClurg MR, Lazarchik SB, Clemm DL, Kerner SA, Elgort MG, Boehm MF, White SK, Pike JW, Heyman RA (1993) Transactivation properties of retinoic acid and retinoid-X receptors in mammalian cells and yeast – correlation with hormone binding and effects of metabolism. J Biol Chem 268: 26625–26633

Alles AJ, Sulik KK (1990) Retinoic acid-induced spina bifida: evidence for a pathogenetic mechanism. Development 108: 73–81

Altaba ARI, Jessell T (1991) Retinoic acid modifies mesodermal patterning in early xenopus embryos. Gene Dev 5: 175–187

Andersen B, Rosenfeld MG (1995) Intracellular receptors – new wrinkles in retinoids. Nature 374: 118–119

Arends MJ, Wyllie AH (1991) Apoptosis – mechanisms and roles in pathology. Int Rev Exp Pathol 32: 223–254
Bamforth JS (1994) Are abnormalities of human organizational genes responsible for causing birth defects? Reprod Toxicol 8: 455–459
Beard RL, Gil DW, Marler DK, Henry E, Colon DF, Gillett SJ, Arefieg T, Breen TS, Krauss H, Davies PJA, Chandraratna RAS (1994) Structural basis for the differential RXR and RAR activity of stilbene retinoid analogs. Bioorg Med Chem Lett 4: 1447–1452
Bendich A, Langseth L (1989) Safety of vitamin A. Am J Clin Nutr 49: 358–371
Benke PJ (1984) The isotretinoin teratogen syndrome. JAMA 251: 2367–2369
Blake KO, Wyse RKH (1988) Embryopathy in infant conceived year after termination of maternal etretinate: a reappraisal. Lancet 2: 1254
Blomhoff R (1994) Transport and metabolism of vitamin A. Nutr Rev 52: 13–23
Boehm MF, Zhang L, Badea BA, White SK, Mais DE, Berger E, Suto CM, Goldman ME, Heyman RA (1994) Synthesis and structure-activity relationships of novel retinoid X receptor-selective retinoids. J Med Chem 37: 2930–2941
Bollag W (1985) New retinoids with potential use in humans. In Laurat JH (ed) Retinoids: new trends in research and therapy. Karger, Basel, pp 274–288
Boylan JF, Gudas LJ (1991) Overexpression of the cellular retinoic acid binding protein-I (CRABP-I) results in a reduction in differentiation-specific gene expression in F9 teratocarcinoma cells. J Cell Biol 112: 965–979
Braun JT, Franciosi RA, Mastri AR et al (1984) Isotretinoin dysmorphic syndrome. Lancet 8375: 506–507
Brickell PM, Tickle C (1989) Morphogens in chick limb development. Bioessays 11: 145–149
Brockes JP (1989) Retinoids, homeobox genes, and limb morphogenesis: specification of the limb and its axes in the chick limb bud and urodele blastema. Neuron 2: 1285–1294
Camera G, Pregliasco P (1992) Ear malformation in baby born to mother using tretinoin cream. Lancet 339: 687
Chambon P (1994) The retinoid signaling pathway: molecular and genetic analyses. Semin Cell Biol 5: 115–125
Chang DT, Lopez A, Vonkessler DP, Chiang C, Simandl BK, Zhao RB, Seldin MF, Fallon JF, Beachy PA (1994) Products, genetic linkage and limb patterning activity of a murine hedgehog gene. Development 120: 3339–3353
Charite J, Degraaff W, Shen SB, and Deschamps J (1994) Ectopic expression of Hoxb-8 causes duplication of the ZPA in the forelimb and homeotic transformation of axial structures. Cell 78: 589–601
Chiocca EA, Davies PJA, Stein JP (1988) The molecular basis of retinoic acid action: transcriptional regulation of tissue transglutaminase gene expression in macrophages. J Biol Chem 263: 11584–11589
Cohlan SQ (1953) Excessive intake of vitamin A as a cause of congenital anomalies in rat. Science 117: 535–536
Cohlan SQ (1954) Congenital anomalies in the rat produced by excessive intake of vitamin A during pregnancy. Pediatrics 13: 556–567
Conlon RA, Rossant J (1992) Exogenous retinoic acid rapidly induces anterior ectopic expression of murine hox-2 genes invivo. Development 116: 357
Creech-Kraft J, Kochhar DM, Scott WJ, Nau H (1987) Low teratogenicity of 13-cis-retinoic acid (isotretinoin) in the mouse corresponds to low embryo concentrations during organogenesis: comparison to the all-trans isomer. Toxicol Appl Pharmacol 87: 474–482
Creech-Kraft J, Lofberg B, Chahoud I, Bochert G, Nau H (1989) Teratogenicity and placental transfer of all-*trans*-, 13-cis-, 4-oxo-all-*trans*-, and 4-oxo-13-cis-retinoic acid after administration of a low oral dose during organogenesis in mice. Toxicol Appl Pharmacol 100: 162–176

Crettaz M, Baron A, Siegenthaler G, Hunziker W (1990) Ligand specificities of recombinant retinoic acid receptors RAR-α and RAR-β. Biochem J 272: 391–397
Cruz Dl (1984) Multiple congenital malformation associated with maternal isotretinoin therapy. Pediatrics 74: 428–430
Dencker L, Annerwall E, Busch C, Eriksson U (1990) Localization of specific retinoid-binding sites and expression of cellular retinoic-acid-binding protein (CRABP) in the early mouse embryo. Development 110: 343–352
Dolle P, Izpisua-Belmonte J, Falkenstein H, Renucci A, Duboule D (1989a) Coordinate expression of the murine Hox-5 complex homoeobox-containing genes during limb pattern formation. Nature 342: 767–772
Dolle P, Ruberte E, Kastner P, Petkovich M, Stoner CM, Gudas LJ, Chambon P (1989b) Differential expression of genes encoding α, β and γ retinoic acid receptors and CRABP in the developing limbs of the mouse. Nature 342: 702–705
Dolle P, Ruberte E, Leroy P, Morriss-Kay G, Chambon P (1990) Retinoic acid receptors and cellular retinoid binding proteins 1: a systematic study of their differential pattern of transcription during mouse organogenesis. Development 110: 1133–1151
Eichele G (1989) Retinoids and vertebrate limb pattern formation. TIG 5: 246–251
Fallon JF, Saunders JW (1968) In vitro analysis of the control of cell death in a zone of prospective necrosis from the chick wing bud. Dev Biol 18: 553–570
Fantel AG, Shepard TH, Newell-Morris LL, Moffett BC (1977) Teratogenic effects of retinoic acid in pigtail monkeys (Macaca nemestrina). I. General features. Teratology 15: 65–72
Farber JL (1985) The biochemical pathology of toxic cell death. Monogr Pathol 26: 19–31
Fawcett D, Pasceri P, Fraser R, Colbert M, Rossant J, Giguere V (1995) Postaxial polydactyly in forelimbs of CRABP-II mutant mice. Development 121: 671–679
Fernhoff PM, Lammer EJ (1984) Craniofacial features of isotretinoin embryopathy. J Pediatrics 105: 595–597
Fesus L, Thomazy V, Falus A (1987) Induction and activation of tissue transglutaminase during programmed cell death. FEBS Lett 224: 104–108
Fiorella PD, Napoli JL (1994) Microsomal retinoic acid metabolism – effects of cellular retinoic acid-binding protein (type I) and C18-hydroxylation as an initial step. J Biol Chem 269: 10538–10544
Flanagan JL, Willhite CC, Ferm VH (1987) Comparative teratogenic activity of cancer chemopreventive retinoidal benzoic acid congeners (arotinoids). JNCI 78: 533–538
Geelen JAG (1978) Hypervitaminosis A induced teratogenesis. Crit Rev Toxicol 6: 351–375
Giroud A, Martinet M (1959) Extension à plusiers espèces de mammifères des malformations embryonnaires par hypervitaminose A. C R Soc Biol 153: 201–202
Glucksmann A (1951) Cell deaths in normal vertebrate ontogency. Biol Rev 26: 59–86
Gorry P, Lufkin T, Dierich A, Rochetteegly C, Decimo D, Dolle P, Mark M, Durand B, Chambon P (1994) The cellular retinoic acid binding protein I is dispensable. Proc Natl Acad Sci USA 91: 9032–9036
Grote W, Harms D, Janig U, Kietzmann H, Ravens U, Schwarze I (1985) Malformation of fetus conceived 4 months after termination of maternal etretinate treatment. Lancet 2: 1276
Hale F (1933) Pigs born without eyeballs. J Hered 24: 105–106
Happle R, Traupe T, Bounameaux T, Fisch T (1984) Teratogene Wirkung von Etretinate beim Menschen. Dtsch Med Wochenschr 109: 1476–1480
Harnish DC, Barua AB, Soprano KJ, Soprano DR (1990) Induction of beta-retinoic acid receptor messenger RNA by teratogenic doses of retinoids in murine fetuses. Differentiation 45: 103–108
Harnish DC, Jiang H, Soprano KJ, Kochhar DM, Soprano DR (1992) Retinoic acid receptor β2 mRNA is elevated by retinoic acid in vivo in susceptible regions of mid-gestation mouse embryos. Dev Dynam 194: 239–246

Hendrickx AG, Hummler H (1992) Teratogenicity of all-trans retinoic acid during early embryonic development in the cynomolgus monkey (Macaca-Fascicularis). Teratology 45: 65–74

Hendrickx AG, Silverman S, Pellegrini M, Steffek AJ (1980) Teratological and radiocephalometric analysis of craniofacial malformations induced with retinoic acid in rhesus monkeys (Macaca nemestrina). Teratology 22: 13–22

Heyman RA, Mangelsdorf DJ, Dyck JA, Stein RB, Eichele G, Evans RM, Thaller C (1992) 9-cis retinoic acid is a high affinity ligand for the retinoid X receptor. Cell 68: 397–406

Hill RM (1984) Isotretinoin teratogenicity. Lancet I: 1465

Hofmann C, Eichele G (1994). Retinoids in development. In: Sporn MB,Roberts AB, Goodman DS (eds) The retinoids: biology, chemistry, and medicine. Raven, New York, pp 387–435

Holder N, Hill J (1991) Retinoic acid modifies development of the midbrain-hindbrain border and affects cranial ganglion formation in zebrafish embryos. Development 113: 1159

Howard WB, Willhite CC, Omaye ST, Sharma RP (1989) Pharmacokinetics, tissue distribution, and placental permeability of all-trans- and 13-cis-N-ethyl retinamides in pregnant hamsters. Fund Appl Toxicol 12: 621–627

Hummler H, Korte R, Hendrickx AG (1990) Induction of malformations in the cynomolgus monkey with 13-Cis retinoic acid. Teratology 42: 263–272

Imakado S, Bickenbach JR, Bundman DS, Rothnagel JA, Attar PS, Wang XJ, Walczak VR, Wisniewski S, Pote J, Gordon JS, Heyman RA, Evans RM, Roop DR (1995) Targeting expression of a dominant-negative retinoic acid receptor mutant in the epidermis of transgenic mice results in loss of barrier function. Gene Dev 9: 317–329

Izpisua-Belmonte J, Dolle P, Duboule D (1991a) Hox genes and the molecular basis of vertebrate limb pattern formation. Semin Dev Biol 2: 385–391

Izpisua-Belmonte JC, Tickle C, Dolle P, Wolpert L, Duboule D (1991b) Expression of the homeobox Hox-4 genes and the specification of position in chick wing development. Nature 350: 585–589

Jiang H, Kochhar DM (1992) Induction of tissue transglutaminase and apoptosis by retinoic acid in the limb bud. Teratology 46: 333–340

Jiang H, Gyda M, Harnish DC, Chandraratna RA, Soprano KJ, Kochhar DM, Soprano DR (1994) Teratogenesis by retinoic acid analogs positively correlates with elevation of retinoic acid receptor-beta 2 mRNA levels in treated embryos. Teratology 50: 38–43

Jiang H, Penner JD, Beard RL, Chandraratna RAS, Kochhar DM (1995) Diminished teratogenicity of retinoid X receptor-selective synthetic retinoids. Biochem Pharmacol 50: 669–676

Jick SS, Terris BZ, Jick H (1993) First trimester topical tretinoin and congenital disorders. Lancet 341: 1181–1182

Jong L, Lehmann JM, Hobbs PD, Harlev E, Huffman JC, Pfahl M, Dawson MI (1993) Conformational effects on retinoid receptor selectivity 1: effect of 9-double bond geometry on retinoid-X receptor activity. J Med Chem 36: 2605–2613

Kalter H (1960) The teratogenic effects of hypervitaminosis A upon the face and mouth of inbred mice. Ann NY Acad Sci 85: 42–55

Kalter H, Warkany J (1961) Experimental production of congenital malformations in strains of inbred mice by maternal treatment with hypervitaminosis A. Am J Pathol 38: 1–21

Kastner P, Grondona JM, Mark M, Gansmuller A, LeMeur M, Decimo D, Vonesch J, Dolle P, Chambon P (1994) Genetic analysis of RXRα developmental function: convergence of RXR and RAR singnaling pathways in heart and eye morphogenesis. Cell, 78: 987–1003

Kessel M (1992) Respecification of vertebral identities by retinoic acid. Development 115: 487–501

Kessel M, Gruss P (1991) Homeotic transformations of murine vertebrae and concomitant alteration of hox codes induced by retinoic acid. Cell 67: 89–104
Kingsley DM (1994) What do BMPs do in mammals? Clues from the mouse short-ear mutation. Trends Genet 10: 16–21
Kistler A (1981) Teratogenesis of retinoic acid in rats: susceptible stages and suppression of retinoic acid-induced limb malformations by cycloheximide. Teratology 23: 25–31
Kistler A (1987) Limb bud cell cultures for estimating the teratogenic potential of compounds: validation of the test system with retinoids. Arch Toxicol 60: 403–414
Kistler A, Galli B, Howard WB (1990) Comparative teratogenicity of three retinoids: the arotinoids Ro 13–7410, Ro 13–6298 and Ro 15–1570. Arch Toxicol 64: 43–48
Kliewer SA, Umesono K, Heyman RA, Mangelsdorf DJ, Dyck JA, Evans RM (1992a) Retinoid X-receptor COUP-TF interactions modulate retinoic acid signaling. Proc Natl Acad Sci USA 89: 1448–1452
Kliewer SA, Umesono K, Mangelsdorf DJ, Evans RM (1992b) Retinoid X receptor interacts with nuclear receptors in retinoic acid, thyroid hormone and vitamin-D3 signalling. Nature 355: 446–449
Knudsen TB, Kochhar DM (1981) Limb development in mouse embryos. III. Cellular events underlying the determination of altered skeletal patterns following treatment with 5'fluoro-2-deoxyuridine. Teratology 23: 241–251
Kochhar DM (1967) Teratogenic activity of retinoic acid. Acta Pathol Microbiol Scand 70: 393–404
Kochhar DM (1968) Studies of vitamin A-induced teratogenesis: effects on embryonic mesenchyme and epithelium, and on incorporation of thymidine-H3. Teratology 1: 299–310
Kochhar DM (1973) Limb development in mouse embryos. I. Analysis of teratogenic effects of retinoic acid. Teratology 7: 289–298
Kochhar DM (1977) Cellular basis of congenital limb deformity induced in mice by vitamin A. In: Bergsma D, Lenz W (eds) Morphogenesis and malformation of the limb. Liss, New York, pp 111–154
Kochhar DM (1987) Mechanisms by which retinoids intercept developmental events of the mammalian limb. Hemisphere, New York
Kochhar DM, Penner JD (1987) Developmental effects of isotretinoin and 4-oxo-isotretinoin: the role of metabolism in teratogenicity. Teratology 36: 67–75
Kochhar DM, Penner JD (1992) Analysis of high dysmorphogenic activity of Ro 13–6307, a tetramethylated tetralin analog of retinoic acid. Teratology 45: 637–645
Kochhar DM, Satre MA (1993) Retinoids and fetal malformations. In: Sharma RP (ed) Dietary factors and birth defects. Pacific Division AAAS, San Francisco, pp 134–230
Kochhar DM, Penner JD, Tellone CI (1984) Comparative teratogenic activities of two retinoids: effects on palate and limb development. Teratogen Carcinogen Mutagen 4: 377–387
Kochhar DM, Kraft JC, Nau H (1987) Teratogenicity and disposition of various retinoids in vivo and In vitro. In: Nau H, Scott WJ (eds) Pharmacokinetic in teratogenesis. CRC Press, Boca Raton, pp 173–186
Kochhar DM, Jiang H, Penner JD, Heyman RA (1995) Placental transfer and developmental effects of 9-cis retinoic acid in mice. Teratology 51: 257–265
Kurokawa R, Direnzo J, Boehm M, Sugarman J, Gloss B, Rosenfeld MG, Heyman RA, Glass CK (1994) Regulation of retinoid signalling by receptor polarity and allosteric control of ligand binding. Nature 371: 528–531
Lammer E, Armstrong D (1991) Malformations of derivatives of thombencephalic alar plate in human retinoic acid embryopathy. 8th international congress on human genetics, 6–11 October 1991, Washington (abstr)
Lammer EJ (1988) A phenocopy of retinoic acid embryopathy following maternal use of etretinate that ended a year before conception. Teratology 37: 472

Lammer EJ, Opitz (1986) The Di George anomaly as a developmental field defect. Am J Med Genet [Suppl] 2: 113–127
Lammer EJ, Chen DT, Hoar RM, Agnish ND, Benke PJ, Braun JT, Curry CJ, Fernhoff PM, Grix AW, Lott IT, Richard JM, Sun SC (1985) Retinoic acid embryopathy. N Engl J Med 313: 837–841
Lammer EJ, Hayes AM, Schunior A, Holmes LB (1988) Unusually high risk for adverse outcomes of pregnancy following fetal isotretinoin exposure. Am J Hum Genet 43: A58
Lampron C, Rochetteegly C, Gorry P, Dolle P, Mark M, Lufkin T, Lemeur M, Chambon P (1995) Mice deficient in cellular retinoic acid binding protein II (CRABPII) or in both CRABPI and CRABPII are essentially normal. Development 121: 539–548
Langman J, Welch GW (1967) Excess vitamin A and development of the cerebral cortex. J Comp Neurol 131: 15–26
Laufer E, Nelson CE, Johnson RL, Morgan BA, Tabin C (1994) Sonic hedgehog and Fgf-4 act through a signaling cascade and feedback loop to integrate growth and patterning of the developing limb bud. Cell 79: 993–1003
Lehmann HW, Bodo M, Frohn C, Nerlich A, Rimek D, Notbohm H, Muller PK (1992) Lysyl hydroxylation in collagens from hyperplastic callus and embryonic bones. Biochem J 282: 313–318
Lehmann JM, Hoffmann B, Pfahl M (1991) Genomic organization of the retinoic acid receptor gamma-gene. Nucleic Acids Res 19: 573–578
Lehmann JM, Zhang XK, Pfahl M (1992) RARγ2 Expression is regulated through a retinoic acid response element embedded in Sp1-sites. Mol Cell Biol 12: 2976–2985
Leid M, Kastner P, Chambon P (1992) Multiplicity generates diversity in the retinoic acid signalling pathways. Trends Biochem Sci 17: 427–433
Levin AA, Sturzenbecker LJ, Kazmer S, Bosakowski T, Huselton C, Allenby G, Speck J, Kratzeisen C, Rosenberger M, Lovey A, Grippo JF (1992) 9-cis retinoic acid stereoisomer binds and activates the nuclear receptor RXRα. Nature 355: 359–361
Linney E, LaMantia A (1994) Retinoid signaling in mouse embryos. Adv Dev Biol 3: 73–114
Lockshin RA, Zakeri ZF (1990) Minireview – programmed cell death: new thoughts and relevance to aging. J Gern 45: B135-B140
Lockshin RA, Zakeri ZF (1991) Programmed cell death and apoptosis. In: Tomei LD, Cope FO (eds) Apoptosis: the molecular basis of cell death. Cold Spring Harbor Laboratory Press, Cold spring Harbor, pp 47–60
Lohnes D, Kastner P, Dierich A, Mark M, Lemeur M, Chambon P (1993) Function of retinoic acid receptor-gamma in the mouse. Cell 73: 643–658
Lohnes D, Mark M, Mendelsohn C, Dolle P, Dierich A, Gorry P, Gansmuller A, Chambon P (1994) Function of the retinoic acid receptors (RARs) during development 1: craniofacial and skeletal abnormalities in RAR double mutants. Development 120: 2723–2748
Lott IT (1983) Fetal hydrocephalus and ear anomalies associated with maternal use of isotretinoin. J Pediatrics 105: 597–600
Lufkin T, Lohnes D, Mark M, Dierich A, Gorry P, Gaub MP, Lemeur M, Chambon P (1993) High postnatal lethality and testis degeneration in retinoic acid receptor-alpha mutant mice. Proc Natl Acad Sci USA 90: 7225–7229
Lyons GE, Houzelstein D, Sassoon D, Robert B, Buckingham ME (1992) Multiple sites of hox-7 expression during mouse embryogenesis – comparison with retinoic acid receptor messenger RNA localization. Mol Reprod Dev 32: 303–314
Maden M (1993) The homeotic transformation of tails into limbs in rana-temporaria by retinoids. Dev Biol 159: 379–391
Mangelsdorf DJ (1994) Vitamin A receptors. Nutr Rev 52: 32–44
Mangelsdorf DJ, Ong ES, Dyck JA, Evans RM (1990) Nuclear receptor that identifies a novel retinoic acid response pathway. Nature 345: 224–229

Mangelsdorf DJ, Umesono K, Evans RM (1994) The retinoid receptors. In Sporn MB, Roberts AB, Goodman DS (eds) The retinoids: biology, chemistry, and medicine. Raven, New York, pp 319–344

Martinez-Frias ML, Salvador J (1990) Epidemiological aspects of prenatal exposure to high doses of vitamin A in Spain. Eur J Epidemiol 6: 118–123

Massarella J, Vane F, Bugge C, Rodriguez L, Cunningham WJ, Franz T, Colburn W (1985) Etretinate kinetics during chronic dosing in severe psoriasis. Clin Pharmacol Ther 37: 439–446

Mavilio F, Simeone A, Boncinelli E, Andrews PW (1988) Activation of four homeobox gene clusters in human embryonal carcinoma cells induces to differentiate by retinoic acid. Differentiation 37: 73–79

McCollum EV, Davis M (1913) The necessity of certain lipins in the diet during growth. J Biol Chem 15: 167–175

Mendelsohn C, Ruberte E, LeMeur M, Morriss-Kay G, Chambon P (1991) Developmental analysis of the retinoic acid-inducible RAR-β2 promoter in transgenic animals. Development 113: 723–734

Mendelsohn C, Ruberte E, Chambon P (1992) Retinoid receptors in vertebrate limb development. Dev Biol 152: 50–61

Mendelsohn C, Lohnes D, Decimo D, Lufkin T, Lemeur M, Chambon P, Mark M (1994a) Function of the retinoic acid receptors (RARs) during development 2: multiple abnormalities at various stages of organogenesis in RAR double mutants. Development 120: 2749–2771

Mendelsohn C, Mark M, Dolle P, Dierich A, Gaub MP, Krust A, Lampron C, Chambon P (1994b) Retinoic acid receptor β2 (RARβ2) null mutant mice appear normal. Dev Biol 166: 246–258

Morgan BA, Tabin C (1994) Hox genes and growth: early and late roles in limb bud morphogenesis. Development 181–186

Morriss-Kay G (1991) Retinoic acid, neural crest and craniofacial development. Semin Dev Biol 2: 211–218

Morriss-Kay GM, Murphy P, Hill RE, Davidson DR (1991) Effects of retinoic acid excess on expression of hox-2.9 and krox-20 and on morphological segmentation in the hindbrain of mouse embryos. EMBO J 10: 2985–2995

Motoyama J, Taki K, Osumiyamashita N, Eto K (1994) Retinoic acid treatment induces cell death and the protein expression of retinoic acid receptor beta in the mesenchymal cells of mouse facial primordia in vitro. Dev Growth Differ 36: 281–288

Murphy P, Davidson DR, Hill RE (1989) Segment-specific expression of a homeobox-containing gene in the mouse hindbrain. Nature 341: 156–159

Napoli JL (1994) Retinoic acid homeostasis. Prospective roles of β-carotene, retinol, CRBP and CRABP. In: Blomhoff R (ed) Vitamin A in health and disease. Dekker, New York, pp 135–188

Nau H (1993) Embryotoxicity and teratogenicity of topical retinoic acid. Skin Pharmacol 6[Suppl 1]: 35–44

Nau H, Chahoud I, Dencker L, Lammer EJ, Scott WJ (1994) Teratogenicity of vitamin A and retinoids. In Blomhoff R (ed) Vitamin A in health and disease. Dekker, New York, pp 615–663

Niswander L, Tickle C, Vogel A, Booth I, Martin GR (1993) FGF-4 replaces the apical ectodermal ridge and directs outgrowth and patterning of the limb. Cell 75: 579–587

Niswander L, Tickle C, Vogel A, Martin G (1994) Function of FGF-4 in limb development. Mol Reprod Dev 39: 83–89

Noback CR, Takahashi YI (1978) Micromorphology of the placenta of rats reared on marginal vitamin A deficient diets. Acta Anat 102: 195–202

O'Toole BA, Fradkin R, Warkany J, Wilson JG, Mann GV (1974) Vitamin A deficiency and reproduction in rhesus monkeys. J Nutr 104: 1513–1524

Ong DE, Chytil F (1975) Retinoic acid binding protein in rat tissue. J Biol Chem 250: 6113–6117

Osumi-Yamashita N, Iseki S, Noji S, Nohno T, Koyama E, Taniguchi S, Doi H, Eto K (1992) Retinoic acid treatment induces the ectopic expression of retinoic acid receptor-β gene and excessive cell death in the embryonic mouse face. Dev Growth Differ 34: 199–209

Palludan B (1976) The influence of vitamin A deficiency on fetal development in pigs with special reference to organogenesis. Int J Vitam Nutr Res 46: 223–225

Paravicini U, Stockel K, McNamara PJ, Hanni R, Busslinger A (1981) On the metabolism and pharmacokinetics of an aromatic retinoid. Ann N Y Acad Sci 359: 54–67

Parr BA, McMahon AP (1995) Dorsalizing signal Wnt-7a required for normal polarity of D-V and A-P axes of mouse limb. Nature 374: 350–353

Piacentini M, Fesus L, Farrace MG, Ghibelli L, Piredda L, Melino G (1991) The expression of "tissue" transglutaminase in two human cancer cell lines is related with the programmed cell death (apoptosis). Eur J Cell Biol 54: 246–254

Public Affairs Committee and the Council of the Teratology Society (1991) Recommendations for isotretinoin use in women of childbearing potential. Teratology 44: 1–6

Riddle RD, Johnson RL, Laufer E, Tabin C (1993) Sonic-hedgehog mediates the polarizing activity of the ZPA. Cell 75: 1401–1416

Rizzo R, Lammer EJ, Parano E, Pavone L, Argyle JC (1991) Limb reduction defects in humans associated with prenatal isotretinoin exposure. Teratology 44: 599–604

Roche L (1985) June 1985 presentation to FDA advisory committee and public hearing on etretinate safety. No Roche Laboratories

Rosa FW (1993) Retinoid embryopathy in humans. In: Koren G (ed) Retinoids: the risk benefit ratio. Dekker, New York

Rosa FW, Wilk AL, Kelsey FO (1986) Teratogen update: vitamin A congeners. Teratology 33: 355–364

Rosen V, Nove J, Song JJ, Thies RS, Cox K, Wozney JM (1994) Responsiveness of clonal limb bud cell lines to bone morphogenetic protein 2 reveals a sequential relationship between cartilage and bone cell phenotypes. J Bone Miner Res 9: 1759–1768

Rossant J, Zirngibl R, Cado D, Shago M, Giguere V (1991) Expression of a retinoic acid response element-hsplacZ transgene defines specific domains of transcriptional activity during mouse embryogenesis. Gene Dev 5: 1333–1344

Rowe A, Richman JM, Brickell PM (1991) Retinoic acid treatment alters the distribution of retinoic acid receptor-beta transcripts in the embryonic chick face. Development 111: 1007–1016

Ruberte E, Dolle P, Krust A, Zelent A, Morriss-Kay G, Chambon P (1990) Specific spatial and temporal distribution of retinoic acid receptor gamma transcripts during mouse embryogenesis. Development 108: 213–222

Ruberte E, Dolle P, Chambon P, Morriss-Kay G (1991a) Retinoic acid receptors and cellular retinoid binding proteins 2: their differential pattern of transcription during early morphogenesis in mouse embryos. Development 111: 45–60

Ruberte E, Kastner P, Dolle P, Krust A, Leroy P, Mendelsohn C, Zelent A, Chambon P (1991b) Retinoic acid receptors in the embryo. Semin Dev Biol 2: 153–159

Ruberte E, Friederich V, Chambon P, Morrisskay G (1993) Retinoic acid receptors and cellular retinoid binding proteins 3: their differential transcript distribution during mouse nervous system development. Development 118: 267–282

Saitou M, Sugai S, Tanaka T, Shimouchi K, Fuchs E, Narumiya S, Kakizuka A (1995) Inhibition of skin development by targeted expression of a dominant-negative retinole acid receptor. Nature 374: 159–162

Satre MA, Kochhar DM (1989) Elevations in the endogenous levels of the putative morphogen retinoic acid in embryonic mouse limb-buds associated with limb dysmorphogenesis. Dev Biol 133: 529–536

Satre MA, Penner JD, Kochhar DM (1989) Pharmacokinetic assessment of teratologically effective concentrations of an endogenous retinoic acid metabolite. Teratology 39: 341–348

Saunders JW (1966) Death in embryonic systems. Science 154: 604–612

Saunders JW, Gasseling MT (1968). Ectodermal-mesenchymal interactions in the origins of limb symmetry. In: Fleischmajer R, Billingham RE (eds) Epithelial-mesenchymal interactions. Williams and Wilkins, Baltimore, pp 78–97

Schweichel JU, Merker HJ (1973) The morphology of various types of cell death in prenatal tissues. Teratology 7: 253–266

Scott WJ, Ritter EJ, Wilson JG (1980) Ectodermal and mesodermal cell death patterns in 6-mercaptopurine riboside-induced digital deformities. Teratology 21: 271–279

Shenefelt RE (1972) Morphogenesis of malformations in hamsters caused by retinoic acid: relation to dose and stage at treatment. Teratology 5: 103–118

Solursh M, Singley CT, Reiter RS (1981) The influence of epithelia on cartilage and loose connective tissue formation by limb mesenchyme cultures. Dev Biol 86: 471–482

Sommer A (1989) New imperatives for an old vitamin (A). J Nutr 119: 96–100

Soprano DR, Gyda M, Jiang H, Harnish DC, Ugen K, Satre M, Chen L, Soprano KJ, Kochhar DM (1994) A sustained elevation in retinoic acid receptol-β2 mRNA and protein occurs during retinoic acid-induced fetal dysmorphogenesis. Mech Dev 45: 243–253

Sporn MB, Roberts AB, Goodman DS (eds) (1994) The retinoids – biology, chemistry, and medicine, (2nd edn.) Raven, New York

Studer M, Popperl H, Marshall H, Kuroiwa A, Krumlauf R (1994) Role of a conserved retinoic acid response element in rhombomere restriction of Hoxb-1. Science 265: 1728–1732

Stunnenberg HG (1993) Mechanisms of transactivation by retinoic acid receptors. Bioessays 15: 309–315

Sucov HM, Dyson E, Gumeringer CL, Price J, Chien KR, Evans RM (1994) RXRα mutant mice establish a genetic basis for vitamin A signaling in heart morphogenesis. Gene Dev 8: 1007–1018

Sulik KK, Cook CS, Webster WS (1988) Teratogens and craniofacial malformations: relationships to cell death. Development 103 [Suppl]: 213 232

Summerbell D (1976) A descriptive study of the rate of elongation and differentiation of the skeleton of the developing chick wing. J Embryol Exp Morphol 35: 241–260

Tabin C (1995) The initiation of the limb bud: growth factors, hox genes, and retinoids. Cell, 80: 671–674

Tabin CJ (1991) Retinoids, homeoboxes, and growth factors – toward molecular models for limb development. Cell 66: 199–217

Teratology Society (1987) Position paper: recommendations for vitamin A use during pregnancy. Teratology 35: 269–275

Thompson JN, Howell JM, Pitt GAJ (1964) Vitamin A and reproduction in rats. Proc R Soc Lond [B] 159: 510–535

Tickle C, Brickell PM (1991) Retinoic acid and limb development. Semin Dev Biol 2: 189–197

Valcarcel R, Holz H, Jimenez CG, Barettino D, Stunnenberg HG (1994) Retinoid-dependent in vitro transcription mediated by the RXR/RAR heterodimer. Gene Dev 8: 3068–3079

Vallet HL, Stark AD, Costas K, Thompson S, Davis R, Teresi N (1985) Isotretinoin (accutane), vitamin A, and human teratogenicity. American Public Health Association, Washington DC

Van Pelt AMM, De Rooij DG (1991) Retinoic acid is able to reinitiate spermatogenesis in vitamin-A-deficient rats and high replicate doses support the full development of spermatogenic cells. Endocrinology 128: 697–704

Wald G (1968) Molecular basis of visual excitation. Science 162: 230–239

Wallingford JC, Underwood BA (1986) Vitamin A deficiency in pregnancy, lactation, and the nursing child. In: Bauernfeind JC (ed) Vitamin A deficiency and its control. Academic, Orlando, pp 101–152
Ward A, Brogden RN, Heel RC, Speight TM, Avery GS (1983) Etretinate. A review of its pharmacological properties and therapeutic efficacy in psoriasis and other disorders. Drugs 26: 9–43
Werler MN, Rosenberg L, Mitchell AA (1989) First trimester vitamin A use in relation to birth defects. Teratology 39: 489
White K, Grether ME, Abrams JM, Young L, Farrell K, Steller H (1994) Genetic control of programmed cell death in drosophila. Science 264: 677–683
Wilkinson DG, Bhatt S, Chavrier P, Bravo R, Charnay P (1989) Segment-specific expression of a zinc-finger gene in the developing nervous system of the mouse. Nature 337: 461–464
Willhite CC (1990) Molecular correlates in retinoid pharmacology and toxicology. In: Dawson MI, Okamura WH (eds) Chemistry and biology of synthetic retinoids. CRC Press, Boca Raton, pp 539–573
Willhite CC, Sharma RP, Allen PV, Berry DL (1990) Percutaneous retinoid absorption and embryotoxicity. J Invest Dermatol 95: 523–529
Williams GT (1991) Programmed cell death – apoptosis and oncogenesis. Cell 65: 1097–1098
Wilson JG (1974) Teratologic causation in man and its evaluation in non-human primates. In Motulsky AG (ed) Birth defects. American Elsevier, New York, pp 191–203
Wilson JG (1977) Current status of teratology. In: Wilson JG, Fraser FC (eds) Handbook of teratology. Plenum, New York, pp 47–74
Wilson JG, Warkany J (1950) Cardiac and aortic arch anomalies in the offspring of vitamin A deficient rats correlated with similar human anomalies. Pediatrics 5: 708–725
Wilson JG, Roth CB, Warkany J (1953) An analysis of the syndrome of malformation induced by maternal vitamin A deficiency. Effects of restoration of vitamin A at various times during gestation. Am J Anat, 92: 189–217
Wolbach SB, Howe PR (1925) Tissue changes following deprivation of fat soluble A vitamin. J Exp Med 42: 753–777
Wyllie AH, Morris RG (1982) Hormone-induced cell death. Purification and properties of thymocytes undergoing apoptosis after glucocorticoid treatment. Am J Pathol 109: 78–87
Yang YL, Vacchio MS, Ashwell JD (1993) 9-cis-retinoic acid inhibits activation-driven T-cell apoptosis – implications for retinoid-X receptor involvement in thymocyte development. Proc Natl Acad Sci USA 90: 6170–6174
Yang YZ, Niswander L (1995) Interaction between the signaling molecules WNT7a and SHH during vertebrate limb development: dorsal signals regulate anteroposterior patterning. Cell 80: 939–947
Yu VC, Delsert C, Andersen B, Holloway JM, Devary OV, Naar AM, Kim SY, Boutin JM, Glass CK, Rosenfeld MG (1991) RXRbeta – a coregulator that enhances binding of retinoic acid, thyroid hormone; and vitamin-D receptors to their cognate response elements. Cell 67: 1251–1266
Zbinden G (1975) Investigations on the toxicity of tretinoin administered systemically to animals. Acta Dermatol Venereol 74 [Suppl]: 36–40
Zelent A, Mendelsohn C, Kastner P, Krust A, Garnier JM, Ruffenach F, Leroy P, Chambon P (1991) Differentially expressed isoforms of the mouse retinoic acid receptor-β are generated by usage of two promoters and alternative splicing. EMBO J 10: 71–81
Zhang LX, Mills KJ, Dawson MI, Collins SJ, Jetten AM (1995) Evidence for the involvement of retinoic acid receptor RAR alpha-dependent signaling pathway in the induction of tissue transglutaminase and apoptosis by retinoids. J Biol Chem 270: 6022–6029

Zhang X, Hoffmann B, Tran P, Graupner G, Pfahl M (1992a) Retinoid X receptor is an auxiliary protein for thyroid hormone and retinoic acid receptors. Nature 355: 441–446

Zhang XK, Lehmann J, Hoffmann B, Dawson MI, Cameron J, Graupner G, Hermann T, Tran P, Pfahl M (1992b) Homodimer formation of retinoid X-receptor induced by 9-cis retinoic acid. Nature 358: 587–591

Zimmer A, Zimmer A (1992) Induction of a RARβ2-lacZ transgene by retinoic acid reflects the neuromeric organization of the central nervous system. Development 116: 977–983

Zimmermann B, Tsambaos D, Sturje H (1985) Teratogenicity of arotinoid ethyl ester (Ro 13–6298) in mice. Teratogen Carcinogen Mutagen 5: 415–431

CHAPTER 22

Peculiarities and Possible Mode of Actions of Thalidomide

R. NEUBERT and D. NEUBERT

A. Introduction

Thalidomide is a Janus-faced substance par excellence. It may be argued that a two-sided behavior is typical for all substances, exhibiting beneficial and adverse effects simultaneously or at least at different dose ranges. However, in the case of thalidomide the situation is somewhat different. On the one hand, the substance was marketed and advertised as being "atoxic," as deduced from animal studies performed routinely at that time, and during the very extensive therapeutic use between 1957 and 1962 very few adverse effects were reported in humans; even several attempted suicides with overdoses of thalidomide were unsuccessful. On the other hand, this substance was responsible for the largest adverse health outcome ever caused by an agent at therapeutic doses and during a short marketing period.

B. Historic Overview

Thalidomide was developed and patented by the Chemie Grünenthal Company in the Federal Republic of Germany in 1954. It was a new type of sedative substance which attracted considerable interest and became extremely popular with both patients and doctors. This can only be understood in the context of the substances available at that time: barbiturates (e.g., barbital, phenobarbital, hexobarbital) and derivatives with rather similar chemical structures and pharmacological properties, such as glutethimide (Doriden). Many of these substances led to tolerance and to psychic and physical dependence and, along with household gas, they were frequently used in suicide. Furthermore, the sleep induced by these agents was often not considered refreshing, possibly because of the pronounced depression of the rapid eye movement (REM) phases. Retrospectively, thalidomide may have had greater similarities to modern tranquilizers than to the barbiturate-type sedatives and hypnotics. This may explain its unusual popularity, not unlike the popularity of benzodiazepins today. However, there is little or no evidence of the development of tolerance or of drug dependence in the case of thalidomide, quite in contrast to benzodiazepins.

Thalidomide was tested in clinical trials during the years 1954/1957 and was introduced on the German market as an adjuvant in a cold remedy at the end of 1956 and as a sedative and hypnotic drug in November 1957. From the years 1958 to 1961 onward, thalidomide, was also officially sold in many other European countries as well as in Australia, Canada, Japan, Brazil, and others, largely over the counter. The substance was never officially marketed in the USA because of administrative reasons, but clinical trials were performed with Kevadon (LASAGNA 1960; GRAY et al. 1960) and malformed children were also born in the USA (MELLIN and KATZENSTEIN 1962).

In adult, thalidomide exhibited a remarkably low acute toxicity; overdosing for various reasons up to ten fold or even 100-fold the usual therapeutic dose did not induce more than a moderately deep sleep, and in all cases there was an uneventful recovery without any therapy within 24 h or less (BURLEY 1960; NEUHAUS and IBE 1960; BRESNAHAN 1961). Furthermore, therapeutic daily doses between 1 and 2.8 g were used in psychotic patients with few side effects (COHEN 1960; ROBERTSON 1962). Overdosing with doses of 0.35–1.2 g was even reported in children between 1½ and 5 years of age, who slept soundly but recovered uneventfully (KUNSTMANN 1960; BURLEY 1960).

Thereafter, because of the serious adverse health effects observed, i.e., polyneuropathy and especially congenital malformations, thalidomide was withdrawn from the German market in early November 1961. Thus for the main indication and in the main distribution form (Contergan), the substance was used in Germany for only about 4 years, with increasing sales frequencies over this period. At the peak period, around 10 tons of the substance were produced and marketed per year. In some other countries (e.g., Brazil and Japan) the substance was still sold during a good part of the year 1962, with adverse effects still being induced at that time.

I. Chemical Structure of Thalidomide and Derivatives

Although a great number of derivatives have been synthesized and studied experimentally, thalidomide has been the only substance used clinically to any significant extent.

The full chemical designation of thalidomide (Fig. 1) is: 3-[1,3-dihydro-1,3-dioxo-2H-isoindol-2-yl] 2,6-dioxopiperidine ($C_{13}H_{10}N_2O_4$). Other designations include 2-[2,6-dioxopiperidine-3-yl] phthalimide and (in the older literature) α-*N*-phthalimidoglutarimide.

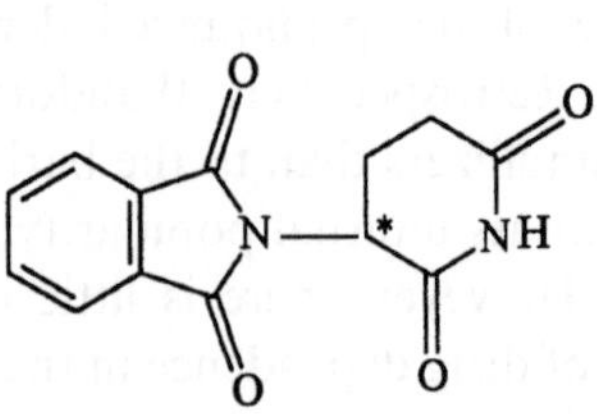

Fig. 1. Chemical structure of thalidomide. The asterisk indicates its asymmetrical carbon atom

Thalidomide is a crystalline white substance very poorly soluble in water and in vegetable oils, but with a good solubility in dimethyl sulfoxide (DMSO). It has an asymmetric carbon atom (asterisk in Fig. 1), and two enantiomers exist, S(–)-thalidomide and R(+)-thalidomide, which can be and have been separated by modern techniques. The enantiomers are unstable in aqueous solution in vitro and in vivo and slowly reracemize. Thalidomide was clinically always used as the racemate.

Of the large number of derivatives synthesized and studied to any extent, two thalidomide analogues deserve special mention: α-EM12 and supidimide (Fig. 2), the former being more potent than thalidomide and the latter exhibiting sedative properties but with a very low or absent teratogenic potential. α-EM12 and β-EM12 form a convenient pair of isomeric derivatives for experimental studies, since α-EM12 is highly teratogenic while β-EM12 (also called EM16) shows an extremely low, if any, teratogenic potency.

In general, the monooxopiperidine derivatives seem to exhibit about the same or slightly lower teratogenic potencies as the corresponding dioxopiperidines (HELM et al. 1981), but the former may be converted in vivo to the dioxo-derivatives.

II. Pharmacokinetics and Metabolism of Thalidomide and Derivatives

Studies with thalidomide are complicated by two characteristics of the substance: (a) the substance is very poorly water soluble (only up to about 100 μg/ml), and (b) the dissolved portion of thalidomide hydrolyzes easily in aqueous solution. Although the water solubility of thalidomide and of many of its derivatives is poor, this is quite sufficient to achieve adequate concentrations for in vitro studies, since therapeutic and embryotoxic (and therefore also experimentally interesting) plasma concentrations are less than 10 μg/ml.

The susceptibility of the substance to spontaneous hydrolysis in aqueous solutions in vivo and in vitro constitutes a greater problem for the inter-

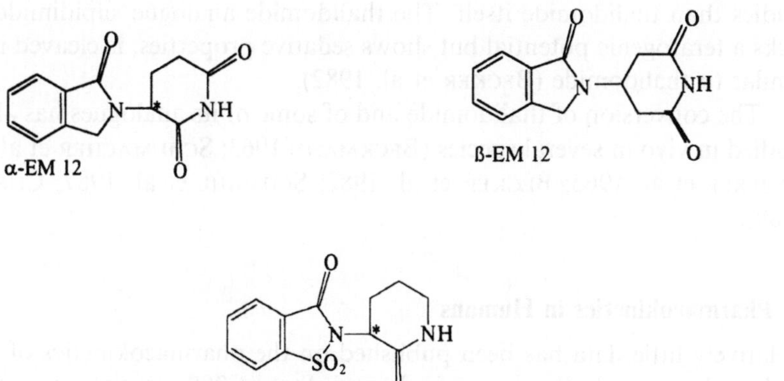

Fig. 2. Derivatives of thalidomide. *Asterisks* indicate asymmetrical carbon atoms

pretation of results. For this reason, it has long been debated whether thalidomide proper or one of its polar hydrolysis products constitutes the biologically active form. Altogether 12 hydrolysis products might be formed, and, taking into account the enantiomers, 22 enantiomeric hydrolysis products are theoretically possible.

As the first step, any one of the four imide bonds in the two ring systems can be cleaved. This gives three primary hydrolysis products, which are monocarbonic acids: (1) cleavage at the phthalimide ring, giving 3-(*O*-carboxybenzamido)-glutarimide, (2) cleavage at the glutarimide ring, giving phthaloylisoglutamine, and (3) phthaloylglutamine.

The rate of hydrolysis in vitro and in vivo was measured in several early studies (BECKMANN 1963; KEBERLE et al. 1965; SCHUMACHER et al. 1965a) and was found to be highly pH dependent. The substance is fairly stable at pH values less than 6 and is very rapidly hydrolyzed at pH 8 or higher. In in vitro studies at pH 7.4 and 37 °C, a physical half-life of thalidomide between 2.4 and 5h was reported. The primary metabolites were found to be more stable than thalidomide. Under in vitro conditions at pH 7.4 and 37 °C, the glutarimide ring seems to be more resistant to hydrolysis than the phthalimide ring, since the formation of 88% of the first product, 8% of the second, and 4% of the third was reported.

A second hydrolysis gives rise to dicarbonic acids, either by opening the second ring system or by deamination of the glutamine. In the latter conversion, metabolic processes might be involved. As a result of a third hydrolysis step, a tricarbonic acid is formed. This may lead to thalidomide proper with extremely small hydrophilic properties entering a cell, and the polar hydrolysis products formed within that cell may be subsequently trapped (KEBERLE et al. 1965).

The thalidomide derivative α-EM12 is much less susceptible to spontaneous hydrolysis, and only a few hydrolysis products occur (JACKSON and SCHUMACHER 1980; SCHMAHL et al. 1987), quite in contrast to thalidomide itself. For this reason, α-EM12 is a much more convenient tool for mechanistic studies than thalidomide itself. The thalidomide analogue supidimide, which lacks a teratogenic potential but shows sedative properties, is cleaved in a way similar to thalidomide (BECKER et al. 1982).

The conversion of thalidomide and of some of its analogues has also been studied in vivo in several species (BECKMANN 1963; SCHUMACHER et al. 1965b; KEBERLE et al. 1965; BECKER et al. 1982; SCHMAHL et al. 1987; CHEN et al. 1989).

1. Pharmacokinetics in Humans

Relatively little data has been published on the pharmacokinetics of thalidomide in humans. Following a single oral dose of 200 mg given to eight male human volunteers weighing 56–88 kg (CHEN et al. 1989), the pharmacological variables measured in blood plasma were reported as follows (mean ± SD):

peak concentration, 1.2 ± 0.2 μg/ml; t_{max}, 4.4 ± 1.3 h; elimination half-life, 8.7 ± 4.1 h; absorption half-life, 1.7 ± 1.1 h. Thus the elimination half-life reported for humans is considerably longer than that reported for experimental animals: approximately 2.7 h in rhesus monkeys and 2–3 h in rabbits and rats (SCHUMACHER et al. 1968a, 1970). Excretion in urine as unchanged thalidomide amounted to only 0.6% ± 0.2% of the total dose administered in the human volunteers.

Since the usual therapeutic dosage during the period of marketing as a hypnotic drug or sedative was 50–100 (200) mg thalidomide once a day (one half to two tablets of Contergan forte or an equivalent preparation), corresponding to about 0.7–3 mg thalidomide/kg body weight, maximum levels of 300–1200 ng thalidomide/ml blood plasma would be expected. Therefore, the peak plasma concentrations of the drug associated with a high probability of a teratogenic action should also be in the range of about 1 μg thalidomide/ml or even less.

Because the doses of thalidomide now used therapeutically in humans to achieve an anti-inflammatory or immunosuppressive effect are higher (3–7 mg/kg body weight), the plasma concentrations resulting from today's therapeutic applications will also be considerably higher than those obtained decades ago with the use as a sedative. It was calculated (CHEN et al. 1989) that peak concentrations of 3–4 μg/ml could be expected at steady state after oral administration of 200 mg thalidomide every 6 h; the minimum plasma concentrations are also much higher with a repeated dose schedule. Therefore, the teratogenic risk, and probably also the risk of polyneuropathy in prolonged treatment, should be even higher with the present therapeutic use when compared with that of 1960/1961.

C. Specific Effects Induced by Thalidomide on Prenatal Development

Thalidomide induces very pronounced effects on the prenatal development of certain species. These effects are unusual, and thalidomide exhibits many peculiarities not shared with other teratogenic agents. Thus in many respects thalidomide may be considered an exception among teratogenic substances. It certainly is not the prototype of all the commonly observed teratogenic actions in experimental research or the known substances that are teratogenic in humans.

I. Recognition of the Teratogenic Potential

The teratogenic effect of thalidomide was first discovered in children after treating the mothers with therapeutic doses of this drug; confirmation in experimental studies was not achieved until considerably later.

1. Experimental Testing in the Mid-1950s

The thalidomide tragedy was, without any doubt, the largest occurrence of a severely adverse health effect induced by a medicinal substance used at therapeutic doses. Many countries were involved, and the shock caused by the birth of several thousand severely malformed babies had a tremendous impact on public concern with respect to the safety of medicinal products and on the legislature and the preclinical testing requirements of drugs from that time on.

Two questions have been hotly debated, namely: (1) whether the tragedy could have been prevented by performing more extensive preclinical studies, and (2) whether a similar tragedy could occur today, even with the more sophisticated testing procedures required by law in most countries.

With respect to the first question, hindsight always makes people wiser decades later. However, there is no doubt whatsoever, despite other claims, that at that time testing to reveal a possible embryotoxic and especially teratogenic potential of newly developed drugs was not carried out. It certainly was not a standard procedure in preclinical safety assessments in those days. The argument that the results of a few experiments on the possible induction of teratogenic effects were already available is beside the point, because these studies concerned vitamin or nutritional deficiencies or used highly reactive substances, such as cytostatics and antimetabolites. They were motivated more by curiosity or the hope of revealing possible mechanisms of action rather than to assess risks. It is clearly wrong that, on the basis of the evidence available at that time it could have been predicted that a substance with such an extremely low general toxicity would induce such drastic malformations at therapeutic doses (WARKANY 1988). We now know, decades later and after extensive experimental experience, that substances with such properties are the exception rather than the rule.

No agreed testing strategies were available at that time. It took years after the thalidomide tragedy to establish standards for reliable experimental testing, and harmonization among different countries was not achieved until only a few years ago. Furthermore, today's tests for revealing potential adverse effects on reproduction and development are performed on rats and a few routine studies (segment II of testing for reproductive toxicity) also on rabbits. Tests on rabbits, which are not used in any other routine toxicological studies, were added to assess teratogenic potential *because* of the information gained with thalidomide. From the experience available in the mid-1950s, only the rat or the chick embryo could have been chosen as experimental models, since some knowledge of the embryology of these species was available. It is well established that in the rat no reproducible teratogenic effects can be induced by thalidomide (SCOTT et al. 1977), and also in the chick embryo no reproducible or specific teratogenic effects of thalidomide were ever established. From these facts it can be concluded that, even if tests for teratogenicity had been performed in the 1950s, there is an overwhelming chance that the teratogenic potential of thalidomide would not have been detected. With the small experience available, this false information might have further obscured the first suspicions of a risk in humans.

With respect to the second question, we can only speculate. Most scientists and regulatory agencies are convinced that the present standard of preclinical testing will enable us to recognize possible teratogenic substances and to initiate measures of primary prevention for women. As a consequence of many of these preventive decisions and risk minimization procedures, we will never find out whether many substances (which humans are not exposed to) really exhibit such a hazardous effect in humans as in experimental models.

It is the peculiarity of adverse health effects that they seldom occur in exactly the same way as experienced before. Therefore, it is most likely that a pronounced, new adverse effect in humans will not be identical to the expected effects. There is no doubt that in the field of reproductive and developmental toxicity there are, even today, areas which are not properly evaluated in the course of the development of a new drug with the routine standards. Possible effects on components or functions of the immune system may be one of these areas.

2. Teratogenic Risk in Humans

The events leading to the revelation that thalidomide is a human teratogenic substance are interesting from both the historic and the medical epidemiological point of view. LENZ (1985) summarized his personal experience with respect to some of the main events. It is quite obvious that a high degree of suspicion and some evidence of a possible causal relationship between the occurrence of reduction limb deformities and thalidomide was not the result of the experience of a single person, but was due to the concerted effort of several clinicians. In retrospective, it is clear from the report by LENZ (1988) and barely understandable today that at least 1800 cases of typical and very unusually deformed babies were reported between 1958 and November 1961 in Germany before the causative agent was recognized with a reasonable degree of certainty (Table 1).

It is fair to state that until the middle of 1961 there was no awareness that a wave of malformations or a serious number of adverse effects induced by a particular drug had occurred. The first person to recognize this and to publish it was WIEDEMANN (1961), and he explicitly suggested an exogenous cause. While he had previously seen only a few cases of severe reduction limb deformities in his clinic, over a decade, 13 cases (including nine cases of phocomelia and amelia) were observed in less than a year in 1960/1961. He also recognized the high mortality rate connected with these types of malformations: seven of the children died at a very young age. Subsequently, he obtained information about this subject in approximately 95 additional cases from several other clinics, indicating that similar observations had been made in many clinics all over the Federal Republic of Germany. Wiedemann published these data in September 1961 (WIEDEMANN 1961; WIEDEMANN and

Table 1. Retrospective assessment of thalidomide-type malformations occurring in the Federal Republic of Germany. (Data from LENZ 1988)

Year	Cases (*n*)
1956	1
1957	1
1958	24
1959	97
1960	450
1961	1233[a]
1961	302[b]
1962	792[c]
1962	135[d]
1963	9
1964	1

[a]January to October.
[b]November and December.
[c]First half of the year.
[d]Second half of the year.

AEISSEN 1961), and the name Wiedemann dysmelie syndrome was used for these defects (e.g., PLIESS 1962). Place of residence (rural or metropolitan), social status and profession of the parents, viral infections, and any contra conceptive agents used by the mother seemed to be irrelevant with respect to the factor inducing the abnormalities. Genetic factors were ruled out by WIEDEMANN, and he recognized that the agent responsible must have been present in the population around the beginning of 1959, and its use must have largely increased over the following years. Thus he assumed a newly introduced "toxic factor." These considerations already greatly limited the number of candidates as a teratogenic factor.

In retrospective, it appears likely that a few cases of typical malformations had already been induced by thalidomide in the premarketing period in 1956/1957, when the new drug was available to doctors, nurses, and members of the Chemie Grünenthal Company (LENZ 1988).

Pinning down the responsible agent as thalidomide was the result of a concerted effort by several clinicians in Germany. Other physicians, besides Lenz, that deserve credit for their extensive involvement in the attempt to analyze the epidemic and the possible cause include WEICKER, WEGERLE, HUNGERLAND, PFEIFFER, BACHMANN, KOSENOW, and JURCZOK, and many others contributed case reports and comments on the possible identity of the responsible agent. On 15 November 1961, LENZ (after obtaining a clue from the father of a malformed child) decided that there was more than a vague indication of a causal relationship between the malformations and thalidomide. He informed Grünenthal of this possibility and urged them to withdraw the drug from the market. It is interesting that just 2 days earlier the evidence of a possible casual relationship had not been particularly clear even for LENZ

(1985). However, researchers were accumulating evidence that thalidomide might be responsible. In December 1961, a short note was published in a widely distributed German medical journal that thalidomide might have caused the malformations (LENZ 1961). At the same time, this was also suggested outside of Germany (MCBRIDE 1961a,b), and several reports were subsequently published in early 1962 (LENZ 1962; PFEIFFER and KOSENOW 1962; WEICKER and HUNGERLAND 1962), followed by an avalanche of publications on this topic from many countries.

In cases in which the fully developed malformation pattern was present, the syndrome seemed well characterized and the diagnosis of thalidomide embryopathy obvious. However, some cases of other causes (e.g., mutations such as the Holt Oram Syndrome) might have been falsely diagnosed as thalidomide-induced disease (VAN REGEMORTER et al. 1982). This most probably also holds true for the report on two cases of malformed children born to malformed fathers, classified as thalidomide victims (MCBRIDE 1994). A speculation that thalidomide might have caused these malformations through mutations is not backed up by any evidence. This problem will be discussed later (see Sect. E.II.1.a).

Credit is due to LENZ for having compiled in very systematic inquiries an impressive amount of evidence to establish without any serious doubt the causal relationship between treatment with thalidomide and the induction of the specific malformations (e.g., LENZ and KNAPPE 1962a,b).

II. Types of Teratogenic Effects and Organotropy

The pattern of malformations induced by thalidomide is almost identical in all non human primates studied so far as well as in children.

In the majority of cases, the defects induced may be classified as reduction abnormalities. Most significant is the finding that not all of the possible defects inducible by thalidomide are present in all of the victims. Thus a typical pattern with a multiplicity of defects is only present in a few patients, in contrast to typical genetic syndromes.

Effects on the developing limbs and on the outer and inner ears were particularly dramatic. Phocomelia and amelia are severe malformations, and it was due to their severity that the epidemic was recognized. However, thalidomide-induced defects were not as severe in all of the cases, and limb malformations often did not consist of an agenesis of all the long bones, but were confined to radius or fibula agenesia and finger defects. This malformation pattern has been systematically described (WILLERT and HENKEL 1969; HENKEL and WILLERT 1969).

Not all limb defects were reduction abnormalities. Polydactyly and a triphalangy of the thumb were also observed. Whether such types of malformations may be considered as overcompensation phenomena is not yet clear.

The malformation pattern induced by thalidomide was by no means restricted to the limbs, and many organ systems were affected in addition to limb defects or alone. Since evaluation was only performed retrospectively and as a collection of case reports, it must have been difficult, if not impossible, to establish a reasonable causal relationship between the exposure to thalidomide and the occurrence of defects already observed at a fairly high rate within a population, such as malformations of the cardiovascular system, if such defects were induced by thalidomide as single abnormalities. For this reason, the best evidence for a causal relationship for these other organotropisms comes from observations of the simultaneous occurrence of additional abnormalities together with typical limb defects.

In an interesting approach, KIDA (1987) grouped thalidomide-affected Japanese children into three groups: those with limb defects without hearing loss, those with hearing loss without limb defects, and those with both hearing loss and limb defects (Table 2). These 137 children (46.7% males) were selected from the Japanese register of a total of 309 thalidomide victims. Of these 137 children, 21 (15%) had amelia or phocomelia, but of these only eight (38%) had no additional ear defects. In contrast, hearing loss occurred frequently without the simultaneous occurrence of limb defects. Facial nerve palsy was strongly correlated with hearing loss, pointing to a similarity in the induction of the defects with respect to time and site. Although defects of the skull and brain defects in terms of lower IQ were not obvious, pathological findings in electroencephalograms (EEG) were frequent. Similarly, about 60% of the children showed abnormal electrocardiograms (ECG), indicating impaired cardiac function.

Gynecological examination of 32 girls aged 15–18 years (and one aged 13) with typical thalidomide-induced limb malformations revealed primary ame-

Table 2. Pattern of abnormalities observed in 137 Japanese children with thalidomide-induced congenital defects (Data from KIDA 1987)

	Upper limb deformities only	Hearing loss only	Limb deformities and hearing loss
Children with these typical main defects	55	35	47
Amelia or phocomelia	8		13
Total hearing loss or severe impairment		23	3
+ Pathological EEG findings	16	18	21
+ Pathological ECG findings[a]	34	16	21
+ Arrhythmia[a]	11	4	8
+ Block[a]	17	10	7
+ Duane's syndrome	1	27	3
+ Facial nerve palsy	3	33	2

EEG, electroencephalogram; ECG, electrocardiogram.
[a]of 118 children examined.

norrhoea in 11 of these cases (MÜHLENSTEDT and SCHWARZ 1984). In nine girls this was associated with genital malformations (seven uterus aplasias, two hypoplastic or rudimentary uteri), eight of them also showing abnormalities of the vagina (three vaginal aplasias, four blind vaginal sacs, one urogenital sinus). In 21 of the girls, puberty developed normally, without any indication of hormonal disturbances. In the United Kingdom, 35 women born affected by thalidomide were identified that gave birth to 64 viable children and had six miscarriages (MAOURIS and HIRSCH 1988).

III. Frequency of Specific Malformations Observed in Some Areas of Germany

There are a number of questions that cannot be answered on the basis of the information available from the observations in humans. These include the following:

- What is the incidence of malformations when the drug is taken during the susceptible period at doses used as sedative or hypnotic?
- What is the incidence of malformations when the drug is taken during the susceptible period at much higher doses for use as an anti-inflammatory or immunosuppressive agent?
- How many abortions are caused by thalidomide?
- Are there more types of malformations induced by thalidomide which could not be detected because they were not compatible with prenatal life?

Most of the statements made on these topics have been pure speculations and cannot be backed up by concrete data.

From a few observations in humans as well as some data from nonhuman primates, there is no doubt that a single dose of thalidomide may be sufficient for the induction of malformations, if the dose is high enough and exposure occurs at the appropriate embryonic stage. From the experimental data, it is also well-known that this is not easily achieved in all animals because of the variability in embryonic development. Furthermore, in humans it is not known whether a low therapeutic dose, e.g., 100 mg thalidomide (1–2 mg thalidomide/kg body weight), is sufficient to induce severe malformations in 100% of the exposed children. With potent thalidomide derivatives (such as α-EM12) this can clearly be achieved in this dose range with nonhuman primates (HEGER et al. 1994). In experimental studies with nonhuman primates, prenatal mortality induced by thalidomide was found to be very low, if not nonexistent. This also excludes the possibility of the occurrence of frequent malformations incompatible with early life, up to the stage at which cesarean section was performed. Of course, a high postnatal mortality is well documented in malformed children.

Some assessments have been made of the frequencies of limb reduction defects probably induced by thalidomide during the epidemic. From the area around Bonn, data from two obstetric clinics were compared with historical reference data from the area around Münster (WEICKER and HUNGERLAND 1962). In the two clinics in Bonn, in a total of approximately 1800 births there were five cases of phocomelia in 1960 (2.8%) and ten cases in 1961. In Münster, in the approximately 302 000 births from 1949 to 1956, there were 60 severe hypo- or aplastic limb defects (0.2%), three phocomelias (0.01%), and two amelias (0.007%). Even if one takes into account the very small number of defects in the reference group and if one combines the data on phocomelias and amelias, these data suggest that the frequency of these defects may have increased about 150- to 330-fold during the epidemic. Remarkably, this would barely increase the total frequency of malformations in this population (assuming a spontaneous rate of 2%–3%). Therefore, the thalidomide epidemic was only detected because of the increase of a very unusual and rare type of gross structural abnormality.

IV. Phase Specificity of Teratogenic Effects

A number of very extensive and careful retrospective inquires were performed with mothers of malformed babies to assess the embryonic period in which the embryo is susceptible to the action of thalidomide. As a result of such studies by LENZ and KNAPP (1962a,b) and by NOWACK (1965) and KREIPE (1967), all from the Institute of Human Genetics in Hamburg, the thalidomide-sensitive phase was determined to be from day 34 to day 50 after the beginning of the last menstrual period or (roughly deduced from such assessments) possibly from day 20 to day 36 of gestation (postovulation or postconception).

However, great caution is required with respect to the assessment of development stages and hence periods during which an embryo is susceptible to the action of a defined teratogenic effect on the basis of days of pregnancy, as judged from the last menstrual period or even ovulation or coitus data. With these serious limitations in mind; the main statements made in several publications are compiled in Tables 3 and 4. From these statements it appears well established that the duration of the sensitive period is 3 weeks or even less.

In addition to the uncertainty of the mothers trying to correctly remember events which happened at least 8 months ago, a serious limitation of such assessments is that the calculations rely on the assumption that a defined embryonic stage can be related to a defined day of pregnancy. This is certainly not the case.

A closer evaluation of the data presented by NOWACK (1965) on the induction of upper limb phocomelias (Fig. 3) reveals that the association of this major malformation in the thalidomide syndrome with the days of drug exposure is rather poor when only the 11 cases are considered with a verified drug medication for 3 days or less. In fact, the range extends from days 37 to 50 of pregnancy (median, day 44). It is quite obvious that the development of an early limb bud in an individual embryo does not extend over a period of 14 days. These data already suggest a considerable variance of the developmental embryonic stages to be expected at a given day of pregnancy.

Table 3. Rough retrospective assessment of days of pregnancy during which the typical malformations may have been induced by thalidomide in children. (From LENZ and KNAPP 1962[a,b], PLIESS 1962; NOWACK 1965; KREIPE 1967)

Defect	Cases (*n*) Total	Cases (*n*) within period stated by author	Sensitive period[c]	Days of exposure[d]
Skeletal defects[a]				
Amelia, upper limbs	10	3	38–43	35 to 55+
Amelia, lower limbs	3	0	41–45?	39 to 50+
Phocomelia, upper limbs	37	19	38–49	–35 to 55+
Phocomelia, lower limbs	18	2	40–47	–35 to 55+
Ray defect, upper limbs	17	6	39–45	–35 to 55+
Ray defect, lower limbs	8	0	45–47	40 to 55+
Agenesia of thumb	6	2	35–40	30 to 45+
Triphalangy of thumb	3	1	46–50	44 to 58
General defects[a]				
Anotia	14	4	34–38	–30 to 50+
Ear deformities	7	1	39–43	37 to 46
Visceral defects[b]				
Cardiovascular defects	17	4	36–48	–30 to 55+
Pylorus hypertrophy	7	2	40–47	35 to 55+
Duodenal atresia	4	0	42–53	40 to 55+
Anal atresia	4	0	41–43	37 to 55+
Various intestinal defects	9	?	??	33 to 55+
Urogenital defects	11	0	41–45	–35 to 55+
Respiratory tract defects	5	0	41–43	–35 to 55+
Pylorus hypertrophy	7	2	40–47	35 to 55+

[a]From NOWACK (1965).
[b]From KREIPE (1967).
[c]As suggested by the author, in days of pregnancy (calculated from last menstrual period).
[d]Minus signs indicate this day and earlier, plus signs this day and later.

The assumptions of definite correlations of embryonic stages with days of pregnancy is also no longer tenable on the basis of our present knowledge of modern embryology. From numerous data it is well known and accepted that there is a considerable variability of embryonic stages at a given day of pregnancy. This is the case even when ovulation or a single coitus date is considered, and it becomes even larger and barely assessable if the last menstrual period is taken as an indicator of the beginning of pregnancy. It would be a serious mistake to rely on such an assumption, since abnormal development might be induced at a period assumed to be "safe." This also has consequences for assessing whether a typical malformation could have been induced when the exposure occurred at a defined period of pregnancy.

From the studies by NISHIMURA et al. (1968) and SHIOTA et al. (1988), it is obvious that a given stage of development may be present at an extended period of pregnancy, and during this period of organogenesis no single day of pregnancy can be assigned to a defined embryonic stage. This has been shown

Table 4. Frequency of malformations observed after exposure to thalidomide on presumed days of pregnancy

Malformation	Frequency of malformations (*n*)						
	Days of pregnancy (after last menstrual period)						
	30–33	34–37	38–41	42–45	46–49	50–53	54–56
Anotia	11	**30**	25	13	14	3[a]	
Agenesia of thumb	5	**15**	13	5	0	0[a]	
Amelia, upper limbs	0	**30**	25	13	14	3[a]	
Phocomelia, upper limbs	0	13	45	**77**	61	14[a]	
Phocomelia, lower limbs	0	14	24	40	**45**	10[a]	
Ray defect, upper limbs	0	4	12	**31**	25	6[a]	
Ray defect, lower limbs	0	0	4	**22**	16	4[a]	
Respiratory tract defects	1	7	7	**10**	9	4	2[b]
Cardiovascular defects	3	8	13	25	**31**	30	14[b]
Anal atresia	0	1	8	7	**10**	8	4[b]
Urogenital defects	1	4	19	31	**33**	30	14[b]
Duodenal atresia	0	0	3	5	8	**9**	6[b]

The number of malformations is given with the days of exposure included in the statements. (Modified from NOWACK 1965; KREIPE 1967). Bold type indicates the period of maximum effect.
[a]Only day 50 evaluated.
[b]Only days 54 and 55 evaluated.

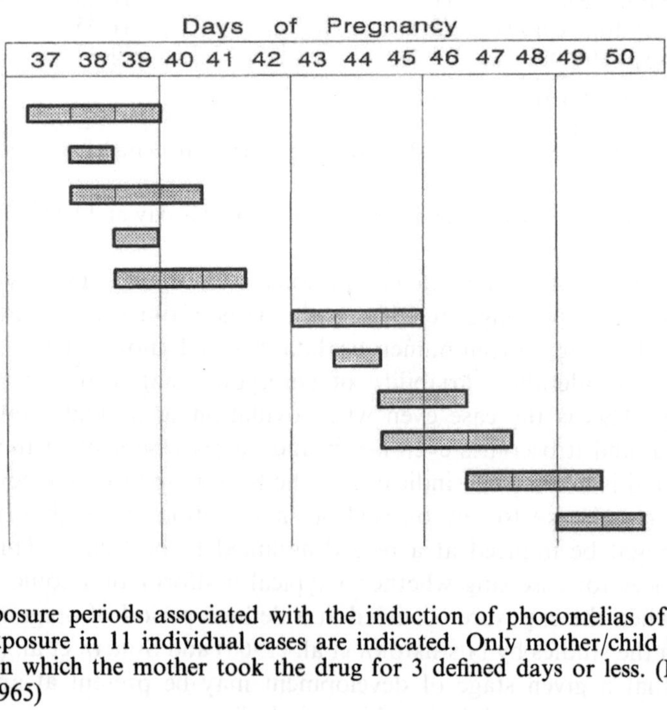

Fig. 3. Exposure periods associated with the induction of phocomelias of the arms. Days of exposure in 11 individual cases are indicated. Only mother/child pairs were evaluated in which the mother took the drug for 3 defined days or less. (Data from NOWACK 1965)

especially convincingly in humans using a single coitus date (SHIOTA et al. 1988), and this result is in complete accord with results obtained on timed pregnancies of nonhuman primates (NEUBERT et al. 1988a; Table 5) and even results obtained in rodents (THIEL et al. 1993). For this reason, developmental stages are no longer correlated with exactly defined days of pregnancy in modern textbooks of embryology (O'RAHILLY and MÜLLER 1992).

Nonetheless, there will be a certain probability that, during a defined period of pregnancy, the embryo passes through defined stages of development. Thus a rough estimation of the period of pregnancy may be made at which susceptible stages of embryonic development occur. These considerations are only important if the possible effect of a short-term exposure is being assessed. With a longer exposure, there is a fair chance that the agent will act during the susceptible embryonic stage.

Assuming from studies in monkeys that the embryonic stages (late) 11 to (early) 14 are the ones in which embryos are susceptible to the induction of upper limb phocomelias by thalidomide (NEUBERT et al. 1988c), it may be expected that the majority of human embryos pass through these stages between days 20 and 50 postconception, with a median around day 30 (SHIOTA et al. 1988). This may correspond to a median of around day 42 of pregnancy when based on the last menstrual period, and this figure is in reasonable agreement with the data presented by NOWACK (1965). The sensitive period may be longer (probably up to about day 60 postmenstruation) than suggested by LENZ and coworkers (LENZ and KNAPP 1962; NOWACK 1965; KREIPE 1967). However, such an assumption is quite compatible with the data of NOWACK (1965). Thus the final conclusion would be similar, but the explanation would

Table 5. Assessment of embryonic developmental stages and number of somite pairs of the individual embryos in each of the litters on defined days of pregnancy in 21 marmosets (*Callithrix jacchus*)

Day of pregnancy							
53		54		55		56	
Stage	Somite pairs (*n*)	Stage	Somite pairs (*n*)	Stage	Somite pairs (*n*)	Stage	Somite pairs (*n*)
11	18/19	10	8/8	10	11/11/12	12	25/25
12	22/22	10	11/11	10	12/12	13	35/35/36
		12	20/23	11	15/15/15	14	37/37
		12	24/24	11	18/18	14	44/44
		12	28/28/28	11	19/19		
		12	28/28	12	22/22/23		
				12	23/24/24		
				12	29/29/29		
				13	35/35		

Each set is one litter of two to three embryos. While interlitter variability is large, intralitter variability is negligible in the marmoset. Stages 11–14 correspond roughly to gestational days 20–50 in humans (SHIOTA et al. 1988).

now be different, based rather on statistical considerations than on stringent correlations between embryonic stages and defined days of pregnancy (postmenstruation). A similar approach was used before by NISHIMURA (1972) to evaluate the possible effect of another teratogenic agent.

V. Species Specificity of Teratogenic Effects

Following the publications on malformed children, worldwide attempts were started to duplicate the teratogenic effect in experimental animals. Since at that time the chick embryo was the main study object of embroyologists, tests in this model and in the rodent species (the species used predominantly in toxicological research) were performed. In the follow-up of the original observations (KEMPER 1962), the results with thalidomide obtained in a considerable number of studies using chick embryos and a variety of techniques and solvents were contradictory. One explanation for the divergent results may be the considerable technical difficulties caused by the poor water-solubility of thalidomide. This species is now usually considered to represent an unsuitable model for studying the teratogenic action of thalidomide or its derivatives.

One of the characteristic features of thalidomide is that it exhibits a very pronounced species and phase specificity. This is unusual for a substance with an extremely high teratogenic potency. As early as the 1960s, studies on a large variety of animal species were performed in order to find a model to mimic the effects observed in children. The data have been compiled by HELM (1966), and no further description of the (mostly negative) results obtained with various more or less exotic species will be given here.

There are three animal species in which considerable research has been carried out, and the results published on these species will be discussed in more detail. These species are rodents, rabbits, and nonhuman primates.

1. Rodents

Since rodents have been, and still are, the best-studied models in toxicology, it is understandable that considerable efforts have been made to reproduce in rodents the thalidomide-induced malformations recognized in children. This would have provided the most convenient test model for systematic studies on a large scale.

Thalidomide certainly crosses the placenta in rodents (BECKMANN 1962; KORANSKY and ULLBERG 1964). However, the transfer to the conceptus has never been quantified at the sensitive embryonic stages (i.e., developmental stages 11–14 in the primate). There is no doubt that serum concentrations exceeding those reached during therapeutic use in humans can easily be reached in rodents. Despite a large interindividual variation, peak concentrations of between 5 and 20 μg/ml were measured after a single oral dose of 50 mg thalidomide/kg in Wistar rats (ERIKSSON et al. 1992).

Overall, the results of teratological studies with rats were disappointing. Multiple strains were tested, but, although some defects were claimed to have been induced (e.g., KING and KENDRICK 1962), the results of the majority of the better studies were completely negative. Although a typical agenesia of the long bones of the limb was never reported after thalidomide exposure in rats and mice, the occurrence of other very divergent (and compared with the malformations in primates quite atypical) abnormalities was reported in a number of early studies performed in the 1960s, as summarized by HELM (1966). We have to take into account the fact that knowledge on testing for reproductive toxicity was limited at that time, and very few of these older studies would meet the standards required today for such experiments. Following high doses of thalidomide (100–500 mg/kg) embryolethality was reported in several studies. However, the effect was variable in different rat strains, e.g., Wistar and BD II rats; even when studied simultaneously, one strain responded to thalidomide and the other not (BROCK and VON KREYBIG 1964). There is also an indication (PALMER 1977) that thalidomide may alter the conception rate and the litter size in rats when given over prolonged periods in a so-called two-litter test (Table 6). The effect increased with subsequent matings. However, secondary-stage testing also suggested an embryolethal effect as a primary cause. Whether this may be considered a specific effect or an action induced by the high doses given over prolonged periods is difficult to decide. The results might be taken as a clue that thalidomide may also induce some reproductive toxicity in certain rat strains, possibly not in all of them. Since the drug was generally not applied before the organogenesis phase, an early embryolethal effect certainly cannot be excluded from the data available. However, the rat is not suitable as a model for studying the typical thalidomide-induced limb abnormalities, since these defects have never been reported.

Table 6. Effect of thalidomide on conception rate and litter size in a two-litter test

	Conception rate (%)		Litter size at birth	
	Control	Thalidomide	Control	Thalidomide
Experiment I				
First mating	79	32	8.3	3.5
Second mating	74	17	8.9	4.7
Third mating	83	6	8.8	3.0
Experiment II				
First mating	50	5	9.2	4.0
Second mating	62	0	9.5	–
Third mating	75	0	–	–

Twenty male and 20 female rats were treated with daily doses of 200 mg thalidomide/kg body wt per day for 60 days before mating and during the subsequent mating, pregnancy, and lactation periods. The pregnancy rate and the mean litter size at birth were evaluated. (Data from PALMER 1977)

Interestingly, these obvious failures at the beginning of the research period on thalidomide to reproducibly induce typical teratogenic effects in rodents have not discouraged investigators from undertaking further attempts to try to induce malformations with thalidomide in rats or mice. Various vehicles have been tested (e.g., DMSO, Tween-20, dimethylformamide) to overcome the extremely poor water solubility of this substance or to achieve high plasma levels via intravenous applications (SCHUMACHER et al. 1968b, 1972b). In a later study, 45 mg thalidomide/kg was given in 0.5 ml dimethylformamide to Wistar rats as single intravenous injections on days 10, 11, or 12 of pregnancy (PARKHIE and WEBB 1983). In addition to an increase in prenatal mortality, the occurrence of malformations was mostly rib and eye abnormalities (Table 7). Again, no limb defects were observed. From our experience, thalidomide comes out of solution in DMSO or dimethylformamide after the first contact with tiny amounts of water. Thus in practice a suspension of thalidomide is injected, with all the possibilities of microembolism etc., that this involves, i.e., the possibility of nonspecific effects. Whether such an experimental set-up can be considered sound may be doubted. Recently, thalidomide derivatives with a higher water solubility have been synthesized (KRENN et al. 1992). We can safely conclude that all these approaches turned out to be failures, and none of the claims of partial success with respect to the induction of limb abnormalities in rats have been substantiated.

Unfortunately, a considerable number of papers have been published by a group in Germany claiming the induction of malformations in mice and rats by a single injection of thalidomide as a suspension in a 25% detergent (Tween-20) solution at a dose of 0.25–1.0 ml intraperitoneally (assuming a body weight of the mice of less than 40 g, this would correspond to more than 1.6–6 ml Tween-20/kg). Using this odd technique, a number of thalidomide derivatives synthesized by various reputable groups were tested, and conclusions on chemical structure–effect relationships were drawn by F. KÖHLER and

Table 7. Embryolethality and teratogenicity of thalidomide in rats

	Dead or resorbed fetuses		Malformations					
			Eye defects		Rib defects		Vertebral column defects	
	(%)	(*n*)	(*n*)	(%)	(*n*)	(%)	(*n*)	(%)
DMF control	3.1	98	0/22	0	9/73	12	4/73	6
Thalidomide (day 10)	23.7	97	4/20	20	9/54	17	4/54	7
Thalidomide (day 11)	10.7	122	1/22	5	29/87	33	0/87	0
Thalidomide (day 12)	15.0	100	4/23	17	28/62	45	7/62	11

Rats were treated with a single dose of 45 mg thalidomide/kg intravenously on days 10, 11, or 12 of pregnancy. Thalidomide was injected in 0.5 ml dimethylformamide (DMF). Evaluation was performed on day 20 of pregnancy. Only some of the fetuses were evaluated for eye defects. No limb defects were observed. (From PARKHIE and WEBB 1983)

coworkers (e.g., KOCH and KÖHLER 1976; FICKENTSCHER et al. 1977). Many of the publications from that laboratory only cause confusion, and there is good reason to believe that the studies were not properly performed. However, results from these studies are often cited, and unjustified conclusions, e.g., on structure–effect relationships, are drawn. Evaluation of this technique in other laboratories (including our own) proved the higher doses of the vehicle itself to be lethal to the pregnant mice, and the procedure to be totally inadequate for any type of testing for reproductive toxicity (SCOTT et al. 1977).

More recently, the possible teratogenic effect of the combination of a hydrolysis product of thalidomide (*N*-phthalyl-glutamic acid, NPGA) and the detergent Tween-20 was investigated in more detail in NMRI mice (KOCHER-BECKER et al. 1992) with a similar experimental design. It was demonstrated that the detergent alone was teratogenic when administered at a dose of 2.5 ml/kg intraperitoneally (a route of administration better avoided in studies on reproductive toxicity). This dose also induced a resorption rate of about 20% and in some cases maternal mortality. It is noteworthy that the authors reported for the first time the induction of a very low incidence of very severe limb abnormalities (including amelia of all the limbs) when the combination of these two substances (NPGA at 350 mg/kg) was administered on one day (day 8 or 9) of pregnancy. However, the embryomortality increased to 45% under these experimental conditions, and a considerable maternal mortality was obvious. It is difficult to judge whether this was predominantly an effect of the intraperitoneally administered high dose of the detergent and whether the thalidomide hydrolysis product, otherwise reported to be nonteratogenic even in a sensitive species (e.g., the rabbit), played any specific part in this outcome. Overall, these results confirm that the mouse cannot be recommended as a species suitable for reproducibly inducing thalidomide-typical limb defects.

2. Rabbits

Besides nonhuman primates, certain rabbit strains (the large New Zealand white and the rather small Himalayan) are the only species suitable for evaluating certain aspects of the teratogenicity of thalidomide derivatives. The first indication of a species responding to thalidomide with clear-cut malformations, came from studies by SOMERS (1962). Using New Zealand white rabbits and a dose of 150 mg thalidomide/kg given on days 8–16 of pregnancy, stillbirths and limb reduction deformities were induced in three animals. These results were subsequently confirmed in several laboratories, but not all rabbit strains are suitable as a model for studying this effect. Furthermore, no study has managed to induce a 100% effect, probably due to the simultaneously occurring embryomortality. No phocomelias or amelias were reported, and compartively high doses are required to induce the adverse effects (LEHMANN and NIGGESCHULZE 1971; STERZ et al. 1987). However, a suitable rabbit strain can be considered a convenient model for studies with thalidomide derivatives if the limitations mentioned are taken into account and if large enough groups of rabbits are evaluated.

After the first report by SOMERS (1962), the New Zealand white rabbit species was widely used for this purpose, and a number of systematic studies were performed (STAPLES and HOLTKAMP 1963; PEARN and VICKERS 1967). In several of the studies with rabbits, the Himalayan strain was used successfully (LEHMANN and NIGGESCHULZE 1971; STERZ et al. 1987).

Examples shown in Table 8 are fairly typical for the outcome of such studies in rabbits:

- Fairly high doses of thalidomide are needed to induce a teratogenic response.
- Pronounced decreases in maternal body weight occurred during the treatment period.
- Thalidomide treatment leads to high resorption rates; at the lowest dose used embryomortality was about 33%, and at the highest dose 72%. Because of this high resorption rate, a reliable evaluation of the dose-related malformation rate is not possible.
- With 50 mg thalidomide/kg, only 10% of the fetuses were malformed. This is typical for most of the studies published, and in only a few studies were effects observed with this or an even lower dose. With thalidomide deri-

Table 8. Teratogenicity and embryomortality induced by thalidomide in Himalayan rabbits

	Daily doses of thalidomide (mg/kg)			
	Controls	50	150	450
Pregnant does	9	10	9	9
Litters with malformed fetuses	1	4	8	8
Dead fetuses (total)	1	7	0	0
Resorptions (total)	7	20	30	49
Viable fetuses (total)	**65**	**55**	**36**	**19**
Malformed fetuses (total)	1	6	23	14
Paw, flexura		4	14	9
Thumb, aplasia of			10	9
Cleft palate			6	1
Radius, aplasia or hypoplasia			5/24[a]	2/14[a]
Tibia, aplasia or hypoplasia			2/25[a]	4/14[a]
Fibula, aplasia or hypoplasia			2/25[a]	2/14[a]
Os pubis, aplasia				1/14[a]
Sternebrae, defects		4/31[a]	9/25[a]	5/14[a]
Vertebral column, disorientations		1/31[a]	13/25[a]	6/14[a]
Vertebral column, synostoses		1/31[a]	1/25[a]	2/14[a]

The drug was given from the sixth to the 18th day of pregnancy; only those examples are shown in which some dose response was obvious. Doses of 150 mg/kg or less caused more than 45% embryomortality; data were difficult to evaluate. The number of viable fetuses evaluated is indicated in bold type.
(Examples of data from LEHMANN and NIGGESCHULZE 1971)
[a]The skeletons were not cleared and stained in all fetuses. The two figures given indicate the number affected/number evaluated.

vatives with a very high teratogenic potency (in primates), such as α-EM12, malformations can only be induced in rabbits within a very small dose range, because the embryolethal effect (not observed in nonhuman primates over a dose range of several orders of magnitude) predominates.
- The types of malformations have few similarities with those observed in nonhuman primates or in children. The most prominent and frequent abnormality was a fixation of the paw in an inwardly rotated position without bone defects. This was confirmed in studies from several laboratories, including our own. It is not a frequent or typical abnormality seen in children with the dysmelia syndrome.
- Although an assessment of a teratogenic potential of thalidomide derivatives is certainly possible in studies performed in rabbits by an experienced laboratory and with a large enough number of animals, the relevance to the situation that might exist in humans is limited. It would be impossible to rely solely on data obtained in this species to market a thalidomide derivative. Studies with primates are essential.

It was reported that in some rabbit strains (e.g., JW-NIBS) very atypical abnormalities were inducible by thalidomide which were not seen in other strains and especially not in humans or nonhuman primates. Such malformations included anencephaly, holoprosencephaly, hydrocephalus, and cleft palate (Matsubara and Mikami 1985). No explanation for such a strain peculiarity is currently available.

3. Nonhuman Primates

The typical malformation pattern first seen in children can be completely reproduced in all Old World and New World monkeys studied so far. The dose range is similar to the one apparently effective in humans, i.e., 10 mg thalidomide/kg or less. However, a teratogenic action of thalidomide could not be induced (Butler 1977; Hendrickx 1972) in the prosimian (bush baby or greater galago, *Galago crassicaudatus*).

Nonhuman primate species responding with the typical pattern of thalidomide abnormalities include the cynomolgus monkey (Delehunt and Lassen et al. 1964), rhesus monkey (Wilson 1966; Barrow et al. 1969), baboon (Hendrickx et al. 1966), bonnet monkey (Hendrickx and Newman 1973; Newman and Hendrickx 1985), crab-eating monkey (Hendrickx 1973), and green monkey (Hendrickx and Sawyer 1978).

First indications that the typical pattern of abnormalities can also be induced by thalidomide in marmosets (*Callithrix jacchus*) were published by Poswillo et al. (1972) and by Hiddleston and Siddall (1978). More recently, very extensive studies with the potent thalidomide derivative α-EM12 have been performed in Berlin (Merker et al. 1988; Neubert et al. 1988c, 1992a; Heger et al. 1988, 1994). Since thalidomide derivatives such as α-EM12 induce typical malformations in this species in almost 100% of the animals at daily doses between 0.1 and 1 mg α-EM12/kg, very few marmosets are necessary to obtain a rough estimation of the possible teratogenic potential of a thalidomide derivative (KLUG et al. 1994).

Although some quantitative differences between rats and marmosets in the hydrolysis and possibly the metabolism of thalidomide and particularly of α-EM12 in vivo were noted (NAU et al. 1988), these do not explain the pronounced species differences with respect to teratogenicity. Thus the cause of the species specificity seems to be more pharmacodynamic than pharmacokinetic.

Less reliable information is available on a possible embryolethal action of thalidomide. Using *Callithrix*, embryomortality was negligible even at high dose levels when thalidomide was administered during organogenesis. In an early attempt to induce malformations in rhesus monkeys, thalidomide was given for 33–45 days beginning right after mating (LUCEY and BEHRMAN 1963). In this insufficiently documented short communication, it was reported that in 44 treated monkeys no live births occurred, while 11 births were recorded in 57 untreated animals. Although an early embryolethal effect or an interference with implantation by thalidomide may be suggested from the results of this study, because of limitations in the experimental design these data do not prove such a causal relationship.

4. Effects Reported from In Vitro Studies

For elucidating the mode of the teratogenic action of thalidomide and for testing the potency of various derivatives, it would be a great advantage to have a simple in vitro system. Many attempts have been made to this end, but largely without success. Difficulties in the interpretation of results arise, especially when attempts are made to include systems for metabolic activation of the test substance. The chance of inducing artifacts irrelevant to the situation in vivo is substantial. Furthermore, for most of the studies the explants were used from rodents, although these species are well known to respond very poorly or not at all to the teratogenic action of thalidomide. The rationale for such attempts lay in considerations that the species difference might be a quantitative problem which could be overcome by using high concentrations in vitro. Since the peak plasma concentrations after therapeutic doses of thalidomide in women giving birth to malformed children were approximately 1 μg/ml, effects induced at this concentration are of special importance.

Using an organoid "micromass" culture of isolated limb bud cells from 13-day-old (34–36 somite pairs) rat embryos, an inhibition of chondrogenesis was reported (FLINT et al. 1985). However, the effect was not significant at concentrations lower than 300 μg thalidomide/ml (a concentration close to general cytotoxicity in this system), and the inhibitory effect could apparently be overcome by increasing the cell density. Apparently no metabolic activation was required for this effect. It should be mentioned that the effective concentrations needed were about two orders of magnitude higher than serum concentrations in humans during the induction of teratogenicity.

Further extensive attempts were made with mouse limb bud cultures, a system allowing the evaluation of morphogenetic differentiation of the cartilaginous bone primordia. In hundreds of studies with various embryonic stages and types of preparations, neither thalidomide nor α-EM12 were able to induce abnormal development in such organ cultures. There were some indications that, when using limb buds from embryos exposed in utero to rifampicine (known to induce specific cytochrome P450 isozymes), abnormal development was observed in the presence of α-EM12 (NEUBERT and BLUTH 1981). However, again high concentrations were used (100 μg α-EM12/ml), and the results, although clear-cut in several experimental series, could not be reconfirmed later in the same laboratory; thus this system also seems to be inadequate for the purposes intended.

According to a small meeting abstract, some studies were also performed using the well-established rat whole embryo culture system. Using concentrations of thalidomide between 1 and 13 μg/ml, no effect on embryonic development was observed. However, it was reported that after preincubation of the drug for 2 h at 37 °C with S9 mix (rat liver homogenate 9000 *g* centrifugation supernatant), abnormal development was induced in the culture (SAEKI et al. 1986). As mentioned before, the formation of artificial metabolites cannot be excluded in such studies, and it is impossible to interpret the significance of such data for the situation that probably exists in vivo. 5-Fluorouracil was found to induce clear-cut abnormal development in this system. Interference with development was also not seen in the rat whole embryo culture with 50 μg thalidomide/ml in the culture medium (HALES and JAIN 1987). It is confusing that, again, abnormal development was reported to occur when S9 mix was added, but the preparation from rabbit liver was inactive, and those from rats and mice were found to be active. Thus the rat whole embryo culture seems to be unsuitable for evaluating abnormal development induced by thalidomide or its derivatives.

An inhibitory effect of thalidomide was observed in an in vitro system not related to any embryonic model, namely the attachment of mouse ovarian tumor (MOT) cells to concanavalin A-coated polyethylene surfaces (BRAUN and WEINREB 1985), but again, the presence of a drug-metabolizing system was required, and the action of teratogenic and nonteratogenic thalidomide derivatives could not be distinguished. Problems related with such systems will be discussed below (see Sect. E.II.2.a.).

Although not representing an in vitro system, it may be worth mentioning that subcutaneous transplantation of small pieces of limb buds of 14-day-old rat fetuses into athymic (nude) mice was followed by good differentiation of the precartilage tissue, but no impaired differentiation was seen in mice treated with 30–240 mg thalidomide/kg body weight (SHIOTA et al. 1990). A significant effect was observed with cyclophosphamide.

VI. Relationship Between Structure and Teratogenic Effect of Thalidomide-Type Substances

Evaluation of the data from the literature on the relationship between the structure and teratogenic effect of thalidomide type substances is extremely hampered by the fact that many data, e.g., on the teratogenic potential of certain thalidomide-derivatives, were obtained from species known to have a very low or no susceptibility to this class of substances. This applies to the numerous results on malformations that were claimed to have been observed after exposure of rodents or chick embryos to thalidomide. These data on the susceptibility of the chick embryo to thalidomide are controversial; the interpretation of the data is hampered by the insolubility of the substance and nonspecific artifacts induced, and it is thus highly questionable whether this species might be suitable as a model for the action of this drug. In a great number of well-performed studies, rats and mice have been found not to respond to thalidomide, despite other claims, and many of the data obtained in these species with various thalidomide analogues are completely worthless. In addition, many of the earlier studies in rabbits, a species exhibiting some susceptibility to the teratogenic action of thalidomide at high doses, are of doubtful quality, and several of the results have not been confirmed in studies, e.g., on nonhuman primates. For this reason, many of the speculations on chemical structure–teratogenic potency relationships cannot be supported, and many wrong conclusions were drawn and widely communicated in the literature from such false data. This has led to the situation that, in spite of a large list of publications, much of the data and several reviews are greatly misleading. Very little is actually known on the requirements for the chemical structure of thalidomide derivatives necessary to induce the teratogenic pattern typical for this type of substance.

It is well documented that the derivative α-EM12 (see Sect. B.I) is highly teratogenic in nonhuman primates and rabbits (SCHUMACHER et al. 1972b; HELM et al. 1981), its potency being clearly higher than that of thalidomide. No other analogue has been reported in well-documented studies to be more potent than thalidomide.

On the other hand, supidimide (see Fig. 2) was convincingly shown to exhibit an extremely low or no teratogenic potency in nonhuman primates (HENDRICKX and HELM 1980), which also holds for a human metabolite of this substance (SCOTT et al. 1980).

The possible teratogenic potential of the numerous polar hydrolysis products of thalidomide has been tested in some older papers in rabbits (FABRO 1967a; WUEST et al. 1966). There is no evidence that the teratogenic action of thalidomide is mediated via one of these hydrolysis products, although testing may be difficult because of the high polarity of some of the di- and tricarbonic acids.

A typical example of the discrepancy between data in the literature is phthalimidophthalimide. This substance is interesting since it does not contain a glutarimide ring. In an early meeting abstract it was mentioned that this substance is a potent teratogen in rabbits (SCHUMACHER et al. 1972a). However, a recent reevaluation in nonhuman primates has shown this substance to exhibit little if any teratogenic potency in marmosets (KLUG et al. 1994).

1. Teratogenicity of Enantiomers of Thalidomide or Derivatives

Since thalidomide is a racemate, it was of interest whether one of the enantiomers is more active than the other. Omitting the dubious studies performed with mice, in early tests using rabbits some evidence was reported (Table 9) that the *d*-enantiomer is more active than the *l*-form (SMITH et al. 1965).

However, this assumption was based on the two to three pregnant rabbits used for these studies. Furthermore, in a later report from the same laboratory (FABRO 1967b) it was stated that no difference in the teratogenicity of the thalidomide enantiomers exists. A meeting abstract briefly mentioned that in all five rhesus monkeys treated, D-thalidomide induced skeletal defects, while L-thalidomide produced less severe skeletal abnormalities in only three out of five monkeys, but severe internal defects (GIACONE and SCHMIDT 1970).

The first clear-cut results were obtained with the enantiomers of α-EM12 using nonhuman primates. Although some degree of reracemization occurred in vivo (SCHMAHL et al. 1988, 1989), the S(–) enantiomer was shown to exhibit a much more potent teratogenic effect in marmosets than the R(+) form (HEGER et al. 1988, 1994). Thus for the potent thalidomide derivative α-EM12, the effectiveness of the S(–) form is clearly established (Tables 10,11). There is good reason to assume that thalidomide behaves in the same way. It is noteworthy that in this primate species teratogenic effects in almost 100% of the marmosets were obtained with daily doses as low as 100 μg/kg of the α-EM12 racemate.

It is difficult to speculate on the reason for this enantiomer specificity. Both EM12 enantiomers are transferred to the rodent or marmoset embryo (SCHMAHL et al. 1996). Although some differences in the pharmacokinetics were found between the enantiomers, these appear insufficient to explain the large difference in biological potency. In the marmoset, R(+) α-EM12 re-

Table 9. Teratogenic potency of thalidomide enantiomers in rabbits. (Data from SMITH et al. 1965)

Derivative	Number of rabbits	Implantations	Resorptions	Malformed	Normal
DL-3′-Phthalimidoglutarimide	3	23	14	**1**	8
D-3′-Phthalimidoglutarimide	3	21	3	**3**	15
L-3′-Phthalimidoglutarimide	2	19	0	0	19

The number of malformed fetuses is indicated in bold type. The number of animals included in the study was too small for a reliable evaluation.

Table 10. Teratogenic potency of EM12 enantiomers in marmosets (Data from HEGER et al. 1994)

EM12 enantiomer	Dose (μg/kg)	Viable fetuses (*n*)	Litters (*n*)	Abnormal fetuses (*n*)
Racemate	100	12	6	12
S(–) EM12	100	13	8	13
R(+) EM12	100	11	7	4

racemizes slightly faster and is slightly faster hydrolyzed in vivo than S(–)α-EM12. For this reason, the S-enantiomer is excreted more slowly and persists longer in the monkey than the R-enantiomer (NAU et al. 1988; SCHMAHL et al. 1989, 1996). Most of these findings on the stability of the enantiomers and on reracemization have been confirmed for thalidomide and the human organism, but in contrast to EM12 and the marmoset, the S(–)-thalidomide is eliminated considerably faster than the R-enantiomer in humans (ERIKSSON et al. 1995). Thus it is very unlikely that the higher teratogenicity of the S-form is due to metabolic differences compared with the R-enantiomer. On the other hand, it is quite feasible that the biological effects require a defined stereospecificity for the interaction with the crucial cellular components or receptors.

Some attempts were also made to prevent reracemization by using thalidomide derivatives substituted at the asymmetric carbon atom. However, the teratogenic potency of the racemate was no longer reliable enough (Table 12) subsequent to, e.g., methylation of thalidomide at this carbon atom (NEUBERT et al. 1988b). It is surprising that methyl-α-EM12 (EL1010) was even less

Table 11. Teratogenic potency of EM12 enantiomers in marmosets. (Data from HEGER et al. 1994)

Defect	Malformed fetuses (*n*)		
	Racemate	S(–) EM12	R(+) EM12
Amelia, upper limb	–	1	–
Amelia, lower limb	4	1	–
Phocomelia, upper limb	–	1	–
Phocomelia, lower limb	–	1	–
Radius aplasia	11	10	2
Radius hypoplasia	–	–	1
Tibia aplasia	6	3	–
Femur aplasia/hypoplasia	4	3	1
Polydactyly	–	–	1
Finger rays, reduced	1[a]	8	2
Mandibula aplasia/hypoplasia	3	6	–
Temporomandibular joint, fused	2	3	–
Fetuses without defects	0	0	7
Total number of fetuses	12	13	11

Dose of the enantiomers or the racemate, 100 μg/kg on days 48–60 of pregnancy
[a] 11 fetuses were evaluated.

Table 12. Teratogenic potency of thalidomide derivatives methylated at the asymmetric carbon atom

Thalidomide derivative	Dose (mg/kg)	Amelia or phocomelia	Abnormal mandibula	Abnormal fetuses
Methyl-thalidomide (GR 2358)	60	3/15	3/13	11/15
α -EM12	5–10	21/40	16/38	40/40
Methyl-α-EM12 (EL1010)	50	0/8	0/8	0/8

Marmosets were treated orally on days 51–57 of pregnancy. Evaluation is based on the total number of fetuses (twin or triplet pregnancies). (Data from NEUBERT et al. 1988b; MERKER et al. 1988; and unpublished results)

teratogenic than the corresponding thalidomide derivative, despite the much higher teratogenic potency of α-EM12.

D. Effects other than Sedative/Hypnotic Ones Induced in Adults

Although thalidomide was used initially and during the official marketing period as a sedative and hypnotic drug, it was tried thereafter for numerous additional therapeutic indications. An extensive discussion is beyond the scope of this handbook, and these data will only be briefly mentioned. The main emphasis will be put on the question of how these effects may relate to the teratogeic potential.

I. Anti-inflammatory and Immunosuppressive Effects

The majority of further therapeutic applications of thalidomide can be classified as treatment of effects of inflammatory diseases and certain indications of immunosuppression. Thalidomide is considered a special type of anti-inflammatory or immunosuppressive agent, since its mechanism of action seems to differ considerably from that of other agents of this class, e.g., glucocorticoids, cyclosporin A, azathioprine, or nonsteroidal antiphlogistic substances.

Since thalidomide is increasingly used clinically for the purpose mentioned, a short discussion of the therapeutic effects achieved and the doses required seems justified here, especially because it is almost impossible to completely avoid adverse effects, including teratogenic ones, in all countries during long-term therapy.

The first hint to possible anti-inflammatory effects of thalidomide in patients was reported quite early on in the therapeutic use of this drug as a sedative. Prevention of edema after surgery, especially in 17 patients after hemorrhoidectomy, was found with a treatment of 100 mg thalidomide given every 8–12 h (MILLER et al. 1960). Apparently, not much attention was given to this observation, and the issue was not continued by further studies.

1. Clinical Effects Observed in Humans

Despite serious concern about severe adverse effects, especially neuropathy, and the danger of inducing malformations if insufficient precautions are taken, more than 10 000 individuals have probably been treated with thalidomide since 1961/1962, and certainly more than 7000 for leprosy. Although frequent warnings on the adverse effects have been issued, it has been claimed that a number of typical malformations have occurred in Brazil subsequent to the treatment of young women with thalidomide for leprosy after the original use of the drug was ended (CUTLER 1994). However, since no scientific reports have been published so far and the possibility of genetic syndromes was not excluded, the causal relationship is difficult to assess.

While, with respect to polyneuropathy, a short-term treatment of a few days or even weeks seems to be medically acceptable, one of the problems with respect to the therapeutic use of thalidomide for inflammatory or immunological disorders is that most of the diseases clearly responding to this drug are chronic and require long-term medication. Without any doubt this magnifies the danger of inducing pathological changes such as polyneuropathy. Therefore, further research is required to develop a thalidomide-type drug which retains the beneficial effects but is devoid of the adverse actions. Although no clues have yet been published, it is feasible that such a goal can be achieved.

The first observation on thereapeutic effects, in addition to sedative ones, was made by chance when SHESKIN (1965) administered thalidomide to a patient with leprosy to achieve a sedative or hypnotic effect. He observed a pronounced beneficial effect on erythema nodosum leprosum, which prompted worldwide trials with this drug for this indication. Medication with thalidomide is now an established therapeutic procedure to lepromatous leprosy, i.e., the so-called leprosy reaction (e.g., IYER et al. 1971; SHESKIN 1980; PARIKH et al. 1986; JADHAV et al. 1990). A critical review of the adequacy and limitations of the various studies is given by JEW and MIDDLETON (1990). When used in combination therapy, there may also be the chance of thalidomide reducing therapeutic steroid requirements (WATERS 1971).

Since the first observation by Sheskin, numerous reports have been published on beneficial therapeutic trials attempting to treat various other inflammatory diseases with thalidomide. The results of more than 40 papers are compiled in Table 13, indicating that ulcerations of various causes, including human immunodeficiency virus (HIV) infection, seem to respond best to thalidomide therapy. Especially impressive are the success rates for certain symptoms in cases of the otherwise therapy-resistant Behçet's syndrome, which has several of the following symptoms: aphthous oral mucosal ulcerations, genital ulcerations, ocular symptoms (uveitis, chorioretinitis, or irridocyclitis), skin lesions (erythema nodosum), arthritic symptoms, gastrointestinal lesions (severe colitis), epididymitis, vascular lesions (including thrombophlebitis), and central nervous system symptoms. Apparently not all of these diverse symptoms respond to thalidomide therapy equally well (see Table 14). Although the etiology and pathogenesis of this complex disease remains unclear, a particular reponse to infections (viral or bacterial, possibly via animals) was assumed as a possible cause, and evidence for autoimmune mechanisms with circulating immune complexes was presented. The latter phenomenon may show similarities with the leprosy reaction. Assuming such etiologies, the therapeutic effects of thalidomide could be explained by its immunosuppressive or anti-inflammatory effects (see below).

While in most of the therapeutic attempts daily doses of around 200 mg thalidomide and sometimes even higher (up to 500 mg) were used, in some of the cases much lower doses (25–100 mg) were reported to be sufficient. Interestingly, the drug was found to provide fast relief when ulcerations and inflammation were associated with pain. However, it was also commented (YOULE et al. 1990) that steroid therapy may be effective in some of these patients, and because of side effects (especially polyneuropathy) thalidomide should be reserved for patients who do not respond to first-line treatment with steroids.

There is also one report on beneficial effects in the treatment of refractory or severe rheumatoid arthritis (GUTIÉRREZ-RODRÍGUEZ et al. 1989) with daily doses of 513 ± 63 mg thalidomide for 18 ± 9 weeks. Although seven out of 17 patients were reported to have shown complete remission and five partial remission, the total number of cases is too small for far-reaching therapeutic conclusions with respect to this important disease.

It is noteworthy that in some of these patients adverse effects were noted during the thalidomide therapy not mentioned in previous reports. These included the new occurrence of erythema nodosum in patients with Behçet's disease (SAYLAN and SALTIK 1982) and hypersensitivity reaction (erythematous macular rush, fever, tachycardia, hypotension) in three of six patients with stage IV HIV infection treated for oropharyngeal ulceration (WILLIAMS et al. 1991).

2. Effects Demonstrated in Experimental Animals

When compared with the quantity of human data on anti-inflammatory and possibly immunosuppressive effects of thalidomide, the number of corresponding publications on animal studies is small. In contrast to the situation in humans, more data is available from animal studies on immunosuppression than on the antiphlogistic effects of thalidomide.

a) Anti-inflammatory Effects

Using endotoxin-induced uveitis in Lewis rats as an experimental model (100 µg *Salmonella lipopolysaccaride*, LPS, injected into the footpad), a single dose of 400 mg thalidomide/kg significantly reduced the number of accumulated cells and the amount of protein in the anterior chamber of the eye when evaluated 20 h after LPS injection (GUEX-CROSIER et al. 1995). This dose was still effective when given 4 h after the LPS challenge. A dose of 300 mg thalidomide/

Table 13. Selected reports on inflammatory dermatological diseases successfully treated with thalidomide

Diagnosis (as reported)	Number of patients			Reference
	Remissions		Failure	
	Complete	Partial		
Aphthae, mucocutaneous, necrotic, recurrent	6	0	0	MASCARO et al. 1979
Aphthae, mucocutaneous, necrotic, recurrent	5	0	0	KÜRKÇÜOGLU et al. 1985
Aphthae, mucocutaneous, necrotic, recurrent	8	0	1	TORRAS et al. 1982
Aphthosis, oral, major symptomes	15	29	0	GRINSPAN et al. 1989
Aphthosis, oral, minor symptomes	19	37	0	GRINSPAN et al. 1989
Behçet's syndrome	20	0	1	BENNOUNA-BIAZ et al. 1983
Behçet's syndrome	0	1	0	COUZIGOU et al. 1983
Behçet's syndrome	1	0	0	EISENBUD et al. 1987
Behçet's syndrome, buccal ulcerations	20	6	0	HAMZA 1986
Behçet's syndrome, genital ulcerations	20	4	1	HAMZA 1986
Behçet's syndrome, erythema nodosum	4	0	0	HAMZA 1986
Behçet's syndrome, uveitis	3	4	4	HAMZA 1986
Behçet's syndrome, active arthritis	13	1	0	HAMZA 1986
Behçet's syndrome	4	0	0	HAMZA et al. 1987
Behçet's syndrome, palmoplantar, pustolosis	1	0	0	HAMZA 1990
Behçet's syndrome	4	0	0	JORIZZO et al. 1986
Behçet's syndrome, severe colitis	1	0	0	LARSSON 1990
Behçet's syndrome, pyoderma gangrenosum	1	0	0	MUNRO and COX 1988
Behçet's syndrome	1	0	0	PATEY et al. 1989
Behçet's syndrome	1	0	0	ROGÉ and TESTAS 1984
Behçet's (Neuro-) syndrome	1	0	0	RAMSELAAR et al. 1989
Behçet's syndrome, pyoderma gangrenosum	1	0	0	RUSTIN et al. 1990
Behçet's syndrome, oral aphthae	22	0	0	SAYLAN and SALTIK 1982
Behçet's syndrome, genital ulcerations	16	0	0	SAYLAN and SALTIK 1982
Behçet's syndrome, iridocyclitis	0	0	4	SAYLAN and SALTIK 1982
Light eruption, polymorphous	14	8	3	SAUL et al. 1976
Lupus erythematosus, cutaneous, articular	0	3	0	BESSIS et al. 1992
Lupus erythematosus, chronic cutaneous	7	2	1	HASPER 1983
Lupus erythematosus, subacute cutaneous	2	0	0	NAAFS et al. 1982
Lupus erythematosus, chronic discoid	?	?	?	KNOP et al. 1983

Lupus erythematosus, chronic discoid	18	0	0	LEVI et al. 1980
Lupus erythematosus, chronic	10	8	5	SAMSOEN et al. 1980
Lupus erythematosus, chronic	10	0	2	SCOLARI et al. 1982
Lupus erythematosus, systemic, cutaneous lesions	18	2	0	ATRA and SATO 1993
Proctitis (HIV-associated)	2	0	0	GEORGHIOU and ALLWORTH 1992
Prurigo, actinic	32	1	1	LONDOÑO 1973
Prurigo, actinic	11	2	0	LOVELL et al. 1983
Prurigo nodularis, genital hemorrhagy	1	0	0	NOUGUDÉ et al. 1983
Prurigo nodularis Hyde	3	0	0	SHESKIN 1975
Prurigo nodularis	1	0	0	VAN DEN BROEK 1980
Prurigo nodularis	4	0	0	WINKELMANN et al. 1984
Pruritus uremic	10	0	8	SILVA et al. 1994
Stomatitis, recurrent, aphthous	32	0	0	REVUZ et al. 1990
Stomatitis, recurrent, aphthous (HIV-associated)	2	0	0	NICOLAU and WEST 1990
Ulceration, orogenital	2	0	0	BOWERS and POWELL 1983
Ulceration, orogenital	14	1	0	JENKINS et al. 1984
Ulceration, orogenital	14	1	0	POWELL et al. 1985
Ulcers, aphthous (HIV-associated)	2	0	0	GHIGLIOTTI et al. 1993
Ulcers, pharyngeal, hyperalgic (HIV-associated)	3	0	0	GORIN et al. 1990
Ulcers, pharyngeal, hyperalgic (HIV-associated)	?	?	?	YOULE et al. 1990
Ulcers, oesophageal (HIV-associated)	1	0	0	GEORGHIOU and KEMP 1990
Ulcers, oesophageal (HIV-associated)	1	0	0	RYAN et al. 1992
Ulcers, recurrent, aphthous (HIV-associated)	1	0	0	RADEFF et al. 1990
Ulcers, oropharyngeal, aphthous (HIV-associated)	3	2	0	WILLIAMS et al. 1991
Ulcer, resistant, aphthous (HIV-associated)	7	0	0	YOULE et al. 1989
Ulcerative colitis	1	0	0	WATERS et al. 1979
Lepromatous leprosy	57	0	0	JADHAV et al. 1990
Lepromatous leprosy	94	0	0	PARIKH et al. 1986
Lepromatous leprosy		4500[a]	?	SHESKIN 1980

Some cases may be reported more than once, in different papers.
HIV, human immunodeficiency virus.
[a]The percentage of remissions is not clearly stated. SHESKIN (1975b) mentions 99%.

Table 14. Therapeutic success rate in 35 patients with different dermatological diseases after treatment with thalidomide. (Data from NAAFS and FABER 1985)

Disease treated	Success (*n*)	Failure (*n*)
Lupus erythematosus and mixed connective tissue disease	17	2
Immune complex vasculitis	2	1
Erythema multiforme	2	0
Lichen planus	1	1
Psoriasis vulgaris and balanitis circinata	0	4
Pemphigoid	2	1
Prurigo nodularis (Hyde)	1	1

kg significantly reduced the protein content, but not the number of cells, while 150 mg/kg had no effect in this inflammation model. A possible effect of 400 mg supidimide/kg was not statistically significant.

b) Transplant Rejection

Almost 30 years ago, indications of the possible inhibition by thalidomide of homograft rejection in mice were published (HELLMANN et al. 1965; MOUZAS 1966), and such an effect was even suggested as a mechanism of the teratogenic action of this drug (HELLMANN 1966). Although such an explanation for the teratogenic potential of thalidomide was never seriously accepted by the scientific community, these data provided the first hint of a possible immunosuppressive action of thalidomide. Unfortunately, these studies and the subsequent experiments were performed in rodents, which are known not to respond to the teratogenic action of thalidomide. Thus the reported enhancement of graft survival was not very impressive (DUKOR and DIETRICH 1967; MOUZAS and GERSHON 1968). Many of the results were contradictory, and the effect was not confirmed in other laboratories using mice or rabbits (PLAYFAIR et al. 1963; FLOERSHEIM 1966; BORE and SCOTHORNE 1966).

Although not relevant to the explanation of the teratogenic action of thalidomide, the idea of the rejection of malformed fetuses being a fundamental homeostatic mechanism essential for the survival of the species (HELLMAN 1966) is certainly valid in the case of primates, as has since been shown (NISHIMURA et al. 1966, 1968) in humans.

Prolongation of canine renal allografts was reported when thalidomide was given 1 week before transplantation and continued after the operation, but no effect was seen when the treatment was started at the time of the transplantation (MURPHY et al. 1970a). However, the dosage regimen, (10 mg thalidomide/kg three times a week) was unusual and probably insufficient. Renal allograft survival was also found to be significantly prolonged in baboons after intramuscular dosing with 10 mg thalidomide/kg per day, and several renal functions of the transplant were clearly improved (MURPHY et al.

1970b). This was quite in contrast to results obtained with azathioprine (Imuran; 3 mg/kg per day) or cortisone (60 mg per day).

When evaluating single lung transplantations in mongrel dogs (five per group), substituting thalidomide (2 × 50 mg/kg per day) for prednisone in a standard triple-drug therapy (2 × 20 mg cyclosporin A/kg plus 2 × 2.5 mg azathioprine/kg, plus 2 × 2 mg prednisone/kg per day) seemed to improve the results. However, the triple combination with thalidomide was unable to prevent rejection when the dose of cyclosporin A was reduced to 2 × 10 mg/kg per day, indicating that even this extremely high dose of thalidomide alone was insufficient to prevent transplant rejection (UTHOFF et al. 1995).

Single doses of up to 100 mg thalidomide/kg were unable to prevent transplant rejection within up to 10 days in Lewis rats transplanted with ACI rat skin grafts. However, a twice daily dose of 50 mg thalidomide/kg resulted in prolongation of the skin grafts up to day 21. In comparative studies, graft survival was considerably longer (52 days) after treatment with daily doses of 15 mg cyclosporin A/kg (VOGELSANG et al. 1986a).

Somewhat prolonged graft survival and fewer necroses were also reported for rat cardiac transplantations after treatment with thalidomide alone or in combination with cyclosporin A (TAMURA et al. 1990; EMRE et al. 1990; DÖSTRAAT et al. 1992).

Since no human data are available, the usefulness of thalidomide therapy for preventing rejection in organ transplantation cannot be assessed. From the animal data available at present, it appears unlikely that thalidomide as a single-drug therapy will be very effective in replacing the standard drug cyclosporin A. However, there may be some place for thalidomide in a (possibly short-term) combination therapy, perhaps superseding glucocorticoids.

c) Graft Versus Host Disease

A possible marginal effect of thalidomide in a graft versus host disease (GvHD) rat model was also suggested in the mid-1960s (FIELD et al. 1966). The experimental procedure consisted in injecting spleen cells of Marshall rats into F1 hybrids and then measuring spleen weight after 9–10 days. The significance of the data is hampered by the fact that huge doses of thalidomide (1.25 g/kg) were given over a period of 9 days intraperitoneally.

About 20 years later, extensive and well-controlled studies were performed by VOGELSANG and coworkers on the same topic, using the Lewis ($RT1^{l}$)/ACI($RT1^{a}$) rat major histocompatibility complex (MHC) mismatch GvHD model (BESCHORNER et al. 1982), which provided clear evidence of a reduction of acute GvHD by thalidomide (VOGELSANG et al. 1986a,b, 1987a). Total body-irradiated Lewis rats were injected intravenously with RT1-incompatible ACI rat bone marrow. Clinical symptoms were evaluated during the following 2 weeks. Prophylactic treatment or treatment of established GvHD, using daily doses of 50–100 mg thalidomide/kg, provided complete protection for 100 days, while untreated rats did not survive for 2 weeks. In a later publication, a beneficial effect on this type of GvHD in the rat was reported even after doses as low as 1 mg thalidomide/kg (VOGELSANG et al. 1987a), but in further studies doses up to 25 mg/kg were ineffective, and dosing with 50 mg thalidomide/kg was required to achieve 100% protection (VOGELSANG et al. 1988a). Chimerism in the Lewis rats was demonstrated by permanent (> 100 days) acceptance of ACI skin grafts but rejection of BN rat grafts. Furthermore, in mixed lymphocyte cultures no proliferative response was induced in the cells from these grafted Lewis rats by ACI rat lymphocytes, but a clear-cut response was seen with BN rat lymphocytes (VOGELSANG et al. 1988b). Combination of subthreshold doses of cyclosporin A and thalidomide seemed to enhance the beneficial effect on GvHD (VOGELSANG et al. 1988a).

VOGELSANG and her group also studied the influence of various immunosuppressive agents on chronic GvHD in the same rat model, in animals developing a clinical picture similar to that of humans. All 18 animals used showed a complete response when treated orally with 50 mg thalidomide/kg (but not with 10 mg/kg). This result was more convincing than effects obtained with azathioprine plus prednisone (1.5 mg/kg plus 1 mg/kg), or cyclosporin A (30 mg/kg). A low-dose combination of thalidomide and cyclosporin A also seemed to be effective (VOGELSANG et al. 1989).

Similar to the effects in animal experiments that have been published, there have also been case reports of successful treatment with thalidomide of human acute GvHD after allogenic bone marrow transplantation (LIM et al. 1988; SAURAT et al. 1988; COLE et al. 1994). However, failures were also reported (RINGDÉN et al. 1988). Furthermore, thalidomide was shown to counteract chronic GvHD (MCCARTHY et al. 1988; VOGELSANG et al. 1987b). From more recent and extensive studies (HENEY et al. 1990, 1991; VOGELSANG et al. 1992; PARKER et al. 1995), it appears that treatment with thalidomide alone shows only limited, if any, benefit in acute GvHD. However, patients with chronic GvHD seem to benefit from such a treatment in perhaps 20%–60% of the cases (Table 15). It seems worthwhile to consider the use of thalidomide in a combination treatment.

d) Other Immunological Reactions

Some positive and mostly negative results were reported when a possible effect of thalidomide was studied on a variety of other immunological systems.

Table 15. Therapeutic success rate in 80 patients with chronic graft-versus-host disease (GvHD) after treatment with thalidomide. (Data from PARKER et al. 1995)

	Patients		Survival	
	(*n*)	(%)	(*n*)	(%)
Complete remission	9	11	9	100
Partial remission	7	9	4	57
Total responders	16	20	13	81
Nonresponders	64	80	30	47

A passive Arthus-type reaction, i.e., an immunologically induced inflammatory response (possibly due to immune complex deposition) with complement fixation, polymorphonuclear leukocyte infiltration, and tissue damage, was reported to be inhibited by oral administration of 25 mg thalidomide/kg in rats, when induced after i.v. injection of chicken ovalbumin and intradermal challenge with the immunoglobulin G (IgG) fraction of rabbit anti-ovalbumin (LAWRENCE et al. 1979). No effect of thalidomide (50 mg/kg) on passive Arthus reactions, on cutaneous anaphylaxis, or on antibody formation was found earlier in guinea pigs sensitized with egg albumin in complete Freund's adjuvant (ULRICH et al. 1971).

At the high dose level of 3 × 100 mg thalidomide/kg given interperitoneally within a period of 24 h, the local Shwartzman skin reaction (acute thrombohemorrhagic vasculitis with increased permeability and leukocyte accumulation) after intradermal LPS injection and intravenous zymosan challenge was found to be reduced in rabbits to about the same extent as after treatment with 4 × 4.6 mg dexamethasone/kg, but 3 × 50 mg thalidomide/kg was inactive under the same experimental conditions (WÖHRMANN et al. 1995). There was no convincing difference between the two thalidomide enantiomers, which may not be expected because of the well-known reracemization. A statistically significant, but biologically not too convincing effect of (±)-thalidomide was also reported with a similar experimental setup with mice (GEIST et al. 1995). The experimental setup of the Shwartzman phenomenon is interesting, because it supposedly shows some similarities with the clinical situations in erythema nodosum or Behçet's disease, known to respond to the therapy with thalidomide.

On the other hand, no inhibition by thalidomide of an experimental allergic encephalomyelitis or allergic neuritis (induced by injection of brain homogenate in Freund's complete adjuvant) in guinea pigs or rats was found (VLADUTIU 1966; GOIHMAN-YAHR et al. 1972, 1974), although in some of the studies daily doses as high as 400 mg thalidomide/kg were used. No effects of thalidomide treatment (up to 2200 mg/kg per day) were observed in guinea pigs with respect to active delayed hypersensitivity, passively induced delayed hypersensitivity, hemagglutinating antibody production, passive cutaneous anaphylaxis, or Arthus reactions (OGILVIE and KANTOR 1968). No effect of thalidomide treatment (orally, daily 100 mg/kg) was observed on the experimental autoimmune myastenia gravis, induced by the injection of acetylcholine receptor of *Torpedo californicus* into rats (CRAIN et al. 1989).

It cannot be judged whether this ineffectiveness of thalidomide is due to the choice of an inappropriate species or a lack of a corresponding potential of the drug. The experimental allergic models apparently respond to the action of other immunosuppressive agents such as cyclophosphamide, 6-mercaptopurine, and azathioprine or methotrexate. It should be noted that the teratogenic effects of thalidomide cannot reproducibly and convincingly be induced in these rodent species. Although rodents seem to respond to some of the effects of thalidomide (at least at quite high dose levels), it is still difficult to extrapolate such effects to the situation in primates.

After dosing Swiss mice with 25 mg thalidomide/kg for 4 weeks, anti-sheep erythrocyte plaque-forming cells and hemagglutinin titers were increased (DESCOTES et al. 1988). There was no significant change when compared with controls in mice treated with 5 mg thalidomide/kg. With the higher dose, but not with the lower one, an enhancement of contact hypersensitivity to picryl chloride was also reported. Similarly, the number of IgM plaque-forming cells against sheep red blood cells was increased in the spleen of C57BL/6 mice treated with 150 mg thalidomide/kg intraperitoneally (ARALA-CHAVES et al. 1994). The absolute number of spleen B lymphocytes was also increased. The authors suggest that thalidomide stimulates both peripheral and central immune systems in mice.

II. White Blood Cells of Primates

There have been a number of attempts to obtain information on the possible mode of action of the clinically verified anti-inflammatory and possible immunosuppressive effect of thalidomide by studying alterations in white blood cells.

1. Number and Function of Leukocytes

A few indications exist that thalidomide might influence the number of the main types of blood cells. In a number of publications, more or less convincing evidence has been reported on alterations in leukocyte functions in thalidomide-treated animals or humans. However, as almost typical for studies on thalidomide, many of the results are contradictory or are not reproducible.

Slight reductions in the number of helper T cells ($CD4^+$) have been reported as well as reduction of the $CD4^+$ to $CD8^+$ ratio in healthy volunteers treated with 100 mg thalidomide every 12 h for 4 days (GAD et al. 1985). Similar effects were reported in two patients with erythema nodosum leprosum (SHANNON et al. 1992), but no corresponding deviations were seen in another report with similar patients (MONCADA et al. 1985). Overall, the reductions reported were not very impressive. Furthermore, no data on the intraindividual variability for these cell subpopulations were given. Thus if changes on $CD4^+$ cells did occur, they are certainly not sufficient to explain pronounced effects on inflammatory and immunological processes. No effect of thalidomide treatment of patients with lepromatous leprosy was found on overall immunoglobulin (Ig) levels (IgM and IgA) in blood plasma (ARRUDA et al. 1986).

There are also some publications on functional changes of white blood cells. Inhibition of chemotaxis of human leukocytes in vitro by thalidomide was reported (FAURE et al. 1980). Interestingly, in contrast to erythromycin, direct addition of thalidomide to the attractant medium did not impair chemotaxis. However, when the cells were preincubated with thalidomide for 30 min at 37 °C, a pronounced but dose-independent effect was observed, and 1 μg thalidomide/ml was sufficient to cause an almost 50% inhibition of chemotaxis. No inhibition of monocyte chemotaxis was found in thalidomide-treated patients with lepromatous leprosy (SCHULLER-LEVIS et al. 1987). Chemotaxis was also not found to be influenced using the blood of guinea pigs treated with 12.5 or 25 mg thalidomide/kg 2 days (MIYACHI and OZAKI 1981).

While no effect whatsoever on chemotaxis was found in vitro in human polymorphonuclear leukocytes (PMN) after preincubation with 1–100 μg thalidomide/ml (NIELSEN and VALERIUS 1986), and no change in the oxygen consumption was observed, a moderate but significant increase in the superoxide anion release was found in both PMN and monocytes in the presence of thalidomide after stimulation with *N-f*-methionyl-leucyl-phenylalanine or with phorbol myristate acetate. There was a poor dose dependency. In contrast to these findings, a decrease in the generation of OH· and O_2^- by PMN was claimed to occur in the presence of thalidomide (MIYACHI et al. 1982). However, the interpretation of the results is hampered by the fact that thalidomide was dissolved with 1 N NaOH and was certainly completely hydrolysed. Only a mixture of undefined hydrolysis products could have been present. This exemplifies difficulties in the interpretation of some studies on thalidomide. Significant enhancement of phagocytosis was reported in a short note (DOLL et al. 1983) in the presence of 1 μg thalidomide/ml, but there was a significant depression with 10 μg/ml. The data are insufficiently recorded to clearly demonstrate a biphasic action. No inhibitory effect of thalidomide on prostaglandin and leukotriene biosynthesis by human platelets or PMN was found (MAURICE et al. 1986).

2. Lymphocyte Proliferation

Several attempts have been made to establish an effect of thalidomide on lymphocyte proliferation in vitro or in vivo, mostly using human lymphocytes. Again, the data are controversial and many of the results reported could not be confirmed by other investigators.

Using phytohemagglutinin (PHA) as a proliferation stimulator and a smear technique, thalidomide was reported to inhibit the "blast" formation in vitro at a concentration of 25–50 μg/ml (ROATH et al. 1962). No effect of thalidomide on lymphocyte proliferation was found using PHA, concanavalin A (ConA), pokeweed mitogen (PWM), or a mixed lymphocyte culture to induce proliferation, and no effect on IgG production of lymphocytes stimulated with PWM was observed (ARONSON et al. 1986). When testing thalidomide and a series of its hydrolysis products, no effect was observed on ConA-stimulated human lymphocyte proliferation in vitro, measured as ^{3}H-thymidine incorporation, regardless of the substance used or the duration of stimulation (GÜNZLER et al. 1986).

However, it would be wrong to conclude that thalidomide treatment does not cause any effect on proliferation of isolated lymphocytes on the basis of these in vitro studies. In fact, a very pronounced effect can be demonstrated when using the lymphocytes of volunteers treated with 5–7 mg thalidomide/kg (NEUBERT et al. 1996b). The extent of the inhibition of lymphocyte proliferation depends on the stimulating agent.

3. Blood Cell Surface Receptors

Within the last 5 years, a series of papers has been published on effects of thalidomide and some of its teratogenic and nonteratogenic derivatives on surface receptors of white blood cells of primates. These effects are the most pronounced and the best reproducible ones reported with respect to thalidomide-induced alterations in the white blood cells and the immune system. This information is the result of the attempt to reveal substance-induced effects in nonhuman primates and subsequently to verify or dismiss the hints and suspicions aroused by studies in humans, using the same testing procedure (including the same antihuman monoclonal antibodies).

In order to find an indication of possible alterations induced by thalidomide in white blood cells, and to obtain the basis for explaining the anti-inflammatory and immunosuppressive actions of thalidomide, surface receptors were analyzed in thalidomide-treated nonhuman primates (marmosets) using antihuman monoclonal antibodies and flow cytometry. It was found that treatment with thalidomide drastically reduced the expression of several surface receptors involved in similar functions, namely those serving as adhesion molecules in cell–cell and cell–extracellular matrix interactions (NEUBERT et al. 1992, 1993). Such adhesion molecules are involved in many immunological reactions, e.g., antigen presentation, "helper" functions, cytotoxic actions, but most importantly also in "homing" of lymphocytes "rolling" and migration of leukocytes into areas of inflammation, and many more functions (SPRINGER 1990). A function of adhesion receptors has also been demonstrated within the intact organism, e.g., with respect to leukocyte rolling (LEY et al. 1991). There is good reason to believe that the effects on the surface receptors are not confined to white blood cells, but may also be induced at other cell types (see Sect. E.II.3.b).

The effects observed in *Callithrix jacchus* were largely confirmed in studies performed in human volunteers. Treatment with daily doses of 5–8 mg thalidomide/kg divided into three doses induced the following changes, as revealed with monoclonal antibodies and flow cytometry (NEUBERT et al. 1992c, 1994; NOGUEIRA et al. 1994):

- Downregulation of members of the β_1-integrin family. CD49d (VLA-α4) was found to be drastically reduced on lymphocytes and monocytes, but there was no effect on CD49b (VLA-α2). The common β-chain of these heterodimers (CD29) was also expressed on much fewer cells after thalidomide treatment.

- Downregulation of members of the β_2-integrin family. CD11a (LFA-1α) was found to be drastically reduced on lymphocytes and monocytes. There was also a pronounced downregulation of the CD11b receptor (Mac-1) on monocytes and granulocytes. The common β-chain of these heterodimers (CD18) was also expressed on fewer cells after thalidomide treatment.
- Downregulation of the "homing receptor" (CD44), especially on monocytes.
- Downregulation of dipeptidyl peptidase IV (CD26), a plasma membrane serine protease, especially on $CD4^+$ cells.
- Downregulation of L-selectin (CD62L) on lymphocytes, monocytes, and granulocytes.
- A rather complex effect on intercellular adhesion moelcule (ICAM)-1 (CD54), the natural ligand for the β_2-integrins. This receptor is downregulated on monocytes, but upregulated on granulocytes. While many $CD4^+$ cells lose this receptor, it is also upregulated on $CD8^+$ cells.

The effect of thalidomide on adhesion receptors is not confined to white blood cells. As will be discussed later (see Sect. E.II.3.a), drastic changes were induced on cells of primate embryos. A clear-cut, but not very pronounced downregulation by thalidomide of ICAM-1 (CD54), E selectin (CD62E), and VCAM (CD106) was also observed in vitro on endothelial cells (stimulated with TNF-α or IL-1) isolated from human umbilical cord veins, but a pronounced inhibition of lymphocyte adhesion to these cells was detectable by 1–10 μg thalidomide/ml or 1 μg EM12/ml (Nogueira et al. 1995a). This more extensive effect was probably due to a dual action of thalidomide on both endothelial cells and lymphocytes.

4. Correlation of Effects on White Blood Cells and Teratogenic Potency

A number of thalidomide derivatives with different teratogenic potentials or potencies were also studied in the marmoset.

With the derivatives showing a high chemical similarity to thalidomide (see Fig. 1, 2), in the majority of cases a good correlation between the capacity to induce the thalidomide-typical alterations on the surface receptors of white blood cells and the teratogenic potential was found (Neubert et al. 1993; Nogueira et al. 1995a,b). Here, only the results obtained on a selection of substances are discussed. In addition to thalidomide, these derivatives include the following (for chemical structure, see Fig. 2):

- *α-EM12*. This substance, differing from thalidomide only in one carbonyl group (phthalimidine instead of phthalimide), exhibits a higher teratogenic potency and is also clearly more potent than thalidomide with respect to its effects on the white blood cells.
- *Supidimide*. No teratogenic potency was found with this substance (in monkeys), which carries a SO_2 group instead of one of the carbonyl groups in the phthalimide ring. This substance also exhibits a much lower potency

with respect to downregulation of the adhesion receptors of white blood cells, but it is not completely ineffective.

- *β-EM12*. This isomer of α-EM12, which differs from α-EM12 in the positioning of the glutarimide ring, has no teratogenic potential (or no more than an extremely low potency, as assessed in the monkey), and no thalidomide-typical effects on white blood cell receptors were found even at high doses.
- *Phthalimidophthalimide*. This substance has only limited chemical resemblance to thalidomide, since it contains two phthalimide rings but lacks the glutarimide moiety. No teratogenic potential could be revealed in the monkey (KLUG et al. 1994), and the substance does not induce thalidomide-typical effects on receptors of white blood cells (NOGUEIRA et al. 1995a,b).
- *Methylthalidomide* and *methyl-α-EM12*. In these derivatives, the hydrogen at the asymmetric carbon atom is substituted by a methyl group. This substitution largely reduces or even completely abolishes the teratogenic potential (evaluated in monkeys). Methyl-α-EM12 (EL1010), which was not found to be teratogenic in the marmoset, shows a weak tendency to induce some of the thalidomide-typical effects on white blood cells in the same species.
- A number of further substances have been tested in this respect. However, since the interesting data have not been published, they are not discussed here.

Using this set of "classical" thalidomide derivatives, there are excellent parallels between the ability to induce the downregulation of integrin receptors on white blood cells and the teratogenic potential. Among these substances, none that has a high potency to react with blood cell receptors is devoid of a teratogenic potential. It seems that all substances with a strong thalidomide-type teratogenic potency also have pronounced effects on white blood cell integrins. However, the selection of analogues mentioned above is too small to completely rule out the general possibility of a dissociation of the two types of effects for any derivative. Recently, evidence has been accumulated that a downregulation of integrin receptors on blood cells is possible with no apparent teratogenicity.

It has been revealed that thalidomide and its active derivatives are capable of triggering at least four types of actions (NOGUEIRA et al. 1994), which may also occur independently of one another (Table 16).

1. A downregulation of several adhesion surface receptors on a variety of white blood cells (lymphocytes, monocytes, granulocytes). Simultaneously, a few receptors (e.g., CD54) are also upregulated on defined cell types.
2. A change in the pattern of $CD8^+$ subtypes of cells, the total number of $CD8^+$ cells remaining fairly constant. The cytotoxic-type cells ($CD8^+CD56^+$) are drastically reduced, while the suppressor-type cells ($CD8^+CD56^-$) are moderately increased.

Table 16. Effects induced by thalidomide on white blood cells of marmosets and humans

	Lymphocytes	Monocytes	Granulocytes	Other cells[c]
β^2-Integrins				
CD11a	Loss	Loss	–	Loss
CD11b	–	Loss	Loss	?
CD18	Loss	Loss	Loss	Loss
β^1–Integrins				
CD49d	Loss	Loss	–	Loss
CD29	Loss	Loss	–	Loss
Homing receptor				
CD44	Loss	(Loss)	Loss	Loss
Selectins				
CD62L	Loss	Loss	Loss	Loss
ICAM–1				
CD54	Variable[a]	Loss	*Increase*	*Increase*
$CD8^+CD56^+$	*Increase*	–	–	–
$CD8^+CD56^-$	Loss	–	–	–
$CD4^+CD45RA^+$ $CD45R0^+$	Loss	–	–	–
$CD4^+CD45RA^-$ $CD45R0^-$	*Appearance*	–	–	–
IL–1 formation and release	–	Reduction	–	–
TNF–α formation and release	–	Reduction[b]	–	–

Upregulations are indicated in italics.
ICAM, intercellular adhesion molecule; IL, interleukin; TNF, tumor necrosis factor.
[a]$CD4^+CD54^+$ downregulated, $CD8^+CD54^+$ upregulated.
[b]Only under certain conditions; apparently not with (or difficult with) pure monocytes.
[c]Other than blood cells.

3. A change in the maturation process of $CD4^+$ cells. The apparent conversion of $CD45RA^+CD45R0^-$ cells to $CD45RA^-CD45R0^+$ cells is affected, as is the occurrence of $CD45RA^-CD45R0^-$ cells with a concomitant loss of $CD45RA^+CD45R0^+$ cells. This phenomenon may be due to altered distribution processes.
4. A change in the formation and release of certain cytokines, e.g., interleukin (IL)-1α and IL-1β, and to a lesser extent possibly also tumor necrosis factor-α (TNF-α), by a defined cell line, monocytes, or mixed monocyte/lymphocyte preparations in vitro.

Not all of the effects occur simultaneously and/or on all of the cell types. It is this divergence which is thought to be responsible for the multiplicity of actions triggered by thalidomide or its active derivatives on the different cell types, e.g., monocytes, lymphocytes, granulocytes, embryonic cells. The first mechanism, i.e., the effect on adhesion receptors, seems to be sufficient to explain the anti-inflammatory effects of thalidomide observed in clinical trials. As an overall result of these studies, the suggestion of an altered cell–cell interaction, resulting in a different tissue distribution of cells involved in inflammatory processes, is very attractive.

Numerous secondary reactions are bound to result from the thalidomide-induced alteration of the adhesion receptors on the various types of white blood cells. Since these receptors are involved in many types of reactions, the cascade of subsequent secondary events must critically depend on the role the specific receptors play in a given cell type or tissue. Since many adhesion receptors are believed to be linked to the cytoskeleton of the cells, direct signal transductions and direct alterations of cell functions via this mechanism are quite feasible. Within the cascade of secondary events, modifications in the release of specific cytokines or other (growth) factors may play an important role. Such changes in cytokine release and function must then again alter cell functions, thus forming a vicious circle.

III. Neurotoxic Effects

Peripheral neuropathy was associated with the therapeutic use of thalidomide. Since, as in the case of malformations, no systematic epidemiological studies are available, the exact incidence cannot be assessed. The (mostly sensory) neuropathy apparently occurred predominantly in older individuals and after prolonged periods of treatment. Because of the multiplicity of the reports and the uniformity of the symptoms described, it is highly likely that a causal relationship exists.

Since the information available on neurotoxic effects has been compiled and extensively discussed elsewhere (NEUBERT 1996), the interested reader is referred to this review.

E. Possible Mechanisms of Action

It is not our intention here to give a complete survey of the numerous hypotheses put forward to try to explain the teratogenic action of thalidomide. These have been compiled before (STEPHENS 1988), and other earlier reviews also exist (e.g., KEBERLE et al. 1965; HELM et al. 1981). None of these hypotheses have proven satisfactory, and many have either turned out to be wrong or the evidence presented so far is extremely meager or even absent (Table 17). We shall only briefly discuss some of the hypotheses that have gained some degree of popularity and such suggestions that cannot be completely dismissed on the basis of the evidence presently available.

I. Tissue of Adult Organisms

Almost all studies performed to unravel the possible modes of action of thalidomide have been performed with tissue of adult organisms or with other simple model systems. Experience during the last few decades has revealed that thalidomide is not a simple "atoxic" sedative substance with pronounced

Table 17. Critical evaluation of some hypotheses on the possible mechanism of the teratogenic action of thalidomide and likelihood of providing a satisfactory explanation for the abnormal development

Hypothesis	Satisfactory explanation for specific teratogenic effects
Interference with glutamine or glutamic acid metabolism	No convincing evidence
Chelation of cations (Ca, Mg, Zn, etc.)	No convincing evidence
Intercalation into DNA	No convincing evidence
Covalent binding to nucleic acids	No convincing evidence
Functional interaction with nucleic acids	May still be possible
Direct interference with collagen metabolism	No convincing evidence
Direct interference with cartilage formation	No convincing evidence
Direct effects on isolated limb buds	No convincing evidence
Induction of cell necrosis or inhibition of cell proliferation	Some evidence
Metabolic conversion of thalidomide is required	No convincing evidence
Thalidomide acts as a mutagen	No convincing evidence
Interference with neural structures in or outside the limbs	No convincing evidence
Interference with early blood vessel formation	No convincing evidence
Interference with cytokine formation, e.g. TNF-α	No convincing evidence
Interference with induction by mesonephros or somites	Not very strong evidence
Interference with cell–cell or cell–extracellular matrix interactions	An attractive hypothesis
Interference with specific migration processes	An attractive hypothesis
Primary interference with adhesion receptors	Best evidence available
Interference with specific signal transductions	Quite feasible

TNF, tumor necrosis factor.

teratogenic properties, but that it exhibits a large variety of additional effects on the adult human organism. Therefore, attempts to reveal modes of action of thalidomide cannot be confined to the teratogenic effect; it is equally important to learn more about the biochemical and cytobiological actions responsible for the pronounced anti-inflammatory and immunosuppressive action of this drug.

1. Importance of Metabolic Activation

The problem has still not been completely solved as to whether thalidomide itself is responsible for the various pharmacological and toxic actions or whether these effects are mediated via an active metabolite or hydrolysis product. There has been no experimental support for the assumption that the biological effects of thalidomide are induced by one of the hydrolysis products and not by thalidomide proper. Thus such an assumption can be largely ruled out on the basis of the existing evidence. However, the possibility cannot be excluded that some of the biological activity of thalidomide may be retained in a defined hydrolysis product or a metabolite.

A number of experimental studies have been performed to obtain evidence for an active metabolite being the ultimate reacting agent responsible. In most of these studies, model systems have been used with little relevance to the end points of the actions of thalidomide. Interpretation of the results is therefore difficult, if not impossible, with respect to a defined action of the drug.

With respect to the formation of biologically active metabolites from thalidomide, there are only two possibilities:

1. Opening of the phthalimide ring and acylation of a diamine
2. Oxidation of the benzene ring via the classical cytochrome P450 (CYP) pathway, giving rise to reactive epoxide species

The first of these possibilities does not represent a true metabolic conversion, but an adduct to physiologically important amine groups may be formed. While such an acylation of diamine groups, as in spermidine or putrescine, is possible (FABRO et al. 1965), there is presently no evidence that such a reaction with protein components or a reduction in the concentration of biologically important aliphatic diamines occurs in vivo and may be responsible for typical effects induced by thalidomide.

With respect to the second possibility, there is no doubt that metabolic oxidation via the CYP system is possible under certain experimental conditions. No indication exists of the CYP isozyme involved. All of the relevant studies were performed in vitro, and the formation of metabolites in such artificial systems with no significance for the situation in vivo cannot be excluded; indeed, it is even likely to occur. At present, there is no evidence whatsoever that such a metabolic conversion occurs in vivo to a sufficient extent that such a metabolite would be capable of inducing the typical effects or that it possesses the required properties (i.e., labile enough for a high reactivity and stable enough to be transferred to various target tissues). Although the possibility of the participation of an oxidation product cannot be completely ruled out, results from in vitro studies in which thalidomide is active in the absence of a metabolizing system, as well as results from studies with various thalidomide derivatives oxidized by CYP systems but biologically inactive (e.g., β-EM12), argue strongly against the necessity of metabolic activation of thalidomide.

2. Fibroblast Growth Factor-Induced Angiogenesis in the Rabbit Cornea

Using a rabbit cornea model, a high dose of 200 mg thalidomide/kg was reported to induce a 30%–50% inhibition of angiogenic protein basic fibroblast growth factor (bFGF)-triggered neovascularization of the cornea (D'AMATO et al. 1994). This finding of an antiangiogenic effect of thalidomide adds an interesting aspect to the spectrum of actions inducible by thalidomide. Unfortunately, this publication is more of a preliminary communication, since no data on dose–response relationships and on representative thalidomide derivatives are provided which would allow conclusions to be drawn about the drug specificity and the relevance of the effect at nearly therapeutic doses. No convincing evidence for an effect (neither by thalidomide nor by α-EM12) was found in another experimental model, the developing vasculature within the chlorioallantoic membrane of the chick. Because of this discrepancy, the authors speculate that metabolic activation of thalidomide is required for the action observed at the rabbit cornea (but see Sect. E.II.2.b).

An inhibitory effect on the pathological growth of vessels into the cornea was also observed after treatment with α-EM12, but not after treatment with supidimide, a substance lacking the anti-inflammatory and immunosuppressive effects of thalidomide. Since cyclosporin A or total body irradiation did not exhibit a similar potency on artificial angiogenesis, the authors concluded that the effect was not mediated by immunosuppression.

Without any evidence of similar effects occurring in embryonic tissue, and especially in the developing limb bud, the authors speculated that the anti-angiogenic effect observed in the cornea might be responsible for the teratogenic action of this class of substances. The only argument in favor of this is that the two teratogenic substances (thalidomide and α-EM12) are active in this assay, while supidimide is not. However, this is a poor argument, because the same holds true for the anti-inflammatory and immunosuppressive potencies of all three substances. Even more confusing in this context is the mention of similar, but weaker inhibitory effects on angiogenesis by two derivatives which are not generally considered to be teratogenic, phthaloylglutamic acid and phthaloylglutamic anhydride (SMITH et al. 1965). The argument that similar changes were reported on the vasculature of the limbs from chick embryos treated with thalidomide is not at all convincing either, since the chick is not a clear-cut model for the teratogenic action of thalidomide, the literature on this subject is contradictory, and nonspecific artifacts are likely to occur (RICKENBACHER 1963).

Furthermore, it appears extremely unlikely that such a broad and common mechanism such as inhibition of angiogenesis would be responsible for such well-defined defects as those induced by thalidomide. From our own experience, there is no indication that general angiogenesis is disturbed in thalidomide-exposed primate embryos. Additionally, according to such a hypothesis, pronounced necroses would be expected. In a short abstract, the occurrence of more cell deaths than normal was reported in the posterior base of the limb bud during its outgrowth and regression of the apical ectodermal ridge in rhesus embryos 3–7 days after maternal treatment with 10 mg thalidomide/kg on days 26–28 of pregnancy (THEISEN and MCGREGOR 1981). However, according to our own studies, thalidomide-induced teratogenesis in nonhuman primates (rhesus monkey as well as marmoset) is characterized by surprisingly few necroses, quite in contrast to abnormalities induced by many other teratogenic agents. A similar observation was published for the rabbit (VICKERS 1967). Thus, while possibly being an interesting pharmacological effect, inhibition of angiogenesis does not seem to be a convincing mechanism for thalidomide-induced teratogenesis.

3. Cytokine Formation In Vitro

It has been reported by KAPLAN and coworkers that in cultures of isolated human monocytes thalidomide at concentrations between 2 and 10 000 ng/ml inhibits the lipopolysaccharide-induced formation of TNF-α and its release into the culture medium (SAMPAIO et al. 1991). There was a very poor dose-response relationship, and it was claimed that no inhibition of the formation and release of IL-1β or IL-6 occurred in the same system. In a subsequent paper, it was reported by the authors that the effect of thalidomide is mediated by an enhancement of TNF-α mRNA degradation (MOREIRA et al. 1993). When these experiments were repeated in several other laboratories, it was difficult to reproducibly confirm the results reported by the group in New York on TNF-α. No effect of thalidomide was observed in a monocytoid cell line (MonoMac-6) with respect to TNF-α formation and release (HELGE et al. 1995). Furthermore, with some preparations of isolated human monocytes, a very slight effect with little or no dose-dependency was inducible, but the effect was highly variable and did not occur at all with preparations from other normal volunteers, or even with further blood samples from the same person (FOERSTER et al. 1995a).

In addition, an *opposite* effect of thalidomide on TNF-α production was observed in three human leukemic cell lines (Hl-60, K562, and U937). Thalidomide clearly enhanced TNF-α production induced by phorbol esters (NISHIMURA et al. 1994). An identical result was obtained with the racemate and the two enantiomers of thalidomide, but reracemization is likely during the long culture period (24 h). Recently, a variety of phthalimide derivatives have been evaluated with these Hl-60 cells, and the chemical structure of many of these derivatives deviates quite considerably from that of thalidomide (SHIBATA et al. 1995). Although no reliable data on the teratogenic potential are available for most of the derivatives tested, it does not appear likely that the stimulating effect on TNF-α formation runs parallel with the teratogenic potency (e.g., methyl-thalidomide is more active on TNF-α formation than thalidomide itself).

Further the claimed selectivity of the effect on the TNF-α was not confirmed. In contrast to the data presented by SAMPAIO et al. (1991), an inhibition of the formation of IL-1α and IL-1β by thalidomide and by other active derivatives can reproducibly be observed with isolated human monocytes and with the human monocytoid cell line MonoMac-6, even in the absence of an effect on TNF-α (HELGE et al. 1995).

The most reliable inhibition by thalidomide of TNF-α formation seems to be found in complex in vitro systems containing cell fractions of peripheral mononuclear blood cells, i.e., monocytes as well as lymphocytes. However, the inhibition again rarely exceeds 50%, and the dose response is always very poor

(FOERSTER et al. 1995). It is still difficult to reproduce the effect of thalidomide on TNF-α formation in vitro regularly, and variations from one cell preparation to another are quite common. Thus the effect of thalidomide on TNF-α formation and release is not very impressive, and the significance of this effect, if it occurs at all in vivo; cannot be assessed on the basis of the presently available data. It is difficult to conceive that as variable an action as the effect of thalidomide on TNF-α production and release is responsible for the clinical effects of this drug on inflammation and immunological reactions, as suggested by Kaplan and her coworkers.

On the other hand, the inhibition of IL-1α and IL-1β formation and release by thalidomide is highly reproducible in MonoMac-6 cells or isolated human monocytes (FOERSTER et al. 1995; HELGE et al. 1995). However, the extent of inhibition achievable was also regularly only 70% or even much less. An effect of thalidomide was also reported in a quite different experimental setup (WEGLICKI et al. 1993): magnesium deficiency induces in young rats an increase of blood plasma TNF-α and other mediators, such as IL-1, and this increase in vivo was also reported to be inhibited by treatment of the animals with thalidomide. No indication of a possible mode of action is available.

Recently, an inhibition of TNF-α production by thalidomide was reported to occur in cells other than those from peripheral blood. Such a dose-dependent decrease in TNF-α production and release into the culture medium was observed with human fetal microglia cells (PETERSON et al. 1995). This is the first indication of a thalidomide-induced effect on TNF-α using fetal cells of a primate. However, there is no indication that such an efffect is correlated with teratogenicity, since indications of thalidomide causing impaired brain development arc scarce. According to these data, the microglial system may prove to be more reliable for assessing thalidomide-type effects on TNF-α than the monocyte cultures, but for inducing a convincing effect, concentrations of 10 μg thalidomide/ml or higher were required. Such constant serum levels are hardly reached under therapeutic conditions.

The question remains whether the variable effect of thalidomide on TNF-α formation and release in monocyte preparations in vitro may also be demonstrated in the intact organism and whether any protective action of the drug can be demonstrated in experimental or pathological conditions in which TNF-α is assumed to play an important or even a crucial role. Using lethal endotoxemia in sensitized mice as a model system, a large dose of thalidomide (400 mg/kg) was unable to protect the mice from a lethal dose of *Escherichia coli* LPS, but chlorpromazine was effective (NETEA et al. 1995). In contrast to chlorpromazine and pentoxifylline, the effect of this thalidomide dose on serum TNF-α was rather unimpressive and not significantly different from controls. The authors concluded that thalidomide did not appear to be an inhibitor of TNF-α production after in vivo administration of endotoxin. It is unknown whether species differences are responsible for this discrepancy. Because of the huge intraindividual variability in normal serum concentrations, no decrease of TNF-α was found in healthy volunteers during treatment with daily doses of 5 mg thalidomide/kg for several days (NEUBERT et al. 1996a). More extensive and carefully controlled studies from different laboratories are necessary to answer the question of whether a convincing effect of thalidomide on TNF-α levels can be obtained in humans under pathological conditions.

In phytohemagglutinin (PHA)-stimulated human peripheral blood mononuclear cells, the production of IL-4 and IL-5 was enhanced, and at the same time that of interferon-γ significantly inhibited (MCHUGH et al. 1995). This effect was obtained in the concentration range of 0.3–10 μg thalidomide/ml. A less pronounced effect was seen after stimulation with recall antigen (streptokinase/streptodornase). The results are compatible with thalidomide switching the response from a Th1 to a Th2 type. It is noteworthy that cyclosporin A, in contrast, inhibited both Th1 and Th2 cytokine production by PHA-stimulated cells.

An additional aspect of the postulated inhibiton of TNF-α-induced effects by thalidomide was the suggestion of a suppression of the activation of latent HIV-1 in a monocytoid cell line (MAKONKAWKEYOON et al. 1993). However, such an inhibitory effect of thalidomide on HIV-1 activation in U1 cells was not restricted to an activation induced by TNF-α, but was also reported to occur in the presence of phorbol myristate acetate, PS, and other cytokine combinations. A pronouced effect was induced in these in vitro studies only at fairly high thalidomide concentrations (50 μg/ml), but it was claimed that HIV activation was also inhibited in vitro in the peripheral blood cells of HIV-infected patients. Another aspect of thalidomide interfering with HIV–blood cell interactions may be based on an interaction of thalidomide with the lymphocyte surface receptor CD26 at therapeutic doses in humans (NEUBERT et al. 1994), although such an interaction with HIV is also speculative at the present time.

II. Teratogenic Action

In contrast and in addition to the hypotheses attempting to explain the action of thalidomide on the adult organism, hypotheses on the possible mode of the teratogenic action of thalidomide must fulfill certain criteria if they are to be taken seriously and considered more than mere speculation. The criteria the mechanism has to meet are as follows:

1. It must be studied in embryonic tissue susceptible to the action of thalidomide.
2. It must be plausible and convincing with respect to possible adverse effects on processes known to be relevant to embryonic development and with respect to the resulting malformation pattern.
3. It must explain the high susceptibility of the embryo in relation to the extremely low general toxicity towards the adult organism.
4. It must explain the pattern of malformations observed in primates and the lack of potential to induce a high incidence of other typical malformations (e.g., in skull and brain, vertebral column, cleft palate).

5. It must be in accord with different teratogenic potentials induced by a variety of chemically closely related derivatives of thalidomide in a susceptible species (e.g., nonhuman primates or possibly rabbits).
6. It must explain the marked species differences (extreme difference in the susceptibility between primates and rodents).

Some of the hypothesis are compiled in Table 17, and it is easy to recognize that none of the older hypotheses and very few of the more modern ones have been developed to the extent that they comply with these simple criteria.

1. Older Speculations

Because some of the hydrolysis products are glutamine or isoglutamine derivatives, it has been suggested that thalidomide may act as a glutamate or glutamine antimetabolite or even as a folic acid antagonist. In view of the chemical structure and the possible formation of isoglutamine derivatives on the one hand, and the fact that folic acid antagonists are known teratogens in humans on the other, this speculation seems attractive. However, no convinving evidence for such antagonism has been presented, and the typical malformation pattern, the species specificity, and the stereospecificity cannot be explained by such an assumption.

In a number of papers (JÖFNSSON 1972; KOCH and CZEJKA 1986), it has been speculated that thalidomide might be intercalated into DNA. Such an interaction with DNA might be of some interest, since some arguments have been provided that only L- thalidomide may be capable of such an interaction. However, such an assumption has remained a speculation, and there are numerous arguments against it (see Sects. E.II.1.a-e); no evidence was presented that such an effect really does occur in vivo and in particular that it takes place within the embryo. Furthermore, no evidence for characteristic mutational effects induced by thalidomide was ever presented, and such a mechanism would explain neither the pattern of malformations nor the species specificity. In addition, from other intercalating agents there is no indication that intercalation into DNA would induce specific malformations. Besides, intercalation (if it takes place in vivo) would predominantly be expected to occur in mito-DNA, and no evidence was ever provided that such an effect on the mitochondrial genome would induce malformations. There are even good arguments against such an assumption, since inhibition of mitochondrial protein synthesis was shown to induce embryomortality but not malformations (BASS 1975; OERTER and BASS 1975; BASS and OERTER 1977).

Thalidomide-induced malformations as a result of neurotoxic effects were postulated in a number of papers from Australia (MCBRIDE 1974; MCCREDIE 1975, 1980; MCBRIDE and VARDY 1983). Besides serious objections against the studies and interpretations of the morphological pictures, further arguments against this hypothesis were advanced and the matter was critically discussed (STEPHENS and STRECKER 1983; STRECKER and STEPHENS 1983; THEISEN 1983).

Although such a mechanism cannot be completely ruled out, it is not very convincing or likely.

a) Mutagenic Effects

More recently, very old data (BAKAY and NYHAN 1968) on an association of radioactivity from ^{14}C-labeled thalidomide with DNA were revived (HUANG and McBRIDE 1990), structural changes of DNA were postulated (HUANG and McBRIDE 1990), and such data were used to postulate a mutagenic effect of the drug, together with observations that children of "thalidomide victims" were reported to show the same malformations as their father (McBRIDE 1994). In a comment in the same journal invited by the journal's editor, this was considered almost certainly unfounded. However, since the reports containing speculations on mutagenic effects of thalidomide received undue attention (induced by the authors) in the nonscientific media, a discussion seems necessary. Although no data on a mutagenic potential of thalidomide were published, an obscure claim of such an effect was made in a semiscientific journal (MACKENZIE 1983), but not confirmed since.

Using Norway LEW/BN rats expressing the *lx* gene (polydactyly-luxate) in heterozygotes, thalidomide given on day 11 of pregnancy intraperitoneally as a suspension with Tween 20 in saline (1:3) was reported to induce polydactyly (BÍLÁ and KREN 1989). No effect was reported to occur when the solvent was given alone. It was claimed that thalidomide may interact with the mutant gene. However, in addition to the limitation of evaluating only one to nine litters in the experimental groups, the obscure procedure of injecting a liquid detergent at a concentration of 3 ml/kg intraperitoneally was used, thus making an interpretation of the results almost impossible.

In a very brief abstract containing no data, an association of radioactivity from ^{3}H-thalidomide with DNA was claimed to occur in rabbits (SCHUMACHER et al. 1967), but, from studies in rat fetuses with the drug ^{14}C-labeled in the phthalic carbonyl, apparently little evidence for such binding was found, although considerable radioactivity was detected in DNA fractions contaminated with highly acidic proteins (BAKAY and NYHAN 1986). In a meeting abstract, results of studies were mentioned in which rats were injected intravenously on day 12 of gestation with ^{14} C-thalidomide labeled either in the phthalimide or in the glutarimide ring using dimethylformamide as a solvent. When DNA was isolated from the embryos, associated radioactivity was only found after the injection of the preparation labeled in the glutarimide ring (HUANG and McBRIDE 1995). Since in the hydrolysis products the carbon atoms of the ring systems remained attached to each other even after both rings were opened, an extensive breakdown of the molecule must have occurred in this study either in vivo or during the isolation procedure. Therefore, the difference in the labeling of the DNA fraction reported for thalidomide preparations with the ^{14}C label in the different rings might also be explained by an incorporation of small ^{14}C-labeled thalidomide fragments via nucleotide precursors into the DNA of the rapidly growing cells. Such an incorporation would have nothing to do with binding of a xenobiotic to DNA. This possible explanation should certainly have been excluded by the authors. Introduction of radioactivity into the salvage pathway might be expected from the glutarimide fragments, but not from the phthalimide moiety, thus explaining the data presented. Furthermore, if an adduct formation is postulated, the investigators must prove its existence. Additionally, it is well known that most adducts do not cause mutagenicity, and again mutations occurring subsequent to the postulated adduct formation would have to be demonstrated. In the rather simple studies, not all these prerequisties and possibilities have been considered, and they are thus no more than mere speculations. It had been reported previously that, with the injection procedure used, no limb deformities can be induced in the rats (PARKHIE and WEBB 1983; see Table 7), thus making the significance of this rat model questionable.

As already pointed out by READ (1994), the observation of a malformed child born to a malformed father is most likely explained by a genetic syndrome. Furthermore, if a mutation is considered as the cause of the malformation of the child, the same cause could be assumed for the father with the similar malformations. It is well known that during the period of thalidomide-induced malformations several misclassifications occurred. In fact, it was reported that a Holt-Oram syndrome, with an autosomal dominant inheritance, was mistaken for a thalidomide embryopathy (VAN REGEMORTER et al. 1982). A wide range of varying abnormalities has been reported for Holt-Oram and similar syndromes. During the period in which thalidomide-induced abnormalities were very frequent, the diagnosis of a teratogenic effect may have been reached rather easily if an exposure to thalidomide was likely: *in dubio pro reo*. This was particularly possible if no pronouced abnormality syndrome (amelia of phocomelia occurred in the child).

Drug-induced, paternally mediated malformations in the offspring are as such very unlikely events, especially when the father has the same type of gross structural abnormalities. The extremely far-fetched assumption would have to be made that the father's malformations were caused by a teratogenic action of the drug in question. Simultaneously, the substance must have a mutagenic property (which so far has not been shown for thalidomide), inducing a mutagenic defect in the germ cells of the father that leads to exactly the same abnormalities as those induced by the teratogenic action. Since mutations occur randomly and they are rare events, such a constellation is extremely unlikely. We should instead keep to the simple and convincing explanation of a mutagenic syndrome in both the parent and the child, as mentioned above.

b) Interference with Formation and Effects of Tumor Necrosis Factor-α on Prenatal Development

Recently, it was reported that expression of TNF-α, of similar proteins, and of corresponding receptors was detected at various stages of prenatal development in chick and rodent embryos (OHSAWA and NATORI 1989; WRIDE and

SANDERS 1993; KOHCHI et al. 1994; JASKOLL et al. 1994), as early as in the blastocyst (PAMPFER et al. 1994), and speculations were made about a possible role of this factor in embryonic development (WRIDE and SANDERS 1995).

Based on the reports of a (not very well founded) thalidomide-triggered inhibition of the formation of TNF-α and its release in monocyte cultures (see Sect. E.I.3), the possibility was raised that this drug might exhibit its teratogenic effect by inhibiting developmentally regulated TNF-α expression (WRIDE and SANDERS 1995), e.g., through a modification of apoptotic processes. At present, no data whatsoever are available to substantiate such a claim, and it has not yet been demonstrated that thalidomide would alter TNF-α expression or modify effects of this cytokine in the embryo. However, it cannot be excluded that for certain effects of high doses of thalidomide, e.g., embryomortality in rats, TNF-α might play a role.

Furthermore, some interesting evidence was summarized that TNF-α may modulate the induction of certain adhesion molecules, such as integrins, involved in inflammatory situations (WRIDE and SANDERS 1995). Thus an interplay of TNF-α concentrations and the expression of certain adhesion molecules seems quite feasible. This again might be of interest when attempting to find an explanation for the downregulation of adhesion receptors, believed to be causally linked to the teratogenic action of thalidomide in primates (see Sect. E.II.3.b). However, it should be remembered that such a link between TNF-α and integrins, if it exists, might be triggered in both directions, and alteration of cell adhesion must certainly be expected to alter the cytokine expression in this tissue. Additionally, no regulatory effect of TNF-α is known to exist in the case of many of the defined adhesion receptors which are substantially altered during the action of thalidomide of blood cells and on cells of the primate embryo.

2. Cell Adhesion in Nonembryonic Model Systems

It is an attractive hypothesis that thalidomide may interfere with embryonic development by interacting with specific cell–cell or cell–extracellularmatrix interactions. Such interactions have long been known to represent key events during morphogenetic differentiation.

In this respect, results of some studies are worth discussion (BRAUN and WEINREB 1985) that indicate that thalidomide can interfere with cell attachment phenomena in nonembryonic model systems. The experimental setup used was the attachment of MOT cells to con A-coated polyethylene surfaces. Although the outcome of these studies, performed entirely in vitro, was virtually negative with respect to explaining the teratogenicity of thalidomide and also the effects on the adult organism (see Sect. E.II.2.a), in our opinion the studies on interference with cell adhesion pointed in the right direction for the first time.

a) Significance of Metabolic Activation for Adhesion of thalidomide of the inhibition of Tumor Cells on Coated Disks

However, there are a number of serious drawbacks in these studies as they were reported: (a) metabolic activation was found to be required for a thalidomide-induced interference with such an attachment of tumor cells to ConA-coated plastic discs, (b) a comparatively high concentration of DMSO (1% final concentration) was used, (c) the concentrations used were quite high (> 100 μg thalidomide/ml), and (d) there is no indication of whether mouse tumor cells behave like cells of primate embryos.

The presence of DMSO complicates the experimental design and the interpretaion of the data, even though it was stated that DMSO alone did not influence the attachment of the tumor cells. Since thalidomide and most of its derivatives are water soluble at concentrations relevant for in vitro studies (< 1–50 μg/ml), the addition of DMSO is quite unnecessary.

The main emphasis in these studies was put on providing evidence for the hypothesis that metabolic activation is a prerequisite for the teratogenic action of thalidomide or its active derivatives (Braun and Weinreb 1985). In order to achieve the mentioned inhibition of attachment of the tumor cells, a preincubation of the thalidomide derivatives with hepatic microsomal fractions was essential, and "activation" products were suggested to be responsible for this interference. Apparently, the well-known hydrolysis products of thalidomide were inactive in this system. The bioactivated intermediates were apparently amazingly stable in the case of thalidomide and α-EM12, inhibiting tumor cell attachment for a couple of hours. The possible nature of the postulated metabolites was not elucidated, but they could apparently be extracted with hexane and chloroform. According to indirect evidence, the reaction was catalyzed by a cytochrome P450-type enzyme system, but no defined CPY isozyme could be specified (Braun et al. 1986). When some thalidomide derivatives were tested in this system (α-EM12 as an active derivative and supidimide as a derivative with low biological potency), surprisingly small differences were found in the ability to inhibit attachment, but the stability of the active metabolites was found to differ considerably. This is not surprising, since the rate of hydrolysis of the three unmetabolized derivatives is already quite different, α-EM12 being the most stable one. The authors speculated that the lack of teratogenicity of supidimide may be due to low embryonic concentrations of the active metabolite rather than an intrinsic lack of teratogenic activity.

The concentrations used for these studies were quite high (> 100 μg/ml), but the experimental approach was not aimed at drawing conclusions on the situation possibly existing in vivo. Furthermore, no indication could be obtained of the rate and the amount of the active metabolite formed under the experimental conditions chosen.

Most importantly, the authors themselves (Braun and Weinreb 1984) were skeptical on the significance of their findings for explaining the teratogenicity of thalidomide. The main argument against the assumption that the effects observed in this model were relevant to explaining the teratogenic mechanism is the fact that the model could not distinguish between teratogenic and nonteratogenic thalidomide analogues: several derivatives with a very low or even absent teratogenic potency (e.g., β-EM12, phthalimidophthalimide, phthalimide, supidimide) gave the same "positive" response on adhesion as thalidomide in this model system. Furthermore, microsomes from thalidomide-sensitive and insensitive species were able to form the inhibitory metabolites. These metabolites were apparently not areneoxides. It also should be remembered that incubation of a substance with hepatic microsomes and cofactors may give rise to "artifacts" with no relevance to the situation in vivo. Therefore, it has to be proven that the metabolites in question really are formed in vivo. This has not been achieved for these hypothetical metabolites. Thus the data presented neither suggest adhesion as a mechanism to explain the teratogenicity of thalidomide, nor can they provide convincing evidence in the eternal question of whether thalidomide itself or a metabolite is the teratogenic species.

b) A Thalidomide Metabolite as the Active Agent

The speculation that an areneoxide may be the ultimate teratogenic substance is not really a hypothesis for explaining a teratogenic mode of action. It may, at best, be a pharmacokinetic suggestion of an active metabolite; such a metabolite has never been convincingly proven to exist, nor has any evidence whatsoever been provided that such a metabolite might be teratogenic. The possibility has also not been excluded that such a metabolite, as produced in vitro, is an artifact (as they are well known to occur during incubation of substances with liver microsomes). Furthermore, if such an areneoxide were formed within the maternal organism, it must be stable enough to reach the embryo, since a CYP-mediated metabolic activation within the early embryo is highly unlikely according to our present knowledge (SCHULZ and NEUBERT 1993).

Suggestions that an areneoxide might be the active teratogenic intermediate were solely based on theoretical considerations and on indirect evidence from studies with an experimental in vitro system that differed greatly from any model relevant to embryology. The experimental design was similar to the one discussed before (BRAUN and WEINREB 1985). Thalidomide was incubated with the usual hepatic microsomal drug-metabolizing systems, and only the target tissues were different; in this case human lymphocytes were used (GORDON et al. 1981). Thus this model system also shows little or no relevance to embryonic development. It may be interesting, however, with respect to effects known today to be induced on human lymphocytes in vivo and in vitro (NEUBERT et al. 1992c).

It was reported (GORDON et al. 1981) that incubation of thalidomide with microsomes from some species (rabbit, monkey, humans) mediated the formation of a metabolite toxic to lymphocytes in vitro, while microsomes from rats were inactive. The formation of an areneoxide was hypothesized from indirect evidence, namely the finding that the cytotoxic effect was abolished by adding epoxide hydrolase and enhanced by 1,2-epoxy-3,3,3-trichloropropane (TCPO), an inhibitor of this enzyme system. Furthermore, an apparent correlation of an effect in this system with the presumed teratogenicity of a few thalidomide derivatives was suggested: α-EM12 and phthalimidophthalimide were "active" in this system, while phthalimide and hexahydrothalidomide were not. The selection of these derivatives was rather unfortunate, since one of the selected "positive" substances, phthalimidophthalimide, is certainly not teratogenic in the monkey (KLUG et al. 1994), and the two "nonteratogenic" substances are also not a good choice. Studies in our own laboratory have also shown β-EM12 to be nonteratogenic (NEUBERT et al. 1987), and there are no good reasons to assume that an areneoxide cannot be formed from this substrate. Thus the areneoxide hypothesis is not really convincing on the basis of the data presented. The evidence available actually argues against it, and these metabolites have never been detected in vivo, although they should be stable enough to reach the embryo. No evidence is available that such a general cytotoxicity would be able to induce the typical malformations induced by thalidomide.

The studies by GORDON et al. (1981) might be of some interest in early hypothetical effects of thalidomide on lymphocytes, effects that are now well established (NEUBERT et al. 1992c, 1994). However, despite the cytotoxicity postulated by GORDON et al. (1981), no lymphopenia occurs in humans exposed to therapeutic doses of thalidomide, and there is no evidence that the drastic effects reproducibly induced on lymphocyte surface (adhesion) receptors require metabolic activation (see Sect. E.II.2.a).

In summary, it is quite feasible that biologically active metabolites are formed from thalidomide in vitro. Whether such metabolites might be formed from thalidomide and certain analogues within the intact organism has not been proven, but cannot be excluded. However, there is no evidence that such reactive metabolites, if they occur in vivo, reach the embryo in sufficient concentrations to induce teratogenic effects. Furthermore, reports on the formation of such biologically active metabolites are controversial (with regard to the possibility of areneoxides), and no convincing correlation with the teratogenic potential has been established in any of these studies when using thalidomide derivatives.

3. Primate Embryos

Of special significance are, of course, results of studies performed with embryonic tissue of primates. Therefore, these merit special attention.

a) Inductive Functions of the Mesonephros of Human Embryos

An especially fascinating topic within the field of morphogenetic differentiation is the inductive influence of two tissue types or of different cell types on each other during organogenesis. Although much of the information on limb development is based on studies with chick embryos, there is little doubt that cell–cell contact and cell–extracellular matrix interactions are essential prerequisites for morphogenesis also in mammals, including at least three fundamental processes of morphogenesis i.e., cellular differentiations, migration processes, and participation in the "zone of polarizing activity" (SUMMERBELL 1979). Cartilage formation in the chick wing primordium requires the inductive interaction of the ectoderm with the mesenchymal cells, and a direct cell contact seems to be necessary (GUMPEL-PINOT 1980). In the chick embryo, and probably also in mammalian embryos, limb bud mesenchymal cells originate from migration from both the somatopleure and the somites, and these two different cell lineages seem to be predetermined when reaching the limb bud: cells from the somatopleure differentiate into cartilage, soft connective tissue, tendons, and smooth muscle, while somite cells form skeletal muscle (CHRIST and JACOB 1980; KIENY 1980). For other tissues, further theoretical and experimentally underpinned considerations on morphogenetic (inductive) tissue interaction between two cell populations were published by GROBSTEIN (1955, 1956, 1967), and the concept was further developed by SAXÉN et al. (1976, 1985; SAXÉN and WARTIOVAARA 1966), using the early development of the metanephric nephron as a model system. This cluster of studies from these and several other laboratories on direct cell–cell interactions as well as on inductions by transmissible substances (using transfilter methods) are now classical publications on this topic.

Using organ cultures of human embryonic tissue as in vitro model systems, LASH and SAXÉN (1971, 1972) studied the influence of thalidomide (8 μg/ml) on induction phenomena occurring on mesonephros and other chondrogenic tissue, including limb buds. The mesonephros culture is a somewhat artificial model system with little relevance for the situation in vivo, because the mesonephros mesenchyme forms cartilage in vitro, which does not happen in vivo. Nevertheless, this is a model system with human tissue exhibiting obvious inhibitory effects induced by thalidomide. The extent of cartilage formation was used as a criterion, and glutethimide (Doriden) served as a control substance at the same concentration. Although this cannot be considered an ideal pair of substances, the results of morphological evaluations (alcian blue staining) using a scoring system were clear-cut (see Tables 18, 19), induction of cartilage matrix within the mesonephros was inhibited by thalidomide when compared with the cultures in the presence of glutethimide (LASH and SAXÉN 1971). Interestingly, the inhibitory effect on cartilage formation was only pronounced with tissue from 6- to 7-week-old embryos, and there was no difference between the cultures in the presence of thalidomide when mesonephros from older embryos was used. However, these explants from older embryos also showed a reduced capacity for forming cartilage in vitro. Similar results were obtained when an explant containing mesonephros with the adjacent limb primordium was used (LASH and SAXÉN 1972).

Table 18. Inhibition by thalidomide of cartilage formation in mesonephron explants of human embryos in vitro

Age of embryo and type of tissue cultured	Formation of cartilage matrix[a] in the presence of	
	Thalidomide	Gluthetimide
Mesonephros		
6 weeks	7	14
6 weeks	6	11
7 weeks	4	8
8 weeks	5	4

Glutethimide (Doriden) was used as a negative control. Substance concentrations: 8 μg/ml culture medium. (From LASH and SAXÉN 1971)
[a]A scoring system was used for estimating the amount of cartilage (alcian blue material) formed in culture. Larger numbers indicate that more cartilage was formed.

No noticeable effect of thalidomide was observed when limb buds or other chondrogenic tissues were cultured alone (Tables 18, 19). This is in agreement with similar attempts using mouse limb buds (see Sect. C.III.1.d). A significant decrease of mitotic figures was only observed in 4-week-old human limb buds cultured in the presence of 1–2 μg thalidomide/ml (YASUDA and NISHIMURA 1989).

Thus LASH and SAXÉN (1972) suggested from the results of their studies that one major site of the action of thalidomide might be tissue associated with the mesonephric mesenchyme, this tissue having a positive influence on limb chondrogenesis. By inhibiting this interaction, thalidomide would inhibit chondrogenesis and induce abnormal development. Regardless of whether this attractive hypothesis is true, the authors seem to have provided good evidence that thalidomide fails to inhibit chondrogenesis directly in the isolated limb

Table 19. Inhibition by thalidomide of cartilage formation in mesonephron explants of human embryos in vitro

	Morphological evaluation (compared with glutethimide)			
	Mesonephros alone	Mesonephros + limb bud	Limb bud alone	Somites + spinal cord
5.5 weeks	±	±	–	–
6 weeks	+	+	–	–
7 weeks	+	+	–	–
7.5 weeks	±	±	–	–
8 weeks	–	–	–	–

Glutethimide (Doriden) was used as a negative control. Substance concentrations: 8 μg/ml culture medium. (From LASH and SAXÉN 1972)
+, marked inhibition; ±, slight inhibition; –, no inhibition.

tissue and that interactions with adjacent tissue or cell migrations into the limb seem to be of importance for the teratogenic effect.

The main argument against the significance of this model system and the hypotheses advanced is that the embryonic stages investigated (6- and 7-week-old embryos) are too advanced for evaluating a thalidomide-induced effect. Assessment in humans and large nonhuman primates indicates the thalidomide-susceptible period to be developmental stages 11-14, which corresponds to an embryonic age of 3–4 weeks (MERKER et al. 1988; NEUBERT et al. 1988c). At 7 weeks of gestation (about 9 weeks of pregnancy), major limb abnormalities (amelia and phocomelia) can no longer be induced in humans and in larger non-human primates (see Sect. C.III.2). Thus, again, the considerations seem to point in the right direction, but the data are inconclusive.

b) Surface Receptors of the Embryo

After it was established that thalidomide is a potent agent that interferes with the expression of surface adhesion receptors on white blood cells in marmoset monkeys (NEUBERT et al. 1991, 1992b, 1993) and in humans (NEUBERT et al. 1992c; NOGUEIRA et al. 1994), the hypothesis was advanced that the teratogenic action of thalidomide might also be correlated with a downregulation of certain adhesion molecules, such as integrins. Such a hypothesis is attractive, since these adhesion receptors are involved in numerous cell–cell interactions and in interactions between cells and the extracellular matrix. Cell interactions have long been recognized as key events in the course of morphogenesis (See above).

In order to prove such a hypothesis on the possible primary mechanism of the teratogenic action of thalidomide, a number of aspects have to be taken into account. These are the same as hold true for establishing any other convincing hypothesis on the teratogenic mode of action of thalidomide:

- The studies must be performed with a susceptible species. Only primates or possibly rabbits can be used.
- It must be shown that these adhesion receptors are present and preferably that they change during the embryonic stage susceptible to thalidomide.
- The studies must be performed under conditions relevant to the teratogenic action of thalidomide, i.e., at the developmental stage of the embryo susceptible to the action of thalidomide (embryonic stages 11–13).
- A teratogenic effect of nearly 100% must be achieved under the experimental conditions, because one has to be sure that a malformation would have been induced in every case. This is only possible with primate embryos and cannot be achieved with rabbits.
- Typical primordia must be used to ensure that a tissue known to respond to the teratogenic action of thalidomide is chosen. Limb buds are a good choice for this purpose.
- The mechanism of action proposed must be plausible. Organotropism and stage specificities should be explained.

- It must be possible to explain the species specificity which is so very typical for the teratogenic action of thalidomide.
- It must be demostrated that nonteratogenic thalidomide derivatives (with very similar chemical structures, such as β-EM12 or the R[+]-enantiomers) do not induce the changes on embryonic cells assumed to be responsible for the teratogenic action.

These types of studies are very difficult to perform, because the whole embryo is only about 4–5 mm long at this susceptible stage and the limb buds are only just starting to be recognizable. Furthermore, the number of embryos that can be examined is limited in studies with primates.

Such studies were initiated in our laboratory, choosing the marmoset as an experimental model and EM12 as a convenient thalidomide derivative. The marmoset was used because of extensive experience and because it seemed to be the most suitable species for these kinds of studies (easy to breed, easy to operate on, carries twins and triplets, etc.). EM12 was chosen because it is even more active than thalidomide itself, and it has the great advantage for experimental studies that only a few hydrolysis products are formed. This substance is an ideal thalidomide-type substance in all respects.

After having tested a number of different procedures, the most suitable technique was to dissect tissue from the different primordia, to separate and isolate the cells, and to analyze the surface receptors with the use of monoclonal antibodies and flow cytometry. It had already been shown that a good number of antihuman monoclonal antibodies against epitopes of adhesion receptors cross-react with the epitopes of these marmoset receptors (NEUBERT et al. 1995a).

In the initial phase of the studies, it was demonstrated that various adhesion receptors are already expressed on cells of typical primordia at these embryonic stages (Figs. 4,5). When analyzing cells from embryos with 18–35 somite pairs, all the integrins tested were found to be already expressed on the surface of some of these cells at these early stages of development. Although the cells isolated varied considerably in their size (as assessed by flow cytometry), for convenience and simplicity the cells of the different primordia were designated in the figures as "smaller" and "larger" cells. The adhesion receptors found to be expressed on the surface of the embryonic cells include the following:

- β_1-*Integrins*. These are heterodimers containing a common β-chain (CD29) and different α-chains. The α-chains expressed in the embryo include CD49d (VLA-α_4), CD49c (VLA-α_3), CD49e (VLA-α_5), and CD49f (VLA-α_6) (NEUBERT et al. 1995b, c, 1996a).
- β_2-*Integrins*. These are heterodimers containing a common β-chain (CD18) and three different α-chains. The α-chains expressed in the embryo include CD11a (LFA-1_α), and CD11b (Mac-1).
- β_3-*Integrins*. These are heterodimers containing a common β-chain (CD61) and three different α-chains. The α-chains are not as well characterized as

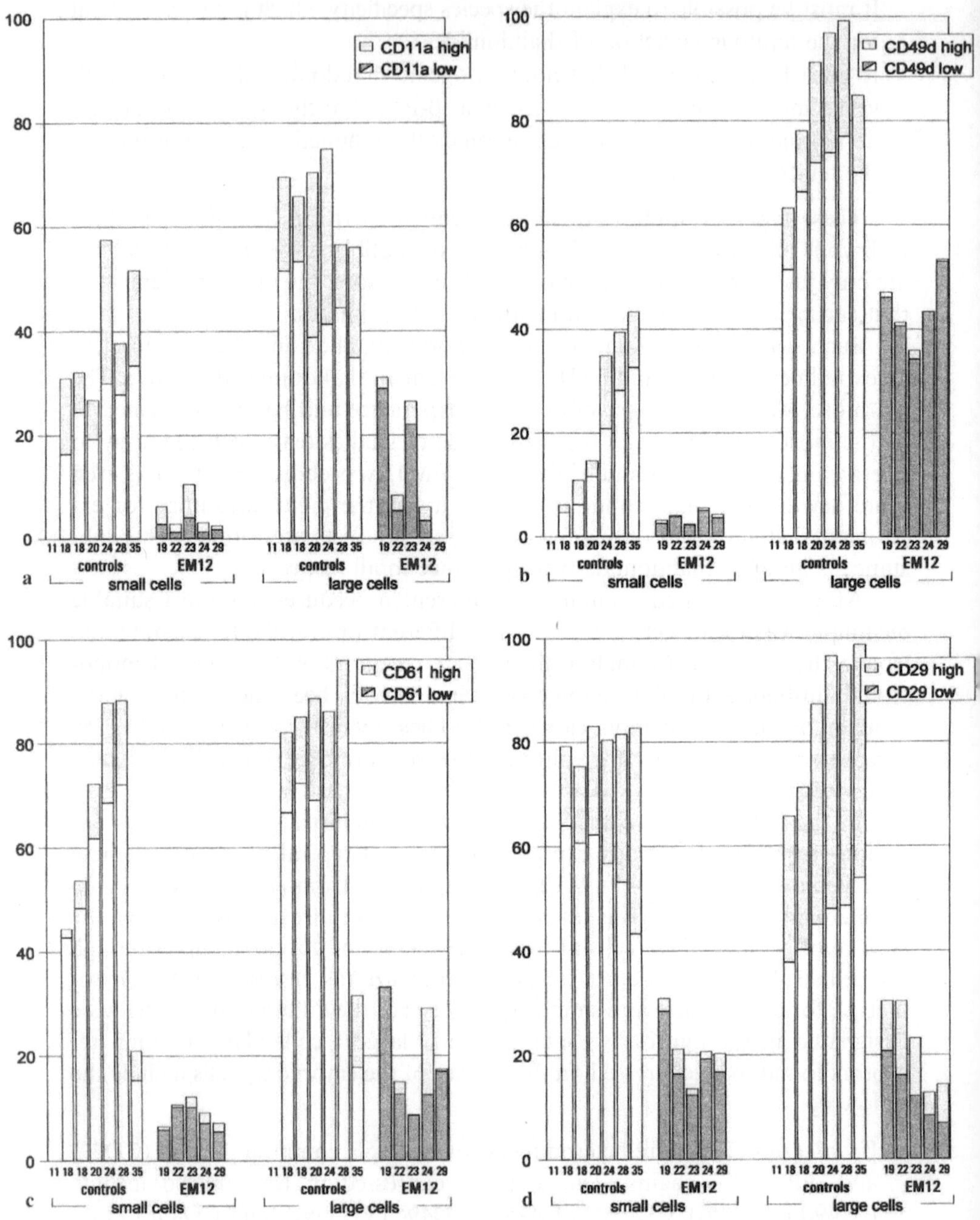
CD11a high
CD11a low
CD49d high
CD49d low
CD61 high
CD61 low
CD29 high
CD29 low
0
20
40
60
80
100
11 18 18 20 24 28 35
19 22 23 24 29
controls
EM12
small cells
large cells
a
b
c
d

for the β_1- and β_2- integrins. Therefore, only the β-chain was measured and shown to be expressed in the embryo.

- ICAM-1 (CD54). This single-chain transmembrane molecule, also shown to be present on some primordial cells, is a member of the immunoglobulin gene superfamily, and in cell–cell adhesion it is the counterpart binding to CD11a/CD18.
- L-Selectin (CD62L). This single-chain transmembrane protein has an N-terminal lectin-like domain, an epidermal growth factor (EGF) repeat, and it also occurred on some cells of the embryo.

The most important finding in these studies was a dramatic downregulation of several of the integrin receptors on the surface of cells from limb bud primordia of EM12-treated marmosets (NEUBERT et al. 1995b,c, 1996a). Some examples are shown in Fig. 4: the percentage of cells expressing the CD11a receptor is greatly reduced, both on small and on large cell types, in the limb bud cells of embryos exposed to α-EM12 in utero. Similarly, the percentage of cells with the CD49d receptor was greatly decreased. Furthermore, the β-chain of the β_3-integrins (CD61) was considerably affected by the exposure of the embryos to α-EM12 in utero.

A similar trend to that seen in the limb buds was also observed in cells of the heart primordium (Fig. 5), another target for the teratogenic action of thalidomide. The decrease in the percentage of cells carrying the CD11a receptor was not as pronounced as found in the limb primordium. However, a marked reduction in expression was observed for the CD49d receptor, especially in the smaller cell type. Similarly, the CD61 receptor was also found to be downregulated in the heart primordium of α-EM12-exposed marmoset embryos. Doses as small as 1 μg α-EM12/kg were effective, an equivalent level of exposure similar to that in mothers of malformed children. It seems quite feasible that the expression of even more receptors is changed than the ones studied so far in embryos exposed to a thalidomide-type substance in vivo.

From the results of these studies, there is no doubt that treatment of marmosets with α-EM12 induces drastic changes in the expression of several

Fig. 4a-d. Percentage of isolated marmoset limb bud cells expressing defined adhesion receptors. **a** CD11a (LFA-1_α). **b** CD49d (VLA-α_4). **C** CD61 (β_3-chain). **d** CD29 (β_1-chain). *Light colored columns* show data of embryonic cells from six untreated marmosets; the *dark-colored columns* show data from five litters exposed in utero to the thalidomide-derivative α-EM12. Embryos at the developmental stages 11–13 were studied. Receptor expression was measured with flow cytometry using fluorescence-tagged antihuman monoclonal antibodies cross-reacting with the marmoset epitopes. According to the flow cytometric light scatter characteristics, the cells in the cell suspension were divided into "smaller" and "larger" cell types. *"High"* and *"low"* indicate the epitope density on this cell fraction. Each column represents the data of the pooled primordia from one litter. The *numbers* under the columns give the number of somite pairs, indicative of the developmental stage of the embryo. In the limb buds of the control animals, the expression of some of the receptors increases during this phase of development (e.g., CD49d, smaller cells). There is a very clear-cut downregulation of cells carrying the adhesion receptors studied (β_1-, β_2-, and β_3-integrins) by EM12

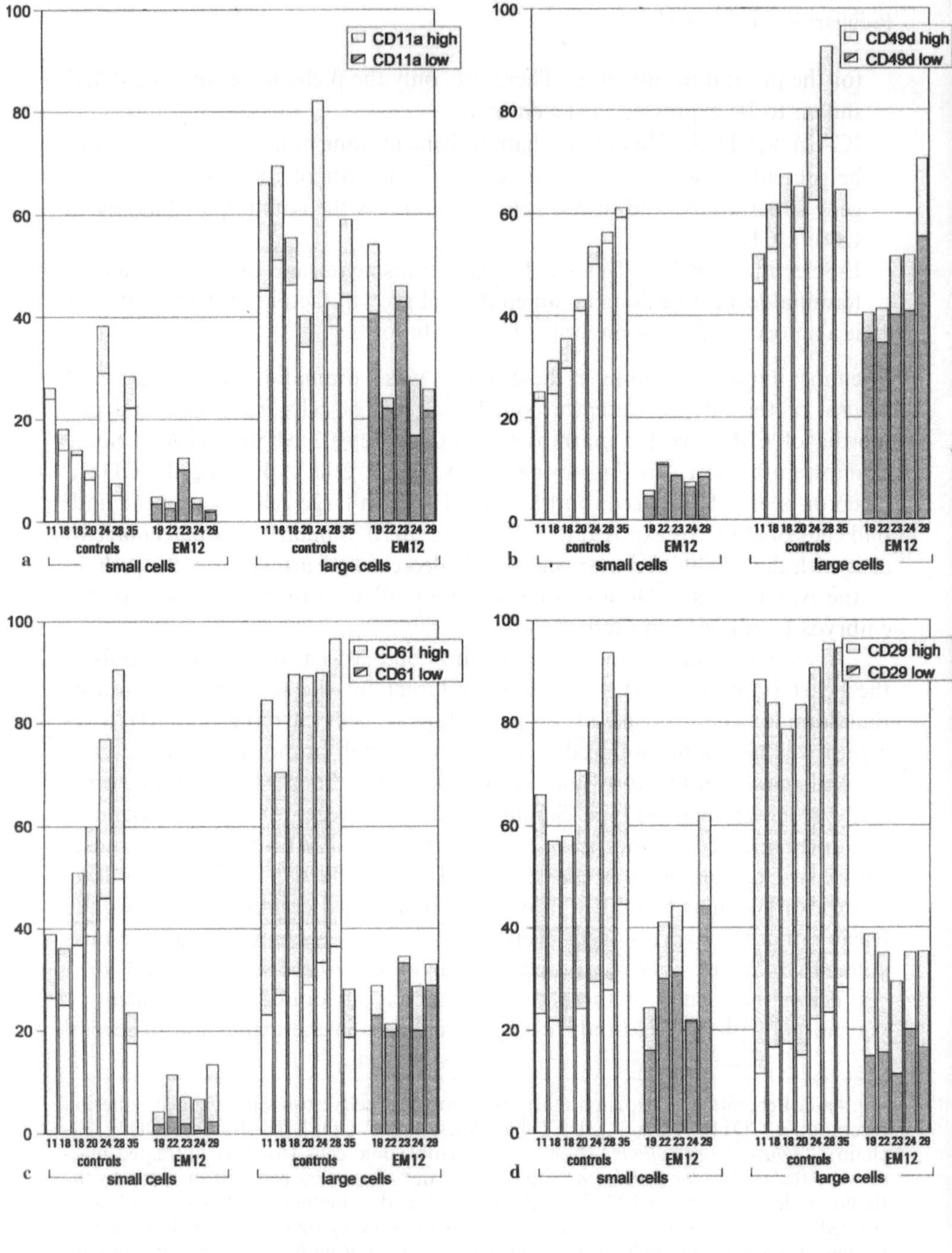

Fig. 5a-d. Percentages of isolated marmoset heart cells expressing defined adhesion receptors **a** CD11a (LFA-1α). **b** CD49d (VLA-α_4). **c** CD61 (β_3-chain). **d** CD29 (β_1-chain). *Light-colored columns* represent cells of embryos from seven untreated marmosets; the *dark-colored columns* give the values from five litters exposed in utero to α-EM12. Other experimental conditions as indicated Fig. 4. There is a clear-cut reduction of cells carrying some of the adhesion receptors (e.g., CD49d or CD61 on "smaller" cells) after dosing with EM12. For other receptors, the effect is not as pronounced

adhesion surface receptors in the primordia of exposed embryos. This is the only hypothesis that has been tested in the most susceptible species and in embryos at developmental stages sensitive to thalidomide. This is also the first time that drastic biological alterations of cells in primordia at the period of pathological development were detected in the course of the action of a thalidomide-type substance.

Whether the alteration observed is a "downregulation" or an inhibition or delay of the development-related expression of these receptors has not yet been established. While these results represent the first step towards understanding the possible mechanism of the teratogenic action of thalidomide, there are still a number of further questions to be answered, including the following:

- Which cells and which cell–cell and cell–extracellular matrix interactions are involved in these changes, and which developmental processes are disturbed?
- Does a downregulation of adhesion receptors always lead to pathological development, or might such effects be compensated for in certain primordia, allowing normal development?
- Are there other substances with different chemical structures deviating from that of thalidomide which induce a similar effect?

While these questions cannot be answered as yet, speculations may be made about the possible consequences of an inhibition of cell–cell or cell–extracellular matrix interactions. According to information on limb development (see above), three types of interference with morphogenesis are feasible:

1. Interference with migration of cells from the somatopleure into the future limb bud. When evaluating the early stages of development, assessment of such migration processes would be included in the experimental setup, since the tissue adjacent to the presumptive limb was investigated. In this process, interactions between the same and between different cell types, and between cells and the extracellular matrix, may be essential.
2. Interference with the differentiation of somatopleuric cells to blastema cells forming cartilage. Although it has been shown that thalidomide does not inhibit cartilage formation in the developed and predetermined limb bud (see above), such an effect may still be important at the early period studied by us.
3. Interference with processes within the zone of polarizing activity, i.e., the early determination of the formation of the specific cartilaginous bone primordia. The process of pattern formation must be initiated among the cells of the early limb bud.

The somatopleuric mesenchyme of the prelimb or early limb region at about stages 12 and 13 was already suggested as a target for the primary lesion induced by thalidomide by Theisen and McGregor (1981) on the basis of their in vivo studies on rhesus monkeys. At later stages of limb development, a regression of the apical ectodermal ridge was observed. However, in contrast to these authors, in numerous similar studies we have never seen extensive necrosis in the early stages of the thalidomide action on nonhuman primate limb primordia (H.J. MERKER, unpublished results).

The most important question is why thalidomide is not teratogenic in rodents, e.g., rats, at reasonable doses and why reduction deformities of the long bones of the limb have never been observed. This problem can now be tackled with rat embryos, using a similar technique as in marmosets.

Two explanations for the apparent lack of thalidomide-inducible teratogenicity in rats may be put forward:

- The corresponding adhesion receptors are not downregulated by thalidomide in the embryos of this species, in contrast to primates.
- The integrin receptors are downregulated, e.g., at very high dose levels. However, the outcome is not teratogenicity but embryomortality, as observed in some of the studies with rodents. In the rabbit, a combination of these two effects might occur.

For studies on rat embryos, antirat monoclonal antibodies must be used, since commonly available antihuman ones do not cross-react. The pool of such antirat monoclonal antibodies is still much smaller than the one for antihuman ones. While results of such studies with embryonic tissue are not yet available, preliminary data suggest that the α-EM12-induced downregulation of adhesion receptors on white blood cells of the adult rat is not very pronounced: while after high doses (100 mg α-EM12/kg for 2 weeks) a slight effect on the CD11a marker (less than 20% reduction) was observed, no effect at all occurred under the same experimental conditions on the CD49d receptor (THIEL et al. 1995).

While the interference with adhesion receptors and functions appears quite plausible as a cause of teratogenicity, it remains to be established how this drastic change in the receptor expression is accomplished by thalidomide and its active derivatives. Since so many quite different surface receptors are involved, a common mechanism of regulation must be affected by thalidomide. Interestingly, such regulatory mechanisms cannot be confined to lymphocytes or white blood cells, but must also be operative in other cells of the primate organism. On the other hand, cells of other species, such as rodents, seem to be rather insusceptible to this action. This is a particularly interesting aspect, since mechanisms such as those involved in signal transduction have long been thought to occur almost identically in different animals, at least as far as mammalian species are concerned.

It is quite feasible, and even very likely, that more effects than those revealed so far are affected by thalidomide. Only a selection of receptors have been studied to date, and interference with the regulatory mechanism involved may still affect other receptors. Furthermore, the change in the receptor expression must inevitably trigger many secondary reactions and changes in cell behavior as well as in the properties of the cells, e.g., with respect to their capability to release mediator substances and to respond to such substances.

Like any other pharmacological effect, this will result in a complex situation with respect to the whole organism, in this case the embryo.

F. Possible Implications of Recent Findings for Prenatal Toxicology

It may be argued that the teratogenicity of thalidomide is an antiquated problem. However, this is a poor argument for several reasons. First of all, it is intriguing that the mechanism of this teratogenic action has not been revealed over the last 30 years. Second, understanding possible mechanisms of substance-induced pathological reactions will not only increase our basic understanding of such processes, but also help to prevent adverse effects induced by substances acting via the same or a similar mechanism. Furthermore, malformed children are still being born (as claimed quite recently in Brazil) to mothers treated with this drug for erythema nodosum leprosum. If this is true, adequate measures for contraception were apparently and incomprehensibly not taken during the treatment period.

If the data on the substance-induced modification of adhesion receptors turns out to point to a causal relationship with pathological development, this will be the first example of an interference with adhesion receptors as a mechanism of not only of pharmacological effects, but also of pronounced and specific toxicity. Integrins and other adhesion receptors are now some of the most interesting types of receptors, since they take part in numerous cell–cell and cell–extracellular matrix interactions, which are certainly key events not only in embryology but in biology in general. This has especially been shown for many interactions in the immune system. We may have just opened the door to a fascinating new field of toxicology, not only concerning developmental toxicology, but also carcinogenicity and immuntoxicology.

Acknowledgement. We thank Jane Klein and Doris Webb for their help in preparing the manuscript. Original work in the author's laboratory was supported by grants from the Deutsche Forschungsgemeinschaft (DFG) awarded to the Sonderforschungsbereich 174 (Sfb 174) at the Free University, Berlin.

References

Arala-Chaves M, Vilanova M, Ribero A, Pinto J (1994) Immunostimulatory effect of thalidomide in normal C57BL/6 mice is compatible with stimulation of a highly connected central immune system. Scand J Immunol 40: 535–542

Aronson IK, Weber L, West D, Buys C, Reder A, Antel J (1986) Thalidomide and lymphocyte function. J Am Acad Dermatol 14: 282–283

Arruda WO, Hacbarth E, Döi E, Santa-Maria JR, Barbosa J, Kajdacsy-Balla AA (1986) Efeito da talidomida sobre os níveis séricos de immunoglobulinas IgM e IgA, fator reumatóide e isohemaglutininas anti-A e anti-B em pacientes com lepra lepromatosa. Um estudo duplo-cego. Rev Inst Med Trop São Paulo 28: 12–14

Atra E, Sato EI (1993) Treatment of the cutaneous lesions of systemic lupus erythematosus with thalidomide. Clin Exp Rheumatol 11: 487–493

Bakay B, Nyhan WL (1968) Binding of thalidomide by macromolecules in the fetal and maternal rat. J Pharmacol Exp Ther 161: 348–360
Barrow MV, Steffek AJ, King CTG (1969) Thalidomide syndrome in rhesus monkeys (macaca mulatta). Folia Primatol (Basel) 10: 195–203
Bass R (1975) Significance of mitochondrial function for embryonic development: proposal of a mechanism for the induction of embryolethal effects. In: Neubert D, Merker H-J (eds) New approaches to the evaluation of abnormal embryonic development. Thieme, Stuttgart, pp 524–541
Bass R, Oerter D (1977) Embryonic development and mitochondrial function. Naunyn Schmiedebergs Arch Pharmacol 296: 191–197
Becker R, Frankus E, Graudums I, Günzler WA, Helm F-C, Flohé L (1982) The metabolic fate of supidimide in the rat. Arzneimittelforschung 32: 1101–1111
Beckmann R (1962) Über das Verhalten von Thalidomid im Organismus. Arzneimittelforschung 12: 1095–1098
Beckmann R (1963) Über die Verteilung im Organismus und den biologischen Abbau von N-Phthalyl-glutaminsäure-imid-[^{14}C] (Thalidomid-[^{14}C]). Naunyn Schmiedebergs Arch Exp Pathol Pharmakol 245: 90–91
Bennouna-Biaz F, Benharbit R, Heid E, Senouci-Belkhadir K, Lazrak B (1986) Le maladie de Behcet. Essai de traitement par la thalidomide. Maroc Med 8: 301–308
Beschorner WE, Tutschka PJ, Santol GW (1982) Chronic graft-versus-host disease in the rat radiation chimera. Transplantation 33: 393–399
Bílá V, Kren V (1989) The significance of genetic factors for thalidomide teratogenicity in the Norway rat. Acta Univ Carolinae Med 35: 3–10
Bessis D, Guillot B, Monpoint S, Dandurand M, Guilhou JJ (1992) Thalidomide for systemic lupus erythematosus. Lancet 339: 549–550
Bore PJ, Scothorne RJ (1966) Effect of thalidomide on survival of skin homografts in rabbits. Lancet i: 1240–1241
Bowers PW, Powell RJ (1983) Effects of thalidomide on orogenital ulceration. Br Med J 287: 799–800
Braun AG, Weinreb SL (1984) Teratogen metabolism: activation of thalidomide and thalidomide analogues to products that inhibit the attachment of cells to concanavalin A coated plastic surfaces. Biochem Pharmacol 33: 1471–1477
Braun AG, Weinreb SL (1985) Teratogen metabolism: spontaneous decay products of thalidomide and thalidomide analogues are not bioactivated by liver microsomes. Teratogenesis Carcinog Mutagen 5: 149–158
Braun AG, Harding AG, Weinreb SL (1986) Teratogen metabolism: thalidomide activation is mediated by cytochrome P-450. Toxicol Appl Pharmacol 82: 175–179
Bresnahan TJ (1961) Attempted suicide with thalidomide. A case report. Appl Ther 3: 621–622
Brock N, von Kreybig T (1964) Experimenteller Beitrag zur Prüfung teratogener Wirkungen von Arzneimitteln an der Laboratoriumsratte. Naunyn Schmiedebergs Arch Exp Pathol Pharmak 249: 117–145
Burley DM (1960) Overdosage with thalidomide ("Distaval"). Med World 93: 26–28
Butler H (1977) The effect of thalidomide on a prosimian: the greater galago (Galago crassicaudatus). J Med Primatol 6: 319–324
Chen T-L, Vogelsang GB, Petty BG, Brundrett RB, Noe DA, Santos GW, Colvin OM (1989) Plasma pharmacokinetics and urinary excretion of thalidomide after oral dosing in healthy male volunteers. Drug Metab Dispos 17: 402–405
Christ B, Jacob HJ (1980) Origin, distribution and determination of chick limb mesenchymal cells. In: Merker H-J, Nau H, Neubert D (eds) Teratology of the limbs. de Gruyter, Berlin, pp 67–77
Cohen S (1960) Sleep regulation with thalidomide. Am J Psychiatry 116: 1030
Cole CH, Rogers PCJ, Pritchard S, Phillips G, Chan KW (1994) Thalidomide in the management of chronic graft-versus-host disease in children following bone marrow transplantation. Bone Marrow Transplant 14: 937–942

Couzigou P, Doutre MS, Houdée G, Richard-Molard B, Fleury B, Amouretti M, Beylot C, Béraud C (1983) Thalidomide et traitement du syndrome de Behcét. Gastroenterol Clin Biol 7: 751–752
Crain E, McIntosh KR, Gordon G, Pestronk A, Drachman DB (1989) The effect of thalidomide on experimental autoimmune Myastenia gravis. J Autoimmun 2: 197–202
Cutler J (1994) Thalidomide revisited. Lancet i: 795–796
D'Amato RJ, Loughnan MS, Flynn E, Folkman J (1994) Thalidomide is an inhibitor of angiogenesis. Proc Natl Acad Sci USA 91: 4082–4085
Delahunt CS, Lassen LJ (1964) Thalidomide syndrome in monkeys. Science 146: 1300–1301
Descotes J, Tedone R, Evreux J-C (1988) Enhancement of antibody response and delayed-type hypersensitivity by thalidomide in mice. Fundam Clin Pharmacol 2: 493–497
Doll NJ, Barnhill RL, Hastings RC, Salvaggio JE, Millikan L (1983) Thalidomide induced alteration of human peripheral blood monocyte activity. Clin Res 31: 920A
Dukor P, Dietrich FM (1967) Immunosuppression by thalidomide? Lancet 1: 569–570
Eisenbud L, Horowitz I, Kay B (1987) Recurrent aphthous stomatitis of the Behcet's type: successful treatment with thalidomide. Oral Surg Oral Med Oral Pathol 64: 289–292
Emre S, Sumrani N, Hong J (1990) Beneficial effect of thalidomide and ciclosporin combination in heterotopic cardiac transplantation in rats. Eur Surg Res 22: 336–339
Eriksson T, Riesbeck K, Östraat Ö, Ekberg H, Björkman S (1992) Drug exposure and flow cytometry analyses in a thalidomide treatment schedule that prolongs rat cardiac graft survival. Transplant Proc 24: 2560–2561
Eriksson T, Björkman S, Roth B, Fyge Å, Höglund P (1995) Stereospecific determination, chiral inversion in vitro and pharmacokinetics in humans of the enantiomers of thalidomide. Chirality 7: 44–52
Fabro S (1967a) Sul meccanismo dell'azione teratogena della talidomide. Acta Vitaminol Enzymol 21: 123–132
Fabro S (1967b) Toxicity and teratogenicity of optical isomers of thalidomide. Nature 215: 296
Fabro S, Smith RL, Williams RT (1965) Thalidomide as a possible biological acylating agent. Nature 208: 1208–1209
Faure M, Thivolet J, Gaucherand M (1980) Inhibition of PMN leukocyte chemotaxis by thalidomide. Arch Dermatol Res 269: 275–280
Fickentscher K, Kirfel A, Will G, Köhler F (1977) Stereochemical properties and teratogenic activity of some tetrahydrophthalimides. Mol Pharmacol 13: 133–141
Field EO, Gibbs JE, Tucker DF, Hellmann K (1966) Effect of thalidomide on the graft versus host reaction. Nature 211: 1308–1310
Flint OP, Orton TC, Girling LR (1985) In vitro inhibition of chondrogenesis in cultured rat embryo limb cells by thalidomide. Br J Pharmacol 84: 122p
Floersheim GL (1966) Another chance for thalidomide. Lancet i: 207
Foerster M, Neubert R (1995) Lack of effects on fetal thymus and lymphocyte development in marmoset fetuses and newborn following treatment with thalidomide. Teratology 51: 23A-24A (P15)
Foerster M, Helge T, Neubert R (1995) Effects of thalidomide and EM12 on the synthesis of TNF-α in cocultures of human monocytes and lymphocytes. Naunyn - Schmiedebergs Arch Pharmacol 351: R130(517)
Gad SM, Shannon EJ, Krotoski WA, Hastings RC (1985) Thalidomide induces imbalances in T-lymphocyte sub-populations in the circulating blood of healthy males. Lepr Rev 56: 35–39

Geist C, Wöhrmann T, Schneider J, Zwingenberger K (1995) Effects of thalidomide on the local Swartzman reaction in mice and rabbits. FEMS Immunol Med Microbiol 12: 165–174
Georghiou PR, Allworth AM (1992) Thalidomide in painful AIDS-associated proctitis. J Infect Dis 166: 939–940
Georghiou PR, Kemp RJ (1990) HIV-associated oesophageal ulcers treated with thalidomide. Med J Aust 152: 382–383
Ghigliotti G, Repetto T, Farris A, Roy MT, De Marchi R (1993) Thalidomide: treatment of choice for aphthous ulcers in patients seropositive for human immunodeficiency virus. J Am Acad Dermatol 28: 271–272
Giacone J, Schmidt HL (1970) Internal malformations produced by thalidomide and related compounds in rhesus monkeys. Anat Rec 166: 306
Goihman-Yahr M, Requena MA, Vallecalle-Suegart E, Convit J (1972) Autoimmune diseases and thalidomide. I. Experimental allergic encephalomyelitis and experimental allergic neuritis of the guinea pig. Int J Lepr 40: 133–141
Goihman-Yahr M, Requena MA, Vallecalle-Suegart E, Convit J (1974) Autoimmune diseases and thalidomide. II. Experimental allergic encephalomyelitis and experimental allergic neuritis of the rat. Int J Lepr 42: 266–275
Gordon GB, Spielberg SP, Blake DA, Balasubramanian V (1981) Thalidomide teratogenesis: evidence for a toxic arene oxide metabolite. Proc Natl Acad Sci USA 78: 2545–2548
Gorin I, Vilette B, Gehanno P, Escande JP (1990) Thalidomide in hyperalgic pharyngeal ulceration of AIDS. Lancet 335: 1343
Gray RW, Coats EA, Pogge RC, MacQuiddy EL (1960) Kevadon: new, safe, sleep-inducing agent. Dis Nerv Syst 21: 392–394
Grinspan D (1985) Significant response of oral aphthosis to thalidomide treatment. J Am Acad Dermatol 12: 85–90
Grinspan D, Fernández Blanco G, Agüero S (1989) Treatment of aphthae with thalidomide. J Am Acad Dermatol 20: 1060–1063
Grobstein C (1955) Inductive interaction in the development of the mouse metanephros. J Exp Zool 130: 319–340
Grobstein C (1956) Inductive tissue interactions in development. Adv Cancer Res 4: 187–236
Grobstein C (1967) Mechanism of organogenetic tissue interaction. Natl Cancer Inst Monogr 26: 279–299
Guex-Crosier Y, Pittet N, Herbort CP (1995) The effect of thalidomide and supidimide on endotoxin-induced uveitis in rats. Graefes Arch Clin Exp Ophthalmol 233: 90–93
Gumpel-Pinot M (1980) Limb chondrogenesis: interaction between ectoderm and mesoderm. In: Merker H-J, Nau H, Neubert D (eds) Teratology of the limbs. de Gruyter, Berlin, pp 43–51
Günzler V, Hanauske-Abel HM, Tschank G, Schulte-Wissermann H (1986) Immunological effects of thalidomide. Inactivity of the drug and several of its hydrolysis products in mononucleocyte proliferation tests. Arzneimittelforschung 36: 1138–1141
Gutiérrez-Rodríguez O, Starusta-Bacal P, Gutiérrez-Montes O (1989) Treatment of refractory rheumatoid arthritis – the thalidomide experience. J Rheumatol 16: 158–163
Hales BF, Jain R (1987) Species differences in the effects of drug metabolizing enzymes on the teratogenicity of thalidomide in rat embryo cultures. Teratology 35: 73A (78)
Hamza MH (1986) Treatment of Behcet's disease with thalidomide. Clin Rheumatol 5: 365–371
Hamza M (1990) Behcet's disease, plamoplantar pustulosis and HLA-B27 treatment with thalidomide. Clin Exp Rheumatol 8: 427

Hamza M, Chaffai M, Ben Ayed H (1982) Le traitement de la maladie de Behcèt par la thalidomide. Nouv Presse Med 11: 1080–1081
Hamza M, Hamzaoui K, Ayed K, Eleuch M, Zribi A (1987) Thalidomide and cell mediated immunity in Behcet's disease. Clin Rheumatol 6: 608–609
Hasper MF (1983) Chronic cutaneous lupus erythematosus. Thalidomide treatment of 11 patients. Arch Dermatol 119: 812–815
Heger W, Klug S, Schmahl H-J, Nau H, Merker H-J, Neubert D (1988) Embryotoxic effects of thalidomide derivatives in the non-human primate Callithrix jacchus. 3. Teratogenic potency of the EM12 enantiomers. Arch Toxicol 62: 205–208
Heger W, Schmahl H-J, Klug S, Felies A, Nau H, Merker H-J, Neubert D (1994) Embryotoxic effects of thalidomide derivatives in the non-human primate Callithrix jacchus. IV. Teratogenicity of μg doses of the EM12 enantiomers. Teratogenesis Carcinog Mutagen 14: 115–120
Helge T, Foerster M, Neubert R (1995) Influence of thalidomide and the derivative EM12 on cytokine release by the monocytoid cell line Monomac 6. Naunyn - Schmiedebergs Arch Pharmacol 351: R129 (516)
Hellmann K (1966) Immunosuppression by thalidomide: implications for teratology. Lancet i : 1136–1137
Hellmann K, Duke DI, Tucker DF (1965) Prolongation of skin homograft survival by thalidomide. Br Med J 2: 687–689
Helm F (1966) Tierexperimentelle Untersuchungen und Dysmeliesyndrom. Arzneimittelforschung 16: 1232–1244
Helm F, Frankus E, Friderichs E, Graudums I, Flohé L (1981) Comparative teratological investigation of compounds structurally and pharmacologically related to thalidomide. Arzneimittelforschung 31: 941–949
Hendrickx AG (1972) A comparison of temporal factors in the embryological development of man, old world monkeys and galagos, and craniofacial malformations induced by thalidomide and triamcinolone. In: Goldsmith EI, Moor-Jankowski J (eds) Medical primatology. Karger, Basel, pp 259–269
Hendrickx AG (1973) The sensitive period and malformation syndrome produced by thalidomide in the crab-eating monkey (Macaca fascicularis). J Med Primatol 2: 267–276
Hendrickx AG, Helm F-C (1980) Nonteratogenicity of a structural analog of thalidomide in pregnant baboons (papio cynocephalus). Teratology 22: 179–182
Hendrickx AG, Newman L (1973) Appendicular skeletal and visceral malformations induced by thalidomide in bonnet monkeys. Teratology 7: 151–160
Hendrickx AG, Sawyer RH (1978) Developmental staging and thalidomide teratogenicity in the green monkey (Cercopithecus aethiops). Teratology 18: 393–404
Hendrickx AG, Axelrod LR, Clayborn LD (1966) Thalidomide syndrome in the baboon. Nature 210: 958–959
Heney D, Lewis IJ, Bailey CC (1988) Thalidomide for chronic graft-versus-host disease in children. Lancet ii : 1317
Heney D, Bailey CC, Lewis IJ (1990) Thalidomide in the treatment of graft-versus-host disease. Biomed Pharmacother 44: 199–204
Heney D, Norfolk R, Wheeldon J, Bailey CC, Lewis IJ, Barnard DL (1991) Thalidomide treatment for chronic graft-versus-host disease. Br J Haematol 78: 23–27
Henkel L, Willert HE (1969) Dismelia. A classification and pattern of malformations in a group of congenital defects of the limbs. J Bone Joint Surg [Br] 51: 399–414
Hiddleston WA, Siddall RA (1978) The production and use of Callithrix jacchus for safety evaluation studies In: Neubert D, Merker HJ, Nau H, Langman J (eds) Role of pharmacokinetics in prenatal and perinatal toxicology. Thieme, Stuttgart, pp 289–297
Huang PHT, McBride WG (1990) Thalidomide induced alterations in secondary structure of rat embryonic DNA in vivo. Teratogenesis Carcinog Mutagen 10: 281–294

Huang PHT, McBride WG (1995) Evidence of interaction of thalidomide with DNA. 23rd ETS Meeting, Dublin, September 1995 (abstract)
Iyer CGS, Languillon J, Ramanujam K, Tarabini-Castellani G, Terencio de las Aguas J, Bechelli LM, Uemura K, Martinez Dominguez V, Sundaresan T (1971) WHO co-ordinated short-term double-blind trial with thalidomide in the treatment of acute lepra reactions in male lepromatous patients. Bull WHO 45: 719–732
Jackson AJ, Schumacher HJ (1980) Identification of urinary metabolites of EM12 N-(2', 6'-dioxopiperiden-3'-yl)-phthalimidine, a teratogenic analogue of thalidomide in rats and rabbits. Can J Pharma Sci 15: 21–23
Jadhav VH, Patki AH, Mehta JM (1990) Thalidomide in type-2 lepra reactions – a clinical experience. Indian J Lepr 62: 316–320
Jaskoll T, Boyer PD, Melnick M (1994) Tumor necrosis factor-α and embryonic mouse lung development. Dev Dyn 201: 137–150
Jenkins JS, Powell RJ, Allen BR, Littlewood SM, Maurice PDL, Smith NJ (1984) Thalidomide in severe orogenital ulceration. Lancet ii: 1424–1426
Jew LJ, Middleton RK (1990) Thalidomide in erythema nodosum leprosum. DICP Ann Pharmacother 24: 482–483
Jönsson NÅ (1972)Chemical structure and teratogenic properties. IV. An outline of a chemical hypothesis for the teratogenic action of thalidomide. Acta Pharm Suecica 9: 543–562
Jorizzo JL, Schmalstieg FC, Solomon AR Jr, Cavallo T, Taylor RS, Rudloff HB, Schmalstieg EJ, Daniels JC (1986) Thalidomide effects in Behcet's syndrome and pustular vasculitis. Arch Intern Med 146: 878–881
Keberle H, Faigle JW, Fritz H, Knüsel F, Loustalot P, Schmid K (1965) Theories on the mechanism of action of thalidomide. In: Robson JM, Sullivan FM, Smith RL (eds) Embryopathic activity of drugs. Churchill, London, pp 210–233
Kemper F (1962) Thalidomide and congenital abnormalities. Lancet ii: 836
Kida M (1987) Thalidomide embryopathy in Japan. Kodansha, Tokyo
Kieny M (1980) The concept of a myogenic cell line in developing avian limb buds. In: Merker H-J, Nau H, Neubert D (eds). Teratology of the limbs. de Gruyter, Berlin, pp 79–88
King CTG, Kendrick FJ (1962) Teratogenic effects of thalidomide in the Sprague Dawley rat. Lancet II 2: 1116
Klug S, Felies A, Stürje H, Nogueira AC, Neubert R, Frankus E (1994) Embryotoxic effects of thalidomide derivatives in the non-human primate Callithrix jacchus. 5. Lack of teratogenic effects of phthalimidophthalimide. Arch Toxicol 68: 203–205
Knop J, Bonsmann G, Happle R, Ludolph A, Matz DR, Mifsud EJ, Macher E (1983) Thalidomide in the treatment of sixty cases of chronic discoid lupus erythematosus. Br J Dermatol 108: 461–466
Koch HP, Czejka MJ (1986) Evidence for the intercalation of thalidomide into DNA: clue to the molecular mechanism of thalidomide teratogenicity. Z Naturforsch 41c: 1057–1061
Koch HP, Köhler F (1976) Teratology study of two isoglutamine derivatives. Arch Toxicol 35: 63–68
Kocher-Becker U, Kocher W, Ockenfels H (1992) Embryotoxische Wechselwirkungen zwischen einem Hydrolyseprodukt des Thalidomids und dem Tensid Tween 20 bei der Maus. Z Naturforsch 47c: 155–169
Kohchi C, Nogushi K, Tanabe Y, Mizuno D-I, Soma G-I (1994) Constitutive expression of TNF-α and TNF-β genes in mouse embryo: roles of cytokines as regulator and effector of development. Int J Biochem 26: 111–119
Koransky W, Ullberg S (1964) Autoradiographic investigations of ^{14}C-labeled thalidomide and glutethimide in pregnant mice. Proc Soc Exp Biol Med 116: 512–516
Kreipe U (1967) Mißbildungen innerer Organe bei Thalidomidembryopathie. Arch Kinderheilk 176: 33–61

Krenn M, Gamcsik MP, Vogelsang GB, Colvin OM, Leong KW (1992) Improvements in solubility and stability of thalidomide upon complexation with hydroxypropyl-β-cyclodextrin. J Pharmaceut Sci 81: 685–689
Kunstmann AH (1960) Erfahrungen mit Contergan in der Kinderheilkunde. Therap Gegenw 99: 106–109
Kürkcüoglu N, Atakan N, Eksioglu M (1985) Thalidomide in the treatment of recurrent necrotic mucocutaneous aphthae. Br J Dermatol 112: 632
Larsson H (1990) Case report. Treatment of severe colitis in Bechcet's syndrome with thalidomide (CG-217). J Intern Med 228: 405–407
Lasagna L (1960) Thalidomide – new nonbarbiturate sleep-inducing drug. J Chronic Dis 11: 627–631
Lash JW, Saxén L (1971) Effect of thalidomide on human embryonic tissues. Nature 323: 634–635
Lash JW, Saxén L (1972) Human teratogenesis: in vitro studies on thalidomide-inhibited chondrogenesis. Dev Biol 28: 61–70
Lawrence R, Pflum R, Graeme ML (1979) The Arthus reaction in rats, a possible test for antiinflammatory and antirheumatic drugs. Agents Actions 9: 184–189
Lehmann H, Niggeschulze A (1971) The teratologic effects of thalidomide in Himalayan rabbits. Toxicol Appl Pharmacol 18: 208–219
Lenz W (1961) Kindliche Mißbildungen nach Medikamenten-Einnahme während der Gravidität? Dtsch Med Wochenschr 52: 2555-2556
Lenz W (1962) Thalidomide and congenital abnormalities. Lancet i: 45
Lenz W (1985) Thalidomide embryopathy in Germany, 1959–1961. In: Marois M (ed) Prevention of physical and mental congenital defects. C. Basic and medical science, education, and future strategies. Liss, New York, pp 77–83
Lenz W (1988) A short history of thalidomide embryopathy. Teratology 38: 2-3-215
Lenz W, Knapp K (1962a) Thalidomid-Embryopathie. Dtsch Med Wochenschr 87: 1232–1242
Lenz W, Knapp K (1962b) Thalidomide embryopathy. Arch Environm Health 5: 14–19
Levi L, Saggorato F, Crippa D (1980) Attività e meccanismo d'azione della talidomide in 18 pazienti affetti da lupus eritematoso discoide fisso. G Ital Dermatol Venereol 115: 471–475
Ley K, Gaehtgens P, Fennie C, Singer MS, Lasky LA, Rosen SD (1991) Lectin-like cell adhesion molecule 1 mediates leukocyte rolling in mesenteric venules in vivo. Blood 77: 2553–2555
Lim SH, McWhannell A, Vora AJ, Boughton BJ (1988) Successful treatment with thalidomide of acute graft-versus-host disease after bone-marrow transplantation. Lancet i: 117
Londoño F (1973) Thalidomide in the treatment of actinic prurigo. Int J Dermatol 12: 326–328
Lovell CR, Hawk JLM, Calnan CD, Magnus IA (1983) Thalidomide in actinic prurigo. Br J Dermatol 108: 467–471
Lucey JF, Behrman RE (1963) Thalidomide: effect upon pregnancy in the rhesus monkey. Science 139: 1295–1296
MacKenzie D (1983) Secret tests say thalidomide is mutagenic. New Sci 99: 457
Makonkawkeyoon S, Limson-Pobre RNR, Moreira AL, Schauf V, Kaplan G (1993) Thalidomide inhibits the replication of human immunodeficiency virus type 1. Proc Natl Sci Acad USA 90: 5974–5978
Maouris PG, Hirsch PJ (1988) Pregnancy in women with thalidomide induced disabilities. Br J Obstet Gynaecol 95: 717–719
Mascaro JM, Lecha M, Torras H (1979) Thalidomide in the treatment of recurrent, necrotic, and giant mucocutaneous aphthae and aphthosis. Arch Dermatol 115: 636–637
Matsubara Y, Mikami T (1985) Teratogenic potentiality of single dose thalidomide in JW-NIBS rabbits. Exp Anim 34: 295–302

Maurice PDL, Barkley ASJ, Allen BR (1986) The effect of thalidomide on arachidonic acid metabolism in human polymorphonuclear leukocytes and platelets. Br J Dermatol 115: 677–680
McBride WG (1961a) Thalidomide and congenital abnormalities. Lancet ii: 1358
McBride WG (1961b) Congenital abnormalities and thalidomide. Med J Aust 2: 1030
McBride WG (1974) Fetal nerve cell degeneration produced by thalidomide in rabbits. Teratology 10: 283–292
McBride WG (1994) Thalidomide may be a mutagen. Br Med J 308: 1635–1636
McBride WG, Vardy P (1983) Pathogenesis of thalidomide teratogenesis in the marmoset (Callithrix jacchus): evidence suggesting a possible trophic influence of cholinergic nerves in limb morphogenesis. Dev Growth Diff 25: 361–373
McCarthy DM, Kanfer E, Taylor J, Barrett AJ (1988) Thalidomide for graft-versus-host disease. Lancet i: 1135
McCredie J (1975) The action of thalidomide on the peripheral nervous system of the embryo. Proc Aust Assoc Neurol 12: 135–140
McCredie J (1980) Thalidomide and the neural crest. In: Merker H-J, Nau H, Neubert D (eds) Teratology of the limbs. de Gruyter, Berlin, pp 431–448
McHugh SM, Rifkin IR, Deighton J, Wilson AB, Lachmann PJ, Lockwood CM, Ewan PW (1995) The immunosuppressive drug thalidomide induces T helper cell type 2 (Th2) and concomitantly inhibits Th1 cytokine production in mitogen- and antigen-stimulated human peripheral blood mononuclear cell cultures. Clin Exp Immunol 99: 160–167
Mellin GW, Katzenstein M (1962) The saga of thalidomide. Neuropathy to embryopathy, with case reports of congenital anomalies. N Engl J Med 267: 1184–1193
Merker H-J, Heger W, Sames K, Stürje H, Neubert D (1988) Embryotoxic effects of thalidomide derivatives in the non-human primate Callithrix jacchus. I Effects of 3-(1,3-dihydro-1-oxo-2H-isoanisol-2-yl)-2,6-dioxopiperidine (EM12) on skeletal development. Arch Toxicol 61: 165–179
Miller JM, Ginsberg M, McElfatrick GC, Shonberg IL (1960) The anti-inflammatory effect and the analgesic property of Contergan-268. Antibiotic Med Clin Ther 7: 743–746
Miyachi Y, Ozaki M (1981) Thalidomide does not affect polymorphonuclear leukocyte (PMN) chemotaxis. Jpn J Lepr 50: 62–64
Miyachi Y, Ozaki M, Uchida K, Niwa Y (1982) Effects of thalidomide on the generation of oxygen intermediates by Zymosan-stimulated normal polymorphonuclear leukocytes. Arch Dermatol 274: 363–367
Moncada B, González-Amaro BR, Urbina R, Loredo CE (1985) Thalidomide – effect on T cell subsets as a possible mechanism of action. Int J Lepr 53: 201–205
Moreira AL, Sampaio EP, Zmuidzinas A, Frindt P, Smith KA, Kaplan G (1993) Thalidomide exerts its inhibitory action on tumor necrosis factor-α by enhancing mRNA degradation. J Exp Med 177: 1657–1680
Mouzas GL (1966) Another chance for thalidomide? Lancet i: 882
Mouzas GL, Gershon RK (1968) The effect of thalidomide on skin allografts in mice: survival of the graft. Transplantation 6: 476–478
Müchlenstedt D, Schwarz M (1984) Gynäkologisch-endokrinologische Untersuchungen bei thalidomidgeschädigten Mädchen. Geburtsh Frauenheilkd 44: 243–248
Munro CS, Cox NH (1988) Pyoderma gangrenosum associated with Behcet's syndrome response to thalidomide. Clin Exp Dermatol 13: 408–410
Murphy GP, Brede HD, Weber HW, Groenewald JH, Williams PD (1970a) Thalidomide immunosuppression in canine renal allotransplantation. J Med 1: 66–73
Murphy GP, Mirand EA, Groenewald JH, Schooness R, Weber HW, Brede HD, DeKlerk JN, vanZyl JJW (1970b) Erythropoietin release in renal allografted baboons, with and without immunosuppression. Invest Urology 7: 271–282
Naafs B, Faber WR (1985) Thalidomide therapy. An open trial. Int J Dermatol 24: 131–134

Naafs B, Bakkers EJM, Flinterman J, Faber WR (1982) Thalidomide treatment of subacute cutaneous lupus erythematosus. Br J Dermatol 107: 83–86
Nau H, Schmahl H-J, Heger W, Neubert D (1988) Pharmacokinetics and chiral aspects of thalidomide and EM12 teratogenesis in monkey and rat. In: Neubert D, Merker HJ, Hendrickx AG (eds) Non-human primates. Developmental biology and toxicology. Ueberreiter, Berlin, pp 461–476
Netea MG, Blok WL, Kullberg B-J, Bemelmans M, Vogels MTE, Buurman WA, van der Meer JWM (1995) Pharmacologic inhibitors of tumor necrosis factor production exert differential effects in lethal endotoxemia and in infection with live microorganisms in mice. J Infect Dis 171: 393–399
Neubert R (1996) Thalidomide. In: Spencer PS, Schaumburg HS (eds) Experimental and clinical neurotoxicology, 2nd edn. Oxford University Press, Oxford (in press)
Neubert D, Bluth U (1981) Limb bud organ cultures from mouse embryos after apparent induction of monooxygenases in utero. Effects of cyclophosphamide, dimethylnitrosamine and some thalidomide derivatives. In: Neubert D, Merker H-J (eds) Culture techniques. Applicability for studies on prenatal differentiation and toxicity. de Gruyter, Berlin, pp 175–195
Neubert D, Heger W, Bremer D, Frankus E, Helm F-C, Merker H-J (1987) Some studies on the areneoxide hypothesis of the teratogenic action of thalidomide In: Nau H, Scott WJ (eds) Pharmacokinetics in teratogenesis vol 1. CRC Press, Boca Raton, pp 177–179
Neubert D, Heger W, Fischer B, Merker H-J (1988a) Variability and speed of development in marmosets and man. In: Neubert D, Merker HJ, Hendrickx AG (eds) Non-human primates. Developmental biology and toxicology. Ueberreiter, Berlin, pp 205–215
Neubert D, Heger W, Klug S, Merker H-J, Nau H, Frankus E (1988b) Exploratory studies on the teratogenicity of two methylated thalidomide derivatives in marmosets. In: Neubert D, Merker HJ, Hendrickx AG (eds) Non-human primates. Developmental biology and toxicology. Ueberreiter, Berlin, pp 477–482
Neubert D, Heger W, Merker H-J, Sames K, Meister R (1988c) Embryotoxic effects of thalidomide derivatives in the non-human primate Callithrix jacchus. II. Elucidation of the susceptible period and of the variability of embryonic stages. Arch Toxicol 61: 180–191
Neubert R, Nogueira AC, Neubert D (1991) Effect of thalidomide-derivatives on the pattern of lymphocyte subpopulations in marmoset blood. Naunyn -Schmiedebergs Arch Pharmacol 344 [Suppl 2]: R123(77)
Neubert D, Klug S, Golor G, Neubert R, Felies A (1992a) Marmosets (Callithrix jacchus) as useful species for toxicological risk assessments. In: Neubert D, Kavlock RJ, Merker H-J, Klein J (eds) Risk assessment of prenatally-induced adverse health effects. Springer, Berlin Heidelberg New York, pp 347–376
Neubert R, Nogueira AC, Helge H, Stahlmann R, Neubert D (1992b) Feasibility of studying effects on the immune system in non-human primates. In: Neubert D, Kavlock RJ, Merker H-J, Klein J (eds) Risk assessment of prenatally-induced adverse health effects. Springer, Berlin Heidelberg New York, pp 313–346
Neubert R, Nogueira AC, Neubert D (1992c) Thalidomide and the immune system. 2. Changes in receptors on blood cells of a healthy volunteer. Life Sci 51: 2107–2117
Neubert R, Nogueira AC, Neubert D (1993) Thalidomide derivatives and the immune system. 1. Changes in the pattern of integrin receptors and other surface markers on T lymphocyte subpopulations of marmoset blood. Arch Toxicol 67: 1–17
Neubert R, Helge H, Neubert D (1994) Thalidomide and the immune system. 4. Downregulation of the CD26 receptor, probably involved in the binding of HIV components to T cells in primates. Life Sci 56: 407–420
Neubert R, Foerster M, Nogueira AC, Helge H (1995a) Cross-reactivity of antihuman monoclonal antibodies with cell surface receptors in the common marmoset. Life Sci 58: 317–324

Neubert R, Hinz N, Helge H, Neubert D (1995b) Occurrence of adhesion receptors on cells of primate embryos, and its down-regulation as the possible primary mode of action of thalidomide. Teratology 51: 14A (S12)
Neubert R, Hinz N, Thiel R, Neubert D (1995c) Down-regulation of adhesion receptors on cells of primate embryos as a probable mechanism of the teratogenic action of thalidomide. Life Sci 58: 295–316
Neubert R, Hinz N, Thiel R (1996a) Embryonic development and surface integrin receptors. Methods Dev Toxicol Biol (in press)
Neubert R, Nogueira AC, Helge H, Neubert D (1996b) Thalidomide and the immune system. 7. Proliferative response of lymphocytes from thalidomide treated volunteers. Life Sci (in preparation)
Neuhaus G, Ibe K (1960) Klinische Beobachtungen über einen Suicidversuch mit 144 Tabletten Contergan-forte (N-Phthalyl-glutaminsäureamid). Med Klin 55: 544–545
Newman LM, Hendrickx AG (1985) Temporomandibular malformations in the bonnet monkey (macaca radiata) fetus following maternal ingestion of thalidomide. J Craniofac Genet 5: 147–157
Nicolau DP, West TE (1990) Thalidomide: treatment of severe recurrent aphthous stomatitis in patients with AIDS. DICP Ann Pharmacother 24: 1054–1056
Nielsen H, Valerius NH (1986) Thalidomide enhances superoxide anion release from human polymorphonuclear and mononuclear leukocytes. Acta Pathol Microbiol Immunol Scand Sect C 94: 233–237
Nishimura H (1972) A note on critical periods for drug induced reduction deformities of total forelimbs in man. Cong Anom 12: 191–193
Nishimura H, Takano K, Tanimura T, Yasuda M, Uchida T (1966) High incidence of several malformations in the early human embryos as compared with infants. Biol Neonat 10: 93–107
Nishimura H, Takano K, Tanimura T, Yasuda M (1968) Normal and abnormal development of human embryos: first report of the analysis of 1213 intact embryos. Teratology 1: 281–290
Nishimura K, Hashimoto Y, Iwasaki S (1994) Enhancement of phorbol ester-induced production of tumor necrosis factor-α by thalidomide. Biochem Biophys Res Commun 199: 455–460
Nogueira AC, Neubert R, Helge H, Neubert D (1994) Thalidomide and the immune system. 3. Simultaneous up- and down-regulation of different integrin receptors on human white blood cells. Life Sci 55: 77–92
Nogueira AC, Gräfe M, Neubert R (1995a) Effect of thalidomide and some derivatives on the adhesion of lymphocytes to endothelial cells. Naunyn Schmiedebergs Arch Pharmacol 351 [Suppl]: R130 (518)
Nogueira AC, Neubert R, Felies A, Jacob-Müller U, Frankus E, Neubert D (1995b) Thalidomid derivatives and the immune system. 6. Effects of two derivatives with no obvious teratogenic potency on the pattern of integrins and other surface receptors on blood cells of marmosets. Life Sci 58: 337–348
Nougué J, Nougué M, Bazex J, Hayek JP, Bidouze H (1983) Survenue d'une hémorragie génitale aucours d'un prurigo nodulaire traité par thalidomide. Ann Dermatol Venereol 110: 551–552
Nowack E (1965) Die sensible Phase bei der Thalidomid-Embryopathie. Humangenetik 1: 516–536
Oerter D, Bass R (1975) Embryonic development and mitochondrial function. I. Effects of chloramphenicol infusion on the synthesis of cytochrome oxidase and DNA in rat embryos during late organogenesis. Naunyn Schmiedebergs Arch Pharmacol 290: 175–189
Östraat Ö, Ekberg H, Schatz H, Riesbeck K, Eriksson T (1992) Thalidomide prolongs graft survival in rat cardiac transplants. Transplant Proc 24: 2624–2625
Ogilvie JW, Kantor FS (1968) The effect of thalidomide on the immune response. Fed Proc 27: 494 (1563)

Ohsawa T, Natori S (1989) Expression of tumor necrosis factor at a specific developmental stage of mouse embryos. Dev Biol 135: 459–461
O'Rahilly, Müller F (1992) Human embryology and teratology. Wiley-Liss, New York
Palmer AK (1977) Reproductive toxicology studies and their evaluation. In: Ballantyne B (ed) Current approaches in toxicology. Wright, Bristol, pp 54–55
Pampfer S, Wuu Y-D, Van der Heyden I, De Hertogh R (1994) Expression of tumor necrosis factor-α (TNFα) receptors and selective effects of TNFα on the inner cell mass in mouse blastocysts. Endocrinology 134: 206–212
Parikh DA, Ganapati R, Revankar CR (1986) Thalidomide in leprosy – study of 94 cases. Indian J Lepr 58: 560–566
Parker PM, Chao N, Nademanee A, O'Donnell MR, Schmidt GM, Snyder DS, Stein AS, Smith EP, Molina A, Stepan DE, Kashyap A, Planas I, Spielberger R, Somlo G, Margolin K, Zwingenberger K, Wilsman K, Negrin RS, Long GD, Niland JC, Blume KG, Forman SJ (1995) Thalidomide as salvage therapy for chronic graft-versus-host disease. Blood 86: 3604–3609
Parkhie M, Webb M (1983) Embryotoxicity and teratogenicity of thalidomide in rats. Teratology 27: 327–332
Patey O, Charasz N, Roucayrol AM, Malkin JE, Lafaix C (1989) Thalidomide et colite ulcéreuse dans la maladie de Behcet. Gastroenterol Clin Biol 13: 104–110
Pearn JH, Vickers TH (1966) The rabbit thalidomide embryopathy. Br J Exp Pathol 47: 186–192
Peterson PK, Hu S, Sheng WS, Kravitz FH, Molitor TW, Chatterjee D, Chao CC (1995) Thalidomide inhibits tumor necrosis factor-α production by lipopolysaccharide- and lipoarabinomannan-stimulated human microglial cells. J Infect Dis 172: 1137–1140
Pfeiffer RA, Kosenow W (1962) Thalidomide and congenital abnormalities. Lancet i: 45–46
Playfair JHL, Leuchars E, Davies AJS (1963) Effect of thalidomide on skin-graft survival. Lancet i: 1003–1004
Pliess G (1962) Beitrag zur teratologischen Analyse des neuen Wiedemann-Dysmelie-Syndroms (Thalidomid-Mißbildungen?). Med Klin 37: 1567–1573
Poswillo DE, Hamilton WJ, Sopher D (1972) The marmoset as an animal model for teratological research. Nature 239: 460–462
Powell RJ, Allen BR, Jenkins JS, Steele L, Hunneyball I, Maurice PDL, Littlewood SM (1985) Investigation and treatment of orogenital ulceration; studies on a possible mode of action of thalidomide. Br J Dermatol 113: 141–144
Radeff B, Kuffer R, Samson J (1990) Recurrent aphthous ulcer in patient infected with human immunodeficiency virus: successful treatment with thalidomide. J Am Acad Dermatol 23: 523–525
Ramselaar CG, Boone RM, Kluin-Nelemans HC (1986) Thalidomide in the treatment of neuroBehçet's syndrome. Br J Dermatol 115: 367–370
Read AP (1994) Thalidomide may be a mutagen. Comment. Br Med J 308: 1636
Revuz J, Guillaume J-C, Janier M, Hans P, Marchand C, Souteyrand P, Nonnetblanc J-M, Claudy A, Dallac S, Klene C, Crikx B, Sancho-Garnier H, Chaumeil JC (1990) Crossover study of thalidomide vs placebo in severe recurrent aphthous stomatitis. Arch Dermatol 126: 923–927
Rickenbacher J (1963) Versuche mit Contergan am Hühnerkeimling. Dtsch Med Wochenschr 88: 2252–2260
Ringdén O, Aschan J, Westerberg L (1988) Thalidomide for severe acute graft-versus-host disease. Lancet ii: 568
Roath S, Elves MW, Israëls MCG (1962) Effect of thalidomide on leukocyte cultures. Lancet ii: 812–813
Robertson WF (1962) Thalidomide ("distaval") and vitamin-B deficiency. Br Med J: 792–793
Rogé J, Testas P (1984) Thalidomide et syndrome de Behçét. Gastroenterol Clin Biol 8: 196

Rustin MHA, Gilkes JJH, Robinson TWE (1990) Pyoderma gangrenosum associated with Behcet's disease: treatment with thalidomide. J Am Acad Dermatol 23: 941–944
Ryan J, Colman J, Pedersen J, Benson E (1992) Thalidomide to treat esophageal ulcer in AIDS. N Engl J Med 327: 208–209
Saeki K, Naya M, Deguchi T, Eto K (1986) Teratogenicity of 5-fluorouracil and thalidomide in rat embryo culture. Toxicol Lett 31: 71
Sampaio EP, Sarno EN, Galilly R, Cohn ZA, Kaplan G (1991) Thalidomide selectively inhibits tumor necrosis factor-α production by stimulated human monocytes. J Exp Med 173: 699–703
Samsoen M, Grosshans E, Basset A (1980) La thalidomide dans le traitement du lupus érythémateux chronique (L.E.C.). Ann Dermatol Venereol 107: 515–523
Saul A, Flores O, Novales J (1976) Polymorphous light eruption: treatment with thalidomide. Aust J Dermatol 17: 17–21
Saurat JH, Camenzind M, Helg C, Chapuis B (1988) Thalidomide for graft-versus-host disease after bone marrow transplantation. Lanceti i : 359
Saxén L, Wartiovaara J (1966) Cell contact and cell adhesion during tissue organization. Int J Cancer 1: 271–290
Saxén L, Lehtonen E, Karkinen-Jääskeläinen M, Nordling S, Wartiovaara J (1976) Are morphogenetic tissue interactions mediated by transmissible substances or through cell contracts? Nature 259: 662–663
Saxén L, Ekblom P, Sariola H (1985) Organogenesis. Prog Clin Biol Res 163A: 41–53
Saylan T, Saltik I (1982) Thalidomide in the treatment of Behçét's syndrome. Arch Dermatol 118: 536
Schmahl H-J, Winckler K, Klingmüller K, Heger W, Barrach H-J, Nau H (1987) Pharmacokinetics of the teratogenic and nonteratogenic thalidomide analogs EM12 and supidimide in the rat and marmoset monkey. In: Nau H, Scott WJ (eds) Pharmacokinetics in teratogenesis vol 1. CRC Press, Boca Raton, pp 181–192
Schmahl H-J, Nau H, Neubert D (1988) The enantiomers of the teratogenic thalidomide analogue EM12: 1. Chiral inversion and plasma pharmacokinetics in the marmoset monkey. Arch Toxicol 62: 200–204
Schmahl H-J, Heger W, Nau H (1989) The enantiomers of the teratogenic thalidomide analogue EM12: 2. Chemical stability, stereoselectivity of metabolism and renal excretion in the marmoset monkey. Toxicol Lett 45: 23–33
Schmahl H-J, Dencker L, Plum C, Chahoud I, Nau H (1996) Stereoselective distribution of the teratogenic thalidomide analogue EM12 in the embryo of marmoset monkey, Wistar rat and NMRI mouse. Arch Toxicol (in press)
Schuller-Levis G, Harris D, Cutler E, Meeker HC, Haubenstock H, Levis WR (1987) Defective monocyte chemotaxis in active lepromatous leprosy. Int J Lepr 55: 267–272
Schulz T, Neubert D (1993) Peculiarities of biotransformation in the pre- and postnatal period. In: Ruckpaul K, Rein H (eds) Regulation and control of complex biological processes by biotransformation. Akademie, Berlin, pp 162–214 (Frontiers in Biotransformation 9)
Schumacher HJ, Smith RL, Williams RT (1965a) The metabolism of thalidomide: the spontaneous hydrolysis of thalidomide in solution. Br J Pharmacol 25: 324–337
Schumacher HJ, Smith RL, Williams RT (1965b) The metabolism of thalidomide: the fate of thalidomide and some of its hydrolysis products in various species. Br J Pharmacol 25: 338–351
Schumacher H, Blake DA, Gillette JR (1968a) Disposition of thalidomide in rabbits and rats. J Pharmacol Exp Ther 160: 201–211
Schumacher H, Blake DA, Gurian JM, Gillette JR (1968b) A comparison of the teratogenic activity of thalidomide in rabbits and rats. J Pharmacol Exp Ther 160: 189–200

Schumacher HJ, Wilson JG, Terapane JF, Rosedale SL (1970) Thalidomide: disposition in rhesus monkey and studies of its hydrolysis in tissue of this and other species. J Pharmacol Exp Ther 173: 265–269
Schumacher H, Blake DA, Gillette JR (1972a) Acylation of subcellular components by phthalimidophthalimide as a possible mode of embryotoxic action. Fed Proc 26: 730
Schumacher HJ, Terapane J, Jordan RL, Wilson JG (1972b) The teratogenic activity of a thalidomide analogue, EM12, in rabbits, rats and monkeys. Teratology 5: 233–240
Scolari F, Harms M, Gilardi S (1982) La thalidomide dans le traitement du lupus érythémateux chronique. Dermatologica 165: 355–362
Scott WJ, Fradkin R, Wilson JG (1977) Non-confirmation of thalidomide induced teratogenesis in rats and mice. Teratology 16: 333–336
Scott WJ, Wilson JG, Helm F-C (1980) A metabolite of a structural analog of thalidomide lacks teratogenic effect in pregnant rhesus monkeys. Teratology 22: 183–185
Shannon EJ, Ejigu M, Haile-Mariam HS, Berhan TY, Tasesse G (1992) Thalidomide's effectiveness in erythema nodosum leprosum is associated with a decrease in CD4+ cells in the peripheral blood. Lepr Rev 63: 5–11
Sheskin J (1965) Thalidomide in the treatment of lepra reaction. Cin Pharmacol Ther 6: 303–306
Sheskin J (1975a) Zur Therapie der Prurigo nodularis Hyde mit Thalidomid. Hautarzt 26: 215
Sheskin J (1975b) Thalidomide in lepra reaction. Int J Dermatol 14: 575–576
Sheskin J (1980) The treatment of lepra reaction in lepromatous leprosy. Fifteen year's experience with thalidomide. Int J Dermatol 19: 318–322
Shibata Y, Shichita M, Sasaki K, Nishimura K, Hashimoto Y, Iwasaki S (1995) N-Alkylphthalimides: structural requirements of thalidomidal action on 12-O-tetradecanoylphorbol-13-acetate-induced tumor necrosis factor-α production by human leukemia HL-60 cells. Chem Pharm Bull 43: 177–179
Shiota K, Fischer B, Neubert D (1988) Variability of development in the human embryo. In: Neubert D, Merker HJ, Hendrickx AG (eds) Non-human primates – developmental biology and toxicology. Ueberreiter, Berlin, pp 191–203
Shiota K, Uwabe C, Yamamoto M, Arishima K (1990) Susceptibility to cyclophosphamide and thalidomide of fetal rat limb buds grafted in athymic (nude) mice. Toxicol Lett 50: 309–318
Smith RL, Fabro S, Schumacher H, Williams RT (1965) Studies on the relationship between the chemical structure and embryotoxic activity of thalidomide and related compounds. In: Robson JM, Sullivan FM, Smith RL (eds) Embryopathic activity of drugs. Churchill, London, pp 194–209
Silva SR, Viana PC, Lugon NV, Hoette M, Ruzany F, Lugon JR (1994) Thalidomide for the treatment of uremic pruritus: a crossover randomized double-blind trial. Nephron 67: 270–273
Somers GF (1962) Thalidomide and congenital abnormalities. Lancet i: 912–913
Springer TA (1990) Adhesion receptors of the immune system. Nature 346: 425–434
Staples RE, Holtkamp DE (1963) Effects of parental thalidomide treatment on gestation and fetal development. Exp Mol Pathol [Suppl 7]: 81–106
Stephens TD (1988) Proposed mechanisms of action in thalidomide embryopathy. Teratology 38: 229–239
Stephens TD, Strecker TR (1983) A critical review of the McCredie-McBride hypothesis of neural crest influence on limb morphogenesis. Teratology 28: 287–292
Sterz H, Nothdurft H, Lexa P, Ockenfels H (1987) Thalidomide studies on the Himalayan rabbit: new aspects of thalidomide-induced teratogenesis. Arch Toxicol 60: 376–381
Strecker TR, Stephens TD (1983) Peripheral nerves do not play a trophic role in limb skeletal morphogenesis. Teratology 27: 159–167

Summerbell D (1979) The zone of polarising activity: evidence for a role in normal chick limb morphogenesis. J Embryol Exp Morphol 50: 217–233
Tamura F, Vogelsang GB, Reitz BA, Baumgartner WA, Herskowitz A (1990) Combination thalidomide and cyclosporine for cardiac allograft rejection. Comparison with combination methylprednisolone and cyclosporine. Transplantation 49: 20–25
Theisen CT (1983) Thalidomide and embryonic sensory peripheral neuropathy: a critical appraisal of the McCredie-McBride theory of limb reduction defects. Issues Rev Teratol 1: 215–249
Theisen CT, McGregor EE (1981) Further observations on the development of thalidomide induced limb deformities. Teratology 23: 65A
Thiel R, Chahoud I, Jürgens M, Neubert D (1993) Time-dependent differences in the development of somites of four different mouse strains. Teratogenesis Carcinog Mutagen 13: 247–257
Thiel R, Foerster M, Stahlmann R, Neubert R (1995) Effects of the thalidomide derivative EM12 on surface receptors of peripheral blood lymphocytes in Wistar rats. Teratology 51: 30A–31A (p42)
Torras H, Lecha M, Mascaro JM (1982) Thalidomide treatment of recurrent necrotic giant mucocutaneous aphthae and aphthosis. Arch Dermatol 118: 875
Ulrich M, deSalas B, Convit J (1971) Thalidomide activity in experimental Arthus and anaphylactic reactions. Int J Lepr 39: 131–135
Uthoff K, Zehr KJ, Gaudin PB, Kumar P, Cho PW, Vogelsang G, Hruban RH, Baumgartner WA, Stuart RS (1995) Thalidomide as replacement for steroids in immunosuppression after lung transplantation. Ann Thorac Surg 59: 277–282
van den Broek H (1980) Treatment of prurigo nodularis with thalidomide. Arch Dermatol 116: 571–572
van Regemorter N, Haumont D, Kirkpatrick C, Viseur P, Jeanty P, Dodion J, Milaire J, Rooze M, Rodesch F (1982) Holt Oram syndrome mistaken for thalidomide embryopathy–embryological considerations. Eur J Pediatr 138: 77–80
Vickers TH (1967) Concerning the morphogenesis of thalidomide dysmelia in rabbits. Br J Exp Pathol 48: 579–591
Vladutiu A (1966) Another chance for thalidomide? Lancet i: 981–982
Vogelsang GB, Hess AD, Gordon G, Santos GW (1986a) Treatment and prevention of acute graftversus-host disease with thalidomide in a rat model. Transplantation 41: 644–647
Vogelsang GB, Taylor S, Gordon G, Hess AD (1986b) Thalidomide, a potent agent for the treatment of graft-versus-host disease. Transplant Proc 18: 904–906
Vogelsang GB, Hess AD, Gordon G, Brundrette R, Santos GW (1987a) Thalidomide induction of bone marrow transplantation tolerance. Transplant Proc 19: 2658–2661
Vogelsang GB, Hess AD, Ling C, Brundrett RB, Colvin OM, Wingard JR, Saral R, Santos GW (1987b) Thalidomide therapy of chronic graft-versus-host disease. Blood 70: 315a
Vogelsang GB, Hess AD, Santos GW (1988a) Thalidomide for treatment of graft-versus-host disease. Bone Marrow Transplant 3: 393–398
Vogelsang GB, Wells MC, Santos GW, Chen TL, Hess AD (1988b) Combination low-dose thalidomide and cyclosporine prophylaxis for acute graft-versus-host disease in a rat mismatched model. Transplant Proc 20 (Suppl 2): 226–228
Vogelsang GB, Hess AD, Friedman KJ, Santos GW (1989) Therapy of chronic graft-*v*-host disease in a rat model. Blood 74: 507–511
Vogelsang GB, Farmer ER, Hess AD, Altamonte V, Beschorner WE, Jabs DA, Corio RL, Levin LS, Colvin OM, Wingard JR, Santos GW (1992) Thalidomide for the treatment of chronic graft-versus-host disease. N Engl J Med 326: 1055–1058
Warkany J (1988) Why I doubted that thalidomide was the cause of the epidemic of limb defects of 1959 to 1961. Teratology 38: 217–219

Waters MFR (1971) An internally-controlled double blind trial of thalidomide in severe erythema nodosum leprosum. Lepr Rev 42: 26–42
Waters MFR, Laing ABG, Ambikapathy A, Lennard-Jones JE (1979) Treatment of ulcerative colitis with thalidomide. Br Med J 1: 792
Weglicki WB, Stafford RE, Dickens BF, Mak IT, Cassidy MM, Phillips TM (1993) Inhibition of tumor necrosis factor-alpha by thalidomide in magnesium deficiency. Mol Cell Biochem 129: 195–200
Weicker H, Hungerland H (1962) Thalidomid-Embryopathie. I. Vorkommen inner- und außerhalb Deutschlands. Dtsch Med Wochenschr 87: 992–1002
Weicker H, Bachmann KD, Pfeiffer RA, Gleiß J (1962) Thalidomid-Embryopathie. II. Ergebnisse individueller anamnestischer Erhebungen in den Einzugsgebieten der Universitäts-Kinderkliniken Bonn, Köln, Münster und Düsseldorf. Dtsch Med Wochenschr 87: 1597–1607
Wiedemann H-R (1961) Hinweis auf die derzeitige Häufung hypo- und aplastischer Fehlbildungen der Gliedmaßen. Med Welt: 1863–1866
Wiedemann H-R, Aeissen K (1961) Zur Frage der derzeitigen Häufung von Gliedmaßenfehlbildungen. Med Monatsschr 15: 816–818
Willert HG, Henkel HL (1969) Klinik und Pathologie der Dysmelie. Springer, Berlin Heidelberg New York
Williams I, Weller IVD, Malin A, Anderson J, Waters MFR (1991) Thalidomide hypersensitivity in AIDS. Lancet 337: 436–437
Wilson JG (1966) Teratogenicity in non-human primates. In: Miller CO (ed) Proceedings of the conference on non-human primate toxicology. Department of Health, Education and Welfare, US Food and Drug Administration, Washington, pp 114–123
Winkelmann RK, Connolly SM, Doyle JA, Padilha-Goncalves A (1984) Thalidomide treatment of prurigo nodularis. Acta Derm Venerol 64: 412–417
Wöhrmann T, Geist C, Schneider J, Matthiesen T, Zwingenberger K (1995) Local skin reactivity after induction of Swartzman reaction in rabbits. Exp Toxicol Pathol 47: 167–172
Wride MA, Sanders (1993) Expression of tumor necrosis factor-α (TNFα)-cross-reactive proteins during early chick embryo development. Dev Dyn 198: 225–239
Wride MA, Sanders EJ (1995) Potential roles for tumor necrosis factor-α during embryonic development. Anat Embryol 191: 1–10
Wuest HM, Fratta I, Sigg EB (1966) Teratological studies in the thalidomide field. Life Sci 5: 393–396
Yasuda Y, Nishimura H (1989) Thalidomide may inhibit proliferation of mesenchyme in human limb buds. Anal Cellul Pathol 1: 247–254
Youle M, Clarbour J, Farthing C, Connolly M, Hawkins D, Staughton R, Gazzard B (1989) Treatment of resistant aphthous ulceration with thalidomide in patients positive for HIV antibody. Br J Med 298: 432
Youle M, Hawkins D, Gazzard B (1990) Thalidomide in hyperalgic pharyngeal ulceration of AIDS. Lancet 335: 1591

CHAPTER 23

Anticonvulsant Drugs: Mechanisms and Pathogenesis of Teratogenicity

R.H. FINNELL, B. BIELEC, and H. NAU

A. Introduction

Seizure disorders secondary to epileptic foci currently affect approximately 6.4 people per 1000 members of the population, or no fewer than 1.6 million Americans. Of these, 400 000 are women of child-bearing age, resulting in some 19 000 pregnancies per year in the United States and tens of thousands more worldwide. As the stigma attached to the seizure disorder is diminishing with improved pharmacological intervention, more epileptic women are contemplating pregnancies, resulting in consultations with primary-care physicians and neurologists over the potential risks involved in anticonvulsant drug-complicated pregnancies. Clinical management decisions are largely overshadowed by the fact that pregnancy in women with epilepsy is still considered a high-risk proposition. This is because of the potential increase in seizure frequency throughout pregnancy, labor, and delivery and because of the higher incidence of adverse pregnancy outcomes (TANGANELLI and REGESTA 1992).

A 30-year review of the medical literature quite clearly establishes that children of epileptic mothers have a higher risk for congenital malformations than do children in the general population (DURNER et al. 1992; DRAVET et al. 1992; KOCH et al. 1992). The first report of a malformation associated with antiepileptic drugs (AED) described a child exposed to mephenytoin in utero who developed microcephaly, cleft palate, malrotation of the intestine and a speech defect and who had an intelligence quotient (IQ) of 60 (MÜLLER-KUPPERS 1963). A subsequent report (JANZ and FUCHS 1964) involved a retrospective survey to evaluate the problem of AED-associated malformations at the University of Heidelberg. The rates of miscarriages and stillbirths from 426 pregnancies in 246 mothers were increased, but the malformation rate (2.2%) was not significantly different from that of the general population of West Germany. The authors concluded that AED were not associated with an increased risk of malformations. Since these initial findings, a vast literature has developed concerning pregnancy outcome among women with seizure disorders (for a review, see DANSKY and FINNELL 1991; FINNELL et al. 1995)

Congenital malformations remain the most commonly reported adverse outcomes in the pregnancies of epileptic mothers. Malformation rates in the general population range from 2% to 3% (KALTER and WARKANY 1983;

KELLY 1984). Reports of malformation rates in various populations of infants of mothers with epilepsy (IME) range from 2.3% to 18.6% (JANZ and FUCHS 1964; MEYER 1973; NAKANE et al. 1980; MEADOW 1968; PHILBERT and DAM 1982; for a review, see DANSKY and FINNELL 1991). These combined estimates yield a risk of malformations in an individual epileptic pregnancy of 4% to 8%. Table 1 reviews several studies which compare malformation rates in the offspring of mothers with and without epilepsy. Evaluation of these studies reveals a wide range of methodologies, and it is questionable whether it is even appropriate to combine data from these studies. Some are matched controls (KOCH et al. 1992; TANGANELLI and REGESTA 1992), while others use population registers (DRAVET et al. 1992; ROBERT et al. 1986), and many have no control populations at all (LINDHOUT et al. 1992; KÄLLEN 1986). Increased awareness and interest in the more subtle effects of AED exposure also complicates comparison. Some recent studies have risk estimates in excess of 20% of the AED-exposed pregnancies, but these numbers often reflect minor anomalies as well as major malformations (VAN DYKE et al. 1988; CZEIZEL et al. 1992; GLADSTONE et al. 1992; YAMATOGI et al. 1993). Despite the variety of approaches, a consistent trend is seen, with IME having roughly two to three

Table 1. Malformation rates in the offspring of epileptic and control mothers

Author	Control		Epileptic mothers	
	Malformation rate (%)	Pregnancies (*n*)	Malformation rate (%)	Pregnancies (*n*)
SABIN and OXORN (1956)			5.4	56
JANZ and FUCHS (1964)			2.3	225
GERMAN et al. (1970)			5.3	243
ELSHOVE and VAN ECK (1971)	1.9	12 051	5.0	65
SPEIDEL and MEADOW (1972)	1.6	483	5.2	427
SOUTH (1972)	2.4	7 892	6.4	31
SPELLACY (1972)		50	5.8	51
BJERKEDAL and BAHNA (1973)	2.2	12 530	4.5	311
FEDRICK (1973)	5.6	649	13.8	217
KOPPE et al. (1973)	2.9	12 455	6.6	197
KUENSSBERG and KNOX (1973)	3.0	14 668	10.0	48
LOWE (1973)	2.7	31 877	5.0	245
MEYER (1973)	2.7	110	18.6	593
MILLAR and NEVIN (1973)	3.8	32 227	6.4	110
MONSON et al. (1973)	2.4	50 591	4.7	306
NISWANDER and WERTELECKI (1973)	2.7	347 097	4.1	413
STARREVELD and ZIMMERMAN et al. (1973)			7.0	372
BIALE et al. (1975)			16.0	56

Table 1. (*Contd.*)

Author	Control		Epileptic mothers	
	Malformation rate (%)	Pregnancies (*n*)	Malformation rate (%)	Pregnancies (*n*)
KNIGHT and RHIND (1975)	3.65	69 000	4.3	140
VISSER et al. (1976)	2.3	9 869	3.7	54
WEBER et al. (1977)	2.2	5 011	4.0	731
ANNEGERS et al. (1978)	3.5	748	8.1	259
SEINO and MIYAKOSHI (1979)			13.7	272
DIETERICH et al. (1980)[a]			59.5	37
MAJEWSHI et al. (1980)			16.0	111
NAKANE et al. (1980)			11.5	700
HILLESMAA et al. (1981)	2.0	5 613	7.7	4795
LINDHOUT et al. (1984)			9.9	151
STANLEY et al. (1985)	3.4	62 265	3.7	244
BEAUSSART-DEFAYE et al. (1986)			7.8	295
KÄLLÉN (1986)			4.9	635
KANEKO et al. (1986)			13.5	166
ROBERT et al. (1986)	1.4	20 916	6.7	148
RATING et al. (1987)	3.7	162	5.3	150
VAN DYKE et al. (1988)[a]			24.3	152
GAILY (1990)	2.9	105	9.1	153
BATTINO et al. (1992a)	2.3		9.1	315
CZEIZEL et al. (1992)[a]			58.9	95
DANSKY et al. (1992)			24.1	119
DRAVET et al. (1992)	1.4	117 183	7.0	281
ESKAZAN and ASLAN (1992)			11.5	104
GLADSTONE et al. (1992)[a]			20.5	44
KANEKO et al. (1992)			6.2	145
KOCH et al. (1992)	4.3	116	6.9	116
LINDHOUT et al. (1992)			7.6	172
OGUNI et al. (1992)			8.8	103
TANGANELLI and REGESTA (1992)	2.6	140	3.6	138
KANEKO et al. (1993)			10.1	354
MARTIN and MILLAC (1993)	2.7		2.9	188
MARTIN and MILLAC (1993)[b]	4.1		4.8	160
YAMATOGI et al. (1993)[a]			25.5	75
HOLMES and HARVEY (1994)			5.4	112
HOLMES and HARVEY (1994)[b]			6.8	453
HOLMES et al. (1994)	5.9	373	15.1	106
MCDONALD (1994)			17.9	28
WATERS et al. (1994)	3.4	355	11.3	115

[a]Includes minor anomalies.
[b]Second cohort.

times the number of adverse pregnancy outcomes when compared to the general population (Table 1). When minor anomalies are considered, the increased risk for IME is even higher. While this literature does not indicate causation, it is clear that one or more complicating factors put these pregnancies at a significantly higher risk for an adverse outcome.

B. Human Studies

I. Fetal Antiepileptic Drug Syndromes: The Case for a Single Syndrome Designation

A variety of dysmorphic syndromes consisting of patterns of congenital anomalies which appear to be nonrandomly associated with in utero AED exposure have been reported. Minor anomalies are considered structural deviations from the norm that do not constitute a threat to health, and by definition these occur in less than 4% of the population (MARDEN et al. 1964). To date, six clinical AED syndromes have been reported in the medical literature. A fetal trimethadione syndrome was first described by ZACHAI and colleagues (1975). Children exposed to this drug in utero were more likely to be short in stature, microcephalic, have V-shaped eyebrows, epicanthal folds, low-set ears, anteriorally folded helices, and irregular teeth. Inguinal hernias, hypospadias, and simian creases were also frequently observed. A retrospective study of trimethadione exposures in 53 pregnancies revealed fetal loss or major malformations in 87% of the pregnancies (FELDMAN et al. 1977). Follow-up studies have reported significant rates of mental retardation among the exposed infants (GOLDMAN et al. 1986).

The fetal hydantoin syndrome (FHS) was initially described, in part, by LOUGHNAN et al. (1973) and expanded upon, including formally naming the syndrome, by HANSON and SMITH (1975). Among the many dysmorphic findings associated with this syndrome, hypoplasia and irregular ossification of the distal phalanges was originally believed to be the single most characteristic feature. These infants displayed facial dysmorphism, including epicanthal folds, hypertelorism, broad, flat nasal bridges, an upturned nasal tip, wide prominent lips, and in addition distal digital hypoplasia (DDH), intrauterine growth retardation, and mental retardation. Subsequently, HANSON et al. (1976) reported a prevalence of FHS of 11%, with an additional 30% of the in utero exposed children expressing some of the syndrome's features. VAN DYKE et al. (1988) studied 62 families with phenytoin (PHT) exposure and confirmed the diagnosis of FHS in 17% of the exposed children. The recurrence risk of having a second child with FHS was highly significant (90%), while families with an unaffected first child had a very low risk of FHS occurence in their second child (2%). These observations directly relate to the issue of the genetically determined sensitivity of individual mothers and infants with respect to AED teratogenicity. As with the fetal alcohol syndrome, it is now common nomenclature to consider children presenting with a more limited pattern of anomalies secondary to in utero hydantoin exposure to be expressing fetal hydantoin effects (FHE; HANSON 1986).

A primidone (PRM) embryopathy has also been described. Affected children present with hirsute foreheads, thick nasal roots, anteverted nostrils, long philtrums, straight thin upper lips, and DDH. These children tend to be small for their age and have an increased risk for psychomotor retardation and heart defects (RUDD and FREEDOM 1979; GUSTAVSON and CHEN 1985). A case report of a family in whom all four siblings had clinical features of the FHE demonstrates the complexity involved in evaluating affected children. The first two children in this family were exposed to both PHT and PRM in utero. In an attempt to prevent dysmorphism in subsequent pregnancies, PHT was discontinued. Unfortunately, the third and fourth children had the same dysmorphic features as did their elder siblings, although the younger siblings had only been exposed to PRM monotherapy (KRAUSS et al. 1984).

A syndrome of facial dysmorphism, pre- and postnatal growth deficiency, developmental delay, and minor anomalies has been described in children exposed to phenobarbital (PB). The clinical features were similar to those seen with in utero exposure to both PHT and alcohol; hence the author felt that it should not be classified as a separate syndrome (SEIP 1976). He also noted that all three of these compounds can result in folate deficiency and hypothesized that such a deficiency could be the common mechanism for the dysmorphism seen with such in utero teratogen exposures.

Dysmorphic features in children exposed to valproic acid (VPA) in utero were defined as a syndrome by DI LIBERTI et al. (1984). These children have inferior epicanthal folds, flat nasal bridges, upturned nasal tips, thin vermilion borders, a shallow philtrum, and downturned mouths. Long, thin overlapping fingers and toes and hyperconvex nails have also been described. Recently, several cases of fetal valproate exposure and radial ray aplasia have been reported, but the prevalence of these abnormalities cannot be determined from case reports (BRON et al. 1990; SHARONY et al. 1993; YLAGAN and BUDORICK 1994). Children exposed to valproate in utero also appear to be at greater risk for perinatal distress (43%), low Apgar scores (28%), postnatal growth deficiency, and microcephaly (JÄGER-ROMAN et al. 1986; ARDINGER et al. 1988). Notably, VPA has been implicated in the incidence of neural tube defects, particularly spina bifida (ROBERT and GUIBAUD 1982; LINDHOUT and MEINARDE 1984; for review see ROSA 1991).

A fetal carbamazepine (CBZ) syndrome has been described by a single group of investigators (JONES et al. 1989). Dysmorphic features include upslanting palpebral fissures, epicanthal folds, short nose, long philtrum, DDH and microcephaly. Developmental delay was also found in 20% of exposed children, although the authors used one standard deviation from the mean as the cut off for abnormality, rather than the customary two standard deviations. Applying the conventional definitions to their data eliminates any increased risk for developmental delay in the CBZ-exposed children. This is further supported by SCOLNIK et al. (1994), who reported that children exposed in utero to CBZ did not differ from controls on IQ tests or Reynell developmental language scales. Reduction in fetal head circumference has been reported in CBZ-exposed children (HIILESMAA et al. 1981). Although smaller than controls, the head sizes were still within the normal range, and the differences between the IME and controls disappeared as the children matured. Generally, CBZ is considered less teratogenic than most other AED, but recent studies have noted an association with the occurrence of spina bifida (ROSA 1991; KÄLLEN 1994).

The application of drug-specific syndromes remains controversial in diagnostic terms. Facial dysmorphism can be difficult to quantify, especially in newborns, and is certainly not specific to any one anticonvulsant drug. IME displaying similar dysmorphic features have been described in the pre-anticonvulsant era (BAPTISTI 1938; PHILBERT and DAM 1982). The long-term outcome and hence significance of these anomalies remain unclear. The hypothesized association of facial dysmorphia with mental retardation (HANSON et al. 1976) has not always been confirmed (GRANSTRÖM 1982; HUTCH et al. 1982). In those few cases which have been followed into early childhood, the dysmorphic features tend to disappear as the child grows older in some, but not all instances (JANZ 1982). In all of these syndromes, the primary abnormalities involve the midface. A retrospective Israeli study spanning 10 years found hypertelorism to be the only anomaly seen more often in IME than controls; this was associated with all anticonvulsant drugs except PRM (NERI et al. 1983). In reviewing this literature, it appears difficult to justify the distinction drawn between the adverse pregnancy outcomes observed following in utero exposure to different AED. The overlap between phenotypes and the variable expression among individuals exposed to these drugs is just too extensive. The pediatric and teratological communities might be better served if all of these collections of anomalies were regrouped under the umbrella term "fetal antiepileptic drug syndrome" (FADS), with minor manifestations being referred to as "fetal antiepileptic drug effects" (FADE). This is suggested to replace a half-dozen overlapping drug-specific syndromes and to describe cases where drug exposure was polytherapeutic.

II. Antiepileptic Drugs and Congenital Malformations

There are a number of potential factors which could account for the increased rates of malformations seen in IME: maternal seizures during pregnancy; the epileptic genotype; falls and injuries secondary to the seizures; and lower socioeconomic status and its attendant limited access to prenatal care. There are, however, a series of observations which strongly implicate AED as the cause of the increased malformation rate amongst IME. These are the following: (a) comparisons between the malformation rates in the offspring of epileptic mothers treated with AED as opposed to those with no AED treatment reveal consistently higher rates in the children of the treated group, as illustrated in Table 2 (for a review, see DANSKY and FINNELL 1991); (b) mean plasma AED

Table 2. Malformation rates in the offspring of treated and nontreated epileptic mothers

Authors	Anticonvulsants		No anticonvulsants	
	Malformation rate (%)	Pregnancies (*n*)	Malformation rate (%)	Pregnancies (*n*)
Janz and Fuchs (1964)	2.2	225	0	120
Speidel and Meadow (1972)	5.0	329	0	59
South (1972)	9.0	22	0	9
Lowe (1973)	6.7	134	2.7	111
Monson et al. (1973)	5.4	205	3.0	101
Annegers et al. (1974)	7.1	141	1.8	56
Annegers et al. (1978)	10.7	177	2.4	82
Nakane et al. (1980)	11.5	478	2.3	129
Lindhout et al. (1984)	7.6	172	7.1	14
Kaneko et al. (1986)	16.2	172	10.0	20
Robert et al. (1986)	7.2	112	0	35
Dravet et al. (1992)	7.4	213	0	14
Eskazan and Aslan (1992)	11.5	104	0	66
Koch et al. (1992)	6.9	116	8.0	25
Lindhout et al. (1992)	8.7	323	7.1	28
Martin and Millac (1993)	4.5	133	3.7	27
Holmes et al. (1994)	15.1	106	6.6	76
Waters et al. (1994)	11.3	159	0	15

concentrations tend to be higher in mothers with malformed infants than mothers with healthy children (Dansky et al. 1980; Battino et al. 1992a); (c): infants of mothers on polytherapy have higher malformation rates than those exposed to monotherapy, as illustrated in Table 3; (d) maternal epilepsy type and seizure frequency during pregnancy do not appear to increase the risk of congenital malformations (Fedrick 1973; Battino et al. 1992a; Koch et al. 1992; Waters et al. 1994).

Although it appears logical to assume that polytherapy carries more inherent risks than monotherapy, high drug levels and multiple drugs are usually associated with the severity of the seizure disorder. It has been suggested that seizure frequency or severity may be a confounding factor, and AED therapy may be only associated with, but not causally responsible for, the increased rate of congenital malformations. Majewski and coworkers (1980) described increased malformation rates and central nervous system injury in IME exposed to maternal seizures. More recently, Lindhout and coworkers (1992) described a marked increase in malformations amongst infants exposed to first-trimester seizures (12.3%) compared to fetuses that were not subject to any maternal seizures (4.0%). Malformations were more often observed in infants exposed to partial seizures than to generalized tonic clonic seizures. Nonetheless, most investigators have found that maternal seizures during

Table 3. Malformation rates in the offspring of epileptic mothers treated with antiepileptic drug monotherapy versus polytherapy

Author	Monotherapy (%)	Total pregnancies (*n*)	Polytherapy (%)	Total pregnancies (*n*)
LINDHOUT et al. (1984)	2.4	42	12.8	108
KANEKO et al. (1986)	6.5	31	19.8	111
VAN DYKE et al. (1988)[a]	38.5	39	28.9	76
BATTINO et al. (1992a)	10.1	207	7.8	102
CZEIZEL et al. (1992)[a]	51.7	60	71.4	35
ESKAZAN and ASLAN (1992)	13.0	23	11.1	81
GLADSTONE et al. (1992)[a]	28.6	21	23.1	13
KANEKO et al. (1992)	4.4	92	10.2	49
LINDHOUT et al. (1992)	7.5	80	7.6	186
TANGANELLI and REGESTA (1992)	3.8	78	9.1	44
KANEKO et al. (1993)	5.9	152	13.4	207
MARTIN and MILLAC (1993)	0	94	15.4	39
WATERS et al. (1994)	11.0	82	10.7	28

[a]Includes minor anomalies.

pregnancy and the type of maternal epilepsy had no impact on the frequency of malformations, development of epilepsy, or febrile convulsions (NAKANE et al. 1980; ANNEGERS et al. 1978; BATTINO et al. 1992a; WATERS et al. 1994; for a review, see DANSKY and FINNELL 1991).

A primary question in evaluating the relationship between maternal epilepsy and congenital malformations is whether or not the association is further confounded by the genetics of the epileptic disorder. Several studies comparing malformation rates among the offspring of treated epileptics as opposed to untreated mothers with epilepsy strongly suggest that it is the use of AED that is the cause of the increased malformation rates, not the presence of a seizure disorder. With the exception of two studies (RATING et al. 1987; KOCH et al. 1992), investigations reveal a consistently elevated malformation rate for infants exposed in utero to AED when compared to IME who were not exposed to AED during pregnancy (Table 2).

The association between AED exposure and increased risk of malformation is further supported by studies comparing malformation rates between infants exposed to AED monotherapy and AED polytherapy. In a multi-institutional Japanese study of two chronological cohorts, the malformation rates in 172 IME born between 1978 and 1984 were compared to a second cohort of 145 IME born between 1985 and 1989. The malformation rates fell from 13.5% to 6.2% in these two cohorts. The change was attributed to the fact that between 1978 and 1984, only 16.1% of IME were exposed to AED montherapy, whereas in the second cohort, 63.4% of the women with epilepsy received AED monotherapy (KANEKO et al. 1992). A joint University of Rotterdam–University of Berlin study compared malformation rates in two cohorts of IME. The first group, born between 1972 and 1979, had an average number of 2.2 AED exposures. Of the malformed infants in that group, only 2% were exposed to monotherapy, while 25% were exposed to four AED. In the second cohort, born between 1980 and 1985, the IME were exposed to an average of 1.7 AED. Eight percent were on monotherapy, and none of the IME was exposed to more than three drugs. DRAVET and colleagues (1992) found that IME on polytherapy had higher malformation rates (16%) than monotherapy-exposed infants (6%). A prospective study from Genoa of 138 pregnancies of women with epilepsy found neither AED monotherapy nor high therapeutic plasma concentrations correlated with increased risk for congenital malformations in the offspring. Polytherapy including PB and PHT appeared to be the only risk factor (TANGANELLI and REGESTA 1992). Recently, an English study of two IME cohorts exposed to AED found no malformations among 94 infants exposed to AED monotherapy and six malformations among the 39 infants exposed to AED polytherapy in the second cohort (MARTIN and MILLAC 1993). Table 3 summarizes 14 studies that distinguish between malformation rates for AED monotherapy- and polytherapy-exposed infants. BATTINO et al. (1992a) noted a small insignificant decrease in malformation rates for AED polytherapy-exposed infants, which they attributed to small sample size. Small sample sizes for VAN DYKE et al. (1988), ESKAZAN and ASLAN (1992), GLADSTONE et al. (1992), and WATERS et al. (1994) may also explain insignificant discrepancies. Overall, malformation rates clearly increase with AED polytherapy.

III. Other Teratogenic End Points

The clinical view of a teratogenic outcome is often unnecessarily restrictive. Valid teratogenic end points following in utero exposure to anticonvulsant drugs are not limited to structural malformations. Mental and behavioral deficits, growth retardation, and death are all potential outcomes of teratogenic exposures.

1. Mental and Behavioral Deficits

IME have been reported to have higher rates of mental retardation when compared to nonepileptic controls. This increased risk is twofold to sevenfold according to various authors (SPEIDEL and MEADOW 1972; HILL et al. 1974; NELSON and ELLENBERG 1982; for a review, see DANSKY and FINNELL 1991; GRANSTRÖM and GAILY 1992). None of these studies controlled for parental intelligence, and while IQ scores at age 7 between groups of children exposed (full-scale IQ, FSIQ, 91.7) or unexposed (FSIQ, 96.8) to PHT reached statistical significance, the clinical significance of such difference is unknown (NELSON and ELLENBERG 1982). Children exposed prenatally to PHT have been found to have significantly lower scores for both performance IQ and

FSIQ and visual motor integration test scores (VANOVERLOOP et al. 1992). LEAVITT and colleagues have found that IME display lower scores in measures of verbal acquisition at both 2 and 3 years of age. Although there was no difference in physical growth parameters between IME and controls, IME scored significantly lower in the Bailey scale of infant development's mental developmental index (MDI) at these ages (LEAVITT et al. 1992). These children also performed significantly less well on the Bates Bretherton early language inventory ($p \leq 0.02$), in the Peabody picture vocabulary's scales of verbal reasoning ($p \leq 0.001$), and composite IQ ($p \leq 0.01$) and displayed significantly shorter mean lengths of utterance ($p \leq 0.001$) (LEAVITT et al. 1992). In a blinded study, SCOLNIK et al. (1994) found that children exposed in utero to PHT had a mean IQ that was 10 points lower than controls and also performed significantly lower on Reynell developmental language scales. This contrasted with the results of CBZ-exposed children, who showed no difference in either IQ or language development compared to controls.

2. Growth Retardation

Low birth weight (less than 2500 g) and prematurity have been described in IME. Infants exposed in utero to PHT, CBZ, and many of the other anticonvulsant compounds have an elevated risk (up to 7.2-fold increase) for intrauterine growth retardation when compared to infants of nonepileptic mothers (for a review, see DANSKY and FINNELL 1991). The average rates in these studies range from 7% to 10% and from 4% to 11%, respectively (BJERKDAL and BAHNA 1973; ANNEGERS et al. 1974; TERAMO et al. 1979; NAKANE et al. 1980; NELSON and ELLENBERG 1982; SVIGOS 1984; BATTINO et al. 1992b; HOLMES et al. 1994). Several recent studies have disputed that AED exposure will lower birth weight. MARTIN and MILLAC (1993) found that AED-exposed infants had a slightly higher average birth weight than unexposed controls, and other studies have shown no significant differences in birth weight between AED-exposed infants and unexposed controls (LEAVITT et al. 1992; GLADSTONE et al. 1992; SCOLNIK et al. 1994; WATERS et al. 1994).

In addition to reductions in birth weights, the epidemiological studies consistently reported a decrease in the mean head circumference or an increased frequency of microcephaly among the infants born to epileptic women (for a review, see DANSKY and FINNELL 1991). Microcephaly has been demonstrated in IME and have been associated with all AED (NELSON and ELLENBERG 1982; NERI et al. 1983; BATTINO et al. 1992b; HOLMES et al. 1994). A Finnish study found a stronger association between CBZ exposure in utero and small head circumference than with other anticonvulsant drugs (HIILESMAA et al. 1981).

3. Stillbirths and Neonatal and Infant Mortality

Fetal wastage, defined as fetal loss after 20 weeks gestation, appears to be as common and perhaps as great a problem in women with epilepsy as is the risk for producing infants with congenital malformations. Studies comparing stillbirth rates found higher rates in IME (1.3%–14.0%) compared to infants of mothers without epilepsy (1.2%–7.8%) (Table 4). Some reports fail to make comparisons with rates observed in the general population or other control groups, making it difficult to establish whether IME are at increased risk (ANNEGERS et al. 1974; KNIGHT and RHIND 1975). A large Norwegian study failed to demonstrate increased risks of stillbirth in IME, but clearly demonstrated increased neonatal deaths (BJERKDAL and BAHNA 1973).

Spontaneous abortions, defined as fetal loss occurring prior to 20 weeks of gestation, do not appear to occur more commonly in IME. A history of previous spontaneous abortions was not found to be significantly different between women with epilepsy (24%) and controls (17.8%) (odds ratio, 1.44; 95% confidence interval 1.03–2.02; YERBY et al. 1985). ANNEGERS and coworkers reported that the gestational age-adjusted rate ratio for spontaneous abortions was no higher for women with epilepsy than for the wives of men with epilepsy. Furthermore, there did not appear to be any difference in

Table 4. Stillbirth and neonatal death rates (%) in infants of epileptic mothers

Author	Stillbirths		Neonatal deaths	
	Cases	Controls	Cases	Controls
JANZ and FUCHS (1964)	12.1	7.0	1.3	–
SPIEDEL and MEADOW (1972)	1.3	1.2	2.7	1.0
BJERKEDAL and BAHNA (1973)	15.3	7.8	3.2	1.5
FEDRICK (1973)	2.7	1.1	–	–
HIGGINS and COMERFORD (1974)	5.2	–	7.8	3.9
KNIGHT and RHIND (1975)	2.0	–	2.9	–
NAKANE(1979)	13.5	4.3	–	–
NAKANE et al. (1980)	14.0	6.7	–	–
NELSON and ELLENBERG (1982)	5.1	1.9	3.5	2.7
SVIGOS (1984)	0	1.3	–	–
KÄLLÉN(1986)	2.2	–	2.7	–
GAILY et al. (1988)[a]	–	–	3.8	–
TANGANELLI and REGESTA (1992)	–	–	2.2	1.4
MARTIN and MILLAC (1993)[a]	–	–	4.7	1.4
MARTIN and MILLAC (1993)[ab]	–	–	2.1	0.9
KÄLLÉN (1994)[a]	–	–	1.5	–
WATERS et al. (1994)[c]	–	–	3.8	–

[a]Described as "perinatal" death rate; for KÄLLDÉN, this includes stillbirths and death within 1 week of birth; for MARTIN and MILLAC and GAILY, unspecified.
[b]Second cohort.
[c]Described as "infant death".

spontaneous abortion rates for treated women with epilepsy (14.6%) compared to untreated women (15.7%) (ANNEGERS et al. 1978). These findings contradict a number of other studies which have demonstrated increased rates of neonatal and perinatal death (Table 4). Perinatal death rates range from 1.3% to 7.8% compared to 1.0%–3.9% for controls.

With respect to the final manifestation of a teratogen, i.e., intrauterine or postnatal death, it is difficult to estimate the number of spontaneous abortions, as many of the early ones go undetected. Spontaneous abortion rates in epileptic women have been reported to range from 4.2% to 22% in the various studies (see DANSKY and FINNELL 1991). ANDERMANN and colleagues (1982) reported the spontaneous abortion rate among women with seizure disorders receiving AED therapy (22.8%) to be twice that among untreated women with epilepsy (10%) (DANSKY et al. 1982). Several authors have reported a two to three-fold increased risk of perinatal mortality in the offspring of epileptic women (NISWANDER and GORDON 1972; GÖPFERT-GEYER et al. 1982; DANSKY and FINNELL 1991). When all studies are considered together, the most frequent causes of perinatal mortality cited in the literature are congenital malformations, placentopathies, prenatal and intrapartum obstetric complications, and neonatal spontaneous hemorrhage (HILL and TENNYSON 1982; GÖPFERT-GEYER et al. 1982; DANSKY and FINNELL 1991).

Clearly, there is an interrelationship between many of the potential end points of teratogenesis. Multiple malformations are often incompatible with life, leading to high rates of pre- and perinatal mortality. Developmental insults that compromise the growth of the embryo can also cause impairment in the final histogenesis of morphogenetic systems, leading to both structural and functional abnormalities. What is important to note is that all of these end points are indicative of a teratogenic insult, and any assessment into adverse pregnancy outcomes associated with a specific anticonvulsant medication must take all four possible outcomes of a human teratogen into consideration, i.e., behavioral deficits, growth retardation, neonatal death, and structural malformations; otherwise the study will be incomplete.

C. Application of the General Principles of Teratogenesis to Antiepileptic Drugs

I. Genetic Susceptibility

Susceptibility to teratogen-induced congenital defects is highly dependent upon the genotype of the infant and the interaction of these genes with AED, including drug-induced changes in maternal physiology.

Anticonvulsant drugs behave as typical teratogens in the sense that only a small percentage of infants exposed in utero to these compounds present with congenital malformations. Generally speaking, the incidence of congenital defects among IME receiving AED treatment is increased two- to three-fold over the prevalence rate found in control populations (DURNER et al. 1992;

LINDHOUT et al. 1992; KANEKO et al. 1992). This means that, while up to approximately 10% of the AED-exposed infants show the clinical stigmata of anticonvulsant drug exposure in utero, nearly 90% of AED-complicated pregnancies do not have adverse outcomes. With respect to VPA and CBZ, it is estimated that 1%–2% of all infants exposed in the first trimester will have a neural tube defect compared to a 0.1% incidence in the general population (ROBERT 1982; ROBERT and GUIBAUD 1982; ROSA 1991). Further evidence supporting the concept that different individuals differ in their susceptibility to anticonvulsant drug-induced teratogenesis can be found in the variable expression of the drug-induced malformations. Clinically, there is a wide range of abnormalities that are nonrandomly associated with in utero exposure to PHT (HANSON and SMITH 1975; HANSON et al. 1976; SMITH 1980), VPA (ARDINGER et al. 1988) and CBZ (JONES et al. 1989), and each exposed and affected neonate presents with some variation of the possible entire spectrum of abnormalities. Although there are selected congenital abnormalities that are most often associated with a specific drug exposure, such as neural tube defects, particularly spina bifida, with VPA (JÄGER-ROMAN et al. 1986; ARDINGER et al. 1988; MARTINEZ-FRIAS 1990) or CBZ (ROSA 1991; LINDHOUT et al. 1992; GLADSTONE et al. 1992; KÄLLEN 1994; OAKSHOTT and HUNT 1994), there is a great deal of overlap in the qualitative pattern of congenital defects observed secondary to fetal AED exposure. Since all of the affected children had chronic drug exposure, the wide variety of malformations and combinations of abnormalities present in affected children cannot simply reflect differences in the time of drug exposure (FINNELL and CHERNOFF 1984). What should be readily apparent is that not all exposed infants are equally susceptible to the teratogenic effects of the anticonvulsant compounds and that this difference in susceptibility is in large part genetically determined.

Specific examples of the genetic basis of susceptibility to anticonvulsant drug-induced birth defects can be seen in those rare experiments of nature involving drug exposure in multiple births. BUSTAMANTE and STUMPFF (1978) described trizygotic triplets who were affected to varying extents with growth retardation and midfacial, nail, and digital phalangeal hypoplasia. While all three triplets were affected having been exposed in utero to both PHT and PB, only one had a cleft lip and palate (triplet C), while only triplet B had craniosynostosis, indicative of the differences in susceptibility to specific malformations expressed by genetically diverse siblings who shared approximately comparable uterine environments. Discordant expression of FADS was reported in dizygotic heteropaternal twins exposed in utero to PHT (PHELAN et al. 1982). Only one of the twins presented with the stigmata of this syndrome, being growth and developmentally retarded and having numerous minor anomalies. Interstingly, when the twins were tested as to the activity of their detoxification enzyme epoxide hydrolase, only the mother and the affected twin had significantly reduced activity compared to the other exposed, but unaffected twin (BUEHLER 1984). STRICKLER et al. (1985) reported that, for 14 PHT-exposed children who had increased cell death in a PHT assay, all had a similarly sensitive parent. The occurrence of major malformations was highly correlated with PHT sensitivity.

In terms of genetically determined sensitivity to anticonvulsant drug-induced teratogenesis, it may well be that such individuals have mutations in specific biochemical pathways that alter the metabolism, with a subsequent deleterious effect on embryonic development. In the final analysis, one could view differences in susceptibility to teratogen-induced birth defects as a genetic trait, determined by the individual's load of "liability genes" (FRASER 1976; FINNELL and CHERNOFF 1987). These genes could regulate enzyme activity either by controlling the quantity of enzyme that is produced or by modifying its function. The microsomal epoxide hydrolase (mEH) gene is one example. Recently, HASSETT et al. (1994) demonstrated that a genetic polymorphism identified in the human mEH gene reduced enzymatic activity by 40% in vitro. Theoretically, the reduced enzymatic activity could increase the teratogenic risk.

The importance of understanding genetically determined sensitivity to AED induced congenital malformations cannot be understated. As previously discussed, the number of infants bearing malformations observed in anticonvulsant drug-treated epileptic mothers is usually between 4% and 8% which is approximately two to three times that of the control populations. The message that gets diluted out of the presentation of data in this fashion is that there are certain individuals that are relatively immune to the adverse effects of the drugs, while at the same time there exists a small, but significant number of women and their infants who, for genetically determined reasons, are at very high risk. Whether this is because the infant has an unusually large number of liability genes that predispose it to many different congenital malformations or whether both mother and fetus have deficient detoxification enzymes is currently unknown. What is of concern is that the risk for these infants and families, greatly exceeds the doubling or tripling of the normal risk that is generally quoted. The actual risk for such families may be as high as 100%, and if these happen to be large families the potential for having multiple siblings born with multiple malformations is unacceptably high. A recently reported case of four siblings all exhibiting typical FADE features and developmental delay after in utero AED polytherapeutic exposure illustrates this point (LAU et al. 1994). The recurrence rates of FADS demonstrated by VAN DYKE et al. (1988) and the occurrence of FADE-affected sibling pairs throughout AED literature clearly demonstrate the importance of genetic susceptibility.

II. Teratogenic Timing

Susceptibility to anticonvulsant drug teratogenesis will depend upon the gestational period of drug exposure.

There is no compelling epidemiological evidence that positively correlates the duration of AED therapy prior to conception with an increased risk for an adverse pregnancy outcome (DANSKY and FINNELL 1991). What is important is the period during development when the embryo is being exposed to the anticonvulsant drug or drugs. For the most part, women with seizure disorders have generally been receiving AED therapy prior to pregnancy, and therefore exposure of the conceptus is likely to be chronic and from the very onset of the pregnancy. The clinical relevance of this teratological principal relates to therapeutic manipulations. Clearly, it would be possible to significantly reduce the risk for neural tube defects associated with both VPA and CBZ by eliminating exposure of the embryo to these two compounds during the first 4–6 weeks of gestation. On the other hand, one could contemplate adding on either of these two drugs in a polytherapeutic treatment regimen at the end of the first trimester without a significant risk of producing a structural malformation.

III. Mechanisms of Teratogenesis

Those anticonvulsant drugs with a teratogenic potential will alter the developing cells and tissues of the embryo by specific mechanisms of action to initiate abnormal morphogenesis.

The specific teratogenic mechanism of action of any of the clinically effective anticonvulsant compounds has yet to be definitively determined. There have been several different hypotheses proposed as the mechanism of action underlying the teratogenicity of the various AED; perhaps the earliest of these suggestions was that PHT was a folic acid antagonist. It is well established that chronic hydantoin therapy can result in depressed levels of circulating folates, which can result in a number of major pregnancy complications. These include placental abruptions, spontaneous abortions, perinatal mortality, and congenital malformations (Fraser and Watt 1964; Hibbard 1964; Hibbard et al. 1965; Stone 1968; Rivey et al. 1984). Such observations serve to link a metabolic interaction between folic acid and PHT with an adverse pregnancy outcome (Blake et al. 1978). It is possible that these indirect maternal effects are the result of PHT inhibiting the enzyme folate conjugase, which normally splits dietary folate into more simple monoglutamate forms. As a result, absorption of the polyglutamated folates is impaired, which can ultimately compromise purine biosynthesis. With respect to folic acid metabolism, it has been shown that VPA directly inhibits the enzyme glutamate formyltransferase, which is involved in the interconversion of the biologically active tetrahydrofolate and its formylated forms (Wegner and Nau 1992). It is possible that, even while maternal folate concentrations are within normal limits, the profile of specific folate metabolites may be significantly shifted, and this might contribute to the development of neural tube defects.

In terms of a specific mechanism of teratogenesis, the most likely explanation for PHT, PB, and CBZ involves the formation of chemically reactive species by the cytochrome P450 system which covalently bind to fetal nucleic acids. Should this occur during critical periods of embryonic development, it could compromise DNA repair mechanisms, result in the preferential degradation of selected RNA and protein species, and adversely affect critical cellular housekeeping functions, including transcription and translation. This could ultimately lead to alterations in cell division and migration and no doubt negatively impact the temporal framework of embryogenesis, resulting in the development of congenital abnormalities. These reactive metabolites are normally detoxified by various enzyme systems, including epoxide hydrolase and glutathione-*S*-transferase. BUEHLER and colleagues (1990) demonstrated that infants with low detoxification capacities that were exposed to PHT in utero were at increased risks for FHS compared with similarly exposed individuals with elevated epoxide hydrolase activity.

The clinical picture of anticonvulsant drug teratogenesis is such that the various drugs are clearly capable of producing the same spectrum of congenital anomalies. As such, there may be a shared or common pathway of teratogenic action by which many of these drugs exert their adverse effects. On the other hand, there may be unique mechanisms exploited by certain anticonvulsant compounds. Both scenarios are complicated by AED polytherapy. Recently, both progabide and valnoctamide, used as anticonvulsants in Europe, were shown to inhibit mEH and thus the clearance of CBZ epoxide in epileptic patients (KROETZ et al. 1993; PISANI et al. 1993). LINDHOUT and colleagues (1984) had previously suggested a model in which combined drug metabolism could increase teratogenicity. They described an interaction where PB induces the arene oxide pathway (where epoxide metabolites of CBZ and PHT are formed), and VPA inhibits mEH but not the monooxygenase complex as a whole. Thus the coadministration of PB and VPA would increase the epoxide metabolites of CBZ or PHT. The polytherapeutic administration of PB and VPA with either CBZ, PHT, or both did yield the highest malformation rates of various polytherapeutic regimens supporting this model. Seven out of 12 (58%)infants exposed in utero to CBZ, PB, and VPA with or without PHT had congenital malformations. What is abundantly apparent is that a great deal more effort will need to be expended in order to reveal the initial site of teratogenic activity at either the molecular or cellular levels.

IV. Drug Access

The access of the anticonvulsant drug to the developing embryo/fetus depends upon its chemical nature.

Unlike other classes of teratogenic compounds that could be physical agents, such as maternal hyperthermia or X-rays, all of the clinically effective anticonvulsant drugs are administered orally. Individual differences in the pharmacokinetic profiles of the drugs will have some bearing on individual fetal exposures, as will various biochemical parameters such as protein binding, molecular weight, and bioavailability in general. Anything that influences the absorption, distribution, metabolism, and elimination of the drug will affect the concentration and duration of the fetal exposure. It is important to note that all of these compounds are lipophilic and will persist in the mother for a period of time after cessation or alteration of the AED treatment regimen.

V. Drug Dosage

The manifestations of abnormal development should increase in frequency and degree as the dosage increases; hence AED should exhibit a dose–response teratogenic effect.

This is not easy to demonstrate in human epidemiological or clinical studies. Since higher doses of individual drugs may correlate with polytherapy, it is not clear whether the critical factor is the number of drugs or the individual drug dosages. Furthermore, at any given drug dosage, there is a large inter-individual variation in the plasma AED concentration due to differences in drug metabolism, drug interactions, pregnancy, and noncompliance with the treatment regimen. Several studies have found no evidence of a dose–response relationship for PHT or other drugs (SPIEDEL and MEADOW 1972; FEDRICK 1973; VAN DYKE et al. 1988; ESKAZAN and ASLAN 1992).

NAKANE and colleagues (1980) and KANEKO and coworkers (1988) noted that the frequency of congenital defects increased with the total amount of anticonvulsant drugs consumed, as determined by the "drug score," thereby taking into consideration both the number of drugs and their dosages. In several other investigations, congenital anomalies were found to occur more frequently when the infants had been exposed to higher doses and/or plasma concentrations of VPA, PB, PRM, and PHT (for a review, see DANSKY and FINNELL 1991). Whenever possible, it is recommended that AED plasma concentrations during pregnancy are closely monitored. The measurement should reflect free drug concentrations, and not total concentrations, as the latter can be misleading, since they include bound drug concentrations, which are generally not pharmacologically active (TOMSON et al. 1994).

D. Experimental Animal Studies

I. Overview

Since MASSEY's original study in 1966, laboratory animal experiments to determine the teratogenic potential of anticonvulsant drugs have been valuable counterparts to human clinical and epidemiological investigations. These studies have helped change or clarify our understanding of the teratogenicity of selected antiepileptic compounds. Animal models can help circumvent problems common to human retrospective and prospective studies; indeed, in a laboratory setting, the investigator can control the nutritional and environmental factors that can modify normal embryonic development. Animal models are also important when it comes to developing testable hypotheses and elucidating the mechanisms of teratogenic action, since one is not limited to investigating only the final manifestation of a given disorder or malformation. Finally, animal studies can help develop the experimental data to support alterations in human therapeutic practices that may be successful in limiting or reducing the adverse effects of drug therapy during pregnancy. These considerations, along with the ethical dilemmas one creates when working with human populations, suggest that our knowledge of anticonvulsant drug teratogenesis would be distinctly poorer were it not for the development and use of animal models.

II. Valproic Acid

Structurally unrelated to any of the other anticonvulsant medications, VPA (Depakene, Abbott Laboratories, Chicago, IL) has a demonstrated teratogenic effect in a wide variety of animal species, including the mouse, rat, hamster, rabbit, and rhesus monkey (for a review, see NAU and HENDRICKX 1987). In all of the aforementioned animal species exposed in utero to VPA, abnormalities of the skeletal system were the most commonly reported developmental defect. There was a dose-dependent increase in various skeletal defects of the ribs, vertebrae, digits, and craniofacial bones in rat fetuses exposed to doses between 300 and 600 mg/kg body weight (BINKERD et al. 1988; VORHEES 1987). Similar observations were made in rabbit fetuses exposed in utero to either sodium or calcium valproate, with the increased rate of skeletal defects observed in both the axial and appendicular structures (PETRERE et al. 1986). Dose-dependent skeletal defects of the ribs and vertebrae were commonly reported in CD-1 mouse fetuses following intraperitoneal injections of VPA (200–400 mg/kg), with over 80% of the fetuses expressing the malformations at the higher drug dosages (KAO et al. 1981). Other skeletal defects observed in valproate-exposed mouse fetuses include syndactyly and partial hemimelia (PAULSON et al. 1985), costal fusions, and multiple limb defects (NAU and SCOTT 1987). The observed shortening of the long bones and digital hypoplasia were not limited to murine fetuses, but were similarly observed in the rhesus monkey (MAST et al. 1987). Cardiovascular malformations were induced in Jcl: ICR mice on gestational days 7–9, with the highest incidence (29%) occurring when the mice were injected with 600 mg/kg on day 7 (SONODA et al. 1993). Transposition of the great arteries with perimembraneous ventricular septal defects and endocardial cushion anomalies were the most commonly observed cardiac defects reported in this study.

The principal malformation associated with VPA exposure in utero in experimental animals has been neural tube defects, including exencephaly and spina bifida. NAU (1985) has produced exencephaly in mouse embryos from dams exposed to sufficiently high dosages of VPA to produce maternal plasma concentrations in excess of 230 μg/ml, irrespective of the route of administration. This concentration represents a two-to five fold increase over the recommended human therapeutic level (NIEDERMEYER 1983). Based upon probit analysis, a single subcutaneous injection of the drug administered on gestational day 8 must produce maternal plasma concentrations of 445 μg/ml in order to produce a 10% increase in the rate of exencephaly over that observed in the controls (NAU 1985). A multiple injection treatment regimen will produce the same increased response frequency for neural tube defects at maternal plasma concentrations of only 225 μg/ml, while osmotic minipumps can be used to induce this same neural tube defect response frequency while delivering a steady-state VPA concentration of 248 μg/ml plasma (NAU 1985). These and additional data indicate that maximal concentrations, and not the total area under the concentration–time curve values (AUC), are correlated with the teratogenic response observed. Thus a total daily dose distributed into multiple administrations is much less teratogenic. This information has been used in many clinical applications where the daily dose is now distributed into two or three portions or a slow release or retard formulation is used. It is highly likely that such a regimen is equally protective against seizures, but may carry a significantly lower risk for teratogenicity as compared to the once-per-day regimen.

With respect to posterior neural tube defects, EHLERS and colleagues (1992a,b) demonstrated that VPA administered (200 mg/kg, IP) at 6-h intervals, beginning on gestational day 9, can produce a 10% response frequency of spina bifida occulta, which increases to 95% affected concomitant with a VPA dosage increase to 500 mg/kg body weight. A significant degree of malformations of the ribs and vertebrae was apparent when the former fetuses were examined following alcian blue–alizarin red skeletal staining (EHLERS et al. 1992a,b). A low frequency (4%–6%) of spina bifida aperta was also induced by the same VPA treatment regimen in the Han: NMRI mouse strain, resulting in a highly disorganized and necrotic spinal cord within the vertebral canal in the lumbosacral region of the developing fetus. The absence of neuronal tissue in these affected fetuses indicates an almost completely localized ablation of the neural tube in the VPA-exposed fetuses (EHLERS et al. 1992b).

Having identified neural tube and skeletal defects as the primary malformations induced by in utero exposure to VPA in the mouse, several studies have been conducted to examine gene–teratogen interactions in the development of these malformations. There appear to be wide differences in susceptibility to the induction of anterior neural tube defects by both VPA and 4-propyl-4-pentenoic acid (4-en-VPA) exhibited by different inbred mouse strains (FINNELL et al. 1988). The highly sensitive SWV fetuses had exencephaly response frequencies in excess of 80% affected when the dams received multiple injections of VPA, while DBA/2J mice were completely resistant. This wide range of variability in the response frequencies to VPA-induced exencephaly in the different inbred mouse strains suggests that susceptibility to these malformations has a strong genetic component (FINNELL et al. 1986, 1988; FINNELL 1991; SELLER et al. 1979).

Recent studies investigating changes in coordinate embryonic gene expression in response to teratogen-induced neural tube defects have revealed significant differences in embryos exposed to either hyperthermia or retinoic acid as opposed to those exposed in utero to VPA. In the hyperthermia–

exposed embryos, there is a highly significant shift in the mRNA expression for *c–fos–c–jun* heterodimers, which increase from gestational day 9.5 to 10.0 at the expense of cyclic adenosine monophosphate (cAMP) response element-binding protein (creb) homodimers. This is just the opposite from the VPA treatment, in which embryos demonstrated a marked increase in the percentage of transcription factor population contributed by the creb homodimer and a decrease in that associated with *c–fos–c–jun* heterodimers (FINNELL et al. 1995). These observed differences in gene expression between various teratogen-induced neural tube defects suggests that there might be multiple mechanistic pathways to the development of these malformations. This is consistent with observations made at the histological level, in which VPA-induced spina bifida aperta differs between VPA and retinoic acid treatment. While VPA actually alters the morphology of the spinal cord, the retinoic acid treatment leaves the cord normal, but irregularly shaped, with only a very thin layer of ependymal cells on the surface, while the dorsal horn is displaced laterally (EHLERS et al. 1992b).

The teratogenicity of various analogues and metabolites of VPA have also been examined in murine models. These studies have determined that strict structural requirements must be met in order for the compound to exert a teratogenic effect (NAU and HENDRICKX 1987; NAU 1991). A valproate-like compound must have the following to be teratogenic: a free carboxyl group, an α-hydrogen atom, branching of carbon chains, no double bonds on C-2 or C-3, and alkyl substituents on C-2 larger than methyl groups (NAU 1985,1991). Homologous compounds with both alkyl chains either longer or shorter than propyl groups have a much more limited teratogenic potential. If there are substitutions of the α-hydrogen atom or if double bonds in carbon positions 2 or 3 are present (2-en or 3-en VPA), the teratogenic activity of the compound is either entirely abolished or significantly restricted (NAU and LÖSCHER 1986; NAU 1991). The addition of a double bond in the C-4 position (4-en VPA) does not interfere with the teratogenic potential of the compound (NAU and LÖSCHER 1986; FINNELL et al. 1988). This high degree of specificity for the teratogenic response differs from the broad, generalized specificity of its antiepileptic activity, suggestive of differing mechanisms of action (NAU and LÖSCHER 1986).

These structure–activity relationships have been further extended to demonstrate that the teratogenic activity of valproate-related compounds is highly stereoselective: of a racemic mixture given, one enantiomer proved to be more potent than the other. Such stereoselectivity was demonstrated for 4-en VPA (2-*n*-propyl-4-pentenoic acid) (HAUCK and NAU 1989), for 2-ethylhexanoic acid (HAUCK et al. 1990; COLLINS et al. 1992), and for 4-yn VPA (HAUCK and NAU 1992), among others. It was also demonstrated that the transplacental pharmacokinetics of a given pair of enantiomers are very similar. The differing teratogenic potencies, together with the comparable embryonic exposure of a given pair of enantiomers, suggest that the teratogenic effect is highly stereoselective and possibly mediated by stereoselective interaction with a chiral constituent within the embryo. This conclusion is supported by recent findings that the high stereoselectivity can also be found by adding these pure enantiomers in embryos cultured in vitro (ANDREWS et al. 1994). The nature of this chiral embryonic structure will be important for elucidating the mechanism of VPA teratogenesis.

Several hypotheses have been proposed to explain how VPA may disrupt normal morphogenetic mechanisms leading to the development of a neural tube defect. NAU and SCOTT (1987) have proposed that weak acids such as VPA will accumulate in the developing embryo because of its naturally alkaline medium, which is approximately 0.4 pH units above the maternal blood pH during the period of neural tube closure (NAU and SCOTT 1986, 1987). As only the nonionized form of the drug is able to cross the extraembryonic membranes, the degree of ionization on both sides of the placental membranes will ultimately determine the transplacental concentration gradient of the drug. Given the slightly acidic nature of valproate, it will more readily ionize in the embryo as opposed to the maternal blood, resulting in significantly higher embryonic concentrations when compared to maternal plasma levels of unbound drug (NAU and SCOTT 1987). The means by which the concentrated VPA level alters normal development remains unknown.

One potential mechanism by which VPA may exert its teratogenic effects is via altered folate metabolism. This is of interest, as the mothers of infants with neural tube defects have been reported to have reduced red cell folate levels when compared to mothers of children without these malformations (SMITHELLS et al. 1976; YATES et al. 1987). When folinic acid was coadministered with teratogenic concentrations of VPA, there was a significant reduction in the frequency of neural tube defects in Han:NMRI inbred mice (TROTZ et al. 1987; WEGNER and NAU 1991, 1992). What is important to consider is that the total folate concentrations in either the dam or the embryos remained constant in response to the VPA treatment. There were, however, significant alterations in the concentrations of selected formylated tetrahydrofolates (THF) (WEGNER and NAU 1991, 1992). Specifically, VPA reduced the concentration of the 5- and 10-formyl-THF and 5-CH_3-THF metabolites. This decrease in the formylated THF could be the result of a VPA-induced block in the interconversion of THF and formylated metabolites be the enzyme glutamate formyltransferase (WEGNER and NAU 1991, 1992). It is of interest to note that, when these metabolites were measured in VPA-exposed embryos from the neural tube defect-sensitive (SWV) and -resistant (DBA/2J) strains, there were significant differences in their metabolic profiles. Specifically, there was an inhibition of between 86% and 92% of the 5-CHO-THF and 5-CH_3-THF metabolites in the SWV embryos, while the DBA/2J embryos had no alterations in their 5-CH_3-THF and only a 50% inhibition in the 5-CHO-THF metabolite. This alteration in the production of specific folate metabolites may adversely affect purine biosynthesis, which could have significant consequences for the embryo's ability to synthesize DNA. Similarly, it may also result in a reduction in the rate of DNA methylation, which could lead to a lack of essential gene expression during critical periods on neural tube closure, resulting in the development of the neural tube defect (FINNELL 1991).

In contrast, HANSEN and GRAFTON (1991) have suggested that VPA teratogenesis is not mediated through folic acid inhibition. Using rat embryo cultures they observed that the addition of folic acid, converted to a stable intermediate, did not alter the teratogenic effects of VPA. The dose-related increase in the incidence of open neural tubes did not change when VPA treatment was combined with supplemental folic acid.

A metabolic block rather than a general folate deficiency is therefore suggested to be of crucial importance for the induction of neural tube defects by valproate. This was recently suggested with regard to the "spontaneous" occurrence of human neural tube defects (SCOTT et al. 1994; MILLS et al. 1995). These investigators suggested that a partial block of methionine synthase may be crucially involved in mechanisms leading to neural tube defects. This enzyme, which is vitamin B_{12} and folate dependent, serves to catalyze the metabolism of homocysteine to methionine and is therefore important for numerous methylation reactions mediated by *S*-adenosyl-methionine. Further support for this was provided by experiments in mice which demonstrated that the coadministration of methionine with VPA significantly reduced the incidence of neural tube defects (EHLERS et al. 1994).

It is also possible that VPA teratogenesis is mediated via interference with folate metabolism based on experimental studies in which antifolate agents are administered along with VPA. Both methotrexate and trimethoprim potentiated the VPA-induced neural tube defects in mice. These studies also point out that trimethoprim may be especially dangerous when administered to pregnant epileptic patients receiving AED therapy.

III. Phenytoin

PHT is the most widely studied of all of the clinically effective anticonvulsant compounds, with almost 50 in vivo animal studies exploring all facets of its teratogenic effects (for a review, see FINNELL and DANSKY 1991). Irrespective of the route of administration, PHT has the potential to disrupt normal morphogenesis in a wide variety of species, including mouse, rat, rabbit, cat and monkey. The focused experimental attention on PHT mirrors our understanding of the clinical expression of pregnancy outcomes in human infants exposed to the drug, which ranks among the best understood of any malformation complex. The majority of experimental studies on PHT teratogenicity have been limited to the ability of this compound to induce orofacial clefts, reflecting the early clinical reports of an increase in the incidence of cleft lip with or without cleft palate among the offspring of epileptic patients (MASSEY 1966; SULIK et al. 1979; GIBSON and BECKER 1968; for review see FINNELL and DANSKY 1991). These studies demonstrated that cleft palate could be induced in susceptible mouse strains when doses as low as 12.5 mg/kg per day were administered to the pregnant dam during the critical period for palate closure. Typically, the drug was administered via injection, either subcutaneous, intramuscular, or intraperitoneal, or by gastric intubation on selected gestational days during murine organogenesis.

Malformations other than cleft lip and/or cleft palate were observed in a number of experimental systems. HARBISON and BECKER (1969) were the first to describe a variety of congenital anomalies in mouse fetuses exposed to PHT during gestational days 8–15. The abnormal fetuses were significantly growth retarded, with shortened long bones, and had hydronephrosis along with renal hemorrhaging. Some fetuses had delayed ossification of the axial skeleton, open eyes, ectrodactyly, and internal hydrocephalus. A similar pattern of congenital defects was induced by the same investigators using PHT administered to pregnant rats (HARBISON and BECKER 1972). This spectrum of congenital defects was confirmed and expanded upon by subsequent investigations into PHT's teratogenic potential to include defects such as tracheoesophageal fistulas, cutaneous hemorrhages, and neural tube defects. FINNELL (1981) described a pattern of malformations in the offspring of pregnant mice that were chronically treated prior to and throughout their entire pregnancy with PHT, which was added to the animals drinking water. This pattern of malformations included ossification delays of the axial skeleton and cranial bones, dilated or immaturely developed cerebral ventricles, renal agenesis and hydronephrosis, cutaneous and renal hemorrhage, and cardiac, digital, and ocular anomalies (FINNELL 1981; FINNELL and CHERNOFF 1982; FINNELL et al. 1989). The differences in the pattern of malformations observed were generally attributable to differences in the species or strain of experimental animal, route of administration, and dosages used in the various studies (FRITZ et al. 1976; SULLIVAN and MCELHATTON 1975, 1977; NETZLOFF et al. 1979; PAULSON et al. 1979; MCDEVITT et al. 1981; HANSEN and BILLINGS 1985). PHT can have other developmental effects in addition to inducing structural malformations. In Sprague-Dawley rats, behavioral deficits described as hyperactivity and impaired memory were observed after in utero exposure to PHT. Further, their ability to learn a complex spatial task was reduced (VORHEES 1986).

The experimental studies on PHT have done more than merely address the teratogenic potential of this compound; they have also helped resolve other complex and controversial issues related to the risks involved in anticonvulsant drug therapy during pregnancy. Perhaps the most important issue to be approached initially was the question of whether it was the PHT or the presence of a maternal seizure disorder that was responsible for the increased incidence of congenital malformations among PHT-exposed fetuses. Using inbred mice with the genetically determined spontaneous seizure disorder quaking (*qk*), it was possible to dissect out the role of the seizure disorder from that of the medication in producing the adverse pregnancy outcome. In the homozygous quaking (*qk*/*qk*) mice with seizures that were left untreated and therefore had several seizures per day throughout their pregnancies, the dams produced normal offspring. As these same mice were placed on increasing dosages of PHT, the frequency of seizures decreased, while the incidence of congenital malformations increased in conjunction with the maternal plasma PHT concentration (FINNELL 1981; FINNELL and CHERNOFF 1982; FINNELL et al. 1989). This study clearly demonstrated that seizures alone do not appear to be a risk factor for congenital defects in experimental animals. Rather, it was the chronic, in utero exposure to PHT that produced offspring with birth defects.

Another issue of considerable interest surrounding the teratogenicity of AED is the extensive clinical variability observed in affected infants. FINNELL and CHERNOFF (1984) suggested that the observed variability was the result of gene–teratogen interactions. By analyzing the most frequently observed defects in mouse fetuses from three different inbred strains that were chronically exposed to PHT, it became apparent that the pattern of malformations in each strain differed significantly from each other and from a heterogeneous population of mice. For example, in the C57BL/6J strain, which has a high spontaneous rate of craniofacial defects (STAATS 1976), it takes relatively lower concentrations of PHT to induce these malformations than would be required to cause a comparable frequency of such defects in the other inbred mouse strains (FINNELL and CHERNOFF 1984; FINNELL et al. 1989). This difference must be related to the number of liability genes maintained within the genetic background of each strain for each of the developing organ systems. These studies demonstrate that genetic variation in the susceptibility to specific PHT-induced malformations among different inbred mouse strains results in distinctive strain differences in the pattern of malformations produced by in utero exposure to PHT. Similar underlying genetic differences in the very heterogeneous human population is one of the likely explanations for the variable phenotypes observed in children gestationally exposed to PHT and other anticonvulsant medications (for a review, see DANSKY and FINNELL 1991).

In spite of considerable effort on the part of the teratological community, the mechanism by which PHT exerts its teratogenic effect is still unknown. Of the many different hypotheses that have been set forth, a current favorite proposes that PHT is metabolized to a toxic, reactive intermediate compound that is ultimately responsible for the observed teratogenic effects. The identification of the actual intermediary metabolic species that is the primary teratogen remains to be determined. Some investigators believe that the source of PHT's teratogenicity is an arene oxide metabolite produced during the bioactivation of the parent compound by one of the P450 cytochromes (MARTZ et al. 1977; BLAKE and MARTZ 1980; WELLS and HARBISON 1980; for a review, see FINNELL and DANSKY 1991). These oxidative metabolites are thought to occur prior to the formation of the dihydrodiol metabolite 5-(3,4-dihydroxyl-1,5-cyclohexadien-1-yl-5-phenylhydantoin) from PHT in a reaction catalyzed by the enzyme epoxide hydrolase (E.C. 3.3.2.3.) (CHANG et al. 1970). Arene oxides, in general, are highly reactive compounds, and when the rate of their bioactivation exceeds the detoxification capacity of the organism, the electrophilic center of the molecule is capable of binding covalently to nucleophilic sites found in fetal macromolecules such as nucleic acids (JERINA and DALY 1974). Such irreversible binding covalently to embryonic or fetal nucleic acids at critical periods of development could initiate a cascade of events which alters important cellular housekeeping functions such as transcription/translation, cell division, and migration, which could ultimately lead to abnormal morphogenesis (MARTZ et al. 1977; WELLS and VO 1989; WELLS and HARBISON 1980; STRICKLER et al. 1985; FINNELL et al. 1992). It is entirely possible that the arene oxide metabolite produced during the biotransformation of PHT in the maternal liver may be sufficiently stable to cross the placenta and bind to fetal tissues. The other possibilities for this molecule to be the primary teratogenic agent involve it being tautomerized to a more stable oxepin, which can then cross the placenta and isomerize back to a reactive arene oxide intermediate (WELLS and HARBISON 1980), or the arene oxide could actually be formed in the fetal liver (BLAKE and MARTZ 1980)

Further evidence in support of the reactive metabolite hypothesis includes studies involving the coadministration of PHT along with various inhibitors of either the bioactivating cytochrome P450 enzymes or the detoxifying enzymes such as epoxide hydrolase. In the presence of TCPO (1,2-epoxy-3,3,3-trichloropropane), the detoxification of PHT by epoxide hydrolase is significantly reduced, and the embryos are exposed to higher concentrations of arene oxide intermediates for longer periods of time, resulting in an enhancement of PHT-induced congenital defects (MARTZ et al. 1977; HARBISON 1978). This is consistent with in vitro studies demonstrating an enhanced binding of radiolabeled PHT in the presence of epoxide hydrolase inhibitors, which include not only TCPO but also cyclohexene oxide (PANTAROTTO et al. 1982). Furthermore, when the mixed-function oxidase enzymes are inhibited by the compound SKF525A, there is an increase in the plasma PHT concentrations and a decrease in its elimination. These combined responses could result in more free PHT being oxidized, which conceivably could elevate the concentration of arene oxides being produced. Under such conditions, both PHT covalent binding and the incidence of fetal defects would be expected to increase (WELLS and HARBISON 1980).

It has also been demonstrated that the coadministration of the cytochrome P450-inhibiting anticonvulsant drug stiripentol (Biocodex Laboratories, Montrouge, France) along with PHT has a profoundly protective effect against PHT-induced congenital defects in a mouse model (FINNELL et al. 1992, 1993, 1994). When teratogenic concentrations of PHT were chronically coadministered with stiripentol to mice of three different inbred strains, the incidence of abnormal fetuses was not significantly different from that observed in the control groups. These findings suggest that stiripentol inhibits the biotransformation of PHT to one or more unknown toxic oxidative metabolites, a suggestion supported by the known inhibitory properties of stiripentol toward the oxidative metabolism of other antiepileptic medications (LEVY et al. 1984; KERR et al. 1991). With the reduction in the production of

PHT-reactive metabolites which could potentially bind to fetal nucleic acids, a significant margin of reproductive safety was afforded the embryos by stiripentol therapy, despite the fetal exposure to teratogenic concentrations of PHT (FINNELL et al. 1992, 1993, 1994). These experimental findings have significant clinical implications, as they suggest that selective polytherapeutic approaches may actually be safer to the developing offspring, which contrasts markedly with the existing epidemiological evidence for multiple drug exposure during pregnancy (LINDHOUT et al. 1984; KANEKO et al. 1988; for a review, see DANSKY and FINNELL 1991).

Finally, it is also possible that the reactive oxidative metabolite of PHT that is the primary teratogenic molecule is one that is not sufficiently detoxified by enzymatic conjugation with reduced glutathione (JOLLOW et al. 1977; SNYNDER et al. 1982; MOLDEUS and JERNSTROM 1983), While a glutathione conjugate of PHT has not as yet been demonstrated, PHT is capable of producing slight, yet significant depletion of hepatic glutathione synthetase in pregnant mice (LUM and WELLS 1986). When mice are pretreated with compounds such as diethyl maleate (HARBISON 1978) or acetaminophen (LUM and WELLS 1986) that deplete normal stores of glutathione, there is a marked increase in the covalent binding of PHT and a subsequent increase in the teratogenic response frequency. The results of these in vivo studies are compatible with similarly studies conducted in vitro using microsomes obtained from either rat (PANTAROTTO et al. 1982) or mouse (KUBOW and WELLS 1986, 1989) hepatocytes, where glutathione reduced the covalent binding of radiolabeled PHT to microsomal proteins.

An alternative model of PHT teratogenicity involves the prostaglandin synthetase enzyme system. WELLS et al. (1989) observed that a combined treatment of CD-1 mice with PHT and acetylsalicylic acid, a strong inhibitor of prostaglandin synthetase, decreased the occurrence of clefting malformations. The model proposes that the teratogenic intermediate of PHT may be a result of cooxidation of prostaglandin synthetase creating teratogenic free radicals. WELLS et al. (1989) describes three apparent paradoxes that favor this model over the P450 arene oxide model. First, HARBISON and BECKER (1970) reported that pretreatment with PB, a P450 inducer, actually decreased PHT teratogenicity, and pretreatment with the P450 inhibitor SKF525A actually increased PHT teratogenicity. Second, WELLS et al. (1982) observed the different behavior of isomers of two PHT-related compounds. Only the l-isomers of mephenytoin and nirvanol, which are not metabolized through an arene oxide intermediate, demonstrated fetal toxicity. Third, WELLS et al. (1989) noted that many similar hydantoin anticonvulsants cause similar teratogenic effects despite the fact that some of them lack a phenyl group, needed in the arene oxide model of teratogenesis.

Recently, MUSSELMAN and colleagues (1994) have conducted preliminary studies into the molecular basis of PHT-induced congenital malformations, utilizing a chronic oral drug treatment regimen. The PHT treatment produced a significant downregulation in the level of expression of several of the genes examined in the gestational day-9.5 embryos, including the growth factors transforming growth factor (TGF)-β and neurotropin (NT)-3, the proto-oncogene *Wnt-1*, the nicotinic receptor, and the voltage-sensitive calcium channel gene. Additional changes in the coordinate expression of several of the growth and transcription factors were observed gestational day-10.0 fetuses. In utero, PHT exposure significantly altered the normal patterns of gene expression, shifting the normal level of gene expression at gestational day 9.5 to those usually occurring 12 h later at day 10.0. The changes in mRNA levels that are seen suggest that the corresponding protein levels are altered in a similar manner. While this may not be true for all of the molecules examined in this study, for those in which it is true, a change in functional protein levels may result in significant alterations in normal embryonic development. These data also support the hypothesis that no one single gene is responsible for teratogen-induced changes in normal morphogenesis. Rather, it is the coordinate change of several molecules, each of which by itself may not be sufficient to elicit detectable developmental changes, but when combined together produce the adverse phenotypic changes observed in the affected embryos (MUSSELMAN et al. 1994).

In summary, it is clear that PHT has a significant teratogenic potential in most, if not all of the common laboratory animal species. PHT is capable of inducing a wide variety of both minor and major congenital anomalies when administered during critical periods of embryogenesis. Given the existing literature, it appears that oxidative metabolites produced during the biotransformation of PHT may be responsible for the majority of adverse reproductive consequences of PHT therapy (FINNELL and DANSKY 1991). As such, it may be possible to limit the reproductive risks inherent in anticonvulsant drug therapy by avoiding drug therapies that either promote the formation of or inhibit the rapid detoxification of toxic oxidative metabolites.

IV. Phenobarbital/Primidone

Experimental work on the reproductive risks associated with in utero PB exposure is surprisingly limited, given the length of time that this drug has been utilized in clinical medicine. Those data that do exist would support the idea that both PB and PRM have a distinct, but limited teratogenic potential in rodents. SULLIVAN and MCELHATTON (1975) fed female ICI strain mice with up to 150 mg/kg doses of the drug, which was mixed in the animal's feed on gestational days 6–16, and observed orofacial clefts in less than 4% of the exposed fetuses. Subsequently, the same investigators administered very low doses of PB by gastric intubation to CD-1 dams and reported no appreciable teratogenic effects (SULLIVAN and MCELHATTON 1977). In both of these studies, the results obtained with PRM were essentially the same as those observed with PB. FRITZ and colleagues (1976) used comparable experimental protocols and confirmed these findings for cleft palate. Although structural malformations were infrequently observed, when pregnant rats were intubated with PB on gestational days 7–18 the resulting offspring had a high frequency of motor disturbances (VORHEES 1983). GUPTA et al. (1982) reported that exposure of Sprague-Dawley CD rats to PB during late prenatal development resulted in a decreased production of testosterone, measured through adulthood and altering sexual function. However, no differences were seen in birth weight or malformation rates between treated and control animals.

FINNELL and coworkers (1987a) chronically exposed pregnant mice from three different inbred mouse strains to a wide range of PB doses in the animal's drinking water and observed a significant teratogenic response. Skeletal, cardiac, neural, and urogential defects were observed in a dose–dependent manner, with LM/Bc being the most sensitive mouse strain, while the C57BL/6J fetuses were the most resistant of the three strains examined. When the pattern of malformations induced by chronic oral PB treatment was compared with a similar study using PHT and the same three inbred strains, it was clear that PB induced a higher frequency of true malformations (urogenital, cleft palate, and cardiac defects), while PHT produced more anomalies related to impaired growth leading to incomplete fetal development. These latter anomalies included hydronephrosis, skeletal ossification delays, and dilated cerebral ventricles (FINNELL et al. 1987b).

V. Carbamazepine

The literature concerning the teratogenic potential of this front-line antiepileptic compound is much more limited than that which exists for either PHT or VPA (for a review, see FINNELL and DANSKY 1991). The initial teratological investigations were those of FRITZ and colleagues (1976), who failed to observe any increase in the incidence of fetal malformations when pregnant dams were treated by gastric intubation on gestational days 6–16 with CBZ dosages of up to 250 mg/kg per day. Other studies noted small increases in the incidence of cleft palate, dilated cerebral ventricles, and growth retardation in mouse fetuses from similarly treated dams (SULLIVAN and MCELHATTON 1977; PAULSON et al. 1979; ELUMA et al. 1984). When mice were chronically exposed to CBZ in their diet prior to and throughout gestation, a significant number of fetuses were observed with congenital malformations of the central nervous or urogenital systems (FINNELL et al. 1986). Fetal growth retardation was consistently observed in serveral different studies (PAULSON et al. 1979; FINNELL et al. 1986; SUCHESTON et al. 1986). In those studies that compared the teratogenicity of CBZ with other anticonvulsant drugs, invariably it was less teratogenic than either PHT or PRM (FRITZ et al. 1976; SULLIVAN and MCELHATTON 1977; PAULSON et al. 1979; ELUMA et al. 1984; EL-SAYED et al. 1983). In two reproductive toxicology studies that were conducted on rats, CBZ had a very limited teratogenic and embryotoxic potential (EL-SAYED et al. 1983; VORHEES et al. 1990). Of the few developmental defects that were produced when CBZ was administered by gastric intubation throughout the period of organogenesis, fetal edema was the most frequently observed anomaly (Vorhees et al. 1990), although fetuses with gastroschisis, omphaloceles, hydronephrosis, ventricular septal defects, hydrocephaly, and skeletal abnormalities were also reported (El-Sayed et al. 1983; Vorhees et al. 1990).

In summary, CBZ appears to have a far more limited and much less malformation-specific teratogenicity than do previously described antiepileptic compounds. Given the limited response frequency obtained in teratology experiments using CBZ, there have not as yet been any truly mechanistic studies reported in the peer-reviewed literature. It would not be unreasonable to consider reactive oxidative metabolic products produced during the biotransformation of CBZ to be the means by which the drug exerts its teratogenic effects. As with PHT, the precise nature of the primary teratogenic metabolic species remains to be elucidated.

Based upon experimental work with several different rodent models of anticonvulsant drug teratology, it appears as if there are considerable differences in the teratognic potential of each compound. Relative to other clinically available anticonvulsant drugs, PB and PRM appear to be much less deleterious than either PHT or VPA (for a review, see Finnell and Dansky 1991).

E. Conclusions

The consensus today is that more than 90% of women with epilepsy who receive anticonvulsant drug therapy will deliver normal children. Nonetheless, prospective and retrospective epidemiological studies have identified unequivocal risks for major malformations and minor anomalies in a small, but significant percentage of pregnancies complicated by anticonvulsant drugs. Using a conservative risk estimate of 6% for AED-complicated pregnancies, this works out to approximately 1100–1200 adverse outcomes per year in the United States alone. This estimate is for major malformations and may not adequately predict AED effects including neurobehavioral deficits. The risk to fetuses is probably higher when multiple anticonvulsant drugs are used in combination and when there is a genetic predisposition to the development of birth defects. To date, no informaion adequately determines which of the anticonvulsant drugs is the most teratogenic and is responsible for the majority of major malformations. Depending upon the study, the relative teratogenicity varies among the front-line anticonvulsant drugs (Table 5). These results are confounded by the use of polypharmacy, different dosages and combinations of drugs, and the different makeup of the patient populations exposed to the drugs.

The consensus opinion is that the anticonvulsant medication that stops seizures in a given patient is the drug that should be used. The clinical evidence would support the concept that monotherapy is perferable during pregnancy; however, this is not practical for all patients. Unfortunately, none of the available studies to date has examined a sufficiently large number of women with epilepsy exposed to antiepileptic drugs in monotherapy during pregnancy. Consequently, inadequate power has skewed the statistical analysis of

Table 5. Malformation rates in the offspring of epileptic women treated with antiepileptic drug monotherapy

Author	VPA		PHT		CBZ		PB		PRM	
	(%)	Total pregnancies (*n*)	(%)	Total pregnancies (*n*)	(%)	Total pregnancies (*n*)	(%)	Total pregnancies (*n*)	(%)	Total pregnancies (*n*)
VAN DYKE et al. (1988)			38.5	39						
BATTINO et al. (1992a)	22.7	22	11.1	27	9.8	61	5.0	60	12.9	31
GLADSTONE et al. (1992)	–	–	31.3	16	6.7	15	–	–	–	–
TANGENELLI and REGESTA (1992)	0	62	0	9	4.8	63	–	–	–	–
KANEKO et al. (1993)	10.0	36	2.3	43	6.5	43	0	14	0	5
WATERS et al. (1994)	–	–	10.7	28	3.0	33	23.8	21	–	–

VAP,Valproic acid; PHT, phenytoin; CBZ, carbamzepine; PB, phenobarbital; PRM, primidone.

risk estimates for specific forms of major birth defects associated with specific drugs. The denominator used for analysis of each anticonvulsant drug combination in polytherapy is even smaller.

It is clear that there are no completely safe AED. The teratogenic effects of the major AED have been described in the scientific and medical literature for the past 30 years. While the risks for an adverse outcome are very real, they may not be as severe as initially suspected. Clinicians should be aware of the serious risks involved when highly susceptible families proceed to have multiple pregnancies, but the majority of women with epilepsy can and do have normal, healthy offspring. Finally, the only way to completely avoid the teratogenic potential of the currently available anticonvulsant drugs is to withdraw drugs in patients planning pregnancy who have been seizure free for at least 2 years (DELGADO-ESCUETA and JANZ 1992).

Acknowledgement. This work was supported in part by PHS grant DE 11303.

References

Andermann E, Dansky L, Kinch RA (1982) Complications of pregnancy, labor and delivery in epileptic women. In: Janz D, Dam M, Richens A et al (eds) Epilepsy, pregnancy, and the child. Raven, New York, pp 61–74

Andrews JE, Ebron-McCoy M, Bojic U et al (1994) The in vitro evaluation of the stereoselective teratogenicity of the enantiomers of the valproic acid analogue 2-n-propyl-4-pentynoic acid (4-yn-VPA). Teratology 49: 381

Annegers JF, Elveback LR, Hauser WA, Kurland LT (1974) Do anticonvulsants have a teratogenic effect? Arch Neurol 31: 364–373
Annegers JF, Hauser WA, Elveback LR et al (1978) Congenital malformations and seizure disorders in the offspring of parents with epilepsy. Int J Epidemiol 7: 241–247
Ardinger HH, Atkin JF, Blackston RD et al (1988) Verification of the fetal valproate syndrome phenotype. Am J Med Genet 29: 171–185
Baptisti A (1938) Epilepsy and pregnancy. Am J Obstet Gynecol 35: 818–824
Battino D, Binelli S, Caccamo ML et al (1992a) Malformations in offspring of 305 epileptic women: a prospective study. Acta Neurol Scand 85: 204–207
Battino D, Granata T, Binelli S et al (1992b) Intrauterine growth in the offspring of epileptic mothers. Acta Neurol Scand 86: 555–557
Beaussart-Defaye J, Bastin N, Demarcq C (1985) Epilepsies and reproduction. Grine, Lille
Biale Y, Lewenthal H, Aderet ND (1975) Congenital malformations due to anticonvulsant drugs. Obstet Gynecol 45: 439–442
Binkerd PE, Rowland JM, Nau H et al (1988) Evaluation of valproic acid developmental toxicity and pharmacokinetics in Sprague-Dawley rats. Fundam Appl Toxicol 11: 485–493
Bjerkedal T, Bahna L (1973) The course and outcome of pregnancy in women with epilepsy. Acta Obstet Gynecol Scand 52: 245–248
Blake DA, Martz F (1980) Covalent binding of phenytoin metabolites in fetal tissue. In: Hassell TM, Jognston MC, Dudley KH (eds) Phenytoin-induced teratology and gingival pathology. Raven, New York, pp 75–80
Blake DA, Collins JM, Miyasaki BC et al (1978) Influence of pregnancy and folic acid on phenytoin metabolism by rat liver microsomes. Drug Metab Dispos 6: 246–250
Bron JTJ, Van DerHarten HJ, Van Geiin HP (1990) Prenatal ultrasonographic diagnosis of radial-ray reduction malformations. Prenat Diagn 10: 279–288
Buehler BA (1984) Epoxide hydrolase activity and fetal hydantoin syndrome. Proc Greenwood Genet Ctr 3: 109–110
Buehler BA, Delimont D, van Waes M, Finnell RH (1990) Prenatal prediction of risk of the fetal hydantoin syndrome. New Engl J Med 322: 1567–1572
Bustamante SA, Stumpff LC (1978) Fetal hydantoin syndrome in triplets: a unique experiment of nature. Am J Dis Child 132: 978–979
Chang T, Savory A, Glazko AJ (1970) A new metabolite of 5,5-diphenylhydantoin (Dilantin). Biochem Biophys Res Commun 38: 444–449
Collins MD, Scott WJ, Miller SJ et al (1992) Murine teratology and pharmokinetics of the enantiomers of sodium 2-ethylhexanoate. Toxicol Appl Pharmacol 112: 257–265
Czeizel AE, Bod M, Halasz P (1992) Evaluation of anticonvulsant drugs during pregnancy in a population based Hungarian study. Eur J Epidemiol 8(1): 122–127
Dansky LV, Finnell RH (1991) Parental epilepsy, anticonvulsant drugs and reproductive outcome: epidemiological and experimental findings spanning three decades. II. Human studies. Reprod Toxicol-5: 281–299
Dansky LV, Andermann E, Sherwin AL et al (1980) Maternal epilepsy and congenital malformations: a prospective study with monitoring of plasma anticonvulsant levels during pregnancy. Neurology 3:15
Dansky LV, Andermann E, Andermann F et al (1982) Maternal epilepsy and congenital malformations: correlation with maternal plasma anticonvulsants levels during pregnancy. In: Janz D, Dam M, Richens A et al (eds) Epilepsy, pregnancy and the child. Raven, New York, pp 251–258
Dansky LV, Rosenblat DS, Andermann E (1992) Mechanisms of teratogenesis: folic acid and antiepileptic drug therapy. Neurology 42 [Suppl]: 32–42
Delgado-Escueta AV, Janz D (1992) Consensus guidelines: preconception counseling, management, and care of the pregnant epileptic patient. Neurology 42 [Suppl 5]: 149–160

Dieterich E, Steveling A, Lukas A et al (1980) Congenital anomalies in children of epileptic mothers and fathers. Neuropediatrics 11(3): 274–283
DiLiberti JH, Farndon PA, Dennis NR et al (1984) The fetal valproate syndrome. Am J Med Genet 19: 473–481
Dravet C, Julian C, Legras C et al (1992) Epilepsy, antiepileptic drugs, and malformations in children of women with epilepsy: a French prospective cohort study. Neurology 42 [Suppl 5]: 75–82
Durner M, Greenberg DA, Delgado-Escueta AV (1992) Is there a genetic relationship between epilepsy and birth defects? Neurology 42 [Suppl 5]: 63–67
Ehlers K, Sturje J, Merker H, Nau H (1992a) Valproic acid-induced spina bifida: a mouse model. Teratology 45: 145–154
Ehlers K, Sturje H, Merker H, Nau H (1992b) Spina bifida aperta induced by valproic acid and by all-trans-retinoic acid in the mouse: distinct differences in morphology and periods of sensitivity. Teratology 46: 117–130
Ehlers K, Drews E, Nau H (1994) The amino acid methione reduces the valproic acid-induced spina bifida rate in the mouse. Teratology 50: 28
El-Sayed MGA, Aly AE, Kadri M et al (1983) Comparative study on the teratogenicity of some antiepileptics in the rat. East Afr Med J 60: 407–415
Elshove J, Van Eck JHM (1971) Aangeboren misvorminge, met name gespleten lipmet zonder gespleten verhemelte, bij kinderen van moeders met epilepsie. Ned Tijdschr Geneeskd 115(33): 1371–1375
Eluma FO, Sucheston ME, Hayes TG et al (1984) Teratogenic effects of dosage levels and time of administration of carbamazepine, sodium valproate, and diphenylhydantoin on craniofacial development in the CD-1 mouse fetus. J Craniofac Genet Dev Biol 4: 191–210
Eskazan E, Aslan S (1992) Antiepileptic therapy and teratogenesis in Turkey. Int J Clin Pharmacol Ther Toxicol 30(8): 261–264
Fedrick J (1973) Epilepsy and pregnancy: a report from the Oxford record linkage study. Br Med J 2: 442–448
Feldman GL, Weaver DD, Lovrien EW (1977) The fetal trimethadione syndrome: report of an additional family and further delineation of this syndrome. Am J Dis Child 131(13): 89–92
Finnell RH (1981) Phenytoin-induced teratogenesis: a mouse model. Science 211: 483–484
Finnell RH (1991) Genetic differences in susceptibility to anticonvulsant drug-induced developmental defects. Pharmacol Toxicol 69: 223–227
Finnell RH, Chernoff GF (1982) Mouse fetal hydantoin syndrome: effects of maternal seizures. Epilepsia 23: 423–429
Finnell RH, Chernoff GF (1984) Variable patterns of malformation in the mouse fetal hydantioin syndrome. Am J Med Genet 19: 463–471
Finnell RH, Chernoff GF (1987) Gene-teratogen interactions: an approach to understanding the metabolic basis of birth defects. In: Nau H, Scott WJ (eds) Drug disposition in teratogenesis vol II. CRC, New York, pp 97–112
Finnell RH, Dansky LV (1991) Parental epilepsy, anticonvulsant drugs and reproductive outcome: epidemiological and experimental findings spanning three decades. I. Animal studies. Reprod Toxicol 5: 281–299
Finnell RH, Moon SP, Abbott LC et al (1986) Strain differences in heat-induced neural tube defects in mice. Teratology 33: 247–252
Finnell RH, Shields HE, Taylor SM, Chernoff GF (1987a) Strain differences in phenobarbital-induced teratogenesis in mice. Teratology 35: 177–185
Finnell RH, Shields HE, Chernoff GF (1987b) Variable patterns in anticonvulsant drug-induced malformation in mice: comparisons of phenytoin and phenobarbitol. Teratog Carcinog Mutagen 7: 541–549
Finnell RH, Bennett GD, Karras SB et al (1988) Common hierarchies of susceptibility to the induction of neural tube defects in mouse embryos by valproic acid and its 4-propyl-4-pentenoic acid metabolite. Teratology 38: 313–320

Finnell RH, Abbott LC, Taylor SM (1989) The fetal hydantoin syndrome: answers from a mouse model. Reprod Toxicol 3: 127–133
Finnell RH, Buehler BA, Kerr BM et al (1992) Clinical and experimental studies linking oxidative metabolism to phenytoin-induced teratogenesis. Neurology 42 [Suppl 5]: 25–31
Finnell RH, van Waes M, Musselman A, Kerr BM, Levy RH (1993) Strain differences in patterns of phenytoin-induced malformations by stiripentol®. Reprod Toxicol 7: 439: 448
Finnell RH, Kerr BM, van Waes M, Steward RL, Levy RH (1994) Protection from phenytoin-induced teratogenesis by the co-administration of stiripentol in a mouse model. Epilepsia 35: 141–148
Finnell RH, Kerr BM, van Waes M et al (1995) Protection from phenytoin-induced congenital malformations by co-administration of the antiepileptic drug stiripentol in a mouse model. Epilepsia (1994)
Fraser FC (1976) The multifactorial/threshold concept-uses and misuses. Teratology 14: 267–280
Fraser FC, Watt HJ (1964) Megaloblastic anemia in pregnancy and the puerperium. Am J Obstet Gynecol 89: 532–534
Fritz H, Muller D, Hess R (1976) Comparative study of the teratogenicity of phenobarbital, diphenylhydantoin and carbamazepine in mice. Toxicology 6: 159–171
Gaily E, Granström ML, Hiilesmaa V et al (1988) Minor anomalies in offspring of epileptic mothers. J Pediatr 112: 520–529
Gaily E, Kantola-Sorsa E, Granström ML (1990) Specific cognitive dysfunction in children with epileptic mothers. Dev Med Child Neurol 32: 403–414
German J, Kowal A, Ellers KH (1970) Trimethadione and human teratogenesis. Teratology 3: 349–362
Gibson JE, Becker BA (1968) Teratogenic effects of diphenylhydantoin in Swiss-Webster and A/J mice. Proc Soc Exp Biol Med 128: 905–909
Gladstone DJ, Bologna M, Maguire C et al (1992) Course of pregnancy and fetal outcome following maternal exposure to carbamazepine and phenytoin: a prospective study. Reprod Toxicol 6: 257–261
Goldman AS, Zachai EH, Yaffe SJ (1986) Environmentally induced birth defect risks. In: Sever JL, Brent RL (eds) Teratogen update. Liss, New York, pp 35–38
Göpfert-Geyer I, Koch S, Rating D et al (1982) Delivery, gestation, data at birth, and neonatal period in children of epileptic mothers. In: Janz D, Dam M, Richens A, Bossi L et al (eds) Epilepsy, pregnancy, and the child. Raven, New York, pp 179–187
Granström ML (1982) Development of the children of epileptic mothers, preliminary results from the prospective Helsinki study. In: Janz D, Dam M, Richens A, Bossi L et al (eds) Epilepsy pregnancy and the child. Raven, New York, pp 403–408
Granström ML, Gaily E (1992) Psychomotor development in children of mothers with epilepsy. Neurology 42 [Suppl 5]: 144–148
Gupta G, Yaffe SJ, Shapiro BH (1982) Prenatal exposure to phenobarbital permanently decreases testosterone and causes reproductive dysfunction. Science 216(7): 640–642
Gustavson EE, Chen H (1985) Goldenhar syndrome, anteriorencephalocele and aqueductal stenosis following fetal primidone exposure. Teratology 32(1): 13–7
Hansen DK, Billings RE (1985) Phenytoin teratogenicity and effects on embryonic and maternal folate metabolism. Teratology 31: 363–371
Hansen DK, Grafton TF (1991) Lack of attenuation of valproic acid-induced effects by folinic acid in rat embryos in vitro. Teratology 43: 575–582
Hanson JW (1986) Teratogen update: fetal hydantoin effects. Teratology 33: 349–353
Hanson JW, Smith DW (1975) The fetal hydantoin syndrome. J Pediatr 87: 285–290
Hanson JW, Myrianthopoulos NC, Harvery MAS et al (1976) Risks to offspring of women treated with hydantoin anticonvulsants, with emphasis on the fetal hydantoin syndrome. J Pediatr 89: 662–668

Harbison RD (1978) Chemical-biological reactions common to teratogenesis and mutagenesis. Environ Health Perspect 24: 87–100
Harbison RD, Becker BA (1969) Relation of dosage and time of administration of diphenylhydantoin to its teratogenic effect in mice. Teratology 2: 305–312
Harbison RD, Becker BA (1970) Effect of phenobarbital and SKF 525A pretreatment on diphenylhydantoin teratogenicity in mice. J Pharmacol Exp Ther 175: 283–288
Harbison RD, Becker BA (1972) Diphenylhydantoin teratogenicicty in rats. Toxicol Appl Pharmacol 22: 193–200
Hassett C, Aicher L, Sidhu JS et al (1994) Human microsomal epoxide hydrolase: genetic polymorphism and functional expression in vitro of amino acid variants. Hum Mol Genet 3 (3): 421–428
Hauck R-S, Nau H (1989) Asymmetric synthesis and entioselective teratogenicity of (4-en-VPA), an active metabolite of the anticonvulsant drug, valproic acid. Toxicol Lett 49: 41–48
Hauck R-S, Nau H (1992) The enantiomers of the valproic acid analogue 2-n-propyl-4-pentynoic acid (4-yn-VPA): assymmetric synthesis and highly stereoselective teratogenicity in mice. Pharm Res 9 (7): 850–855
Hauck R-S, Wegner C, Blumtritt P et al (1990) Asymmetric synthesis and teratogenic activity of (R)- and (S)-2–ethylhexanoic acid, a metabolite of the plasticizer-Di-(2-ethylhexyl)phthalate. Life Sci 46: 513–518
Hibbard BM (1964) The role of folic acid in pregnancy. J Obstet Gynecol Br Common 71: 529–542
Hibbard BM, Hibbard ED, Jeffcoate TNA (1965) Folic acid and reproduction. Acta Obstet Gynec Scand 44: 375–400
Higgins TA, Comerford JB (1974) Epilepsy in pregnancy. J Ir Med Assoc 67(11): 317–320
Hiilesmaa VK, Teramo K, Granström ML et al (1981) Fetal head growth retardation associated with maternal antiepileptic drugs. Lancet 2: 165–167
Hill RM, Tennyson L (1982) Premature delivery, gestational age, complications of delivery, vital data at birth on newborn infants of epileptic mothers: review of the literature. In: Janz D, Bossi L, Dam M et al (eds) Epilepsy, pregnancy and the child. Raven, New York, pp 167–73
Hill RM, Verniaud WM, Horning MG et al (1974) Infants exposed in utero to antiepileptic drugs. Am J Dis Child 127: 645–653
Holmes LB, Harvey EA (1994) Holoprosencephaly and the teratogenicity of anticonvulsants. Teratology 49: 82
Holmes LB, Harvey EA, Brown KS et al (1994) Anticonvulsant teratogenesis: I. A study design for newborn infants. Teratology 49: 202–204
Hutch HC, Steinhousen HC, Helge H (1982) Mental development in children of epileptic parents. In: Janz D, Bossi L, Dam M et al (eds) Epilepsy, pregnancy and the child. Raven, New York, pp 437–42
Jäger-Roman E, Deichl A, Jakob S et al (1986) Fetal growth , major malformations, and minor anomalies in infants born to women receiving valproic acid. J Pediatr 108: 997–1004
Janz D (1982) Antiepileptic drugs and pregnancy: altered utilization patterns and teratogenesis. Epilepsia 23 [Suppl 1]: 853–863
Janz D, Fuchs U (1964) Are antiepileptic drugs harmful when given during pregnancy? Ger Med Mon 9: 20–22
Jerina DM, Daly JW (1974) Arene oxides: a new aspect of drug metabolism. Science 185: 575–582
Jollow DJ, Kocsis JJ, Snyder R et al (eds) (1977) Biological reactive intermediates: formation, toxicity, and inactivation. Plenum New York
Jones KL, Lacro RV, Johnson KA et al (1989) Pattern of malformations in the children of women treated with carbamazepine during pregnancy. New Engl J Med 320: 1661–1666
Källén B (1986) A register study of maternal epilepsy and delivery outcome with special reference to drug use. Acta Neurol Scand 73(3): 253–259

Källén AJB (1994) Maternal carbamazepine and infant spina bifida. Reprod Toxical 8 (3): 203–205
Kalter H, Warkany J (1983) Congenital malformations. New Engl J Med 308: 491–497
Kaneko S, Fukishima Y, Sato T et al (1986) Teratogenicity of antiepileptic drugs – a prospective study. Jpn J Psychiatry Neurol 40: 447–450
Kaneko S, Otani K, Fukushima Y et al (1988) Teratogenicity of antiepileptic drugs: analysis of possible risk factors. Epilepsia 29: 459–467
Kaneko S, Otani K, Fukushima Y et al (1992) Malformation in infants of mothers with epilepsy receiving antiepiletics drugs. Neurology 42 [Suppl 5]: 68–74
Kaneko S, Otani K, Kondo T et al (1993) Teratogenicity of anticonvulsant drugs and drug specific malformations. Jpn J Psychiatry Neurol 47(2): 306–308
Kao J, Brown NA, Schmidt B et al (1981) Teratogenicity of valproic acid: in vivo and in vitro investigations. Teratog Carcinog Mutagen 1: 367–382
Kelly TE (1984) Teratogenicity of anticonvulsant drugs 1. Review of literature. Am J Med Genet 19: 413–434
Kerr BM, Martinez-Lage JM, Viteri C et al (1991) Carbamazepine dose requirements during stiripentol therapy: influence of cytochrome p450 inhibition by stiripentol. Epilepsia 32: 267–274
Knight AH, Rhind EG (1975) Epilepsy and pregnancy: a study of 153 pregnancies in 59 patients. Epilepsia 16: 99–110
Koch S, Losche G, Jäger-Roman E et al (1992) Major and minor birth malformations and antiepileptic drugs. Neurology 42 [Suppl 5]: 83–88
Koppe JG, Bosmon W, Oppers VM et al (1973) Epilepsie en aangeborn afwijkingen. Ned Tijdschr Genesskd 117: 220–224
Krauss CM, Holmes LB, Van Lang QC, Keith DA (1984) Four siblings with similar malformations after exposure to phenytoin and primidone. J Pediatr 105: 750–755
Kroetz DL, Loiseau P, Guyot M et al (1993) In vivo and in vitro correlation of microsomal epoxide hydrolase inhibition by progabide. Clin Pharmacol Ther 54: 458–497
Kubow S, Wells PG (1986) In vitro evidence for prostaglandin synthetase-catalyzed bioactivation of phenytoin to a free radical intermediate. Pharmacologist 28: 195
Kubow S, Wells PG (1989) In vitro bioactivation of phenytoin to a reactive free radical intermediate by prostaglandin synthetase, horseradish peroxidase and thyroid peroxidase. Mol Pharmacol 35: 504–511
Kuenssberg EV, Knox JDE (1973) Teratogenic effect of anticonvulsants. Lancet 1: 198
Lau K, Lee A, Ch'ien L (1994) Four siblings with malformations due to maternal epilepsy and anticonvulsants. J Tenn Med Assoc 87 (5): 193–194
Leavitt AM, Yerby MS, Robinson N et al (1992) Epilepsy and pregnancy: developmental outcomes at 12 months. Neurology 42 [Suppl 5]: 141–143
Levy RG, Loisseau P, Guyot M et al (1984) Stiripentol kinetics in epilepsy: nonlinearity and interactions. Clin Pharmacol Ther 36: 661–669
Lindhout D, Meinardi H (1984) Spina bifida and in-utero exposure to valproic acid. Lancet 2: 396
Lindhout D, Hoeppener RJEA, Meinardi H (1984) Teratogenicity of antiepileptic drug combinations with special emphasis on epoxidation (of carbamazepine). Epilepsia 25: 77–83
Lindhout D, Meinardi H, Meijer JWA, Nau H (1992) Antiepileptic drugs and teratogenesis in two consecutive cohorts: changes in prescription policy paralleled by changes in pattern of malformations. Neurology 42 [Suppl 5]: 94–110
Loughnan PM, Gold H, Vance JC (1973) Phenytoin teratogenicity in man. Lancet 1: 70–72
Lowe CR (1973) Congenital malformations among infants born to epileptic women. Lancet 1: 9–10
Lum JT, Wells PG (1986) Pharmacological studies on the potentiation of phenytoin teratogenicity by acetaminophen. Teratology 33: 53–72

Majewski F, Raft W, Fischer P et al (1980) Zur Teratogenität von Anticonvulsiva. Dtsch Med Wochenschr 105: 719–723
Marden PM, Smith DW, McDonald MJ (1964) Congenital anomalies in the newborn infant, including minor variations. J Pediatr 64: 357
Martin PJ, Millac PAH (1993) Pregnancy, epilepsy, management and outcome: a 10-year perspective. Seizure 2: 277–280
Martinez-Frias ML (1990) Clinical manifestation of prenatal exposure to valproic acid using case reports and epidemiologic information. Am J Med Genet 37: 277–282
Martz F, Failinger C, Blake D (1977) Phenytoin teratogenesis: correlation between embryopathic effect and covalent binding of putative arene oxide metabolite in gestational tissue. J Pharmacol Exp Ther 203: 231–239
Massey KM (1966) Teratogenic effects of diphenylhydantoin sodium. J Oral Ther Pharmacol 2: 380–385
Mast TJ, Nau H, Wittfoht W, Hendrickx AG (1987) Teratogenicity and pharmacokinetics of valproic acid in the rhesus monkey. In: Nau H, Scott WJ (eds) Pharmacokinetics in teratogenesis. CRC, Boca Raton, pp 131–147
McDevitt JM, Gautieri RF, Mann DE Jr (1981) Comparative teratology of cortisone and phenytoin in mice. J Pharm Sci 70: 631–634
McDonald AD (1994) Therapeutic drugs in early pregnancy and congenital defects. J Clin Epidemiol 47(1): 105–110
Meadow SR (1968) Anticonvulsant drugs and congenital abnormalities. Lancet 2: 1296
Meyer JG (1973) The teratological effects of anticonvulsants and the effects of pregnancy and birth. Eur Neurol 10: 179–180
Millar JHD, Nevin NC (1973) Congenital malformations and anticonvulsant drugs. Lancet 2: 328
Mills JL, McPartlin JM, Kirke DN et al (1995) Homocysteine metabolism in pregnancy complicated by neural-tube defects. Lancet 345: 149–51
Moldeus P, Jernstrom B (1983) Interaction of glutathione with reactive intermediates. In: Larsson A, Orrenius S, Holmgren A et al (eds) Functions of glutathione: biochemical, physiological, toxicological and clinical aspects. Raven, New York, pp 99–108
Monson RR, Rosenberg L, Hartz SC (1973) Diphenyhydantoin and selected congenital malformations. N Engl J Med 289: 1049–1052
Mullers-Kuppers von M (1963) Embryopathy during pregnancy caused by taking anticonvulsants. Acta Paedopsychiatr 30: 401–405
Musselman AC, Bennett GD, Greer KA et al (1994) Preliminary evidence of phenytoin-induced alterations in embryonic gene expression in a mouse model. Reprod Toxicol 8: 383–395
Nakane Y (1979) Congenital malformations among infants of epileptic mothers treated during pregnancy. Folia Psychiatr Neurol Jpn 33: 363–369
Nakane Y, Okuma T, Takahashi R et al (1980) Multi-institutional study on the teratogenicity and fetal toxicity of antiepileptic drugs: report of a collaborative study group in Japan. Epilepsia 21: 663–680
Nau H (1985) Teratogenic valproic acid concentrations: infusion by implanted minipumps vs conventional regimen in the mouse. Toxicol Appl Pharmacol 80: 243–250
Nau H (1991) Valproic acid induced NTDs in mouse and human: aspects of chirality, alternative drug development, pharmacokinetics, and possible mechanisms. Pharmacol Toxicol 69: 310–321
Nau H, Hendrickx AG (1987) Valproic acid teratogenesis Pharmacology, vol 1. Institute of Scientific Information, Philadelphia, pp 52–56 (ISI atlas of science)
Nau H , Loscher W (1986) Pharmacologic evaluation of various metabolites and analogs of valproic acid: teratogenic potencies in mice. Fundam Appl Toxicol 6: 669–676
Nau H, Scott WJ (1986) Weak acids may act as teratogens by accumulating in the basic milieu of the early mammalian embryo. Nature 323: 276–278

Nau H, Scott WJ (1987) Teratogenicity of valproic acid and related substances in the mouse: drug accumulation and pH_i in the embryo during organogenesis and structure-activity considerations. Arch Toxicol [Suppl]11: 128–139

Nelson KB, Ellenberg JH (1982) Maternal seizure disorder outcomes of pregnancy and neurologic abnormalities in the children. Neurology 32: 1247–1254

Neri A, Heifetz L, Nitke S et al (1983) Neonatal outcomes in infants of epileptic mothers. Eur J Obstet Gynecol Reprod Biol 16: 263–268

Netzloff ML, Streiff RR, Frias JL et al (1979) Folate antagonism following teratogenic exposure to diphenylhydantoin. Teratology 19: 45–50

Niedermeyer E (1983) Epilepsy guide: diagnosis and treatment of epileptic seizure disorders. Urban and Schwarzenberg, Baltimore

Niswander JD, Wertelecki W (1973) Congenital malformations among offspring of epileptic women. Lancet 1 (811): 1062

Niswander KR, Gordon E (1972) The women and their pregnancies. U.S. Department of Health, Education and Welfare. Saunders, Philadelphia

Oakeshott P, Hunt G (1994) Prevention of neural tube defects. Lancet 343: 123

Oguni M, Dansky L, Andermann E et al (1992) Improved pregnancy outcome in epileptic women in the last decade: relationship to maternal anticonvulsant therapy. Brain Dev 14(6): 371–380

Pantarotto C, Arboix M, Sezzano P et al (1982) Studies on 5,5-diphenylhydantoin irreversible binding to rat liver microsomal proteins. Biochem Pharmacol 31: 1501–1507

Paulson RB, Paulson GW, Jreissaty S (1979) Phenytoin and carbamazepine in production of cleft palates in mice: comparison of teratogenic effects. Arch Neurol 36: 832–836

Paulson RB, Sucheston ME, Hayes TW et al (1985) Teratogenic effects of valproate in the CD-1 mouse fetus. Arch Neurol 42: 980–983

Petrere JA, Anderson JA, Sakowski R et al (1986) Teratogenesis of calcium valproate in rabbits. Teratology 34: 263–269

Phelan MC, Pellock JM, Nance WE (1982) Discordant expression of fetal hydantoin syndrome in heteropaternal dizygotic twins. N Engl J Med 307: 99–101

Philbert A, Dam M (1982) The epileptic mother and her child. Epilepsia 23: 85–99

Pisani F, HajYehia A, Fazio A et al (1993) Carbamazepine-valnoctamide interaction in epileptic patients: in vitro/in vivo correlation. Epilepsia 34(5): 954–959

Rating D, Jager-Roman E, Koch S et al (1987) Major malformations and minor anomalies in infants exposed to different anticonvulsants during pregnancy. In: Wolf P, Dam M, Janz D et al (eds) Advances in epileptology: the 16th international epilepsy symposium. Raven, New York, 16: 561–565

Rivey MP, Schottelius DD, Berg MJ (1984) Phenyotin-folic acid: a review. Drug Intell Clin Pharm 18: 292–301

Robert E (1982) Valproic acid and spina bifida: a preliminary report-France. MMWR Morb Mortal Wkly Rep 31: 565–566

Robert E, Guibaud P (1982) Maternal valproic acid and congenital neural tube defects. Lancet 11: 937

Robert E, Lofkvist E, Mauguiere F et al (1986) Evaluation of drug therapy and teratogenic risk in a Rhone-Alps district population of pregnant epileptic women. Eur Neurol 25: 436–443

Rosa FW (1991) Spina bifida in infants of women treated with carbamazepine during pregnancy. N Engl J Med 324(10) 647–677

Rudd NL, Freedom RM (1979) A possible primidone embryopathy. J Pediatr 94: 835–837

Sabin M, Oxorn H (1956) Epilepsy and pregnancy. Obstet Gynecol 7(2): 175–179

Scolnik D, Nulman I, Rovet J et al (1994) Neurodevelopment of children exposed in utero to phenytoin and carbamazepine monotherapy. JAMA 271 (10): 767–770

Scott J, Kirke P, Molloy A et al (1994) The role of folate in the prevention of neural-tube defects. Proc Nutrit Soc 53: 631–636

Seino M, Miyakoshi M (1979) Teratogenic risks of antiepileptic drugs in respect to the type of epilepsy. Folia Psychiatri Neurol Jpn 33(3): 379–385
Seller MJ, Embury S, Polani PE et al (1979) Neural tube defects in curly-tail mice. 2. Effect of maternal administration vitamin A. Proc R Soc Lond Biol Sci 206: 95–107
Sharony R, Garber A, Viscochil D et al (1993) Preaxial ray reduction defects as part of valproic acid embryofetopathy. Prenat Diagn 13: 909–918
Smith DW (1980) Hydantoin effects on the fetus. In: Hassell TM, Johnston MC, Dudley KH (eds) Phenytoin-induced teratology and gingival pathology. Raven, New York, pp 35–40
Smithells RW, Sheppard S, Schorah CJ (1976) Vitamin deficiencies and neural tube defects. Arch Dis Child 51: 944–950
Snyder R, Parke DV, Kocsis JJ, Jollow DJ et al (eds) (1982) Biological reactive intermediates. 2. Chemical mechanisms and biological effects, parts A and B Plenum, New York
Sonoda T, Ohdo S, Ohba K et al (1993) Sodium valproate-induced cardiovascular abnormalities in the Jcl: ICR mouse fetus: peak sensitivity of gestational day and dose-dependent effect. Teratology 48: 127–132
South J (1972) Teratogenic effects of anticonvulsants. Lancet 2: 1154
Speidel BD, Meadow SR (1972) Maternal epilepsy and abnormalities of the fetus and newborn. Lancet 2: 839–843
Spellacy WN (1972) Maternal epilepsy and abnormalities of the fetus and newborn. Lancet 2: 1196–1197
Srinivasan G, Seeler RA, Tiruvury A et al (1982) Maternal anticonvulsant therapy and hemorrhagic disease of the newborn. Obstet Gynecol 59: 250–252
Staats J (1976) Standardized nomenclature for inbred strains of mice: sixth listing. Cancer Res 36: 4333–4377
Stanley FJ, Priscott PK, Johnston R et al (1985) Congenital malformations in infants of mothers with diabetes and epilepsy in Western Australia, 1980–82. Med J Aust 143: 440–442
Starreveld-Zimmerman AAE, Van Der Kolk WJ, Meinardi H (1973) Are anticonvulsants teratogenic? Lancet 2: 48–49
Stone ML (1968) Effects on the fetus of folic acid deficiency in pregnancy. Clin Obstet Gynec 11: 1143–1153
Strickler SM, Dansky LV, Miller MA et al (1985) Genetic predisposition to phenytoin-induced birth defects. Lancet 2: 746–749
Sucheston ME, Hayes TG, Eluma FO (1986) Relationship between ossification and body weight of the CD-1 mouse fetus exposed in utero to anticonvulsant drugs. Teratog Carcinog Mutagen 6: 537–546
Sulik KK, Johnston MC, Ambrose LJH et al (1979) Phenytoin-induced cleft lip and palate in A/J mice: a scanning and transmission electron microscopic study. Anat Rec 195: 243–256
Sullivan FM, McElhatton PR (1975) Teratogenic activity of the antiepileptic drugs phenobarbital, phenytoin, and primidone in mice. Toxicol Appl Pharmacol 34: 271–282
Sullivan FM, McElhatton PR (1977) A comparison of the teratogenic activity of the antiepileptic drugs carbamazepine, clonazepam, ethosuximide, phenobarbital, phenytoin, and primidone in mice. Toxicol-Appl Pharmacol 40: 365–378
Svigos JM (1984) Epilepsy and pregnancy. Aust NZ J Obstet Gynaecol 24: 182–85
Tanganelli P, Regesta G (1992) Epilepsy, pregnancy, and major birth anomalies: an Italian prospective, controlled study. Neurology 42 [Suppl 5]: 89–93
Teramo K, Kuusisto AN, Raivio KO (1979) Perinatal outcome of insulin-dependent diabetic pregnancies. Ann Clin Res 11(4): 146–155
Tomson T, Lindborn U, Ekquist B et al (1994) Epilepsy and pregnancy: a prospective study of seizure control in relation to free and total plasma concentrations of carbamazepine and phenytoin. Epilepsia 35 (1): 122–130

Trotz M, Wegner C, Nau H (1987) Valproic acid-induced neural tube defects: reduction by folinic acid in the mouse. Life Sci 41: 103–110
Van Dyke DC, Hodge SE, Helde F et al (1988) Family studies in fetal phenytoin exposure. J Pediatr 113: 301–306
Vanoverloop D, Schnell RR, Harvey EA et al (1992) The effects of prenatal exposure to phenytoin and other anticonvulsants on intellectual function at 4 to 8 years of age. Neurotoxicol Teratol 14(5): 329–335
Visser GH, Hisjes HJ, Elshove J (1976) Anticonvulsants and fetal malformations. Lancet 1: 970
Vorhees CV (1983) Fetal anticonvulsant syndrome in rats: dose- and period- response relationships of prenatal diphenylhydantoin, trimethadione and phenobarbital exposure on the structural and functional development of the offspring. J Pharmacol Exp Ther 227: 274–287
Vorhees CV (1986) Fetal anticonvulsant exposure effects on behavioral and physical development. Ann NY Acad Sci 477: 49–62
Vorhees CV (1987) Teratogenicity and developmental toxicity of valproic acid. Teratology 35: 195–202
Vorhees CV, Acuff KD, Weisenburger WP et al (1990) Teratogenicity of carbamazepine in rats. Teratology 41: 311–317
Waters CH, Belai Y, Gott PS et al (1994) Outcomes of pregnancy associated with antiepileptic drugs. Arch Neurol 51: 250–253
Weber M, Schweitzer M, Mur JM et al (1977) Epilepsie medicaments antiepileptiques et grossesse. Arch Fr Pediatr 34: 374–383
Wegner C, Nau H (1991) Diurnal variation of folate concentrations in mouse embryo and plasma: the protective effect of folinic acid on valproic acid-induced teratogenicity is time-dependent. Reprod Toxicol 5: 465–471
Wegner C, Nau H (1992) Alteration of embryonic folate metabolism by valproic acid during organogenesis: implications for mechanism of teratogenesis. Neurology 42 [Suppl 5]: 17–24
Wells PG, Harbison RD (1980) Significance of the phenytoin reactive arene oxide intermediate, its oxepin tautomer, and clinical factors modifying their roles in phenytoin-induced teratology. In: Hassell TM, Johnston MC, Dudley KH (eds) Phenytoin-induced teratology and gingival pathology. Raven, New York, pp 83–108
Wells PG, Vo HPN (1989) Effects of the tumor promoter 12-O-tetradecanolylphorbol-13-acetate on phenytoin-induced embryopathy in mice. Toxicol Appl Pharmacol 97: 398–405
Wells PG, Kupfer A, Lawson JA et al (1982) Relation of in vivo drug metabolism to stereoselective fetal hydantoin toxicology in the mouse: evaluation of mephenytoin and its metabolite nirvanol. J Phamacol Exp Ther 221: 228–234
Wells PG, Zubovits JT, Wong ST et al (1989) Modulation of phenytoin teratogenicity and embryonic covalent binding by acetylsalicyclic acid, caffeic acid, and a-phenyl-*N*-t-butylnitrone: implications for bioactivation by prostaglandin synthetase. Toxicol Appl Pharmacol 97: 192–202
Yamatogi Y, Oka E, Satoh M et al (1993) A prospective follow-up of the offspring of epileptic patients. Jpn J Psychiatry Neurol 47 (2): 309–311
Yates JR, Ferguson-Smith MA, Shenkin A et al (1987) Is disordered folate metabolism the basis for the genetic predisposition to neural tube defects? Clin Genet 31: 279–287
Yerby MS, Friel PN, Miller DQ (1985) Carbamazepine protein binding and disposition in pregnancy. Ther Drug Monit 7: 269–273
Ylagan LR, Budorick NE (1994) Radial ray aplasia in utero: a prenatal finding associated with valproic acid exposure. J Ultrasound Med 13: 408–411
Zachai EH, Mellman WJ, Neiderer B et al (1975) The fetal trimethadione syndrome. J Pediatr 87: 280–284

CHAPTER 24

Cardiovascular Active Drugs

B.R. DANIELSSON and W.S. WEBSTER

A. Introduction

There are many classes of therapeutic agents whose primary action is on the cardiovascular system. These include antihypertensive drugs (e.g., beta-blocking agents and calcium antagonists), antiarrhythmics (e.g., potassium channel inhibitors), and sympathomimetic drugs (e.g., epinephrine). Other substances, such as ergotamine and vasopressin, are vasoactive and are used for special indications. In addition, other drugs whose primary action is on other organ systems (mainly the central nervous system) such as phenytoin, caffeine, nicotine, and the narcotic drug cocaine also exert pharmacological action on the cardiovascular system as a side effect.

Drugs in all of the above-mentioned classes have been reported to cause adverse effects in fetuses ranging from mild effects, such as transient bradycardia, to severe malformations in animal and/or human studies. There is evidence from mechanistically oriented experimental studies that malformations and other adverse fetal effects induced by these drugs are secondary to hemodynamic alterations causing fetal hypoxia/ischemia (mediated via pharmacological action) rather than to direct toxic effects of the drug or its metabolites. In this chapter, the term hypoxia is used as a common expression for oxygen deprivation, whether caused by lack of perfusion (ischemia) or by aberrant oxygenation (hypoxic hypoxia). The drugs are classified in relation to how they exert their pharmacological effect on the cardiovascular system. The clinical relevance of experimental studies is discussed in relation to doses, plasma concentrations, proposed mechanism of induction of the adverse effect, and similarities (and differences) in malformation pattern/observed adverse fetal effects in animal studies compared with human studies.

B. Uterine Vessel Clamping

Mechanical impairment of the uteroplacental blood flow in the rat by clamping the uterine blood vessels causes fetal bradycardia, hypoxia, and malformations. The early postimplantation rat embryo, i.e., gestational days (GD) 6–11, is relatively resistant to induced hypoxia, and many embryos survive 90–120 min of uterine vessel clamping (FRANKLIN and BRENT 1964;

LEIST and GRAUWILER 1973a). A low incidence of congenital malformations is seen in the surviving fetuses, including situs inversus, omphalocele, gastroschisis, microphthalmia, and heart defects.

By GD 12 the embryos become more sensitive, and on days 13–16 most embryos are unable to survive 60 min of hypoxia caused by vascular clamping. Shorter periods of hypoxia on these days (30–45 min) induced a range of malformations. The most common are distal digital defects (Fig. 1a) such as shortened or absent metatarsals, metacarpals, phalanges, and nails, dislocated digits, and syndactyly. These defects range in severity from loss of the entire footplate to missing nails. Other defects, less frequently seen, include shortened tail, cleft palate, and abnormalities of the nose and genital tubercle (LEIST and GRAUWILER 1974; WEBSTER et al. 1987). Histological examination of the brain from some fetuses after treatment on GD 16 revealed cavitation defects in the cerebral cortex, basal ganglia, brain stem, and eye (WEBSTER et al. 1991). Pathological examination (2–24 h after the clamping procedure) showed that the malformations are preceded by edema, dilated blood vessels, hemorrhage, and blisters (Fig. 2a) and subsequent development of necrosis (LEIST and GRAUWILER 1974; WEBSTER et al. 1987). After GD 16, the fetus appears to become less sensitive to hypoxia induced by uterine vessel clamping (LEIST and GRAUWILER 1973a).

C. Vasoactive Drugs

In addition to hypoxia induced by mechanical means, there is extensive evidence that severe reduction in the uterine blood can be induced pharmacologically by vasoactive drugs. Control of the uterine blood flow is primarily by sympathetic, α-adrenergic vasoconstrictors, with the usual state being a maximally dilated vasculature. Drugs that cause decreased uteroplacental blood flow are potentially both embryotoxic and teratogenic, as discussed below. The reproductive outcome of exposure to these drugs depends on whether the induced hypoxia/ischemia is above threshold in severity and duration for the embryonic day of exposure. As seen in the uterine vessel-clamping experiments, the most sensitive days appear to be in the late organogenic or early fetal period, and the most likely malformations are limb defects, particularly missing or reduced digits and/or nails.

I. Vasoconstrictors

1. Sympathomimetic Drugs

Epinephrine is a sympathomimetic drug that acts on both α- and β-receptors, but is a more potent α-receptor activator. As expected, when administered intravenously to pregnant ewes it is a potent constrictor of the uterine blood vessels (GREISS 1963). Although comprehensive teratology studies of epinephrine are not available, teratogenic effects have been demonstrated (JOST

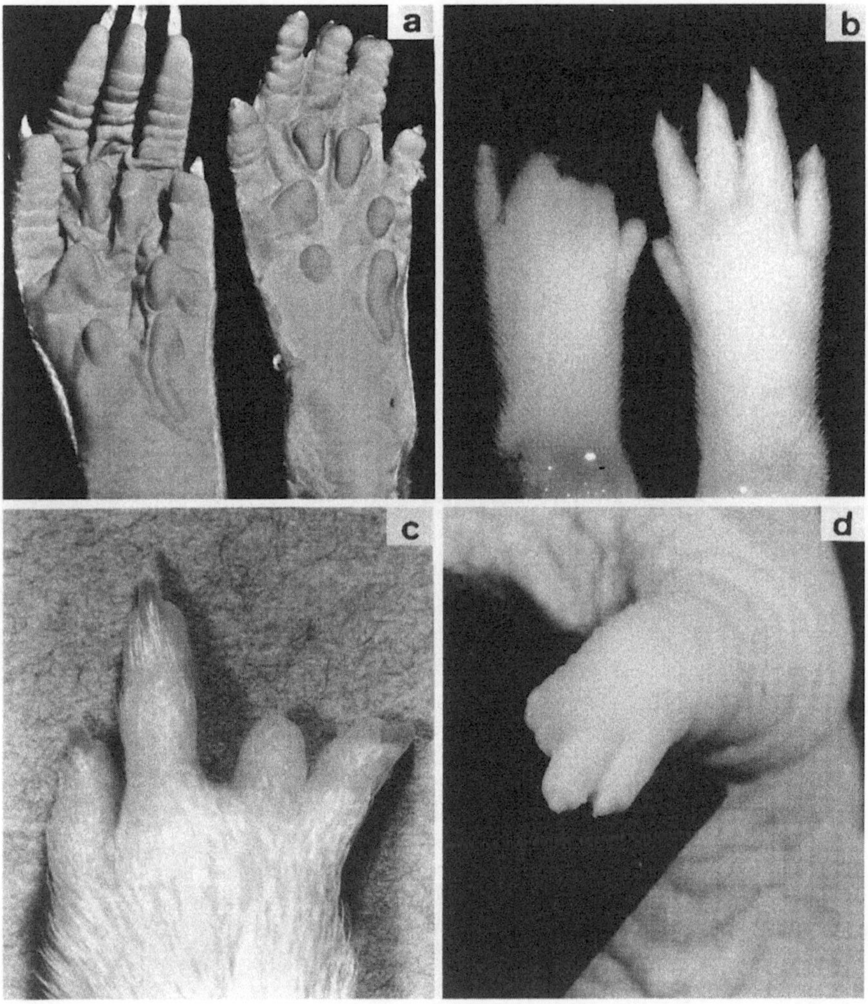

Fig. 1a-d. Digital amputation defects caused by **a** uterine vessel clamping **b** cocaine, **c** almokalant, and **d** dofetilide. **a** The hindlimb on the *right* is from a 6-week-old rat after uterine vessel clamping on day 16 of gestation. Compared with the hindlimb from a control 6-week-old rat on the *left*; note the distal hypoplasia and proximal syndactyly of the digits. **b** Right and left hindlimbs of a 5-week-old from a dam given cocaine (60 mg/kg) on day 16 of gestation. Note the reduction of the second and fourth digits of the left limb. **c** Left forelimb of a 3-week-old rat from a dam given almokalant (25 μmol/kg) during gestation. Note reductions of all digits. **d** Left forelimb from a rat fetus 7 days after a single oral dose of dofetilide (25 μmol/kg) on day 13 of gestation. Note the loss of the second and third digits. (Fig. 1a,b,d from WEBSTER et al. 1987, 1990, 1996 by courtesy of Teratology; Fig. 1c from DANIELSSON 1993 by courtesy of LINFO)

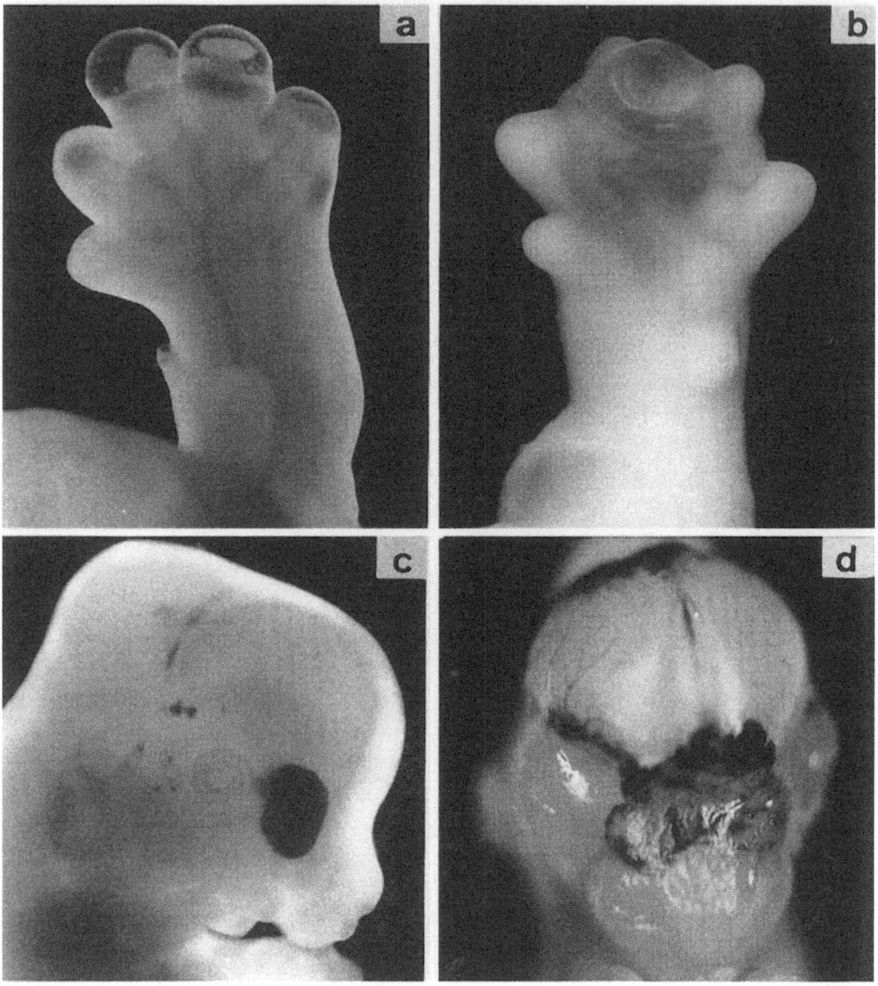

Fig. 2a-d. Early changes after **a** uterine vessel clamping, **b** cocaine, **c** dofetilide, and **d** phenytoin. **a** Hindlimb from an 18-day-old rat fetus 48 h after uterine vessel clamping. Note hemorrhage and edema of the third to fifth digits. **b** Palmar surface of the forelimb of a rat fetus from a dam given a single intraperitoneal dose of cocaine (60 mg/kg) 48 h earlier on gestational day 16. Note the large blood-filled blister affecting the second to fourth digits. **c** Lateral view of a rat fetus 96 h after a single oral dose of dofetilide (50 μmol/kg) on day 11 of gestation. The fetus has a right-sided oblique facial cleft with a large hemorrhage at the upper end of the cleft. The left side of the fetus appeared to be normal. **d** Hemorrhage in the nasal and frontal region on day 18 of a rabbit fetus 48 h after last dose of phenytoin (150 mg/kg orally during days 14–16 of gestation). (Fig. 2a-c from Webster et al. 1987, 1990, 1996; and Fig. 2d from Danielson et al. 1992 by courtesy of Teratology; copyright by Wiley-Liss Inc)

1953). Epinephrine was injected into the peritoneal cavity of pregnant rats or rabbits on GD 17 and the fetuses were examined 2–4 days later. There were hematomas on the head and extremities of the fetuses, and the latter led to distal limb amputation defects, such as are regularly seen following uterine vessel clamping. In an investigation of the mechanism of epinephrine-induced teratogenesis, CHERNOFF and GRABOWSKI (1971) showed that an intraperitoneal or intrajugular injection of epinephrine (25 μg) into a 16-day-pregnant rat caused fetal bradycardia, which began 5–20 min after injection and lasted up to 65 min. The fetal heart rate decreased from 140–180 bpm to a mean of 35 bpm. Similar effects could be produced in rats of 15–19 days gestation. The reduction of the fetal heart rate was accompanied by reduced fetal blood pressure. On day 14 the average blood pressure was 28.3 mm water, rising to 63.5 mm water on day 18. A reduction of fetal heart rate from 145 to 50 reduced blood pressure from 54 mm to 20–27 mm water.

Direct injection of epinephrine into the pericardial cavity of a day-16 fetus did not cause fetal bradycaradia; instead it increased fetal blood pressure. This indicated that the fetal bradycardia induced by intraperitoneal injection of epinephrine into the pregnant rat was secondary to an effect on the maternal system. The authors proposed two possible mechanisms by which epinephrine caused fetal bradycardia: either by contracting the uterus or by producing generalized vasoconstriction of the uterine blood vessels. Vasoconstriction was the favored mechanism, as doses of oxytocin sufficient to cause uterine contractions did not affect fetal heart rate (MARTIN and YOUNG 1960).

Although epinephrine is a widely used drug, human teratogenicity has not been suspected (BRIGGS et al. 1990). Theoretically, a large dose could cause a transient decrease in uterine blood flow. A case has been reported in which epinephrine (1.5 ml of a 1:1000 solution over 1 h) was administered to a 28-week-pregnant patient to reverse severe hypotension secondary to an allergic reaction. Decreased fetal movements occured after treatment, and the infant delivered at 34 weeks had evidence of intracranial hemorrhage and died at 4 days (ENTMAN and MOISE 1984).

Norepinephrine is a potent agonist at α-receptors and, like epinephrine, is a potent constrictor of uterine blood vessels (GREISS 1963). Standard teratology studies are not available for norephinephrine, and its teratogenic potential has not been demonstrated. However, as discussed in a later section, the teratogenic activity of cocaine is thought to be mediated by increased circulating levels of norepinephrine. Intravenous administration of norepinephrine to pregnant rhesus monkeys caused severe asphyxia of the fetus (ADAMSONS et al. 1971). Vasoconstriction of the uterine circulation was identified as the central factor responsible, since direct injection of catecholamines into the fetus produced no adverse effects on fetal acid–base state.

Pseudoephedrine, a sympathomimetic agent with both α- and β-agonist activity, could potentially reduce uterine blood flow. The over-the-counter availability of this drug, usually in conjunction with antihistamines and other ingredients, results in extensive, unsupervised use. However, no relationship between first-trimester use of pseudoephedrine and congenital malformations was noted in 421 pregnancies (Aselton et al. 1985).

2. Vasopressin (Antidiuretic Hormone)

Vasopressin (antidiuretic hormone ADH) is formed in the hypothalamus and released by the posterior pituitary into the circulation in response to increased plasma osmolality and decrease in extracellular volume. ADH acts directly on the kidney, and failure to secrete adequate amounts causes diabetes insipidus. At high concentrations, vasopressin causes general vasoconstriction. This effect is not antagonized by adrenergic blocking agents nor by vascular denervation. Intraperitoneal injection of vasopressin (10–40 mU) into pregnant rats on GD 17 caused hemorrhage in the fetal digits, leading to typical amputation-type defects (Jost 1951). In an investigation of the mechanism of vasopressin-induced teratogenesis, Chernoff and Grabowski (1971) showed that vasopressin (0.5 units) injected intraperitoneally in 16-day-pregnant rats caused fetal bradycardia, which began 5–20 min after injection and lasted up to 65 min. The same decrease in fetal heart beat and fetal blood pressure as occurred after a dose of 25 μg adrenalin was observed (Chernoff and Grabowski 1971).

Vasopressin or synthetic analogues such as desmopressin (which has decreased pressor activity) have been used to treat diabetes insipidus in humans during pregnancy, but no adverse effects have been reported. There have been no reports linking birth defects in the human with the use of vasopressin, desmopressin, or lypressin (Briggs et al. 1990).

3. Ergotamine

Ergotamine is an alkaloid prepared from ergot, the product of a fungus (*Claviceps purpurea*) that grows on rye and other cereals. The predominant action of ergotamine is peripheral vasoconstriction, and its main therapeutic use is in the treatment of migraine; this may be due to its direct vasoconstrictive action on dilated extracerebral arteries. Ergotamine also has uterotonic activity and on this basis is not recommended for use during pregnancy. The effect of ergotamine on vascular and uterine smooth muscle is thought to be mediated by α-adrenergic receptors and/or tryptaminergic receptors (Rall 1991). At high doses in rats, the drug causes increased blood pressure from stimulation of α_2-adrenergic receptors in arterioles. With the exception of the brain, the increased vascular resistance is accompanied by decreased blood flow in various organs.

In a standard teratology study, ergotamine was administered to rats and mice by daily oral doses on GD 6–15 and in rabbits by daily dosing on GD 6–18 (Grauwiler and Schon 1973). At doses that affected maternal weight gain during treatment, ergotamine increased resorptions in rats and caused fetal retardation in all three species. However, no specific teratogenic activity was observed. In a subsequent study (Schon et al. 1975), ergotamine was administered to pregnant rats on a single day between GD 4 and 19. No adverse effects were seen from days 4 to 10, but from day 11 onwards prenatal mortality increased, reaching a maximum on day 14. In addition, 10-mg/kg doses given on days 13, 14, 15, or 16 caused characteristic anomalies, including cleft palate and bilateral limb defects (shortening or absence of nails, phalanges, and entire digits). The authors commented on the similarity between the abnormalities seen after ergotamine and those seen after uterine vessel clamping. They also demonstrated that the effect of ergotamine could be completely antagonized by α-adrenoceptor blockade with phenoxybenzamine (SCHON et al. 1975). Intra-aminotic doses of ergotamine, giving embryonic concentrations 50 times higher than those achieved after oral dosing, were needed to produce a similar embryolethal effect (LEIST and GRAUWILER 1973b), implying that the observed fetal effects were secondary to an effect of ergotamine on the dam. The uterotonic activity of ergotamine was not thought to be involved in the pathogenesis, as ergometrine did not inhibit uteroplacental blood flow despite having uterotonic activity comparable to ergotamine (LEIST and GRAUWILER 1973b).

In human pregnancy, there has been a report of two cases of Poland anomaly in infants born to women who had attempted abortion with ergotamine or derivative (DAVID 1972). Poland anomaly consists of unilateral absence of the sternocostal part of the greater pectoral muscle and ipsilateral syndactyly and may include absence of digits; it is thought to have a vascular-based pathogenesis. Two cases of disruptive vascular etiologies leading to malformations have been associated with intake of tablets containing ergotamine and caffeine (GRAHAM et al. 1983) and after therapy with ergotamine, caffeine, and a beta blocker (HUGES and GOLDSTEIN 1988), suggesting possible additive or synergistic effects.

4. Cocaine

It is well established that cocaine blocks the presynaptic uptake of catecholamines (dopamine and norepinephrine), allowing these neurotransmitters to accumulate in the postsynaptic cleft and in the systemic circulation. Studies in the pregnant ewe (MOORE et al. 1986; WOODS et al. 1987) and the pregnant baboon (MORGAN et al. 1991) have shown that intravenously administered cocaine causes a rise in maternal arterial blood pressure and a dose-dependent fall in uterine blood flow. These vascular changes are related to cocaine-induced increased circulating levels of norepinephrine and possibly dopamine and a postulated direct effect of cocaine on the uterine vasculature (MORGAN et al. 1991).

There is clear evidence that cocaine is teratogenic in the rat. A single large dose of cocaine was teratogenic when administered to pregnant rats between GD 14 and 19, but not when administered at earlier stages (WEBSTER and BROWN-WOODMAN 1990). The most common defects were reduction deformities of the limbs (see Fig. 1b) and tail. Examination of the fetuses 48 h after cocaine dosing revealed that the defects were preceded by hemorrhage, edema, and necrosis in the same structures (see Fig. 2b). Subsequent examination of the brains of the fetuses showed that the cocaine caused hemorrhage, necrosis, and cavitation in the fetal cerebral cortex, corpus striatum, brain stem, and retina (WEBSTER et al. 1991). The lens of the eye showed vacuolization. These defects were not lethal, and many offspring with severe limb defects were easily reared to adulthood. The limb and central nervous system (CNS) defects seen in the rat fetuses following cocaine exposure were identical with the defects seen after uterine vessel clamping on the same GD. It was proposed that cocaine causes severe constriction of the uterine vasculature, leading to an hypoxic response in the placental/fetal unit which causes the observed hemorrhage and edema. The hypoxia induced by uterine vessel constriction in the pregnant rat may be further augmented by a direct cardiotoxic effect of cocaine on the maternal heart, producing a concentration-dependent negative inotrophy (SHARMA et al. 1992).

There is some clinical evidence that cocaine abuse during human pregnancy causes an increased risk of adverse reproductive outcome, including spontaneous abortion, abruptio placentae, prematurity and low birth weight for gestational age, and transitory behavioral disorders in the newborn (CHASNOFF et al. 1985; MACGREGOR et al. 1987; BINGOL et al. 1987). A small number of congenital malformations have also been reported, including prune belly syndrome, ileal atresia, hypospadias, and limb malformations (CHASNOFF et al. 1988; HOYME et al. 1990; CHAVEZ et al. 1988). The limb malformations included two unilateral terminal transverse limb reductions of the upper extremities just distal to the elbow, missing medial ray of the hand, missing third, fourth, and fifth digits of the hand, two asymmetric radial ray anomalies, and one symmetrical bilateral reduction of the arms with only a single forearm bone and digit present (HOYME et al. 1990).

The incidence of these defects or other adverse outcomes in relation to the incidence of cocaine use in pregnancy is difficult to calculate, but has probably been overestimated (HUTCHINGS 1993). There have been a number of studies that have failed to show any increased risk of malformations with cocaine use in pregnancy. This suggests that cocaine-induced malformations are rare and probably only occur following extremely high-dose usage at critical times of pregnancy. There is also the important point that may cocaine users are polydrug users (e.g., alcohol, marijuana, cigarettes); hence the specific contribution of cocaine to the observed adverse outcomes is difficult to isolate (HUTCHINGS 1993).

There have been a number of reports of severe brain damage in human neonates following maternal cocaine use during pregnancy (CHASNOFF et al. 1986; CHASNOFF and GRIFFITH 1989; DIXON and BEJAR 1989; GREENLAND et al. 1989; HOYME et al. 1990). In the most detailed study, out of 32 neonates exposed to continued cocaine use throughout gestation (affirmed by urine toxicology at birth), 15 had CNS abnormalities detected by ultrasonography (DIXON and BEJAR 1989). Some of the lesions were suggestive of prior hemorrhage and ischemic injury with cavitation in the white matter. These 3- to 10-mm cystic lesions were sometimes bilateral and were concentrated anterior and inferior to the lateral ventricles, in the frontal lobes, and in the basal ganglia. These findings of "old" infarction with cavitation were interpreted as evidence of brain injury prior to birth, perhaps due to drug-induced vasoconstriction of the branches of the middle cerebral artery. This pattern of CNS damage is very similar to that seen in the rat teratology study (WEBSTER et al. 1991). There was also evidence of intraventricular, subarachnoid, and subependymal hemorrhage in the human neonates, indicating further vascular reactions nearer the time of birth. The medical report does not mention whether the neonates were examined for eye defects. There has also been a report describing fetuses aborted from cocaine-addicted mothers (KAPUR et al. 1991). Significant pathologic lesions were found only in the brain, with three out of four cases examined showing hemorrhages in the ventricular wall and randomly in other locations of the cerebral parenchyma.

5. Nicotine

Nicotine is an alkaloid and a potent agonist at nicotinic receptors in the CNS. It also stimulates the adrenal medulla, causing the release of catecholamines (epinephrine and nonepinephrine). Subcutaneous injection of nicotine (0.5 or 5 mg/kg) in the pseudopregnant rat constricts the uterine blood vessels, causing a decrease in uterine blood flow of up to 40% for 90 min and reducing intrauterine oxygen tension by over 50% (HAMMER et al. 1981). Studies using pregnant sheep showed that maternal intravenous injection of nicotine at 30 μg/kg per min decreased uterine blood flow by 42%, increased fetal blood pressure by 25%, and decreased fetal heart rate by 12% (CLARK and IRION 1992). The vasoconstrictive effect of nicotine in sheep has been attributed primarily to catecholamine release (RESNICK et al. 1979). Similar studies using pregnant rhesus monkeys with an infusion rate of 100 μg nicotine/kg per min decreased uterine blood flow by as much as 38% and induced acidosis and hypoxia in the fetus (SUZUKI et al. 1980). In humans, cigarette smoking has been reported to increase maternal catecholamine levels and to increase the fetal heart rate (QUIGLEY et al. 1979). Increased catecholamine levels have not been found in all studies (LINDBLAD et al. 1988), and no change in uterine artery vascular resistance has been detected (MORROW et al. 1988; BRUNER and FOROUZAN 1991). Although appropriate measurements are difficult in humans, these results suggest that cigarette smoking does not cause an acute change in uterine blood flow. Since nicotine crosses the placenta, an alternative or additional proposed mechanism is that nicotine affects fetal hemodynamics directly or indirectly by increasing fetal catecholamine levels (LINDBLAD et al. 1988).

There have been relatively few teratology studies using nicotine. NISHIMURA and NAKAI (1958a) gave pregnant mice subcutaneous doses of nicotine (25 mg/kg) on a single day of gestation or on two or three consecutive days.

Limb and digital defects were induced when nicotine was administered on any day from GD 7 to 14. Most of the defects were described as malformation of a joint, but there were a few fetuses with brachydactyly. Higher doses given to rats reduced fetal weight and viable litters, but did not induce malformations (HUDSON and TIMIRAS 1972). There were no reported teratogenic effects in rabbits (VARA and KINNUNEN 1951) or in cattle or swine (KEELER 1980).

In humans, maternal cigarette smoking is associated with abruptio placentae, fetal growth retardation, and increased perinatal mortality but not birth defects (NIEBYL 1988). It is likely that maternal and/or fetal catecholamine release in response to smoking is insufficient to produce the severe hypoxia needed to induce malformations.

II. Vasodilators

1. Calcium Channel Antagonists and Hydralazine

Vasodilators are mainly used as antihypertensive agents by decreasing the resistance in peripheral vessels, resulting in a decrease in mean arterial blood pressure. However, a risk for decreased uteroplacental blood flow and decreased oxygen saturation in fetal blood secondary to maternal hypotension and a diversion of blood flow from the placenta to peripheral vascular beds may exist. Such effects have been observed in animal studies in late pregnancy after administration of hydralazine in the ewe (LADNER et al. 1970) and the calcium antagonists nifedipine in the ewe (HARAKE et al. 1987) and nicardipine in the rhesus monkey and the rabbit (DUCSAY et al. 1987; LIRETTE et al. 1987). Similar findings were observed after administration of another calcium antagonist, felodipine, administered during late organogenesis and the postorganogenic period (GD 14–16) in rabbits. A marked decrease in maternal blood pressure (approximately 30%) and uteroplacental flow (40%–50%) was observed for more than 6 h after dosing; at the same time, the blood flow increased up to sixfold in peripheral vascular beds such as muscles (LUNDGREN et al. 1992).

In experimental studies in different species (rat and/or rabbit), the vasodilating calcium antagonists nifedipine (FUKUNISHI et al. 1980; DANIELSSON et al. 1989), felodipine, nitrendipine (DANIELSSON et al. 1989), nicardipine (YOSHIDA et al. 1989), and the chemically unrelated hydralazine (DANIELSSON et al. 1989) have induced structural abnormalities, mainly characterized as abnormal structure and hypoplasia of the distal parts of the digits (see Fig. 3a). In a comparative study of vasodilators (calcium antagonists and hydralazine), identical phalangeal defects were observed. No digital defects were seen after administration of the inactive first-step metabolite of one of the calcium antogonists (H152/37; see Table 1). Distal phalangeal defects have also been reported for the vasodilators diltiazem (ARIYUKI 1975) and flunarazine (YOSHIDA et al. 1989). The sensitive period for induction of the defects seems to be late organogenesis and early postorganogenesis (GD 13–16 in the rat and GD 14–17 in the rabbit).

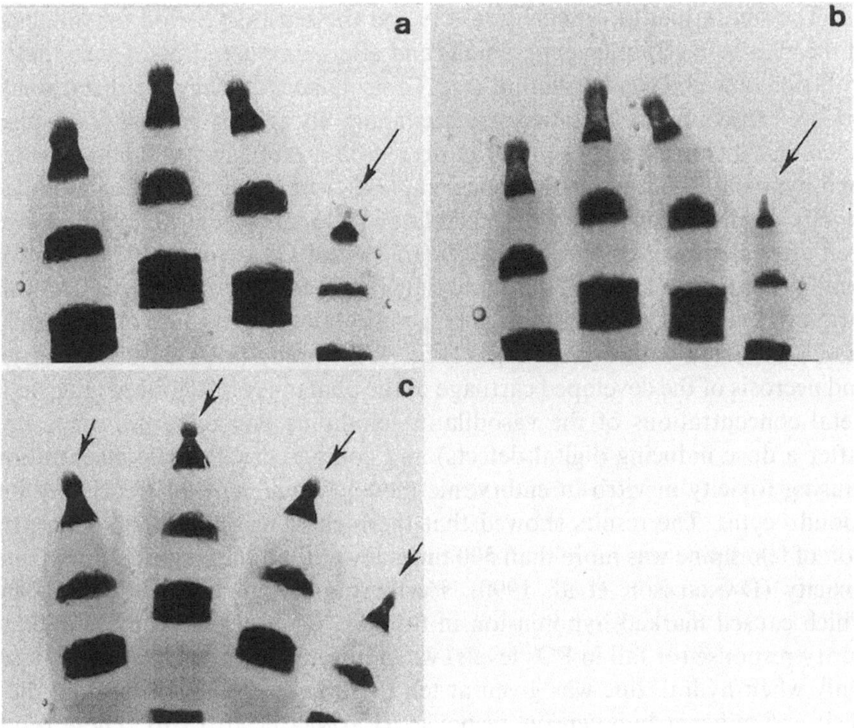

Fig. 3a-c. Identical distal digital hypoplasia (arrows) of rabbit fetuses at term on the fourth digit on the hindpaw after a single oral dose of 33.2 mg nifedipine/kg (**a**) and 150 mg phenytoin/kg (**b**) on day 16. Also note abnormal macroscopical structure of other phalanges, especially on the distal phalanx of the third digit. Hypoplasia on all distal phalanges (*arrow*) of the forepaw after repeated administration of phenytoin (150 mg/kg) on days 14–17 of gestation (**c**). Alizarine-stained skeletons. (From DANIELSSON et al. 1989 and DANIELSON et al. 1992 by courtesy of Teratology; copyright by Wiley-Liss Inc.)

Table 1. Observed defects in term fetuses of the distal phalanx of the fourth digit of the hindpaws after single oral administration of different vasodilators on day 16 in the rabbit

Test substance	Dose (mol/kg)	Animals (*n*)	Fetuses (*n*)	Reduced (%)	Unossified (%)	Abnormal structure (%)
Hydralazine	381	8	51	0	3.9	5.9
	763	7	33	3.0	3.0	9.1
Nifedipine	40, 50	4	25	4.0	0	12.0
	80, 100	17	115	2.2	0	91.3
Nitrendipine	40	6	46	50.0	4.3	100
	80	19	103	44.7	6.8	100
Felodipine	12	5	36	47.2	5.6	100
H152/37	80	12	77	0	0	0
Control	–	18	110	0	0	0

(Modified from DANIELSSON et al. 1989 by courtesy of Teratology).

The similarities in observed defects and the sensitive period for induction of the defects in clamping experiments and after administration of vasodilators with different chemical structure (see Table 1) suggest that the distal digital defects caused by vasodilators are secondary to aggravated maternal pharmacological side effects, resulting in decreased uteroplacental blood flow and fetal hypoxia. The same pathological changes were detected in the limb plates shortly after administration of vasodilator (DANIELSSON et al. 1990) as were seen after clamping of uterine vessels (LEIST and GRAUWILER 1974). Histologically, the digital areas of the limb plates showed extensive edema and dilatation of marginal sinus within 2 h (Fig. 4a). After 8 h, rupture of thin-walled vessels occurred with hemorrhages (Fig. 4b). Finally (after 24–48 h), blisters and necrosis of the developed cartilage of the phalanges was noticed (Fig. 4c,d). Fetal concentrations of the vasodilator felodipine have also been measured (after a dose inducing digital defects) and compared with the concentrations causing toxicity in vitro in embryonic mesenchymal cells differentiating into chondrocytes. The results showed that the highest measured fetal concentration of felodipine was more than 500 times lower than that required for in vitro toxicity (DANIELSSON et al. 1990). Furthermore, the doses of hydralazine, which caused marked hypotension in the ewe, failed to elicit any fetal circulatory response (or fall in PO_2 levels) when injected intravenously in the fetus. Only when hydralazine was given at ten to 20 times the maternal dose did it elicit a significant hypotensive response. (LADNER et al. 1970). These results support the idea that the digital defects caused by vasodilators are secondary to pharmacological action on the maternal side (resulting in a reduction in uteroplacental blood flow causing fetal hypoxia) and are not caused by a direct pharmacological action or toxic effect on the embryo/fetus.

Clinical data suggest that vasodilators in therapeutic doses may cause short-lasting fetal hypoxic episodes secondary to transient maternal hypotension (WIDE-SVENSSON et al. 1990). However, there was no evidence of digital defects (or other fetal defects) in more than 100 pregnant women treated with hydralazine in early pregnancy (SANDSTRÖM 1978; TCHERDAKOFF et al. 1978), nor is there any evidence for other vasodilators. The doses and plasma concentrations of vasodilators have in general been very high in the animal studies when digital defects have been observed. For example, in the rabbit after administration of a dose inducing digital defects, the plasma concentration was greater than 200 nmol/l for more than 12 h and caused severe maternal hypotension; decreased uteroplacental blood flow was also observed (for a review, see DANIELSSON et al. 1990). In humans, the peak therapeutic concentration after a slow-release tablet is around 7 nmol/l (EDGAR et al. 1987). Furthermore, in the clinical situation the purpose of the treatment is to lower an increased blood pressure to normal values, not to induce severe hypotension and decreased uteroplacental blood flow of long duration. An increased vascular resistance (due to vasoconstriction) in the uteroplacental vascular bed has been observed in severe hypertension, resulting in decreased uteroplacental blood flow. In this state, vaso dilation at a moderate dose of

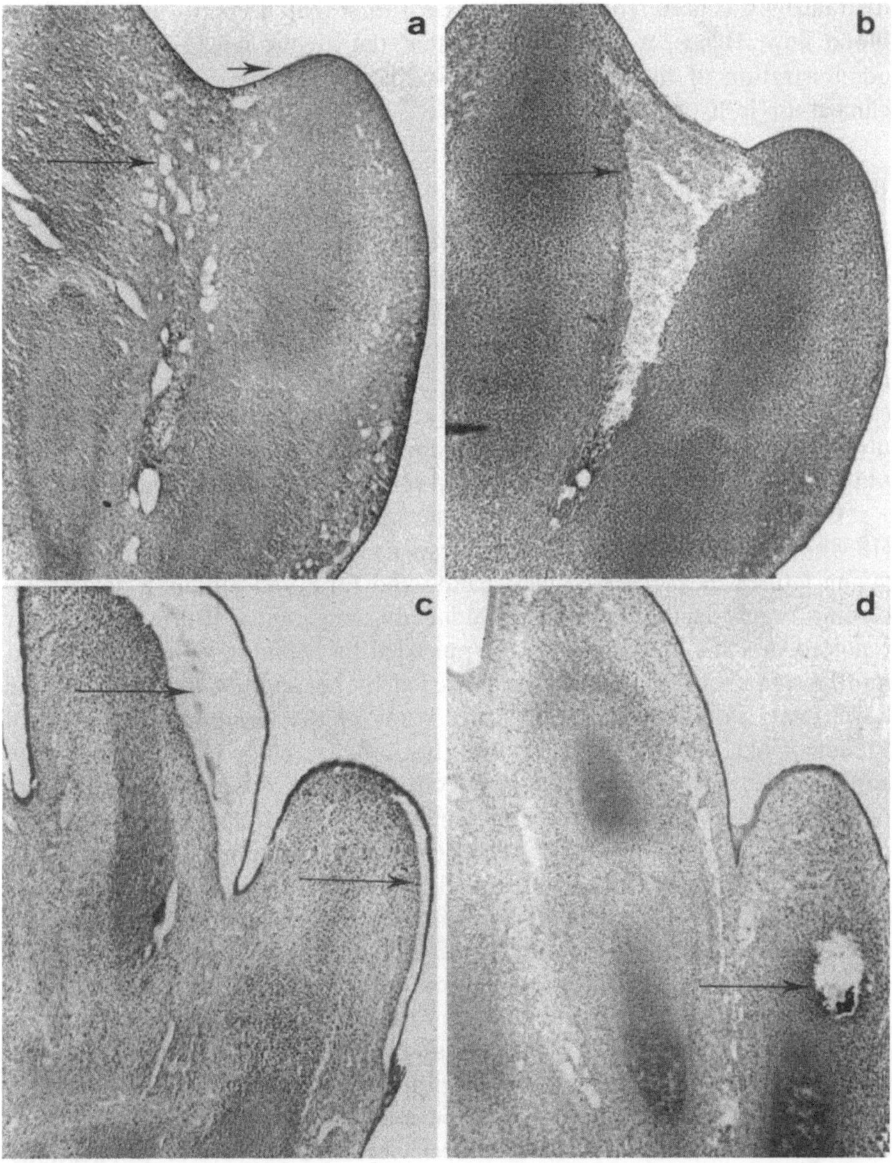

Fig. 4a-d. Histologic sections of the third and fourth right hind digits of rabbit fetuses **a** 2 h **b** 8 h and **c,d** 24 h postadministration of a calcium channel antagonist (felodipine 12 μmol/kg) on day 16. **a** Marked widening of thin-walled blood vessels (*long arrow*) and severe mesenchymal edema. Poorly defined borders of cartilaginous phalangeal primordia and distended, tight ectoderm (*short arrow*).**b** Mesenchymal edema with poorly defined cartilaginous phalangeal primordia. Rupture of the marginal vessels with extensive hemorrhage (*arrow*). **c** Mesenchymal edema with blister formation (*arrows*) and separation between mesenchyme and ecoderm. **d** Mesenchymal edema and distended marginal vessels. Small necrosis at this site of the cartilaginous primordium of the third phalange of the fourth digit (*arrow*). (From Danielsson et al. 1990 by courtesy of Teratology; copyright by Wiley-Liss Inc.)

hydralazine has been reported to increase, rather than decrease, uteroplacental blood flow. Hence, it seems unlikely that the digital defects obtained after administration of high dosages of vasodilators in animal studies relate to clinical use in hypertensive women.

III. Caffeine

The main pharmacological effects of caffeine are exerted on the CNS and the cardiovascular system. Similar to the vasodilators, caffeine can directly dilate peripheral blood vessels, but can also release catecholamines, which may result in vasoconstriction of uterine vessels in the same way as cocaine. Decreased uterine blood flow has been suggested as the cause of the teratogenicity of caffeine observed in animal studies, and caffeine has been reported to decrease uterine and ovarian blood flow in rats in midgestation, while maternal cardiac output was not altered (KIMMEL et al. 1984).

Caffeine was first reported as a teratogen by NISHIMURA and NAKAI (1958b). They injected large doses intraperitoneally in mice, and in the surviving fetuses they observed cleft palate and fore- and hindlimb defects, including brachydactyly, syndactyly, adactyly, and joint malformations. They reported that the limb defects were preceded by hematomas in the limb primordia and that cleft palate was preceded by hematomas in the upper and lower jaws. In a more comprehensive study of the teratogenic potential of caffeine, COLLINS et al. (1981) showed that caffeine (80 or 125 mg/kg per day) administered orally to pregnant rats from GD 0–19 caused increased resorptions and in the surviving fetuses caused edema, hemorrhage, and limb defects, characterized by missing digits. The hemorrhages were mostly found in the limbs, but also in the skull and trunk. The limb defects appear to be the same as those caused by uterine vessel clamping and cocaine. Another reactive species tested has been the rabbit, with 9% of rabbit fetuses showing distal digital amputation defects following oral doses during the first half of gestation (BERTRAND et al. 1970). Preliminary results show that digital amputation defects in the rabbit could be induced after a single dose (150 mg/kg) during GD 14–17 (B.R.G. DANIELSSON et al., unpublished ; see Fig 5).

The effect of caffeine-containing beverages during pregnancy have been investigated in many studies and been reviewed in several publications. Overall, the studies suggest that intake of less than eight cups of coffee per day (one cup corresponds to about 1.4–2.1 mg caffeine/kg) does not result in any increased risk for malformations (for a review, see SCHARDEIN 1993). However, in two of the studies excess consumption of coffee in pregnancy was associated with an increased risk. One study indicated that consumption of more than eight cups per day was associated with an increased frequency (23% vs. 13%) of congenital malformations (BORLEE et al. 1978). In the other report, JACOBSON et al. (1981) attributed excess daily consumption of coffee in pregnancy (19–30 mg/kg per day) with bilateral digital hypoplasia (ectrodactyly) seen in three children. Some human studies have indicated low

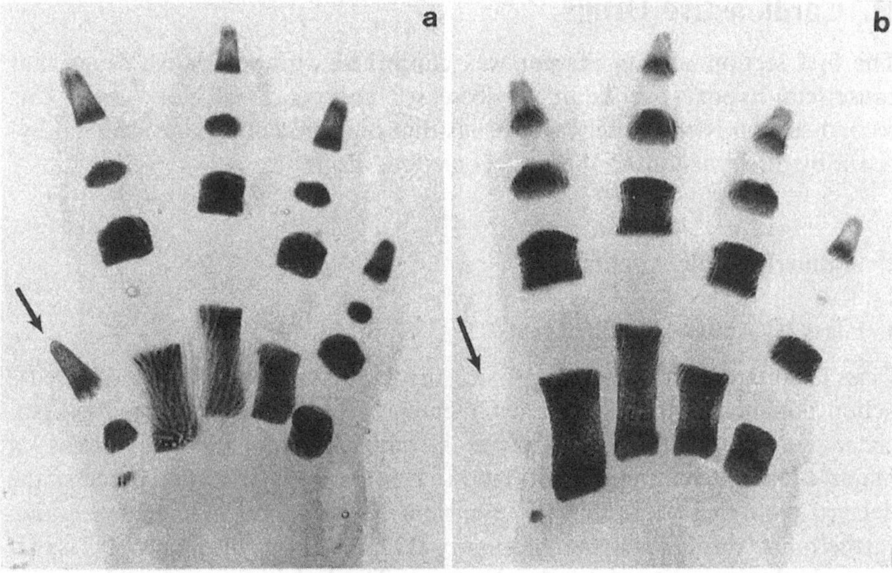

Fig. 5a,b. Amputation defect on the first digit of the forepaw (**b**) after administration of 150 mg caffeine/kg on day 14 of gestation. Compare with control fetus (**a**). Alizarine-stained skeletons

birth weight in offspring of women ingesting coffee in pregnancy and "moderate to heavy" caffeine intake during pregnancy has been associated with increased risk of fetal loss in controlled studies (e.g., INFANTE et al. 1993). On the other hand, there was no evidence that moderate caffeine use increased the risk of abortions or intrauterine growth retardation in a recent prospective, well-controlled study (MILLS et al. 1993).

Caffeine's known pharmacological effects on the cardiovascular system (which may result in fetal hypoxia) and the similarities in malformation pattern and sensitive period for induction of the defects suggest embryonic/fetal hypoxia as the explanation of the observed caffeine-induced malformations in animal studies. Caffeine does not seem to present any teratogenic risk in the human fetus when intake is low or moderate and spread out over the day. However, caffeine may have the potential to act in synergy with other substances with the capacity to cause embryonic/fetal hypoxia. As mentioned before, there are two case reports of disruptive vascular etologies leading to malformations after intake of tablets containing both ergotamine and caffeine (HUGHES and GOLDSTEIN 1988; GRAHAM et al. 1983), and experimentally caffeine potentiated the embryotoxic effect of drugs which may cause fetal hypoxia (NEHLIA and DERBY 1994). The available information regarding caffeine's potential adverse effects in fetuses may warrant avoidance of excessive caffeine intake during pregnancy.

D. Cardioactive Drugs

The first section of this chapter was concerned with vasoactive drugs that cause fetal hypoxia secondary to decreased uteroplacental blood flow. This second section is concerned with drugs that have potential to cause fetal hypoxia by direct action on the embryonic/fetal heart.

I. Antiarrhythmic Agents

1. Class III Antiarrhythmic Agents

Class III antiarrhythmic agents prolong myocardial refractoriness and increase action potential duration. This lengthening of the refractory period is associated with a decrease in the incidence of reentrant ventricular arrhythmias – a major cause of ventricular fibrillation. For the repolarization process, the delayed rectifying potassium (K^+) current (I_k) is of major importance and, furthermore, this channel is the primary target for most of the novel class III antiarrhythmic agents under development. The I_k has two outward components, a rapidly activating current (I_{kr}) and a slowly activating current (I_{ks}). Class III antiarrhythmic drugs have differential effects on these currents, with most agents (e.g., dofetilide, almokalant, L-691,121) inhibiting I_{kr}.

In doses not causing cardiodepressive or any other significantly adverse effects in the dams, a high incidence of embryonic/fetal death has been observed after administration of different class III antiarrhythmic agents. Embryonic/fetal death has been reported in rats for almokalant (DANIELSSON 1993; ABRAHAMSSON et al. 1994) and dofetilide (SPENCE et al. 1994), L-691,121 (BAN et al. 1992; KONOSHI et al. 1992) and in rabbits for sotalol (B.R.G. DANIELSSON et al., unpublished).

In addition to embryonic death, morphological abnormalities (including distal digital amputations, Fig. 1c,d; general edema and hemorrhage) were observed in surviving fetuses in some of the studies (DANIELSSON 1993; BAN et al. 1992). In order to establish the sensitive days for induction of these defects and to fully examine the spectrum of external malformations, rats were given an oral dose of almokalant or dofetilide on a single day during the organogenic period. No defects or embryolethality were seen after the highest doses on GD 8, 15, or 16 and these doses had no apparent effect on the dam. On days 9 and 10, the drugs caused embryolethality but no external malformations in surviving fetuses. On days 11 and 12, it caused facial defects, including right-sided oblique facial clefts; these defects were preceded by hemorrhage (Fig. 2c). On day 13 it induced forelimb defects, primarily missing digits on the left side (Fig. 1d), and on day 14 mostly hindlimb defects; again, these defects were preceded by hemorrhage. Embryonic day 13 was the most sensitive day with limb defects induced at the lowest dose (WEBSTER et al. 1996). These low doses resulted in plasma concentrations in rats close to the therapeutic concentration range (50–200 nmol/l) for almokalant (ABRAHAMSSON et al. 1994).

The margin of safety also appears to be small for dofetilide when comparing human plasma concentrations (RASMUSSEN et al. 1992) and concentrations in the experimental situation when adverse effects occur (WEBSTER et al. 1996).

Mechanistic studies, including electrophysiological investigation of adult and fetal rat hearts on day 13 of gestation (ABRAHAMSSON et al. 1994) and measurements of embryonic heart rate in whole-embryo cultures on day 10, 13, or 14 (BAN et al. 1992; SPENCE et al. 1994; WEBSTER et al. 1996, have all revealed that almokalant, dofetilide, sotalol, and L-691,121 caused a concentration-dependent slowing of the embryonic heart, arrhythmias, and temporary/permanent cardiac arrest. Sensitivity to these drugs started on GD 9, when the embryonic heart starts beating, and ended on GD 15 (WEBSTER et al. 1996). The effects were reversible, and depressed heart rates recovered to normal after a removal of the drug. The heart rate of embryos on GD 11 was more sensitive to dofetilide than the heart rate of embryos on day 14 of gestation (a 14%–64% decrease in heart rate vs. an 11%–43% decrease in heart rate, respectively), for the same concentrations tested (SPENCE et al. 1994). The results suggest that I_{kr} may be "downregulated" or suppressed in favor of other repolarizing currents with increasing gestational age.

During prenatal development, cardiac tissue undergoes dramatic changes in its electrophysiological properties, as described for the rat (COUCH et al. 1969). The electrophysiological investigations by ABRAHAMSSON et al. (1994) clearly illustrated the dramatic differences in the cardiac action potential configuration between the immature and the adult rat heart. The ventricular as well as the atrial action potential in the spontaneously beating heart of the fetus had a pronounced plateau of considerable duration. In adult papillary muscle, on the other hand, the action potential was short and spike-like. The class III compounds examined (*d*-sotalol, almokalant, and dofetilide) did not influence the action potential duration in the adult myocardium, which is consistent with evidence that the adult rat heart is not dependent on an I_k. However, in the fetal heart all agents induced a concentration-dependent prolongation of the action potential duration, which was accompanied by a marked bradycardia. At the higher drug concentrations, rhythm abnormalities and/or bradycardia-dependent early after-depolarizations occasionally appeared in both ventricular and atrial tissue (Fig. 6). Hence it appeared that the embryonic heart was dependent upon I_k for its normal function (ABRAHAMSSON et al. 1994).

Altogether, the studies show that the critical period for induction of embryonic death/teratogenicity in vivo and direct effects on the embryonic heart in vitro for class III antiarrhythmic agents is between GD 9 and 14 in the rat. It would appear that the teratogenic properties of these drugs represent a class effect. Since recent results show that similar embryotoxic effects can be induced by class III antiarrhythmic agents in other species (mice, ABRAHAMSSON et al. 1994; rabbits, B.R.G. DANIELSSON et al., unpublished), it is to be expected that, at an appropriate serum concentration, they would have the same effect on the human embryonic heart. It is proposed that the embry-

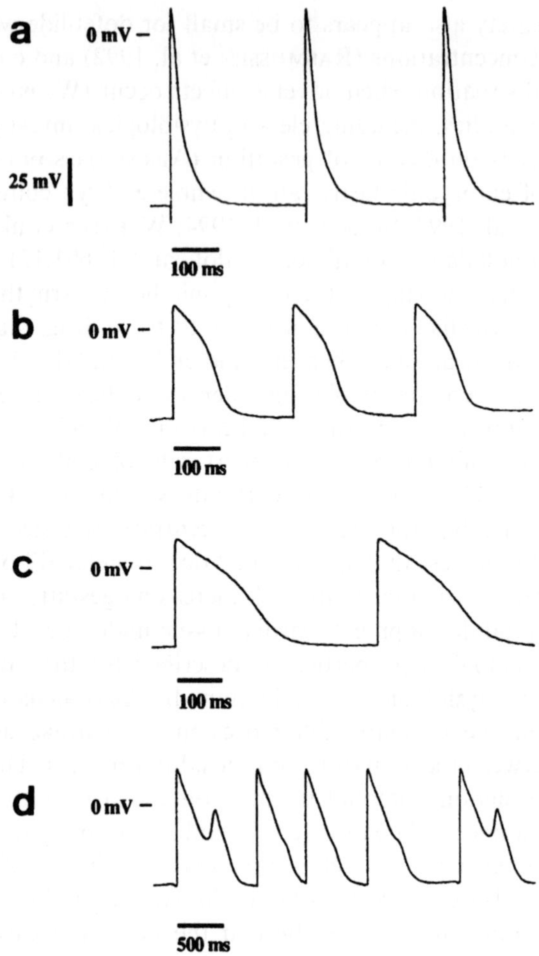

Fig. 6a-d. Transmembrane recordings of action potentials from rat ventricular myocardium **a** Adult myocardium in the control situation paced at 3.3 Hz. **b-d** Recordings from spontaneously beating fetal heart on gestational day 13. **b** In the control situation. **c** After 0.1 μM dofetilide. **d** After 10 μm dofetilide. (From ABRAHAMSSON et al. 1994 by courtesy of Cardiovascular Research; copyright by BNJ Publishing Group)

olethality and teratogenic activity is due to hypoxia/ischemia (as previously described for vasoactive drugs) secondary to bradycardia and arrhythmia in the embryo.

2. Phenytoin

Phenytoin is an effective antiarrhythmic drug and has been used for treatment of ventricular arrhythmias secondary to digitalis intoxication, open-heart surgery, and acute myocardial infarction. However, its main use is as an anticonvulsant in epilepsy. Phenytoin is an established human teratogen. Limb anomalies, consisting of hypoplasia of the nails and distal phalanges, together with midfacial hypoplasia are the most consistent morphological markers (HANSON 1986). Phalangeal hypoplasia is a rare congenital defect in humans (JONES 1988), but a high incidence of this defect has been reported in children (22%–34%) after in utero exposure to phenytoin, as reviewed by GAILY (1990). Less frequently, more severe defects such as orofacial clefts and serious limb reduction defects have been reported. All these manifestations have been induced in animal studies (see Chap. 28).

The mechanism by which phenytoin causes malformations is still unknown, but several hypotheses have been presented (see Chap. 28). The similarities in the clinical and experimental pattern of malformations caused by phenytoin with those caused by clamping of uterine vessels and cardiovascular active drugs suggest that pharmacologically induced fetal hypoxia may be involved in phenytoin teratogenicity. In this chapter, we present recent supporting evidence for this hypothesis. The most common fetal defect after oral dosing with phenytoin (100 mg/kg) during GD 7–18 in rabbits was phalangeal hypoplasia (MCCLAIN and LANGHOFF 1980). Vasodilators have also been reported to cause such defects in rabbits (DANIELSSON et al. 1989,1990). In a comparative study between phenytoin and a vasodilator (nifedipine), both drugs caused identical dose-dependent phalangeal hypoplasia. The distal phalanx on fourth digit on the hindpaw was most easily affected at the lowest doses (Fig. 3a,b). With increasing doses, other phalanges on the hind- and forepaws were involved in an identical manner (Fig. 3c). The histological appearance of the defects and the sensitive period (days 14–16) for induction of the defects was identical for the two drugs (DANIELSSON et al. 1992). Subsequent studies showed that digital defects caused by phenytoin were preceded by edema, dilated blood vessels, hemorrhage, and eventually mesenchymal necrosis in the same way as shown for vasodilators (Fig. 4a-d). Similar hemorrhage lesions were seen in the frontonasal region and in the CNS (DANIELSON et al. 1992; see Fig. 2d).

A dose of phenytoin (150 mg/kg) in rabbits known to cause a high incidence of hypoplasia/aplasia of the distal phalanges (all the distal phalanges on all digits) and other defects resulted in free and total phenytoin plasma concentrations of 9.5–12.7 and 129–176 μmol/l, respectively, during the 24-h period after single oral dosing on days 14–16, (DANIELSSON et al. 1995). A dose of phenytoin (100 mg/kg) causing a moderate incidence of digital hypoplasia in rabbits (25% of the distal phalanges of the first, third, and fifth digit on the forepaw and 50% of the fourth digit on the hindpaw affected) resulted in free and total concentrations of 2.1–9.7 and (20–170) μmol/l respectively. This dose level has also been shown to cause cleft palate in rabbit fetuses (MCCLAIN and LANGHOFF 1980). A lower dose (50 mg/kg), not causing digital hypoplasia or any other adverse fetal effects at all, resulted in free and total concentrations of 0.9–5.0 and 17–112 μmol/l, respectively. The free concentrations of phenytoin in the fetuses and amniotic fluid were of the same magnitude as in maternal plasma at all these dose levels (DANIELSSON et al. 1995). These results suggest an association between phenytoin-induced malformations and free plasma concentrations.

Although numerous studies have related phenytoin therapy with malformations in human pregnancy, very few studies have related plasma concentrations of phenytoin in early pregnancy to fetal outcome. The results from one such study showed that in utero exposure during the first 20 weeks of pregnancy (a total of 75 children) was associated with radiographically defined phalangeal hypoplasia (GAILY 1990). The mothers of seven children in the study had plasma phenytoin levels over 40 μmol/l, and all of their children showed more prominent phalangeal defects than did children exposed to lower levels. One child, who was exposed in utero to particularly high levels (maximum 86 μmol/l), showed complete aplasia or severe hypoplasia of distal phalanges of most fingers and toes (GAILY 1990). A total plasma concentration of 86μmol/l corresponds to a free concentration of 9.5 μmol/l assuming an 11% free fraction, as has been shown in the first and second trimesters (TOMSON et al. 1994). The free plasma concentration (which is the most relevant to compare if a pharmacological effect is assumed) causing digital hypoplasia in humans thus seems very similar to that causing the same type of defects in rabbits.

A teratogenic dose of phenytoin in the rabbit caused hemodynamic alterations on GD 16, manifested as decreased maternal heart rate and blood pressure (both decreased by around 15%) in the awake rabbit, resulting in a significant decrease in maternal PO_2 and increase in PCO_2 (DANIELSON et al. 1992). In the same study, the fetal heart rate was examined in anesthetized rabbits on the same day. A decrease in the fetal heart rate of 12% in phenytoin-treated animals was observed, compared to anesthetized controls. As was seen in the rabbit, a teratogenic dose of phenytoin in A/J mice has also been shown to markedly depress the maternal heart rate for up to 6 h after dosing (WATKINSON and MILLICOVSKY 1983). This may be a contributing factor to embryonic hypoxia in the mouse and rabbit, but a similar effect is not observed with therapeutic dosing in humans.

In order to examine possible effects of phenytoin on the embryonic heart, phenytoin has been administered to day-10 A/J mouse embryos (whole-embryo culture in vitro). In vivo administration of phenytoin to pregnant A/J mice on this day of gestation induces a high incidence of oral clefts, and the incidence of oral clefts was greaetly reduced if the mice were placed in a hyperoxic (50% PO_2) chamber (MILLICOVSKY and JOHNSTON 1981). Similar to the class III antiarrhythmic agents dofetilide and almokalant, which induce oral clefts after a single administration at the same stage of organogenesis in the rat (day 11), phenytoin in vitro caused a decrease in embryonic heart rate (7%–50%) and an increase in arrhythmias and cardiac arrest with increasing concentration (100–300 μM). The no-effect concentration was approximately 50 μM (DANIELSSON et al. 1996).

The observed hemodynamic effects in the rabbit (DANIELSON et al. 1992) and in vitro on the embryonic heart in both mice and rats (DANIELSSON et al.

1996, DANIELSSON and WEBSTER 1996) may be explained by the known pharmacological effects of phenytoin. In addition to inhibition of Na^+ channels, which is the main component of phenytoin's stabilizing effects on excitable membranes of neurons and cardiac myocytes, phenytoin has Ca^{2+}-antagonizing properties (MCLEAN and MACDONALD 1983). These pharmacological properties of phenytoin may explain the observed decreased maternal heart rate and blood pressure in the rabbit. Phenytoin also has the capacity to delay K^+ currents during action potentials (YAARI et al. 1986), which may be of importance in explaining the similarities in cardiodepressive effects on the embryonic heart in vitro between antiarrhythmic K^+ channel blockers and phenytoin.

Overall, the results suggest that phenytoin has a cardiodepressive effect on the embryonic/fetal heart and a smaller, negative effect on the maternal circulation (hypotension and decreased maternal heart rate). These combined effects result in embryonic hypoxia/ischemia. The human relevance of the proposed teratogenic mechanism is strengthened by the fact that the free concentrations of phenytoin in the rabbit are the same as have been associated with the same type of defects in humans. The fetal hypoxia may also explain other manifestations included in fetal hydantoin syndrome, such as pre- and postnatal growth retardation and mental deficiency.

II. β-Adrenergic Antagonists

β-Adrenergic antagonists (beta blockers), such as propranolol, and the selective beta-1 blockers metoprolol and atenolol are widely used in the treatment of hypertension, angina pectoris, and arrhythmias. Treatment with beta blockers decreases the effect of catecholamines in physical and emotional stress and results in decreased heart rate, cardiac output, and blood pressure. These pharmacological effects of beta blockers suggest that these drugs have the potential to cause fetal hypoxia and associated malformations secondary to decreased placental perfusion and fetal bradycardia during sensitive stages.

However, in teratological studies in rats and/or rabbits, metoprolol (BODIN et al. 1975), propranolol (SCHOENFELD et al. 1978), and atenolol (ESAKI and IMAI 1980; ESAKI 1980) have not shown any teratogenic potential. The same hold true for published studies in early human pregnancy. The widely used propanolol was safe in this respect in several reports (O'CONNOR et al. 1981; TCHERDAKOFF et al. 1978). Negative reports with regards to malformations have been reported in over 100 cases with metoprolol (SANDSTRÖM 1978). According to SCHARDEIN (1993), negative reports have also appeared with atenolol, betaxolol, labetalol, oxprenatolol, and beta blockers generally.

In the later stages of human pregnancy, an increased risk for adverse fetal effects has been reported. Growth retardation has been reported in human studies and in animal studies after administration of high doses (SCHOENFELD et al. 1978; REDMOND 1982). An association between treatment with beta blockers and late intrauterine death and premature deliveries has been reported in a few studies, but these findings were not confirmed in other studies. Beta blockers may also cause fetal bradycardia. This can be illustrated by the results from two studies investigating the circulatory effects of treatment with atenolol in pre-eclampsia and pregnancy-induced hypertension in late pregnancy. In the first study (THORLEY et al. 1981), atenolol (5 mg orally) decreased systolic and diastolic pressure (from 171 to 155 mmHg and from 116 to 107 mmHg, respectively). The mean maternal heart rate decreased from 90 to 74 and fetal heart rate from 145 to 138. In the other study, the blood pressure and maternal heart rate decreased significantly, and the mean fetal heart rate decreased by 5% after the same oral dose (LUNELL et al. 1979).

The lack of teratogenicity of beta blockers, despite their potential to cause embryonic/fetal hypoxia, might partly be related to the immaturity of the adrenergic system. In the rat, the heart begins to beat around GD 10, but does not receive the extrinsic autonomic innervation until about GD 16–18 (HOGG 1957; GOMEZ 1958; ADOLPH 1965). The sinus node is innervated on about GD 16. In early embryonic life, there is a slow, fixed heart rate, which progressively increases with age. The absence of an extrinsic innervation in the early heart means that reflex responses to cardiovascular active drugs will be absent, leaving only those responses that result from direct action on the heart.

β-Adrenergic receptors are present in the non-innervated embryonic rodent heart (MARTIN et al. 1973; ROBKIN et al. 1973). In the rat, receptor subtype selectivities (such as β_1-receptors in the fetal heart) are already in place in early development (GD 12), but receptor coupling to adenylate cyclase (indicating some aspect of physiological function) is low at this stage. In contrast, coupling showed a spike at GD 18 and, furthermore, on this day concentration of β-receptors in fetal tissues had climbed fivefold compared to GD 12 (SLOTKIN et al. 1994). These data suggest an association between β-receptor expression and cell differentiation in late fetal stages and may contribute to explain the relative absence of effects of β-agonists and β-antagonists (despite the occurrence of β-receptors) on the embryonic/fetal heart. Thus the chronotrophic response of a β-agonist on the heart was minimal on GD 12–14 in mice compared with GD 15–16, when it was more prominent, and just prior to birth, when peak responsiveness occurred (WILDENTHAL 1973). The beta blocker propranolol did not slow the (preneural) heart after administration to rat embryos (GD 11) in vitro, but blocked the accelerating effect of the β-agonist isoproterenol (ROBKIN et al. 1973). In mammalian embryos, propranolol does not decrease the fetal heart rate until after the establishment of the sympathetic innervation on day 16 (MARTIN et al. 1973). Altogether, these results suggest that the ability of beta-blockers to cause fetal bradycardia by direct pharmacological action is low during susceptible stages for induction of hypoxia-related malformations. Furthermore, the clinical studies indicate that only slight bradycardia may occur at therapeutic doses during later stages of pregnancy.

With regard to beta blockers' potential to cause malformations secondary to decreased placental perfusion, animal studies with vasodilators showed that the mildest form of malformations related to hypoxia only occurred after severe hemodynamic alterations. Thus phalangeal hypoplasia was obtained only at doses causing a fall in maternal blood pressure of 30% and a decrease in uteroplacental blood flow of 40%–50% (secondary to hypotension and diversion of blood flow from the pregnant uterus to peripheral vascular beds) during the entire sensitive period (DANIELSSON et al. 1989, 1990; LUNDGREN et al. 1992). It is uncertain whether beta blockers can cause such pronounced hemodynamic alterations. Beta blockers are highly effective in decreasing the effects of catecholamines in the diseased state (e.g., they normalize blood pressure in hypertension or decrease the heart rate in tachycardia), but these effects under normal conditions are much less pronounced (BIGGER and LEFKOWITZ 1990).

E. Discussion

The results presented in this chapter suggest that vascular disruption resulting in tissue necrosis and malformations seems to be a common response to hypoxic situations of different origin, e.g., mechanical clamping of uterine vessels or pharmacological action of cardiovascular active drugs, such as vasoconstrictors, vasodilators, and antiarrhythmics (see Fig. 7). The same type of malformations, preceded by hemorrhage and necrosis, have also been induced by decreasing PO_2 in the atmosphere, as reviewed by GRABOWSKI (1970) (see Fig. 7). The most sensitive days appear to be in the late organogenic and the early fetal period, and the most likely malformations are limb defects, particularly missing or reduced digits and/or nails. After more severe hypoxia, other malformations may occur, including CNS damage, and they may occur also at stages of pregnancy other than the most sensitive one. Since all observed fetal abnormalities are preceded by edema and hemorrhage in previously established structures, they should be called disruptions rather than malformations. The cellular mechanism by which hypoxia causes injury is not full understood. During recent years, the use of new techniques has revealed in more detail the biochemical events which occur in response to transient hypoxia/ischemia and reperfusion, which may result in local generation of oxygen free radicals in embryonic/fetal tissues (FANTEL et al. 1992).

The reason for the increased susceptibility to hypoxia during the late organogenic and early postorganogenic period is unknown, but might be related to the immaturity of the embryonic/fetal cardiovascular system. The early embryo (before dependence on its own cardiovascular system) is mainly nourished by passive diffusion processes and anaerobic glycolysis. During the sensitive period, the cardiovascular system is morphologically developed and the heart is beating, but the innervation of the heart is sparse and the vascular autonomic nervous system is poorly developed (DOWNING 1960; BARTOLOMÉ et al. 1980). The phase-specific sensitivity could thus be due to a functional deficiency to adequately express compensative responses to hypoxia

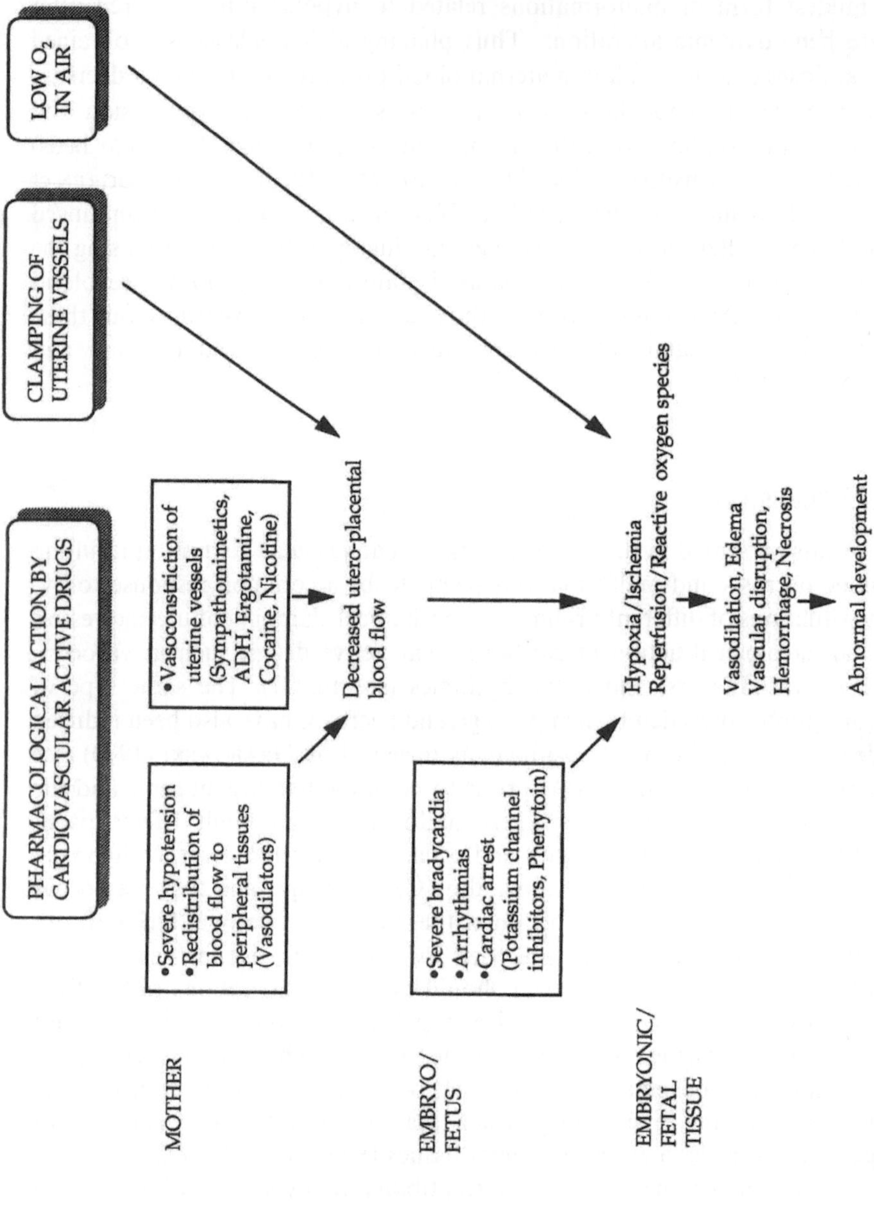

Fig. 7. Conditions related to hypoxia induced abnormal development by a common mechanism. *ADH*, anti-diuretic hormone

(e.g., vasopressor response), resulting in more severe hypoxia than during earlier and later developmental stages. In the case of class III antiarrhythmics, their teratogenicity is clearly related to a different reactivity of the fetal heart, resulting in severe embryonic hypoxia/ischemia, only during the period when malformations are induced.

The similarities in pattern of birth defects observed in man and in animal studies for some of the substances (e.g., cocaine, phenytoin, and maybe ergotamine) suggest that pharmacologically induced hypoxia resulting in vascular disruption is of relevance also in humans. The effect of such drugs seems to depend on whether the induced hypoxia/ischemia is above threshold in severity and duration for the embryonic day of exposure. A final general comment is that complementary hemodynamic studies together with pharmacokinetic data may contribute to a better assessment of the clinical relevance of adverse findings in teratology studies for new drugs.

References

Abrahamsson C, Palmer M, Ljung B, Duker G, Baarnhielm C, Carlsson L, Danielsson B (1994) Induction of rhythm abnormalities in the fetal rat heart. A tentative mechanism for the embryotoxic effect of class III antiarrhythmic agent almokalant. Cardiovasc Res 28: 337–344

Adamsons K, Mueller-Heubach E, Myers RE (1971) Production of fetal asphyxia in the rhesus monkey by administration of catecholamines to the mother. Am J Obstet Gynecol 109: 248–252

Adolph EF (1965) Capacities for regulation of heart rate in fetal, infant and adult rats. Am J Physiol 209: 1095–1105

Ariyuki P (1975) Effects of diltiazem hydro-chloride on embryonic development: species differences in the susceptibility and stage specificity in mice, rats, and rabbits. Okajimas Fol Anat Jpn 52: 103–117

Aselton P, Jick H, Milunsky A, Hunter JR, Stergachis A (1985) First trimester drug use and congenital disorders. Obstet Gynecol 65: 451–455

Ban Y, Konishi R, Kawana K, Nakatsuka T, Fujii T (1992) Embryotoxic effects of class III antiarrhythmic agent, L-691,121, in rats. Teratology 45: 462

Bartolomé J, Mills E, Lau C, Slotkin TA (1980) Maturation of sympathetic neurotransmission in the rat heart. V. Development of baroreceptor control of sympathetic tone. J Pharmacol 215: 596–599

Bertrand M, Girod J, Rigaud MF (1970) Ectrodactylie provoquee par la cafeine chez les rongeurs. Role des facteurs specifiques et genetiques. C R Soc Biol (Paris) 164: 1488–1489

Bigger TJ, Lefkowitz KJ (1990) Adrenergic receptor antagonists. In: Gilman AG, Rall TW, Nies AS, Taylor P (eds) Goodman and Gilman's pharmacological basis of therapeutics, 8th edn, vol 2. Pergamon, New York, pp 221–243

Bingol N, Fuchs M, Diaz V, Stone RK, Gromisch DS (1987) Teratogenicity of cocaine in humans. J Pediatr 110: 93–96

Bodin N-O, Flodh H, Magnusson G, Malmfors T, Nyberg J-A (1975) Toxicological studies on metoprolol. Acta Pharmacol Toxicol. [Suppl 5] 36: 96–103

Borlee I, Lechat MF, Bouckaert A, Misson C (1978) Coffee, risk factor during pregnancy? Louvain Med 98: 279–284

Briggs CG, Freeman RK, Yaffe SJ (1990) Drugs in pregnancy and lactation. Williams and Wilkins, Baltimore

Bruner JP, Forouzan I (1991) Smoking and bucally administered nicotine. Acute effect on uterine artery doppler flow velocity waveforms. J Reprod Med 36: 435–440

Chasnoff IJ, Griffith DR (1989) Cocaine: clinical studies of pregnancy and the newborn. Ann NY Acad Sci 562: 260–266
Chasnoff IJ, Burns WJ, Schnoll SH, Burns KA (1985) Cocaine use in pregnancy. New Engl J Med 313: 666–669
Chasnoff IJ, Bussey ME, Savich R, Stack CM (1986) Perinatal cerebral infarction and maternal cocaine use. J Pediatr 108: 456–459
Chasnoff IJ, Chisum GM, Kaplan WE (1988) Maternal cocaine use and genitourinary tract malformations. Teratology 37: 201–204
Chavez GF, Muliare J, Cordero JF (1988) A population-based study of periconceptional cocaine use and genitourinary tract malformations. Presented at the 9th Annual David W Smith Workshop on Malformations and Morphogenesis, 3–7 August 1988, Oakland, California, p 25
Chernoff N, Grabowski CT (1971) Responses of the rat fetus to maternal injections of adrenalin and vasopressin. Br J Pharmacol 43: 270–278
Clark KE, Irion GL (1992) Fetal hemodynamic response to maternal intravenous nicotine administration. Am J Obstet Gynecol 167: 1624–1631
Collins TFX, Welsh JJ, Black TN, Collins EV (1981) A study of teratogenic potential of caffeine given by oral intubation to rats. Regu Toxicol Pharmacol 1: 355–378
Couch JR, West C, Hoff HE (1969) Development of the action potential of the prenatal rat heart. Circ Res 24: 19–31
Danielson MK, Danielsson BRG, Marchner H, Lundin M, Rundqvist E, Reiland S (1992) Histopathological and hemodynamic studies supporting hypoxia and vascular disruption as explanation to phenytoin teratogenicity. Teratology 46: 485–497
Danielsson BR (1993) Malformations and fetal hypoxia induced by exaggerated pharmacological effects. In: Sundvall A, Danielsson BR, Hagberg O, Lindgren E, Sjöberg P, Victor A (eds) Development toxicology – preclinical and clinical data in retrospect. Tryckgruppen, Stockholm, pp 44–57
Danielsson BRG, Reiland S, Rundqvist E, Danielson M (1989) Digital defects induced by vaso-dilating agents. Relationship to reduction in utero-placental blood flow. Teratology 40: 351–357
Danielsson BRG, Reiland S, Rundqvist E, Danielson M, Dencker L, Regård C-G (1990) Histological and in vitro studies supporting decreased utero-placental blood flow as explanation to digital defects after administration of vasodilators. Teratology 41: 185–193
Danielsson BRG, Danielson M, Rundqvist E, Reiland S (1992) Identical phalangeal defects induced by phenytoin and nifedipine suggest fetal hypoxia and vascular disruption behind phenytoin teratogenicity. Teratology 45: 247–258
Danielsson BRG, Danielson K, Tomson T (1995) Phenytoin causes phalangeal hypoplasia in the rabbit fetus at clinically relevant free concentrations. Teratology 52: 252–259
Danielsson BR, Azarbayani F, Wisén AC (1996), Initiation of phenytoin teratogenesis: Pharmacologically induced embryonic bradycardia and arrhythmia resulting in hypoxia/ischemia and possible free radical damage at reperfusion. Teratology 53: 99
David TJ (1972) Nature and etiology of the Poland anomaly. New Engl J Med 287: 487–489
Dixon SD, Bejar R (1989) Echoencephalographic finding in neonates associated with maternal cocaine and methamphetamine use: incidence and clinical correlates. J Pediatr 115: 577–778
Downing SE (1960) Baroreceptor reflexes in new-born rabbit. J Physiol 150: 201–213
Ducsay CA, Thompson JS, Wu AT et al. (1987) Effects of calcium entry blocker (nicardipine) tocolysis in rhesus macaques: fetal plasma concentrations and cardiorespiratory changes. Am J Obstet Gynecol 157: 1482

Edgar B, Collste P, Haglund K, Regård CG (1987) Pharmacokinetics and haemodynamic effects of felodipine as monotherapy in hypertensive patients. Clin Invest Med 10: 388–394
Entman SS, Moise KJ (1984) Anaphylaxis in pregnancy. South Med J 77: 402
Esaki K (1980) Effects of oral atenolol administration on the rabbit fetus. Preclin Rep Cent Inst Exp Animal 6: 259–264
Esaki K, Imai K (1980) Effects of oral aternolol on reproduction in rats. Preclin Rep Cent Inst Exp Animal 6: 239–258
Fantel AG, Barber CV, MacKler B (1992) Ischemia/reperfusion: a new hypothesis for the developmental toxicity of cocaine. Teratology 46: 285–292
Franklin JB, Brent RL (1964) The effect of uterine vascular clamping on the development of rat embryos three to fourteen days old. J Morphol 115: 273–290
Fukunishi K, Yokoi Y, Yoshida H, Nose T (1980) Effects of nifedipine on rat fetuses (in Japanese). Med Consult N Remed 17: 2245–2256
Gaily E (1990) Distal phalangeal hypoplasia in children with prenatal phenytoin exposure. Am J Med Genet 35: 403–414
Gomez H (1958) The development of the innervation of the heart in the rat embryo. Anat Rec 130: 53–71
Grabowski CT (1970) Embryonic oxygen deficiency – a physiological approach to analysis of teratological mechanisms. In: Woollam DHM (ed) Advances in teratology, vol 9. Logo, London, pp 125–169
Graham JM Jr, Marin-Padilla M, Hoefnagel D (1983) Jejunal atresia associated with cafergot ingestion during pregnancy. Clin Pediatr (Phila) 22: 226–228
Grauwiler J, Schon H (1973) Teratological experiments with ergotamine in mice, rats and rabbits. Teratology 7: 227–235
Greenland VC, Delke I, Minkoff HL (1989) Vaginally administered cocaine overdose in a pregnant women. Obstet Gynecol 74: 476–477
Greiss FC (1963) The uterine vascular bed: effect of adrenergic stimulation. Obstet Gynecol 21: 295–301
Hammer RE, Goldman H, Mitchell JA (1981) Effects of nicotine on uterine blood flow and intrauterine oxygen tension in the rat. J Reprod Fertil 63: 163–168
Hanson JW (1986) Teratogen update: fetal hydantoin effects. Teratology 33: 349–353
Harake B, Gilbert RD, Ashwal S, Power GG (1987) Nifedipine: effects on fetal and maternal hemodynamics in pregnant sheep. Am J Obstet Gynecol 157: 1003
Hogg IE (1957) Developing innervation of the heart in the albino rat. Anat Rec 127: 470 (abst)
Hoyme HE, Jones KL, Dixon SD, Jewett T, Hanson JW, Robinson LK, Msall ME, Allanson JE (1990) Prenatal cocaine exposure and fetal vascular disruption. J Pediatr 85: 743–747
Hudson BB, Timiras PS (1972) Nicotine injection during gestation: impairment of reproduction, fetal viability, and development. Biol Reprod 7: 247–253
Hughes HE, Goldstein DA (1988) Birth defects following maternal exposure to ergotamine, beta blockers, and caffeine. J Med Genet 25: 396–399
Hutchings DE (1993) The puzzle of cocaine's effect following maternal use during pregnancy: are there reconcilable differences? Neurotoxicol Teratol 15: 282–286
Infante-Rivard C, Fernández A, Gauthier R, David M, Rivard G-F (1993) Fetal loss associated with caffeine intake before and during pregnancy. JAMA 270: 2940–2943
Jacobson MF, Goldman AS, Syme RH (1981) Coffee and birth defects. Lancet 1: 1415–1416
Jones KL (1988) Smith's recognizable patterns of human malformation. Saunders, Philadelphia, pp 662–705
Jost A (1951) Sur le role de la vasopressine et de la corticostimuline (ACTH) dans la production experimentale de lesions des extremites foetales (hemorragies, necroses, amputations congenitales). CR Soc Biol (Paris) 145: 1805–1809

Jost MA (1953) Degenerescence des extremites du foetus de rat provoquée par l'adrenaline. CR Soc Biol (Paris) 236: 1510–1513
Kapur RP, Shaw CM, Shepard TH (1991) Brain hemorrhages in cocaine-exposed human fetuses. Teratology 44: 11–18
Keeler RF (1980) Congenital defects in calfs from maternal ingestion of nicotiana glauca of high anabasine content. Clin Toxicol 15: 417–426
Kimmel CA, Kimmel GL, White CG, Grafton TF, Young JF, Nelson CJ (1984) Blood flow changes and conceptal development in pregnant rats in response to caffeine. Fundam Appl Toxicol 4: 240–247
Konoshi R, Ban Y, Kawana T, Nakatsuka, Fujii T (1992) In vitro embryotoxic effects of a class III antiarrhythmic agent L-691,121, in rats using whole embryo culture. Teratology 46: 11B
Ladner C, Weston P, Brinkman CR, Assai NS (1970) Effects of hydralazine on uteroplacental and fetal circulations. Am J Obstet Gynecol 108: 375–381
Leist KH, Grauwiler J (1973a) Influence of the developmental stage on embryotoxicity following uterine vessel clamping in the rat. Teratology 8: 227–228
Leist KH, Grauwiler J (1973b) Transplacental passage of 3H ergotamine in the rat and determination of the intraamniotic embryotoxicity of ergotamine. Experientia 29: 764
Leist KH, Grauwiler J (1974) Fetal pathology in rats following uterine vessel clamping on day 14 of gestation. Teratology 10: 55–68
Lindblad A, Marsal K, Andersson K-E (1988) Effect of nicotine of human fetal blood flow. Obstet Gynecol 72: 371–382
Lirette M, Holbrook RH, Katz M (1987) Cardiovascular and uterine blood flow changes during nicardipine HC J tocolysis in the rabbit. Obstet Gynecol 169: 79
Lundgren Y, Thalen P, Nordlander M (1992) Effects of felodipine on uteroplacental blood flow in normotensive rabbits. Pharmacol Toxicol 71: 361–364
Lunell NO, Persson B, Aragon G, Fredholm BB, Aström H (1979) Circulatory and metabolic effects of acute beta 1-blockade in severe pre-eclampsia. Acta Obstet Gynecol Scand 58: 443–445
MacGregor SN, Keith LG, Chasnoff IJ, Rosner MA, Chisum GM, Shaw P, Minogue JP (1987) Cocaine use during pregnancy: adverse perinatal outcome. Obstet Gynecol 157: 686–690
Martin JD, Young IM (1960) Influence of gestational age and hormones on experimental fetal bradycardia. J Physiol 152: 1–13
Martin S, Levey BA, Levey GS (1973) Development of betaadrenergic receptor in fetal rat heart. Biochem Biophys Res Commun 54: 949–954
McClain RM, Langhoff L (1980) Teratogenicity of diphenylhydantoin in the New Zealand white rabbit. Teratology 21: 371–379
McLean MJ, MacDonald RL (1983) Multiple action of phenytoin on mouse spinal cord neurons in cell culture. J Pharmacol Exp Ther 227: 779–798
Millicovsky G, Johnston MC (1981) Maternal hyperoxia greatly reduces the incidence of phenytoin-induced cleft lip and palate in A/J mice. Science 212: 671–672
Mills JL, Holmes LB, Aarons JH, Simpson JL, Brown ZA, Jovanovic-Peterson LG, Conley MR, Graubard BI, Knopp RH, Metzger BE (1993) Moderate caffeine use and the risk of spontaneous abortion and intrauterine growth retardation. JAMA 269 5: 593–597
Moore TR, Sorg J, Miller L, Key TC, Resnik R (1986) Hemodynamic effects of intravenous cocaine on the pregnant ewe and fetus. Am J Obstet Gynecol 155: 883–888
Morgan MA, Silavin SL, Randolph M, Payne GG, Sheldon RE, Fishburne JI, Wentworth RA, Nathanielsz PW (1991) Effect of intravenous cocaine on uterine blood flow in the gravid baboon. Am J Obstet Gynecol 164: 1021–1030
Morrow RJ, Ritchie JWK, Bull Sb (1988) Maternal cigarette smoking: the effect on umbilical and uterine blood flow velocity. Am J Obstet Gynecol 159: 1069–1071

Nehlia A, Derby G (1994) Potential teratogenic and neurodevelopmental consequences of coffee and caffeine exposure; a review on human and animal data. Neurotoxical Teratol 16: 531–543
Niebyl JR (1988) Drug use in pregnancy, 2nd edn. Lea and Febiger, Philadelphia
Nishimura H, Nakai K (1958a) Developmental anomalies in offspring of pregnant mice treated with nicotine. Science 127: 877–878
Nishimura H, Nakai K (1958b) Congenital malformations in offspring of mice treated with caffeine. Proc Soc Exp Biol Med 104: 140–142
O'Connor PC, Jick H, Hunter JR, Stergachis A, Madsen S (1981) Propranolol and pregnancy outcome. Lancet 2: 1168
Quigley ME, Sheehan KL, Wilkes MM, Yen SSC (1979) Effects of maternal smoking on circulating catecholamine levels and fetal heart rates. Am J Obstet Gynecol 133: 685–690
Rall TW (1991) Oxytocin, prostaglandins, ergot alkaloids, and other drugs; tocolytic agents. Chapter 39 In: Gilman AG, Rall TW, Nies AS, Taylor P (eds) Goodman and Gilman's pharmacological basis of therapeutics, 8th edn, vol 2. Pergamon, New York
Rasmussen HS, Allen MJ, Blackburn KJ, Butrous GS, Dalrymple HW (1992) Dofetilide, a novel class III antiarrhythmic agent. J Cardiovasc Pharmacol 20: S96–S105
Redmond GP (1982) Propranolol and fetal growth retardation. Semin Perinatol 6: 142–147
Resnick R, Brink GW, Wilkes M (1979) Catecholamine-mediated reduction in uterine blood flow after nicotine infusion in the pregnant ewe. J Clin Invest 63: 1133–1136
Robkin M, Shepard TH, Baum D (1973) Autonomic drug effects on the heart rate of rat embryos. Teratology 9: 35–44
Sandström B (1978) Antihypertensive treatment with the adrenergic beta-receptor blocker metoprolol during pregnancy. Gynecol Invest 9: 195–204
Schardein JL (1993) Chemically induced birth defects, 2nd edn. Dekker, New York
Schoenfeld N, Epstein O, Nemesh L, Rosen M, Atsmon A (1978) Effects of propranolol during pregnancy and development of rats. I. Adverse effects during pregnancy. Pediatr Res 12: 747–750
Schön H, Leist KH, Grauwiler J (1975) Single day treatment of pregnant rats with ergotamine. Teratology 11: 32A
Sharma A, Plessinger MA, Sherer DM, Liang C, Miller RK, Woods JR (1992) Pregnancy enhances cardiotoxicity of cocaine: role of progesterone. Toxicol Appl Pharmacol 113: 30–35
Slotkin TA, Lau C, Seidler FJ (1994) β-Adrenergic receptor overexpression in the fetal rats. Distribution, receptor subtypes and coupling to adenylate cyclase activity via G-proteins. Toxicol Appl Pharmacol 129: 223–234
Spence GS, Vetter C, Hoe C-M (1994) Effects of the class III antiarrhythimic, dofetilide (UK-68,798) on the heart rate of midgestation rat embryos in vitro. Teratology 49: 282–292
Suzuki K, Minei LJ, Johnson EE (1980) Effect of nicotine upon uterine blood flow in the pregnant rhesus monkey. Am J Obstet Gynecol 136: 1009–1013
Tcherdakoff PH, Colliard M, Berrard E, Kreft C, Dupay A, Bernaille JM (1978) Propranolol in hypertension during pregnancy. Br Med J 2: 670
Tomson T, Lindbom U, Ekqvist B, Sundqvist A (1994) Epilepsy and pregnancy: a prospective study of seizure control in relation to free and total plasma concentrations of carbamazepine and pheoytoin. Epilepsia 35,1: 122–130
Thorley KJ, McAinsh J, Cruickshank JM (1981) Atenolol in the treatment of pregnancy-induced hypertension. Br J Clin Pharmacol 12: 717–730
Vara P, Kinnunen O (1951) Effect of nicotine on the female rabbit and developing fetus. Ann Med Exp Biol Fenn 29: 202–213
Watkinson WP, Millicovsky G (1983) Effect of phenytoin on maternal heart rate in A/J mice: possible role in teratogenesis. Teratology 28: 1–8

Webster WS, Brown-Woodman PDC (1990) Cocaine as a cause of congenital malformations of vascular origin: experimental evidence in the rat. Teratology 41: 689–697

Webster WS, Lipson AH, Brown-Woodman PDC (1987) Uterine trauma and limb defects. Teratology 35: 253–260

Webster WS, Brown-Woodman PDC, Lipson AH, Ritchie HE (1991) Fetal brain damage in the rat following prenatal exposure to cocaine. Neurotoxicol Teratol 13: 621–626

Webster WS, Danielsson BR (1996) Phenytoin-induced bradycardia in rat embryos. Teratology 53: 123

Webster WS, Brown-Woodman PDC, Snow M, Danielsson BRG (1996) Teratogenic potential of almokalant, dofetilide and d-sotalol: drugs with potassium channel blocking activity. Teratology (in press)

Wide-Swensson D, Ingemarsson I, Arulkumaran S, Andersson K-E (1990) Effects of isradipine, a new calcium antagonist, on maternal cardiovascular system and uterine activity in labour. Br J Obstet Gynaecol 97: 945

Wildenthal K (1973) Maturation of responsiveness to cardioactive drugs. Differential effects of acetylcholine, norepinephrine, theophylline, tyramine, glucagon, and dibutyryl cyclic AMP on atrial rate in hearts of fetal mice. J Clin Invest 52: 2250

Woods JR, Plessinger MA, Clark KE (1987) Effect of cocaine on uterine blood flow and fetal oxygenation. JAMA 257: 957–961

Yoshida T, Takegawa Y, Miyago M, Hasegawa Y (1989) Hyperphalangism induced by Ca-blockers in rat fetuses. Teratology 40: 668

Yaari Y, Selzor ME, Pineus JH (1986) Phenytoin: mechanisms of its anticonvulsant action. Ann Neurol 20: 171–184

CHAPTER 25

Anticoagulants

R.M. PAULI

A. Introduction

Anticoagulants have been used clinically for over half a century and are critical for the adequate treatment of a variety of health- and life-threatening conditions. The recognition that some of them may also pose a risk to the developing embryo and fetus is more recent. This review summarizes and assesses the evidence of harmful intrauterine effects of the two major classes of anticoagulants, coumarin derivatives and heparin, with particular emphasis on postulated pathogenetic mechanisms.

B. Use of Anticoagulants in Pregnancy

Pregnancy results in a hypercoagulable state in the mother secondary to increased levels of coagulation factors and of thrombin generation and decreased fibrinolysis (DEMERS and GINSBERG 1992), resulting in a chronic, low-grade intravascular coagulopathy (RUTHERFORD and PHELAN 1991). These changes presumably prepare the mother's body for the hemostatic challenge of separation of the placenta from the uterine wall (GREAVES 1993). In conjunction with venous stasis and venous outflow obstruction from the legs, this hypercoagulability results in markedly increased risks of thromboembolism in the pregnant woman (DEMERS and GINSBERG 1992). As many as one in 1500 to one in 5000 pregnancies are complicated by maternal thromboembolic disease with a resulting maternal mortality rate of about 1 per 100 000 births (RUTHERFORD and PHELAN 1991; DEMERS and GINSBERG 1992). Anticoagulation is essential when severe thromboembolic disease develops (GINSBERG and HIRSH 1992) and also may be used prophylactically in those with a prior history of serious thromboembolic disease (RUTHERFORD and PHELAN 1991).

There are a number of other indications for anticoagulation in pregnancy (CONARD et al. 1990; DEMERS and GINSBERG 1992; GINSBERG and HIRSH 1992; DAHLMAN 1993; NEERHOF et al. 1993), including maternal mechanical heart valve prostheses, inborn deficiencies of anticoagulation factors (protein S, protein C, and antithrombin III) and prevention of fetal sequelae in patients with antiphospholiopid antibodies (GINSBERG and HIRSH 1992).

Coumarin derivatives and heparin are the only agents available which are effective in the treatment of such complications (Greaves 1993). These will be considered in turn from the prospective of their teratogenic potential.

C. Coumarin Derivatives

I. Historical Overview

Vitamin K (*K*oagulation factor) was discovered by Henrik Dam almost 70 years ago and was introduced into clinical practice shortly thereafter for correction of defective clotting secondary to liver dysfunction. Subsequently, it proved critical in the treatment and prevention of hemorrhagic disease of neonates, which remains one of its most vital clinical functions today (Uotila 1990; Shearer 1992).

The anticoagulant activity of coumarin derivatives was first demonstrated through the investigation of sweet clover disease in cattle (Uotila 1990). These compounds eventually found utility in clinical medicine for the treatment of disorders which result in increased risk of intravascular coagulation, thrombosis formation, and embolization, as well as in the public health arena as potent rodenticides.

Two decades ago it was demonstrated that a specific constellation of abnormalities could arise in some of the offspring of women given these coumarin derivatives during pregnancy (Becker et al. 1975; Shaul et al. 1975; Pettifor and Benson 1975). Simultaneously, a far clearer understanding developed of the essential nature of vitamin K (as a cofactor in post-translational γ-carboxylation of proteins), of the function of γ-carboxyglutamyl residues that result (calcium binding), and of the mechanism of action of oral anticoagulants (as an inhibitor of this post-translational process) (Shearer 1992). Subsequent demonstration of the existence of other, noncoagulation proteins which are vitamin K dependent and warfarin inhibitable (Hauschka et al. 1989) allowed for a confluence of clinical and biochemical observations, resulting in a novel understanding of the pathogenesis of the teratologic effects of coumarin derivatives.

II. Mode of Action

1. Vitamin K

Vitamin K is one of the four fat-soluble vitamins, and most of it is obtained through ingestion of phyloquinone (Shearer 1992). Absorption in the upper portion of the small intestine is dependent on the presence of both bile salts and on pancreatic lipolysis (Shearer 1992). The primary biological function of vitamin K is as a cofactor in post-translational modification of a variety of proteins, which results in the creation of γ-carboxylated glutamyl residues (Stenflo and Suttie 1977). This modification, in turn, results in a conformational change, which allows the modified proteins to bind calcium and to interact with phospholipids (SUTTIE 1980; HIRSH and FUSTER 1994). The biological activity of vitamin K-dependent proteins requires this transformation.

The longest and best known proteins which require such post-translational modification are the vitamin K-dependent coagulation factors i.e., factors II, VII, IX, and X. These are synthesized in the liver in an inactive form (SUTTIE 1980). The carboxylase reaction by which they are made biologically active requires the reduced form of vitamin K (FURIE and FURIE 1990). Therefore, vitamin K is reduced to vitamin K hydroquinone by various vitamin K reductases (FURIE and FURIE 1990). During the reaction which results in γ-carboxylation of proteins, the reduced vitamin K (vitamin K hydroquinone) is simultaneously converted to vitamin K epoxide (FURIE and FURIE 1990), both reactions probably reflecting different actions of the same enzyme (SHEARER 1992). In order for the vitamin K epoxide to be salvaged for further use, it is cycled by conversion back to vitamin K through the action of vitamin K epoxide reductase (FURIE and FURIE 1990) (Fig. 1).

Coagulation factor synthesis and their post-translational modification occurs only in the liver. The liver-localized enzymes have been best studied. However, all tissues in which vitamin K-dependent proteins are synthesized contain all of the machinery necessary both for carboxylation and for vitamin K recycling (HAUSCHKA et al. 1989; FURIE and FURIE 1990).

2. Mechanism of Anticoagulation

Oral anticoagulants exert their effect through inhibition of the post-translational carboxylation of coagulation proteins (STENFLO and SUTTIE 1977;

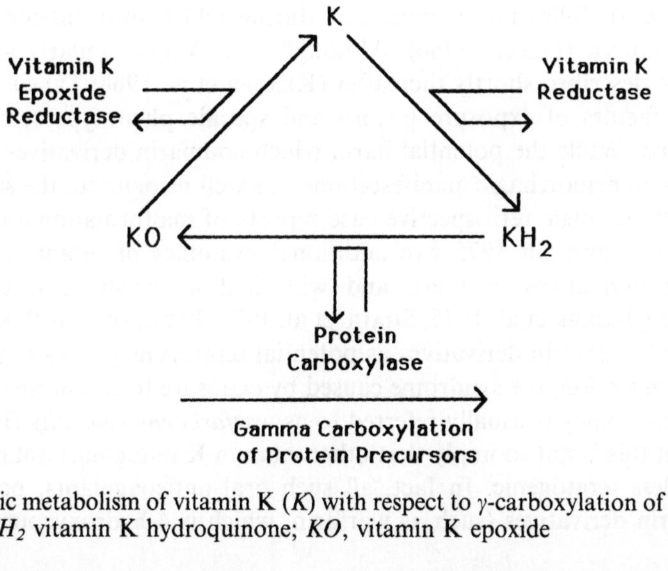

Fig. 1. Cyclic metabolism of vitamin K (*K*) with respect to γ-carboxylation of precurson proteins. KH_2 vitamin K hydroquinone; *KO*, vitamin K epoxide

GALLOP et al. 1980; SUTTIE 1980, 1983). Precursor proteins continue to be formed but, lacking γ-carboxyglutamic acid sites, these proteins are biologically inactive (SUTTIE 1983). The oral anticoagulants accomplish this through interference with the cyclic interconversion of vitamin K and vitamin K epoxide. It should be recalled that vitamin K epoxide, generated in the carboxylation reaction, is recycled through the action of vitamin K epoxide reductase (Fig. 1). All coumarin derivative anticoagulants specifically inhibit vitamin K epoxide reductase, so that this recycling possibility is lost. This, in turn, results in decreased function of all proteins requiring γ-carboxylation, both coagulation factors II, VII, IX, and X and anticoagulation factors protein C and protein S (GREAVES 1993), but, on balance, the net effect is decreased coagulation activity.

These anticoagulant effects, in part, can be overcome by dietary vitamin K, which can enter the vitamin K cycle through a second, warfarin-independent reductase (SHEARER 1992), but a continuing supply of exogenous vitamin K would of course be needed, since no recycling would occur.

One would anticipate that the most potent effects on coagulation (and on other vitamin K-dependent functions) would arise when low vitamin K levels (whether secondary to dietary depletion or to physiologically low levels, as are present in the fetus) co-occurred with exogenous inhibitors. In such circumstances one might anticipate that even weak inhibitors could have a substantial effect (SHEARER 1992).

III. Warfarin Embryopathy

1. Initial Recognition

Three decades ago an infant was described with abnormalities of ossification, nasal hypoplasia, terminal phalangeal abnormalities, and other birth defects who was born following a pregnancy during which oral anticoagulant exposure occurred (DISAIA 1966). Although remarkably similarly affected infants were described shortly thereafter (KERBER et al. 1968; TEJANI 1973), the common factors of exposure history and specific phenotype remained unappreciated. While the potential harm which coumarin derivatives can cause secondary to hemorrhagic manifestations was well recognized, the significance of these three single retrospective case reports of malformational effects was ignored. However, in 1975 five additional examples of infants exposed to coumarin derivatives in utero and who had a specific constellation of anomalies (BECKER et al. 1975; SHAUL et al. 1975; PETTIFOR and BENSON 1975) established coumarin derivatives as potential teratogens (WARKANY 1975).

By convention, the syndrome caused by exposure to coumarin derivatives early in pregnancy is usually referred to as *warfarin embryopathy* (HALL et al, 1980), but this is not to imply that other vitamin K antagonist anticoagulants are any less teratogenic. In fact, all such oral anticoagulants, both hydroxycoumarin derivatives (such as warfarin, which is 4-hydroxycoumarin) and indane-1,3-dione-derived compounds, appear to have a similar teratogenic potential (HALL et al. 1980; GÄRTNER et al. 1993).

2. Phenotypic Manifestations

Early reports established that a specific combination of features sometimes arises in coumarin derivative-exposed infants, principally including nasal hypoplasia (Fig. 2), abnormalities of ossification (Fig. 3), and limb abnormalities. By 1980, when HALL et al. published a review of the effects of anticoagulants in pregnancy, 29 instances of warfarin embryopathy had been identified. Since, then, another three dozen examples have been published. Table 1 summarizes the prominent features in the 64 individuals whose case reports either have been previously published or are known to the author by personal communication or who have been examined by the author. The data in Table 1 do not include all instances of presumed embryopathic effects, since in many series the clinical features of affected infants are insufficiently documented. Furthermore, such unequivocal instances are likely to represent only the more severe end of a continuum of effects. We have assessed neonates and fetuses who were exposed in appropriate periods of gestation who had subtle facial features suggestive of the embryopathy but in whom no other char-

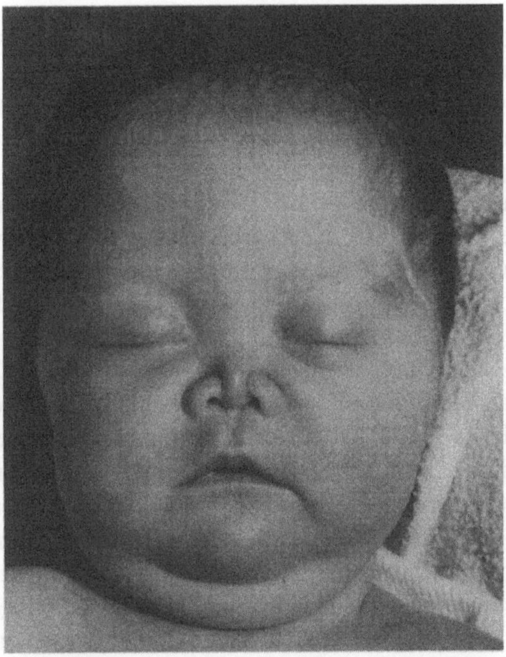

Fig. 2. Craniofacial features of warfarin embryopathy. Note the small nose and septation between the alae nasi and nasal tip. (Reproduced with permission from PAULI et al. 1976)

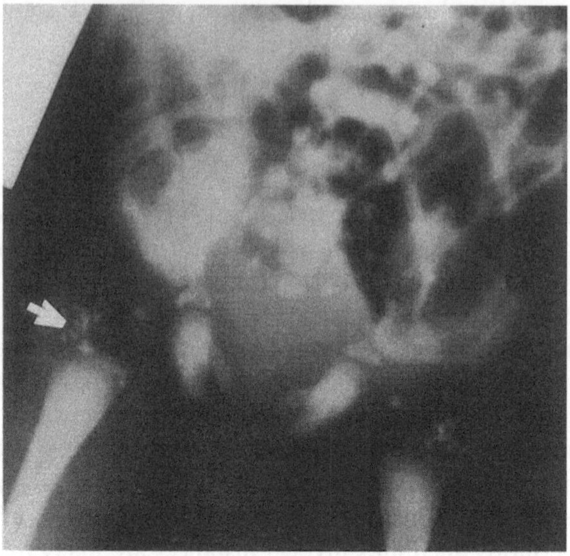

Fig. 3. Radiograph showing so-called stippled epiphyses in infant with warfarin embryopathy. The *arrow* points to one such area of stippling. (Reproduced with permission from PAULI et al. 1976)

acteristics were present, an observation suggested by others as well (WONG et al. 1993). We have used the previously suggested minimal diagnostic criteria for determining whether an example is included (HALL et al. 1980): appropriate intrauterine exposure to a coumarin derivative and either nasal hypoplasia or stippling of epiphyses.

The one constant feature of infants diagnosed as having warfarin embryopathy is the presence of nasal hypoplasia. Note, however, that this statement is somewhat tautologic, since many infants (all those in whom appropriate radiologic evaluation was not completed) would be excluded a priori, given the suggested minimal diagnostic criteria, if nasal hypoplasia were not present. Typically, the nasal bridge is depressed, the nasal septum remarkably diminished in length, and the nose small with anteverted nares. In many affected infants, there is a curious and characteristic septation between the alae nasi and the nasal tip (Fig. 2). Neonatal respiratory distress is exceedingly frequent (estimated to occur in 47%–65% of affected infants, Table 1). Most often this appears to arise as a result of the markedly diminished size of the nasal airways in conjunction with the obligate nose breathing in infants. The consequent obstruction can usually be straightforwardly relieved by use of a nasal airway (HALL et al. 1980). Use of nasal stents or orotracheal intubation likely would be of similar benefit. In a few instances choanal stenosis has also been documented. Although respiratory compromise appears to be limited to infancy, the nose and nasal septum do not show catchup growth (HOSENFELD and WIEDEMANN 1989; R.M. PAULI, personal

Table 1. Major manifestations in 64 reported instances of the warfarin embryopathy

Feature	Total cases reported[a]	Affected (n)	Affected (%)	Minimum risk (%)[b]
Nasal hypoplasia	61	61	100	95
Neonatal respiratory distress	46	30	65	47
Periepiphyseal stippling	56	49	88	77
Limb anomalies (distal digital hypoplasia and/or foreshortening)	44	26	59	41
Small for gestational age (< 10th percentile)	49	21	43	33
CNS abnormality	26	13	50	20
Ophthalmologic abnormality	40	7	18	11
Hearing loss	39	8	21	13
Congenital heart abnormality	35	4	11	6
Death	63	10	16	16

Modified and updated from data summarized in PAULI and HAUN 1993. Data from the following sources: DISAIA 1966; KERBER et al. 1968; TEJANI 1973; BECKER et al. 1975; FOURIE and HAY 1975; PETTIFOR and BENSON 1975; SHAUL et al. 1975; BARR and BURDI 1976; PAULI et al. 1976; RICHMAN and LAHMAN 1976; RAIVIO et al. 1977; VANLAEYS et al. 1977; GOOCH et al. 1978; LUTZ et al. 1978; ROBINSON et al. 1978; HALL et al. 1980; BAILLIE et al. 1980; CURTIN and MULHERN 1980; WHITFIELD 1980; HARROD and SHERROD 1981; WEENINCK et al. 1981; LAFRANCHI et al. 1982; O'NEILL et al. 1982; JULLIAN et al. 1983; LARREA et al. 1983; LAMONTAGNE et al. 1984; JAVARES et al. 1984; SALAZAR et al. 1984; STRÜWE et al. 1984; MATORRAS et al. 1985; PAWLOW and PAWLOW 1985; ITURBE-ALLESIO et al. 1986; VITALI et al. 1986; ZAKZOUK 1986; TAMBURRINNI et al. 1987; ZIPPRICH et al. 1987; HOSENFELD and WIEDEMANN 1989; SARELI et al. 1989; LEICHER-DÜBER et al. 1990; FREUDE et al. 1991; MASON et al. 1992; DE VRIES et al. 1993; GÄRTNER et al. 1993; PAULI and HAUN 1993; BARKER et al. 1994.
CNS, central nervous system.
[a]Shown here are the number of instances in which published cases or unpublished materials allow assignment in each of the feature categories as unaffected or affected.
[b]Data derived from assuming that in each instance in which there is insufficient information about a feature to allow assignment this feature is *absent*. The actual frequency of each of these characteristics in infants with the warfarin embryopathy is likely somewhere between the percentages given for "Affected" and "Minimum risk".

observation); facial features similar to the so-called Binder syndrome (HOWE et al. 1992) persist.

Stippling of periepiphyseal regions is the second virtually constant feature of warfarin embryopathy (Fig. 3). Such irregular calcific deposition has been demonstrated in 88% of those affected infants in whom radiographs were obtained (Table 1). Even this may be an underestimate. Stippling appears to be incorporated into areas which are normally calcified with maturation and, therefore, stippling is usually not identifiable after about 1 year of age. Therefore, if radiographs are not obtained in the neonatal period, the stippling may go undetected. In the warfarin embryopathy, the stippled regions are mainly along the axial skeleton, at the proximal femora, and within the calcanei. Irregularity of the ossification of the calcanei may be the longest persisting radiologic manifestation (R.M. PAULI, personal observation), similar to that seen in mild forms of chondrodysplasia punctata (SCHEFFIELD et al. 1976). Occasionally, calcific stippling will also be seen in the laryngeal or tracheal cartilages (ROBINSON et al. 1978; HALL et al. 1980), but this does not seem to contribute substantially to the observed respiratory distress. Incidentally, MONCADA et al. (1992) report the calcification of cartilaginous rings of the trachea and bronchi in adults after long-term administration of coumarin derivatives, and it is likely that the mechanism of this effect is similar to that resulting in warfarin embryopathy when prenatal exposure occurs.

Other features may also reflect abnormality of osteochondral development. Around half of all described patients have limb abnormalities (Table 1). Most commonly these consist of distal digital hypoplasia with shortening of the terminal phalanges and hypoplastic fingernails (HALL et al. 1980). Varying degrees of rhizomelic shortening are present in a minority of affected individuals (see, e.g., BECKER et al. 1975).

At least a third of recognized infants are small for their gestational age, and smaller proportions have hearing loss, congenital heart disease, central nervous systems anomalies, and ophthalmologic abnormalities (Table 1). It is not certain which of these features are intrinsic to warfarin embryopathy and which arise independently of it. It seems likely that hearing loss is a component of the embryopathy per se, while heart disease may arise either associated with or independent of the other features of the embryopathy (see Sect. IV). The central nervous system and eye defects almost certainly arise through a separate mechanism and by exposure to coumarin derivatives later in pregnancy (see Sect. IV); both studies of the "genocopy" of warfarin embryopathy and information regarding critical periods of exposure support this. Thus those features of greatest long-term consequence are likely not part of warfarin embryopathy per se.

Death, which has occurred in about one sixth of those infants recognized to have warfarin embryopathy, also most often seems to be secondary to central nervous system effects rather than to warfarin embryopathy itself.

3. Critical Period of Exposure

One limit of the critical period of exposure for warfarin embryopathy was demonstrated by its occurrence in an infant whose mother began use of warfarin in the sixth postconceptual week (PAULI et al. 1976). Analysis of exposure periods of all instances known at that time led HALL et al. (1980) to postulate that exposure had to include some period between the sixth and ninth (postconceptual) weeks of pregnancy. This estimate of the critical period has been supported by virtually all subsequent cases of warfarin embryopathy. The one exception (DEVRIES et al. 1993) is said to have been exposed only from conception to the fifth week and again after 12 weeks of gestation; note, however, that there was no independent confirmation of those dates. Nonetheless, this, and one case reported by HARROD and SHERROD (1981), in whom exposure did not begin until the ninth postconceptual week, leads to some uncertainty about the temporal range, during which exposure can cause embryopathic effects.

4. Pathogenetic Mechanism Resulting in Warfarin Embryopathy

It was initially postulated that warfarin embryopathy resulted from inhibition of fetal coagulation factors and consequent hemorrhage (BECKER et al. 1975). This was demonstrably in error based upon the critical period of exposure of 6–9 weeks of gestation and the previous demonstration that vitamin K-dependent coagulation proteins do not appear until 12–14 weeks of gestation (BLEYER et al. 1971). This implied that either warfarin embryopathy results from direct pharmacologic effects on other vitamin K-dependent proteins which appear prior to this gestational age or that the teratogenic effects of coumarin derivatives are unrelated to those primary pharmacologic effects. Evidence now suggests that other vitamin K-dependent proteins are involved in the pathogenesis of warfarin embryopathy.

a) *Pseudowarfarin Embryopathy*

Much of what we understand about the biochemical basis of the warfarin embryopathy is based upon a rare "genocopy" which combines the phenotypic features of warfarin embryopathy and a vitamin K-dependent combined coagulopathy.

α) Vitamin K and Coagulation. As reviewed in Sect. II above, four factors in the clotting cascade require vitamin K for their post-translational modification into active proteins (UOTILA 1990). Calcium binding by factors II, VII, IX, and X is dependent on this vitamin K-dependent post-translational carboxylation of glutamyl residues (STENFLO and SUTTIE 1977). Vitamin K deficiency or coumarin administration results in underdecarboxylation (STENFLO and SUTTIE 1977) and consequent deficiency of function (SUTTIE 1980, 1983) of these factors. Coumarin derivatives exert this effect through inhibition of vitamin K epoxide reductase, which is essential for the cycling of vitamin K (Fig. 1). Inborn abnormalities of vitamin K utilization, of vitamin K cycling, or of protein carboxylation could cause a congenital coagulopathy entirely analogous to that caused by the anticoagulant effects of coumarin derivative administration.

β) Vitamin K-Dependent Coagulopathies. Eight individuals have been recognized in whom there appears to be isolated abnormality of all vitamin K-dependent coagulation factors (NEWCOMB et al. 1956; MCMILLAN and ROBERTS 1966; FISCHER and ZWEYMÜLLER 1966. CHUNG et al. 1979; JOHNSON et al. 1980; GOLDSMITH et al. 1982; BRENNER et al. 1990), a process designated as familial multiple-factor deficiency type III (SOFF and LEVIN 1981). This defect has affected both sexes, has arisen in two sibships (GOLDSMITH et al. 1982; BRENNER et al. 1990), including one with parental consanguinity (BRENNER et al. 1990), and is seen in offspring of parents who have normal vitamin K-dependent clotting factors. Therefore, it is very likely secondary to one or more autosomal recessive alleles. In none of these instances have abnormal external phenotypic features been described. A variety of mechanisms could cause such combined deficiencies. Two enzymopathies are most likely: either a defect in vitamin K epoxide reductase or in protein carboxylase (both of which are present in all tissues with γ-carboxylated glutamyl residue containing proteins ; FURIE and FURIE 1990; SHEARER 1992). Data from BRENNER et al. (1990) suggest that a carboxylase defect is most likely the etiology for this "pure" form of vitamin K-dependent coagulation deficiency (which might arise without other phenotypic manifestations if carboxylase activity is tissue specific).

γ) Association of Vitamin K-Dependent Coagulopathy and the Phenotype of Warfarin Embryopathy. In contrast to the individuals described above, we examined a boy who not only had functional deficiency of all vitamin K-dependent coagulation factors, but who also had nasal hypoplasia (Fig. 4), distal digital hypoplasia (Fig. 5), radiographic stippling, and conductive hearing loss, all of which were quite similar to what is seen in warfarin embryopathy (PAULI et al. 1987). During delivery, this boy incurred a scalp laceration, which continued to bleed until blood products were administered. He bruised easily and had episodes of bleeding in early childhood that were difficult to control. Uncontrollable epistaxis led to coagulation studies, which, then and subsequently, showed selective deficiency of all the vitamin K-dependent coagulation factors. Oral vitamin K resulted in resolution of the bleeding problems and partial correction of the coagulation parameters. Thus this boy had a combined vitamin K-dependent coagulopathy and the phenotype of warfarin embryopathy. Subsequent biochemical assessments (PAULI et al. 1987) showed that the coagulation abnormalities arose secondary to undercarboxylation of glutamyl residues of the coagulation factors, and various assays of prothrombin showed a pattern equivalent to that seen after warfarin therapy. These abnormalities, in turn, were shown to be secondary to a defect in vitamin K epoxide reductase (rather than to a carboxylase defect per se; note that this observation suggests that the same epoxide reductase activity is shared among different tissues).

Subsequent to our description of the first example of pseudowarfarin embryopathy (PAULI et al. 1987), five additional infants have been recognized who appear to have the same biochemical defect resulting in the same unique combination of features as was described in this boy. Only one of these has been published. LEONARD (1988) described a male infant with severe nasal hypoplasia, distal digital hypoplasia, and generalized stippling on radiographs; he also experienced excessive bleeding following circumcision. Coagulation studies demonstrated a vitamin K-dependent combined caogulopathy virtually identical to that seen in the boy reported by us (PAULI et al. 1987) with deficient post-translational carboxylation of vitamin K-dependent factors (PAULI 1988). This infant also developed postnatal hydrocephalus (perhaps secondary to intracranial hemorrhage). A brother of this patient died neo-

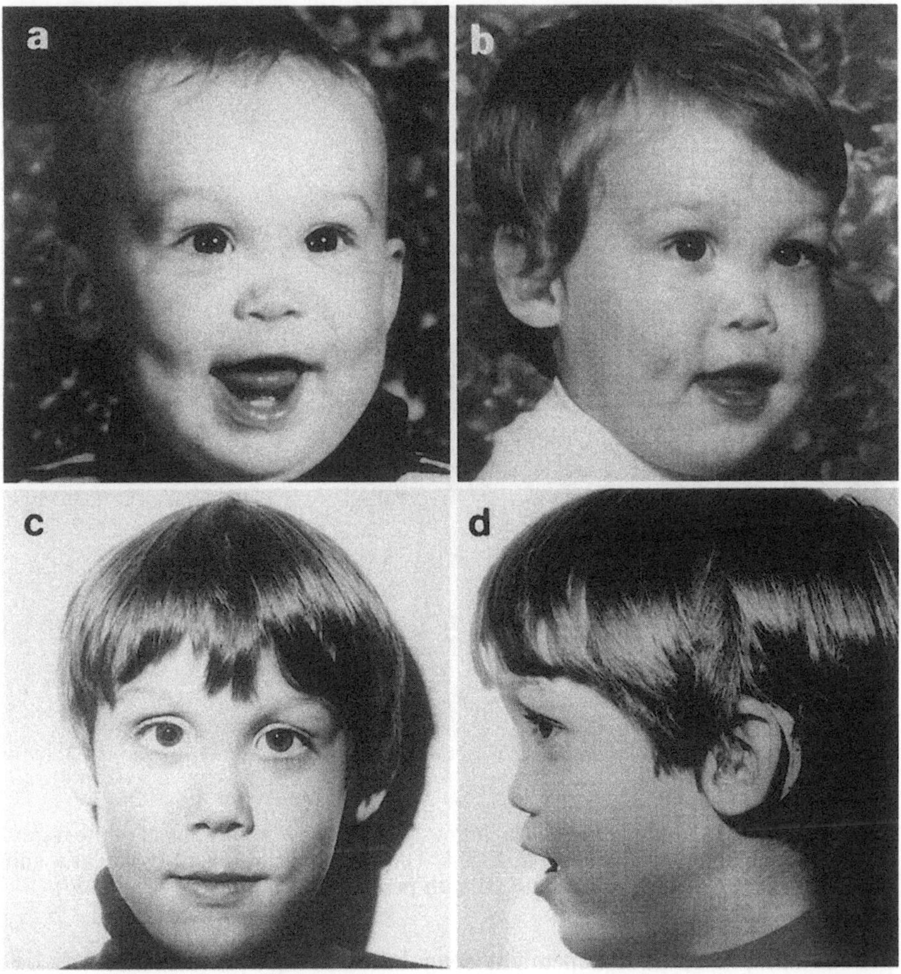

Fig. 4a-d. Facial features at 10 months (**a**), 23 months (**b**), and 7 1/2 years (**c,d**) in the first patient recognized to have pseudowarfarin embryopathy. (Reproduced with permission from PAULI et al. 1987)

natally from hemorrhagic disease (even though vitamin K was administered at birth) and likely was also affected (LEONARD 1988).

Two other unrelated males with both a vitamin K-dependent coagulopathy and clinical features of warfarin embryopathy have shown unequivocal, if indirect, evidence of the same defect in vitamin K epoxide reductase (unpublished observations); one of these died secondary to a central nervous system hemorrhage.

Another male child had both clinical and radiographic features consistent with this diagnosis but, in addition, demonstrated neonatal cerebral and cerebellar atrophy as well as complex congenital heart disease, which ultimately

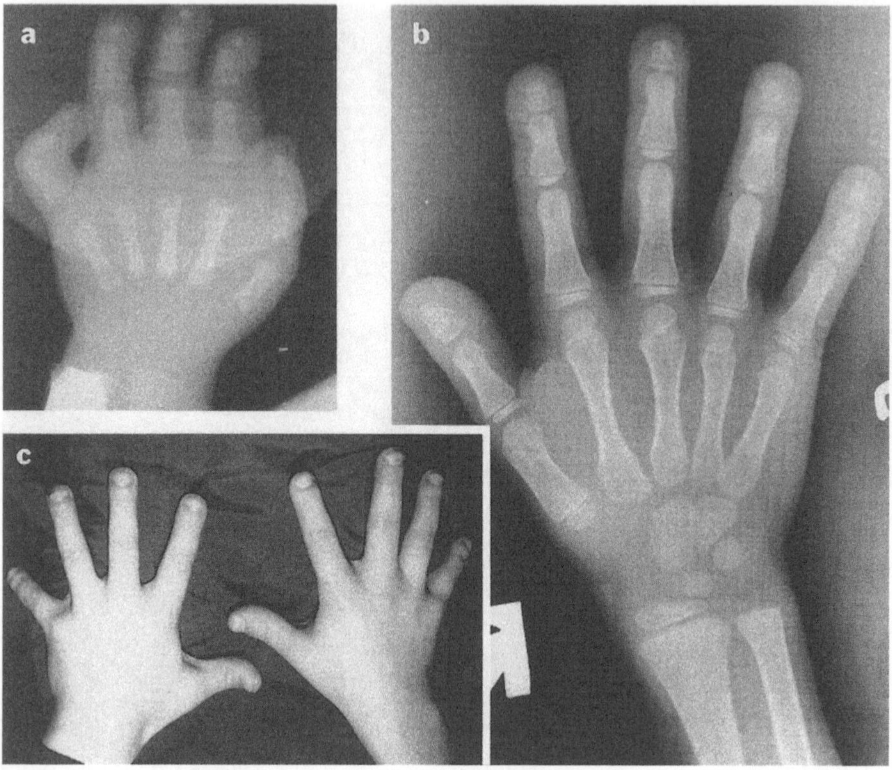

Fig. 5a-c. Infant (**a**) and childhood (**b**) radiographs and childhood clinical photographs (**c**) of the same patient as shown in Fig. 4. They show distal digital hypoplasia and generalized brachydactyly. (Reproduced with permission from PAULI et al. 1987)

proved fatal; vitamin K-dependent coagulation factors were moderately depressed (unpublished observation). Liver samples taken after his death were assayed for vitamin K and vitamin K epoxide levels and for epoxide reductase activity (J.W. SUTTIE, unpublished observation). These assays, when compared with control samples from infants that had died from sudden infant death syndrome, suggest that this patient did have a congenital deficiency of vitamin K epoxide reductase.

Finally, we recently assessed a female infant with similar clinical and radiographic features who had sufficiently severe airway obstruction to necessitate tracheostomy. Detailed assays of vitamin K-dependent function are pending, but this observation is of some importance, since she would be the first recognized female and, hence, would suggest that this defect is likely an autosomal recessive (rather than possibly x-linked) process.

δ) Vitamin K-Dependent Bone and Cartilage Proteins. Proteins which require vitamin K-dependent post-translational carboxylation and hence are warfarin inhibitable have been identified in nonhepatic tissues and apparently have functions independent of coagulation (HAUSCHKA et al. 1989; UOTILA 1990). Two of these are particularly relevant to the likely pathogenesis of warfarin embryopathy. Both osteocalcin (HAUSCHKA et al. 1978) and matrix Gla protein (PRICE et al. 1983) are such vitamin K-dependent proteins (HAUSCHKA et al. 1989). Both arise ontogenetically sufficiently early to be consistent with the known critical period of coumarin derivatives' embryopathic effects (HAUSCHKA and REID 1978; OTAWARA and PRICE 1986). As with other vitamin K-dependent proteins, post-translational modification of these bone and cartilage proteins modify their ability to bind calcium (HAUSCHKA et al. 1989). Both may play roles in calcium homeostasis (PRICE et al. 1982, 1983; SHEARER 1992). Matrix Gla protein, in particular, seems to be an attractive candidate for regulation of calcium deposition in differentiating cartilage. Chondrocytes contain all of the enzymatic machinery for vitamin K-dependent and warfarin-inhibitable post-translational modification of matrix Gla protein (LOESER and WALLIN 1991). Matrix Gla protein is synthesized in the growth plate (HALE et al. 1988), is present in uncalcified cartilage and as bone begins to ossify (PRICE et al. 1981; OTAWARA and PRICE 1986), and is anchored to the organic bone matrix (OTAWARA and PRICE 1986).

b) Proposed Mechanism

The characteristic features of warfarin embryopathy appear to result either from abnormalities of calcium deposition (e.g., stippled epiphyses, conductive hearing loss secondary to ossicular dysfunction) or from abnormalities of cartilage growth (nasal septal hypoplasia, distal digital hypoplasia, limb fore-shortening). One could, then, postulate that warfarin-inhibitable proteins of cartilage might result in the embryopathy through the primary pharmacologic action of coumarin derivatives (inhibition of γ-carboxylation) on matrix Gla protein or a similar vitamin K-dependent protein. *If* this were true and *if* there was an inborn error of metabolism which caused inhibition of all vitamin K-dependent functions (e.g., deficiency of vitamin K epoxide reductase activity), then we would predict that it would cause both a type III combined coagulopathy and a phenotypic copy of warfarin embryopathy as well. Pseudowarfarin embryopathy has demonstrated exactly this (PAULI et al. 1987). By analogy it appears likely that the embryopathic effects of coumarin derivatives are the result of their primary pharmacologic action, inhibition of vitamin K epoxide reductase (Fig. 6; PAULI et al. 1987; PAULI and HAUN 1993).

5. Relationship to Other Disorders Which Cause Radiographic Stippling

a) Processes Which May Share a Common Pathogenesis

Anything that results in abnormal vitamin K function (e.g., through inhibition of vitamin K epoxide reductase) during a critical period of development (6–9 weeks) might result in embryopathic features similar to those seen following exposure to coumarin derivatives. Even weak inhibitors might have a sub-

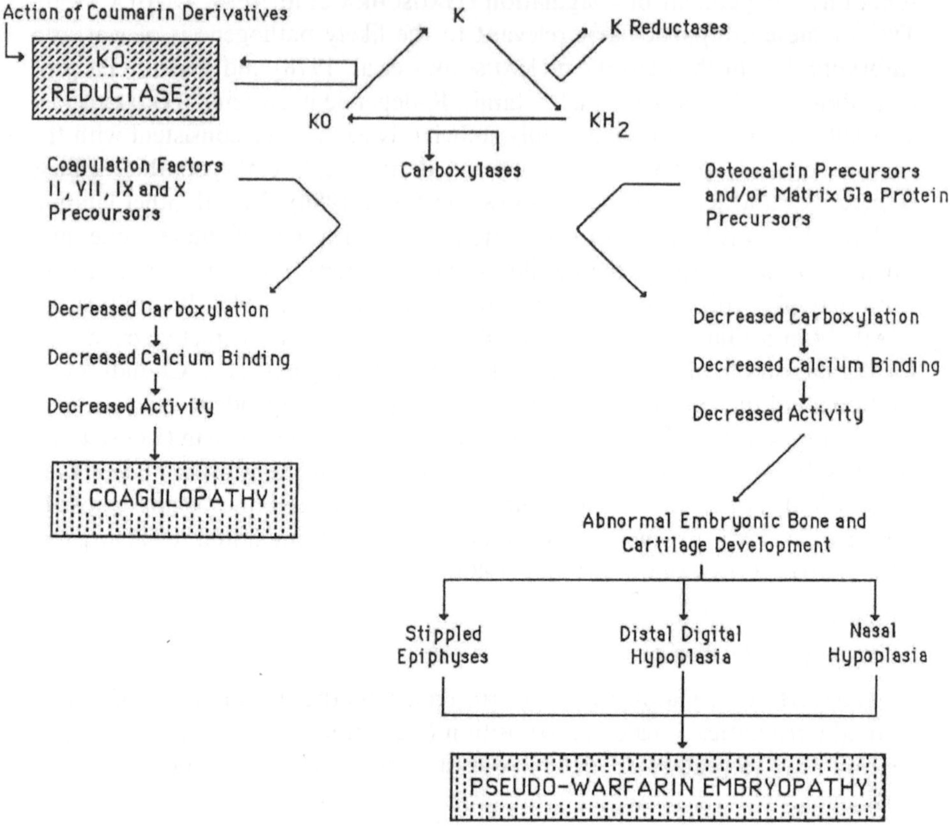

Fig. 6. Proposed mechanism for pseudowarfarin embryopathy. In parallel, this suggests that the underlying mechanism of coumarin derivatives' embryopathic effects are also secondary to inhibiton of vitamin K epoxide reductase. (Reproduced with permission from PAULI and HAUN 1993)

stantial teratogenic effect given the marked maternal–fetal gradient and consequent relative fetal deficiency for vitamin K (SHEARER 1992).

α) Hydantoins. Prior to recognition of a teratogenic risk, the anticonvulsant hydantoins were shown to predispose neonates to hemorrhage secondary to inhibition of vitamin K-dependent coagulation factors (MOUNTAIN et al. 1970; SOLOMON et al. 1972; BLEYER and SKINNER 1976). The fetal hydantoin syndrome shares many characteristics with warfarin embryopathy: growth retardation, nasal and midface hypoplasia, and distal digital hypoplasia (HANSON and SMITH 1975). Anecdotal evidence suggests that prenatal hydantoin exposure can, indeed, result in bony features virtually identical to those seen in warfarin embryopathy. One of the early instances of warfarin embryopathy was observed in an infant born to a mother who had also taken diphenylhydantoin during her pregnancy (Pettifor and Benson 1975), and Sheffield et al. (1976), in their report of 26 individuals with a particularly mild form of chondrodysplasia punctata, included two instances in which babies born with radiographic stippling had been exposed to hydantoins in utero. A similar example was reported more recently (Howe et al. 1992). It seems likely that these observations together are most easily explained by postulating that hydantoins cause many of their in utero effects through inhibition of vitamin K epoxide reductase activity. Incidentally, it has been suggested that administration of vitamin K to the mother can reduce the deleterious hemorrhagic effects of antiseizure medications (Deblay et al. 1982); exogenous supplementation might also abrogate the malformational effects of hydantoins and conceivably might prevent the consequences of inherent deficiency of vitamin K epoxide reductase as well.

β) Alcohol. Occasionally, infants with the fetal alcohol syndrome have been noted to have radiographic stippling (Badois et al. 1983; Leicher-Duber et al. 1990; Howe et al. 1992). Whether this might arise either from direct effects on vitamin K epoxide reductase or, alternatively, secondary to concomitant malnutrition remains entirely speculative.

γ) Maternal Malnutrition. The hypothesis that maternal malnutrition may result in phenotypic features similar to those seen in warfarin embryopathy has received some support from the observation made by Toriello et al. (1990), who describe two patients born after pregnancies complicated by maternal malnutrition (secondary to functional or structural short gut syndrome) and total parenteral nutrition. In each instance the infant had phenotypic features identical to warfarin embryopathy, and Toriello et al. (1990) hypothesized that malabsorption and malnutrition may have resulted in maternal vitamin K deficiency, which, in the face of the marginal vitamin K status of the normal fetus (Mandelbrot et al. 1988), resulted in the observed phenotype.

b) Binder Syndrome

Binder syndrome is a descriptive diagnostic label sometimes applied to individuals with isolated nasal spine hypoplasia and related facial characteristics (Quarrell et al. 1990). Sheffield and his colleagues have recognized the similarity of these and the facial features of chondrodysplasia punctata and have provided clinical evidence that many instances of Binder syndrome do in fact have radiographic stippling and other calcification abnormalities if evaluated early in life (Sheffield et al. 1990; Howe et al. 1992). It now seems likely that many instances of Binder syndrome are identical to mild chondrodysplasia punctata (Sheffield et al. 1976), being identified as one or the other solely dependent upon the age at which diagnosis is sought. In infancy, such individuals usually will be recognized as having chondrodysplasia punctata, while those evaluated in adolescence will be diagnosed as having

Binder syndrome (HOWE et al. 1992). It is conceivable that many or all instances of the Sheffield form of chondrodysplasia punctata and of the Binder syndrome are secondary to in utero abnormalities of vitamin K epoxide reductase activity – some secondary to a variety of exposures and some (probably) secondary either to an inherent transient abnormality of vitamin K epoxide reductase activity or to partial deficiency of matrix Gla protein or to selective abnormality of a bone- and cartilage-specific vitamin K-dependent carboxylase. Investigations of these alternatives in patients with mild chondrodysplasia punctata would be of considerable interest.

c) Other Syndromes with Stippling: Chondrodysplasia Punctata

The chondrodysplasia punctatas are a heterogeneous group of genetic disorders which share stippling of periepiphyseal cartilages on neonatal radiographs (WULFSBERG et al. 1992). These and other disorders in which stippling arises seem to be secondary to one of two primary mechanisms (PAULI 1988; SHEFFIELD et al. 1990; PAULI and HAUN 1993; Fig. 7). One group arises secondary to abnormalities of vitamin K epoxide reductase activity and, perhaps, through other mechanisms affecting vitamin K-dependent post-translational γ-carboxylation, as discussed above. The others are a series of disorders – rhizomelic chondrodysplasia punctata, x-linked dominant chondrodysplasia punctata, and Zellweger syndrome, in particular – which may arise secondary to abnormalities of peroxisomal function.

FRANCO et al. (1995) have characterized another form of chondrodysplasia punctata and, through this, have proposed an alternative mechanism for warfarin's embryopathic effects. X-linked recessive chondrodysplasia punctata arises secondary to nullisomy of a portion of the noninactivated region of the X chromosome in males (CURRY et al. 1984) or secondary to point mutations in the arylsulfatase E gene in this region (FRANCO et al. 1995). In vitro assessments suggest that arylsulfatase E is inhibitable by warfarin, at least in very high concentrations (FRANCO et al. 1995). Given the phenotypic overlap between warfarin embryopathy and X-linked recessive chondrodysplasia punctata and this apparent inhibition by warfarin, FRANCO et al. (1995) proposed that the embryopathic effects of coumarin derivatives may occur through direct inhibition of arylsulfatase E rather than through the mechanism summarized above.

IV. Other Effects of Coumarin Derivative Exposure

1. Hemorrhage

Perinatal hemorrhagic risks of coumarin derivatives were recognized early (see, e.g., QUENNEVILLE et al. 1959). Their use in the last stages of pregnancy is uniformly regarded as contraindicated (DEMERS and GINSBERG 1992; GINSBERG and HIRSH 1992; MATERNAL and NEONATAL HAEMOSTASIS WORKING PARTY 1993; GREAVES 1993; HIRSH and FUSTER 1994).

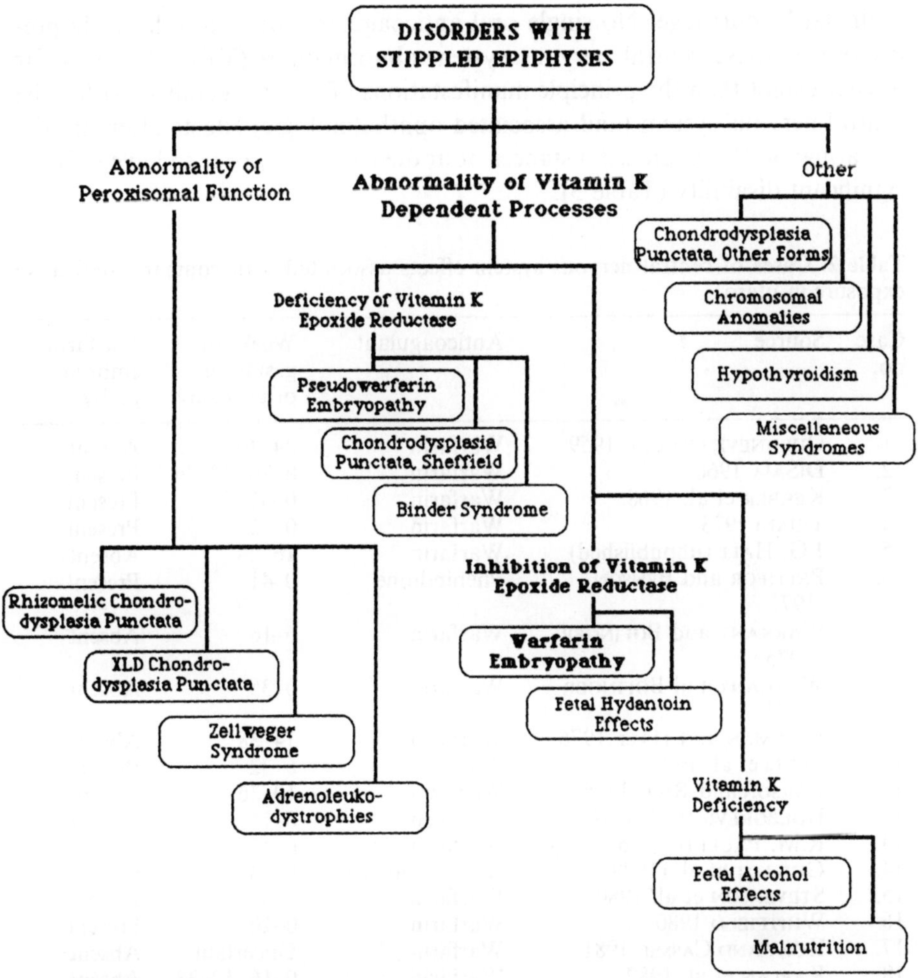

Fig. 7. Relationship of various clinical disorders in which stippled epiphyses are demonstrable

2. Central Nervous System Effects

a) Recognition and Phenotype

Half a century ago, VON SYDOW (1947) reported an instance of hydrocephalus in a child exposed throughout gestation to a coumarin derivative. This example and subsequent similar ones were assumed to be secondary to perinatal hemorrhage. As described in Sect.IV.1, this remains a realistic concern, and thus the recommendation that oral anticoagulants be discontinued prior to onset of labor became incorporated into standard practice (see e.g., HIRSH et al. 1970). Nonetheless, children with central nervous system abnormalities which are not so readily explained in this manner continued to be born.

To date, there are least 31 examples (Table 2) of infants in whom central nervous system problems have arisen which are not explicable on the basis of perinatal hemorrhage. No single oral anticoagulant seems to selectively predispose to these central nervous system abnormalities (Table 2). To a far greater extent than the principle manifestations of warfarin embryopathy, the central nervous system (and associated ophthalmologic) effects often are devastating: of 31 recognized instances, four died and 21 appear to have suffered significant disability (Table 3).

Table 2. Cases of central nervous system effects associated with coumarin derivative exposure in utero

Case no.	Source	Anticoagulant	Weeks of gestation of exposure	Warfarin embryopathy
1.	QUENNEVILLE et al. 1959	Warfarin	24–40	Absent
2.	DISAIA 1966	Warfarin	0–26, 28–36	Present
3.	KERBER et al. 1968	Warfarin	0–31	Present
4.	TEJANI 1973	Warfarin	0–35	Present
5.	J.G. HALL (unpublished)	Warfarin	16–24	Absent
6.	PETTIFOR and BENSON 1975	Phenindione	0–41	Present
7.	WARKANY and BOFINGER 1975	Warfarin	6–16	Absent
8.	WARKANY and BOFINGER 1975	Warfarin	0–35	Absent
9.	SHERMAN and HALL 1976	Warfarin	14–38	Absent
10.	PAULI et al. 1976	Warfarin	6–32	Present
11.	CARSON and REID 1976	Warfarin	10–36	Absent
12.	HOLZGREVE et al. 1976	Warfarin	0–15	Absent
13.	R.M. PAULI (unpublished)	Warfarin	6–40	Absent
14.	GUILLOT et al. (1979)	Acenocoumarin	0–28	Present
15.	STEVENSON et al. 1980	Warfarin	0–31	Present
16.	WHITFIELD 1980	Warfarin	0–20	Present
17.	KORT AND CASSEL 1981	Warfarin	Uncertain	Absent
18.	KAPLAN et al. 1982	Warfarin	0–16, 17–33	Absent
19.	CHEN et al. 1982	Warfarin	0–36	Absent
20.	LARREA et al. 1983	Acenocoumarin	First/second trimester	Present
21.	SHEIKHZADEH et al. 1983	Warfarin	0–32	Absent
22.	KAPLAN 1985	Warfarin	8–12	Absent
23.	PAWLOW AND PAWLOW 1985	Phenprocoumin	0–11	Present
24.	LAPIEDRA et al. 1986	Acenocoumarin	0–6, + uncertain	Undocumented
25.	R.M. PAULI (unpublished)	Warfarin	14–20	Absent
26.	GÄRTNER et al. 1993	Phenprocoumin	9–14	Present
27.	GÄRTNER et al. 1993	Phenprocoumin	0–16	Absent
28.	DEVRIES et al. 1993	Acenocoumarin	0–5, 12–32	Present
29.	VILLE et al. 1993	Warfarin	26–36	Absent
30.	WONG et al. 1993	Warfarin	All trimesters	Absent
31.	OLTHOFF et al. 1994	Acenocoumarin	Second/third trimester	Absent

Modified and updated from data summarized in PAULI and HAUN (1993) Sources are, listed chronologically.

Table 3. Clinical features in 31 identified instances of central nervous system abnormalities associated with exposure to coumarin derivatives

Case no.	CNS structural features	CNS functional features	Severity of CNS effects[a]	Ophthalmologic characteristics
1.	Cerebral agenesis; microcephaly	–	Disabling	Optic atrophy
2.	–	Mild MR; blindness	Disabling	Optic atrophy
3.	–	Moderate to severe MR; seizures; deaf; spasticity	Disabling	–
4.	Hydrocephalus; meningocele	Severe MR	Disabling	Microphthalmia
5.	Cerebellar atrophy; cerebral atrophy	Moderate MR; seizures; spasticity	Disabling	Optic atrophy
6.	–	Mild MR	Disabling	Optic atrophy
7.	Hydrocephalus	–	Lethal	–
8.	DWM; encephalocele; hydrocephalus	–	Disabling	–
9.	Microcephaly	MR; hypotonia	Disabling	Possibly blindness
10.	Cerebral atrophy	MR; hypotonia	Lethal	–
11.	Microcephaly	MR; spasticity	Disabling	Blindness
12.	Possibly DWM (Posterior fossa cyst); possibly ACC	Mild to moderate MR;	Disabling	–
13.	DWM; possibly ACC	Mild MR; spasticity	Disabling	–
14.	Microcephaly	DD; hypotonia	Disabling	–
15.	–	DD	Disabling	Optic atrophy
16.	–	DD; hemiparesis; encephalitis at 9 months	Incidental	–
17.	Meningomyelocele	–	Unknown	–
18.	DWM; absent septum pellucidum; hydrocephalus	–	Disabling	–
19.	Hydrocephalus	MR	Disabling	–
20.	–	Moderate MR	Disabling	–
21.	Anencephaly	–	Lethal	–
22.	DWM; ACC; hydrocephalus	Mild MR; seizures	Disabling	Peter anomaly
23.	ACC; hydrocephalus	seizures; apnea	Unknown	–
24.	Hydrocephalus	–	Unknown	–
25.	Posterior fossa cyst; cerebral atrophy; microcephaly	Moderate MR; seizures; spasticity	Disabling	–
26.	DWM; hydrocephalus	–	Unknown	–
27.	Meningocele	–	Unknown	–
28.	Microcephaly; hydrocephalus	DD	Disabling	–
29.	Cerebellar hypoplasia; cerebral atrophy	–	Lethal	–
30.	Hydrocephalus	DD	Disabling	Microphthalmia; cataract
31.	–	MR; seizures; spasticity	Disabling	Optic nerve hypoplasia

Adapted and updated from material originally published in PAULI and HAUN 1993. Case numbers as in Table 2. MR, mental retardation; DWM, Dandy–Walker malformation; ACC, agenesis of the corpus callosum; DD, developmental delays; CNS, central nervous system.

[a]Best estimate of functional effect based upon descriptions in original publications.

b) Distinction from Warfarin Embryopathy

In only 12 of 30 children with central nervous system abnormalities associated with coumarin derivative exposure were features of warfarin embryopathy documented (Table 2). Likewise, only a minority of those with the embryopathy have any evidence of damage to their central nervous system. This suggests that independent mechanisms may result in these two constellations of abnormality. Furthermore, in six instances no exposure occurred between 6 and 9 weeks, the apparently critical period for embryopathic effects (Table 2). Instead, in most instances exposure occurred in the second and third trimesters. This, too, suggests an independent origin of the embryopathic and central nervous system sequelae. If there is a separate critical period for these central nervous system effects, it must be quite broad, extending from at least 11 weeks of gestation – the latest point of exposure in a case described by PAWLOW and PAWLOW (1985) – to about 26 weeks – the earliest date of exposure reported by VILLE et al. (1993).

Analysis of patterns of central nervous system anomalies also suggests that central nervous system effects arise through a pathogenetic mechanism distinct from the proposed cause of warfarin embryopathy.

c) Patterns of Abnormality and Likely Pathogenetic Mechanism

Despite considerable variation among affected individuals, the data presented in Table 3 suggest that patterns of central nervous system damage are discernible. Indeed, three, possibly four such patterns are suggested (PAULI and HAUN 1993).

Eleven of 31 instances of central nervous system effects have characteristics consistent with the *septo-optic dysplasia* spectrum (JONES 1988). Six had optic atrophy or similar ophthalmologic features (QUENNEVILLE et al. 1959; DISAIA 1966; J.G. HALL, unpublished; PETTIFOR and BENSON 1975; STEVENSON et al. 1980; OLTHOFF et al. 1994). Four instances of probable or certain agenesis of the corpus callosum have been recognized (HOLZGREVE et al. 1976, R.M. PAULI unpublished; KAPLAN 1985; PAWLOW and PAWLOW 1985), as has one example of absence of the septum pellucidum (KAPLAN et al. 1982).

A second pattern encompasses instances in which *Dandy-Walker malformation* (MURRAY et al. 1985; PASCUAL-CASTROVIEJO et al. 1991) and other posterior fossa abnormalities that are difficult to differentiate have been demonstrated. These include five examples of Dandy-Walker malformation (WARKANY and BOFINGER 1975, HOLZGREVE et al. 1976; R.M. PAULI, unpublished; KAPLAN et al. 1982; KAPLAN 1985), one noncharacterized posterior fossa cyst (R.M. PAULI, unpublished), and one with apparent cerebellar atrophy (J.G. HALL, unpublished). In total, then, seven instances of this cluster of abnormalities have been recognized. In this regard it is interesting that, in addition to these seven examples, BITTEL et al. (1993) noted that two of 72 infants and children ascertained through the presence of intracranial cysts had been exposed to coumarin derivatives during pregnancy. Unfortunately, data were not presented which would allow sorting with respect to whether or not these were examples of posterior fossa cysts or Dandy-Walker malformation (Bittel et al. 1993). Some children display features of both the septo-optic dysplasia spectrum and the Dandy-Walker malformation (Table 3).

Thirdly, five instances of *neural tube defects* have been documented. Three of these had a meningocele or meningomyelocele (Tejani 1973; Kort and Cassel 1981; Gärtner et al. 1993), one an encephalocele (Warkany and Bofinger 1975), and one was born with anencephaly (Sheikhzadeh et al. 1983). Although these occurrences might be coincidental rather than causal, the likelihood of the latter is sufficient that women exposed to oral anticoagulants should probably be considered at a higher than average risk for the occurrence of neural tube defects in their offspring.

Finally, the examples of various anterior segment defects of the eye and of microphthalmia (Tejani 1975; Harrod and Sherrod 1981; Kaplan 1985; Wong et al. 1993) might suggest another pattern of ophthalmologic effects.

For all three prominent patterns – Dandy-Walker malformations, septo-optic dysplasia, and neural tube defects – there has been some evidence to suggest that they may arise from a secondary vascular or hemorrhagic disruption rather than from an intrinsic malformational process (Pascual-Castroviejo et al. 1991; Dominguez et al. 1991, and Stevenson et al. 1987, respectively). Conceivably, then, the central nervous system effects might arise secondary to early (neural tube defects) or later (Dandy-Walker malformation, septo-optic dysplasia) vascular or hemorrhagic disruptions.

Thus there is considerable evidence to suggest not only that the central nervous system effects are independently determined from the embryopathic effects of coumarin derivatives, but also that they may be disruptional in origin. First, most instances have arisen in association with second and third trimester exposures. Secondly, pattern analysis suggests that vascular or hemorrhagic disruptions may account for the specific central nervous systems features seen. In addition, direct demonstration of prenatal intracranial hemorrhage (Ville et al. 1993) and evidence of damage in the distribution of the middle cerebral arteries (Pati and Helmbrecht 1994) support the notion that some or all of the central nervous system effects are vascular and disruptive in origin.

The cases reported by Kaplan (1985) and Pawlow and Pawlow (1985) are troubling in this regard. If one postulates that the disruptive effects are a consequence of the effects of coumarin derivatives on the coagulation factors of the vulnerable fetus (made more vulnerable because of the relative sensitivity of the fetal vitamin K-dependent coagulation factors; Ville et al. 1993) then central nervous system anomalies should not arise unless exposure extends to at least 14 weeks of gestation or so (Bleyer et al. 1971). Likewise, postulating that neural tube defects arise through such a late-gestation effect is inconsistent with generally accepted theory. Whether multiple mechanisms may give rise to central nervous system anomalies remains unproven.

Sundaram and Lev (1988, 1990) provide an alternative mechanism that could give rise to central nervous system effects. They demonstrated that vitamin K may have a role in sphingolipid biosynthesis in the fetal brain (SUNDARAM and LEV 1988) and that abnormalities of this sphingolipid biosynthesis may arise secondary to coumarin derivative administration (SUNDARAM and LEV 1988) through inhibition of sulfotransferase activity (SUNDARAM and LEV 1990). They point out that, since sulfatides have an important role as components of the myelin sheath, inhibition of their synthesis by vitamin K inhibitors could result in significant abnormalities of brain morphogenesis (SUNDARAM and LEV 1990). These observations would seem more relevant if some of the brain abnormalities described in humans were more obviously linked with anomalies of myelination.

3. Other Malformations

A variety of other malformations have been reported in association with in utero exposure to coumarin derivatives. These include cleft lip and/or cleft palate (KORT and CASSEL 1981; CHEN et al. 1982; VITALI et al. 1986; AYHAN et al. 1991; WONG et al. 1993), congenital hip dysplasia (KORT and CASSEL 1981), growth failure (COTRUFO et al. 1991), diaphragmatic hernia (NORMAN and STRAY-PEDERSON 1989), renal anomalies (WARKANY and BOFINGER 1975; LUTZ et al. 1987; HALL 1989; AYHAN et al. 1991), and others (CASANEGRA et al. 1975; DEAN et al. 1981; CHEN et al. 1982; TAMBURRINNI et al. 1987; RUTHNUM and TOLMIE 1987; AYHAN et al. 1991). Many of these may have arisen coincidentally in association with such exposure. Alternatively, some may reflect additonal, low-frequency teratogenic effects which are more difficult to identify simply because of their low-frequency occurrence.

Anomalies of sidedness determination may be one such low-frequency, but specific association. I was struck during the preparation of this review that a pattern of abnormal sidedness determination may be enmeshed within the various case reports and series. Among all exposures reported, about a dozen infants have had documented congenital heart disease (BRAMBEL et al. 1951; AARO and JUERGENS 1971; PAULI et al. 1976; COX et al. 1977; DEAN et al. 1981; BALDE et al. 1988; COTRUFO et al. 1991; BORN et al. 1992; GÄRTNER et al. 1993; BARKER et al. 1994; LEE et al. 1994). Given that over a thousand exposures have been described, one would anticipate occasional coincidental co-occurrence of coumarin derivative exposure and congenital heart disease. That would be particularly true for relatively common lesions such as septal defects (COTRUFO et al. 1991; GÄRTNER et al. 1993; LEE et al. 1994). Less expected is the occurrence of at least four instances of congenital heart disease associated with abnormalities of sidedness determination (although in one, exposure appears not to have occurred at appropriate gestation to be causal; BRAMBEL et al. 1951). DEAN et al. (1981) described an infant with pulmonic atresia, ventricular septal defect, and anomalous venous return; COX et al. (1977) detailed an instance of single ventricle, L transposition of a single great artery, pulmonary atresia, and partial anomalous venous return in association with asplenia and partial malrotation of the gut; BARKER et al. (1994) described dextrocardia and abdominal situs inversus; while a patient reported by BRAMBEL et al. (1951) had asplenia, a two-chamber heart, and pulmonary artery artesia. In none was a related defect reported in any family members. If anomalies of sidedness determination are the result of exposure to coumarin derivatives, then this would need be secondary to deleterious effects earlier in gestation than the critical period for development of warfarin embryopathy (HUTCHINS et al. 1983; YOKOYAMA et al. 1993; PHOON and NEILL 1994) and would have to arise by a completely independent mechanism than that posited for warfarin embryopathy, since vitamin K-dependent, warfarin-inhibitable proteins are not known to be in any way involved in development of appropriate sidedness.

V. Animal Models

Most attempts at creating an animal model of the deleterious effects of coumarin derivatives on human embryos and fetuses have been remarkably un successful. In particular, efforts to generate features similar to warfarin embryopathy failed in mice (ROLL and BAER 1967; MCCALLION et al. 1971; J.N. KRONICK et al. 1974; R.M. PAULI unpublished observations), rats (BECKMAN et al. 1982), chicks (LAVELLE et al. 1994), and rabbits (KRAUS et al. 1949; HIRSH et al. 1970; MCCALLION et al. 1971). Often these efforts resulted in fatal hemorrhagic manifestations in the offspring and/or the mother, but not in demonstrable parallels to the nasal hypoplasia or stippled epiphyses or the central nervous system disruptions seen in human offspring similarly exposed. Thus coumarin derivatives seemed unique in being a human teratogen without demonstrable teratogenicity in other species.

Recently, HOWE and WEBSTER (1990, 1992) have succeeded in mimicking certain of the effects of coumarin derivatives in the rat. Their success arose principally because of recognition of the potential to "rescue" dams from the hemorrhagic consequences of warfarin treatment. Rescue, through administration of vitamin K, takes advantage of a second liver-specific, but warfarin-insensitive reductase (FASCO et al. 1982; PRICE and KANEDA 1987; HOWE and WEBSTER 1990). Normally carboxylated coagulation factors can then be made in the dam so long as vitamin K is provided, while little antagonism of the effects of coumarin derivatives on vitamin K-dependent fetal protein modification occurs (probably because of a large maternal–fetal gradient for vitamin K; MANDELBROT et al. 1988). In their first set of experiments, HOWE and WEBSTER (1990) demonstrated that warfarin administration (with dam rescue using vitamin K) between pregnancy days 9 and 20 resulted in hemorrhage in the fetuses. Those hemorrhagic manifestation were primarily of the brain, eyes, and face. Had pups been allowed to proceed to term, such hemorrhages likely would be recognized as brain disruptions similar to the central nervous system effects identified in human offspring similarly exposed.

HOWE and WEBSTER (1992) further reasoned that absence of recognition of stippling and nasal hypoplasia in laboratory animals likely arose because

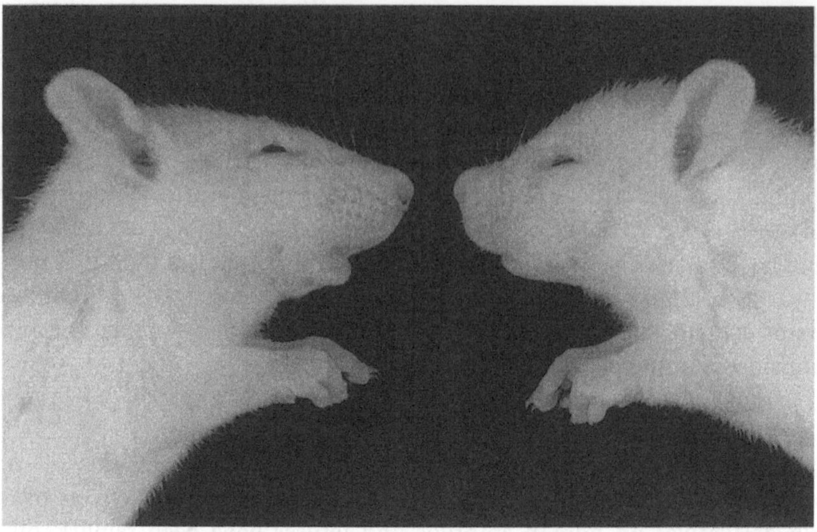

Fig. 8. Control (*left*) and treated (*right*) rats who received warfarin (and vitamin K) from birth to current age of 3 weeks. (Reproduced with permission from HOWE et al. 1992)

much of analogous nasal and ossific development occurs perinatally in rodents rather than being early prenatal processes. Therefore, it seemed reasonable to assess the effects of neonatal administration of coumarin derivatives. Rats treated in this way showed growth retardation, anomalous proportions of craniofacial growth, nasal hypoplasia, and mild foreshortening of the long bones (HOWE and WEBSTER 1992). Abnormalities of the nasal septum were particularly evident, with decreased growth arising apparently secondary to abnormal and abnormally rapid ossification of the nasal cartilage. Gross effects were quite similar to changes seen in human infants exposed to coumarin derivatives early in pregnancy (Fig. 8). No stippling was identified, however, although unusual calcified bridges within growth plates were identified histologically: similar disruptions of bone morphogenesis in the rat secondary to late prenatal exposure to coumarin derivatives have been described by others as well (FETEIH et al. 1990). The rat so treated appears to be an adequate model and could be used to further delineate the molecular mechanisms underlying warfarin embryopathy. Thus far, however, no direct biochemical examination of the pathogenesis of the abnormalities seen in the rat has been published.

VI. Estimation of Risks

As for all other known teratogens, only a minority of appropriately exposed embryos and fetuses demonstrate the expected teratogenic effects of coumarin derivative exposure. Accurate estimates of those risks have proven difficult to generate, however. It appears unlikely that a sufficiently large controlled trial will ever be undertaken (or, ethically, that it could be undertaken now given the unequivocal risks of exposure already demonstrated). Clear ascertainment biases may arise in attempting to estimate the risk resulting from coumarin derivative exposure in pregnancy through literature review. While these are ultimately of undeterminable effect, it seems likely that, overall, such estimates, as attempted by HALL et al. (1980), result in disproportionate ascertainment of abnormal outcomes. Similar biases are present in the efforts of GINSBERG and HIRSH (1988, 1989a, 1989b) and GINSBERG et al. (1989a), which are also based upon such literature-based data sets.

Not surprisingly, risk estimates of these groups are not markedly discrepant. On the basis of 418 exposures, HALL et al. (1980) estimated the risk for embryopathic effects to be about 4% and for central nervous system anomalies to be about 3% (HALL et al. 1980). The GINSBERG and HIRSH series, based upon an updated collection of 970 published instances of exposure, yielded risk estimates of 4.7% for embryopathic manifestations and 1.6% for central nervous system effects.

We reasoned (PAULI and HAUN 1993) that one could limit certain expected biases of literature-based estimates by utilizing only relatively large published series (rather than case reports) and by using only relatively recent investigations (to minimize effects of changes in medical care over time). Nonetheless, such series still present their own difficulties. None is sufficiently large to be used independently for risk estimation of relatively rare events. Few series (ITURBE-ALLESIO et al. 1986; SARELI et al. 1989, COTRUFO et al. 1991; BORN et al. 1992) are prospective. Antecedent characteristics and severity of underlying disease of treated women may vary markedly from study to study. Most series are reported either by adult cardiologists or by obstericians, and it seems likely that considerable under-ascertainment of subtle neonatal features is present in most of them. Given that very few have directly examined neonates (e.g., CHONG et al. 1984; SALAZAR et al. 1984; ITURBE-ALLESIO et al. 1986), such under-ascertainment seems particularly likely. Similarly, virtually uniform lack of long-term follow-up suggests inevitable under-ascertainment of central nervous system effects in particular. The preliminary report by OLTHOF et al. (1994) is a refreshing exception. They provide follow-up information on 21 exposed children at 8–10 years of age using an appropriate control group; one exposed child had severe neurological abnormalities, while the others were cognitively normal and no different from the comparable controls.

Other studies which otherwise might provide additional information on risk had to be eliminated. Some (e.g., GUIDOZZI 1984) are presented in a manner which makes it impossible to sort data appropriately, for example by anticoagulant used. Others are so methodologically weak as to be of little value. The publication by SBAROUNI and OAKLEY (1994) deserves special comment in this regard. It purports to demonstrate absence of risk of cou-

Table 4. Study designs of published series of coumarin derivative exposure in pregnancy

Study no.	Study	Design	Indications for anticoagulation use	Anticoagulant	Weeks used	Cases (n)
I.	Ayhan et al. 1991	Retrospective, sequential, single institution	Valves	Warfarin	Variable	47
II.	Ben Ismail et al. 1986	Retrospective; sequential; single instituion	Valves	"Oral"	Not stated	53
III.	Born et al. 1992	Prospective; sequential; single institution	Valves	"Coumarin-like"	Not stated	40
IV.	Braun et al. 1984	Retrospective; sequential; single institution	Valves	Warfarin	Throughout	24
V.	Chen et al. 1982	Retrospective; sequential;	Valves	Warfarin	Variable	30
VI.	Chong et al. 1984	Retrospective follow-up of 46 pregnancies; 22 of 42 eligible examined; two institutions	Various	Warfarin	Variable	46
VII.	Cotrufo et al. 1991	Prospective; selective (20/29); single institution	Valves	Warfarin	Throughout	20
VIII.	Iturbe-Alessio et al. 1986	Prospective; sequential, 35/49 infants examined; multiple institutions	Valves	Warfarin	Variable	49
IX.	Javares et al. 1984	Retrospective; sequential single institution	Valves	Acenocoumarin	0 to 34–36 weeks	29
X.	Kort and Cassel 1981	Retrospective; sequential single institution	Cardiac disease	Warfarin	Variable	40
XI.	Larrea et al. 1983	Retrospective; sequential; single institution	Valves	Acenocoumarin	Variable	47

XII.	Lee et al. 1986	Retrospective; sequential; single institution	Valves	Warfarin	0 to 6–8 and 13–37 weeks	18
XIII.	Lee et al. 1994	Retrospective; sequential; single institution	Valves	Warfarin	Variable	47
XIV.	Matorras et al. 1985	Retrospective; sequential; single institution	Valves	Warfarin	0 to 37–38 weeks	44
XV.	O'Neill et al. 1982	Retrospective; sequential; four institutions	Valves	Warfarin	Variable	17
XVI.	Pavankumar et al. 1988	Retrospective; sequential; single institution	Valves	Acenoco-umarin	Throughout	47
XVII.	Salazar et al. 1984	Retrospective; sequential; 34/128 infants examined; single institution	Valves	Acenoco-umarin	0–38 weeks	128
XVIII.	Sareli et al. 1989	Prospective; sequential; single institution	Valves	Warfarin	First and second trimesters	50
XIX.	Sheikhzadeh et al. 1983	Retrospective; sequential; single institution	Valves	Warfarin	0–39 weeks	13
XX.	Vitali et al. 1986	Retrospective; sequential; single institution	Valves	Warfarin	Variable	85
XXI.	Wong et al. 1993	Restrospective; sequential; single institution	Various	Warfarin	Variable	53
	Total					927

Updated and modified from information originally published in Pauli and Haun (1993).

marin derivatives when they are used appropriately in pregnancy. This conclusion is based upon responses to a mail questionnaire that was sent to cardiologists treating adults concerning their clinical experience with women so treated. It is imprudent on the basis of such a survey for the authors to conclude that warfarin treatment is not associated with warfarin embryopathy.

With the various caveats in mind, Tables 4 and 5 present data concerning 21 series (ranging in size from 13 to 128 cases) published since 1980. Table 4 provides informaton regarding design, while Table 5 presents a summary of fetal outcome information from those series. A total of 927 pregnancies are reported, including 213 spontaneous miscarriages and 47 stillbirths (the later being more than four times the expected rate for the general population; GREB et al. 1987). Among the 927 pregnancies and 691 live births, 12 had unequivocal features of warfarin embryopathy (1.3% of all pregnancies and 1.7% of live births). Another 25 were reported as having the embryopathy with insufficient documentation for confirmation or had insufficient reported features to conform to the suggested diagnostic criteria. These plus the unequivocally affected cases yield an estimated risk of 3.9% of all pregnancies and 4.9% of all live births.

Central nervous system manifestations are considerably less common, having been identified in only six infants (KORT and CASSEL 1981; CHEN et al. 1982; LARREA et al. 1983; SHEIKHZADEH et al. 1983; WONG et al. 1993), or 0.6% of all pregnancies and 0.9% of liveborn infants.

All such estimates have limitations. Nonetheless, it seems reasonable to estimate risks in liveborn infants born after pregnancies in which coumarin derivative exposure occurs as follows. The warfarin embryopathy will occur in 2%–5% exposed infants, but only when exposure includes some portion of postconceptual weeks 6 through 9. Central nervous system abnormalities will occur in 0.5%–2% of exposed infants and likely arise in most instances from exposures in the second and third trimesters.

D. Heparin

I. Historical Overview

Heparin was discovered about 75 years ago – another of those discoveries first made by a precocious medical student (KANDROTAS 1992). It is a naturally occurring family of glycosaminoglycans extractable from a variety of animal sources (most often porcine or bovine intestinal mucosa; GREAVES 1993) with molecular weights ranging from about 5000 to 35 000 (GREAVES 1993). Heparin has been used extensively since the 1940s (KANDROTAS 1992), particularly in situations in which rapid anticoagulation is needed.

Table 5. Study outcomes of published series of coumarin derivative exposure in pregnancy

Study no.	Cases	Abortion		Stillbirth		Neonatal death		Embryopathy[a] Certain	Embryopathy[a] Certain + possible	CNS/eye
	(*n*)	(*n*)	(%)	(*n*)	(%)	(*n*)	(%)	(*n*)	(*n*)	(*n*)
I.	47	19	40	4	9	2	4	0	0	0
II.	53	8	15	3	6	4	8	0	0	0
III.	40	7	18	1	3	5	13	3	0	0
IV.	24	7	29	5	21	?		0	0	0
V.	30	10	33	0		0		0	0	1
VI.	46	?		2	4	2	4	0	0	0
VII.	20	0		0		0		0	0	0
VIII.	49	9	18	1	2	0		2	8	0
IX.	29	?		?		?		0	1	0
X.	40	2	5	3	7	0		0	0	1
XI.	47	11	23	0		1	2	2	2	1
XII.	18	9	50	0		0		0	0	0
XIII.	47	7	15	6	13	1	2	0	1	0
XIV.	44	8	18	0		1	2	1	2	0
XV.	17	6	35	2	12	0		1	1	0
XVI.	47	3	6	1	2	1	2	0	0	0
XVII.	128	36	28	9	7	3	2	1	3	0
XVIII.	50	9	18	7	14	2	4	1	2	0
XIX.	13	2	15	2	15	2	15	0	1	1
XX.	85	37	44	0		1	1	1	0	0
XXI.	53	23	43	1	2	0	0	0	16	2
Total[b]	927	213	23	47	5	25	3	12[c]	37[d]	6[e]

Updated and modified from information originally published in PAULI and HAUN (1993). Study numbers as in Table 4.
CNS Central nervous system.
[a]See text for description of justification for dividing embryopathy cases into certain and probable ones.
[b]Given the differences in the ways that adverse outcomes were reported in the original studies, the data are often either not strictly comparable or difficult to apportion within the categories used here. Therefore, the totals should be viewed with considerable skepticism.
[c]1.3% of all pregnancies.
[d]4.0% of all pregnancies.
[e]0.6 % of all pregnancies.

II. Mode of Action

The means by which heparin exerts its anti-thrombotic effect is complex and only partially understood. Much of this effect appears to be secondary to specific binding and a resultant conformational change in antithrombin III (HIRSH 1991; RUTHERFORD and PHELAN 1991; KANDROTAS 1992; GREAVES 1993). The consequent conformational change potentiates the effect of antithrombin III (GREAVES 1993), causing more rapid and more marked inactivation of a number of coagulation factors (HIRSH 1991; RUTHERFORD and

PHELAN 1991), but probably most importantly through inhibition of thrombin activation of factors V and VIII (HIRSH 1991; RUTHERFORD and PHELAN 1991). In addition, heparin has effects on platelet function, vascular permeability, etc., which also may play some role in its anticoagulant effects (HIRSH 1991).

III. Use and Effects in Pregnancy

The indications for long-term use of heparin in pregnancy are outlined in Sect. B. Heparin and oral coumarin derivatives are virtually interchangeable for most indications, but have two essential differences: (1) administration of heparin must be either intravenous or subcutaneous, since it is so poorly absorbed from the gut (RUTHERFORD and PHELAN 1991); (2) there is a substantial delay in the anticoagulant effect of coumarin derivatives, since the stores of vitamin K must be exhausted prior to any substantial anticoagulant effect, while heparin's effect is direct potentiation of antithrombotic activity and is immediate.

Despite similar indications, because heparin is usually the drug of choice for initial anticoagulation and for impatient management, comorbidity is likely to be greater, on average, than in individuals treated with coumarin derivatives. This makes analysis and comparison of potentially deleterious effects of the two classes of anticoagulants difficult (GINSBERG and HIRSH 1988, 1989a,b; GINSBERG et al. 1989a).

1. Maternal Risks

In all circumstances heparin shows a high variability of individual response, requiring careful monitoring (HIRSH 1991). Obviously, such a need is at least as important when the patient is pregnant. Collective experience suggests that there are a series of special risks which need to be addressed regarding heparin use in pregnancy.

First, maternal hemorrhage may arise, not only because of excessive anticoagulation, but also because delivery necessarily involves a need to limit bleeding after placental separation. Serious bleeding arises as a complication in pregnancy in around 2%–10% of treated women (RUTHERFORD and PHELAN 1991; GINSBERG and HIRSH 1992). If severe enough, hemorrhage may threaten the pregnancy and may have secondary harmful effects on the fetus.

Maternal thrombocytopenia, which may arise in about 1% of treated pregnant women (RUTHERFORD and PHELAN 1991; GREAVES 1993), may also pose secondary threats to the developing fetus.

Osteoporosis may arise following long-term use of heparin (GRIFFITHS et al. 1965), particularly in pregnant women (GRIFFITHS and LIU 1984; HIRSH 1991; RUTHERFORD and PHELAN 1991; GREAVES 1993; HARAM et al. 1993). In a series of observations, DAHLMAN and colleagues demonstrated that osteopenia is common in pregnant women treated with heparin, occuring in about 17% (DAHLMAN et al. 1990), that osteoporotic spinal fractures are not rare, arising in about 2% of treated women (DAHLMAN 1993), and that these effects arise secondary to changes in calcium homeostasis (DAHLMAN et al. 1992). While only a minority of women experience clinically demonstrable abnormalities, most show reduction in bone mass (DAHLMAN et al. 1994; BARBOUR et al. 1994) a loss which persists long after the end of the pregnancy (BARBOUR et al. 1994) but which ultimately probable is reversible (DAHLMAN et al. 1994). A variety of mechanisms have been proposed (HARAM et al. 1993), but none has been proved (DAHLMAN et al. 1992; DAHLMAN et al. 1994). While of serious concern if heparin is being considered as an alternative to coumarin derivative use, there does not seem to be any harmful fetal effects of these alterations in calcium homeostasis.

Various devices and techniques have been developed to make administration of heparin for extended periods simpler and safer, as throughout pregnancy, including indwelling catheters (ANDERSON et al. 1993), programmable pumps (FLOYD et al. 1991), and use of low molecular weight heparin, which has a longer half-life and can therefore be administered only once daily (STURRIDGE et al. 1994).

2. Placental Barrier

Although indirect means of teratogenicity of heparin have been speculated upon (HALL et al. 1980), it is most likely that heparin will exert no teratogenic effects on the fetus if it cannot reach the fetus. FLESSA et al. (1965) first showed that placental passage of heparin seems not to occur. This has been confirmed by others through various experimental designs (MÄTZSCH et al. 1991, BAJORCA and CONTRACTOR 1992). Certainly, in part, this is secondary to its high molecular weight and its high negative charge (BAJORCA and CONTRACTOR 1991). Although exceedingly small quantities may be transferred to the fetal circulation (BAJORCA and CONTRACTOR 1992), these are insufficient to affect any fetal clotting parameters. Thus, if such small quantities of heparin were teratogenic, these effects would have to occur via a mechanism other than its anticoagulant effects.

Even the lower molecular weight heparins show no substantial fetal transfer (MÄTZSCH et al. 1991), perhaps because of a nonphysical heparin barrier of the placenta which causes neutralization of heparin (USZYNSKI 1992).

3. Fetal Risks

HALL et al. (1980) suggested that there may be considerable fetal risk of heparin use, speculating that this might arise from chelation of calcium and secondary fetal calcium deficiency. However, given that maternal ionized calcium levels *increase* with heparin use (DAHLMAN et al. 1992), this postulate seems untenable. Furthermore, review of literature cases of heparin use, when controlled for maternal comorbidity, seem not to demonstrate an increased risk for fetal death or any other fetal sequelae, except for uncomplicated permaturity (GINSBERG and HIRSH 1988, 1989a,b; GINSBERG et al. 1989a). This conclusion is supported by the results of a retrospective cohort study (Ginsberg et al. 1989b).

E. Recommendations for Anticoagulation in Pregnancy

A general consensus has arisen that heparin is the anticoagulant of choice in pregnancy (Rutherford and Phelan 1991; Demers and Ginsberg 1992; Ginsberg and Hirsh 1992; Maternal and Neonatal Haemostasis Working Party of the Haemostasis and Thrombosis Task 1993; Greaves 1993) and that warfarin is generally contraindicated (Hirsh and Fuster 1994), although this consensus is not without its dissenters (Sbarouni and Oakley 1994). At least from the fetal perspective, such a recommendation appears appropriate: coumarin derivative use can result in significant fetal sequelae (although none arises in a large proportion of exposed fetuses), while no clear demonstration of fetal risks of heparin has been forthcoming.

References

Aaro LA, Juergens JL (1971) Thrombophlebitis associated with pregnancy. Am J Obstet Gynecol 109: 1128–1136
Anderson DR, Ginsberg JS, Brill-Edwards P, Demers C, Burrows RF, Hirsh J (1993) The use of an indwelling teflon catheter for subcutaneous heparin administration during pregnancy. Arch Intern Med 153: 841–844
Ayhan A, Yapar EG, Yüce K, Kisnisci HA, Nazli N, Özmen F (1991) Pregnancy and its complications after cardiac valve replacement. Int J Gynaecol Obstet 35: 117–122
Badois C, Lavollay B, Bomsel F (1983) Maladie des épiphyses ponctuées (MEP) associée a une foetopathie éthyligue (FE). Trois premiers cas rapportés. Ann Radiol (Paris) 26: 244–250
Baillie M, Allen ED, Elkington AR (1980) The congenital warfarin syndrome: A case report. Br J Ophthalmol 64: 633–635
Bajoria R, Contractor SF (1992) Transfer of heparin across the human perfused placental lobule. J Pharm Pharmacol 44: 952–959
Balde MD, Breitbach GP, Wettstein A, Hoffmann W, Bastert G (1988) Fallotsche Tetralogie nach Cumarineinnahme in Frühschwangerschaft – eine Embryopathie? Geburtshilfe Frauenheilkd 48: 182–183
Barbour LA, Kick SD, Steiner JF, LoVerde ME, Heddleston LN, Lear JL, Barón AE, Barton PL (1994) A prospective study of heparin-induced osteoporosis in pregnancy using bone densitometry. Am J Obstet Gynecol 170: 862–869
Barker DP, Konje JC, Richardson JA (1994) Warfarin embryopathy with dextrocardia and situs inversus. Acta Paediatr 83: 411
Barr M, Burdi AR (1976) Warfarin-associated embryopathy in a 17-week-old abortus. Teratology 14: 129–134
Becker MH, Genieser NB, Finegold M, Miranda D, Spackman T (1975) Chondrodysplasia punctata: is maternal warfarin therapy a factor? Am J Dis Child 129: 356–359
Beckman DA, Soloman HM, Brent RL (1982) Third trimester teratogenic effects of sodium warfarin in the rat. Teratology 26: 28A
Ben Ismail M, Abid F, Trabelsk S, Taktak M, Fekih M (1986) Cardiac valve prostheses, anticoagulation, and pregnancy. Br Heart J 55: 101–105
Bittel M, Ehrensberger J, Gysler R, Illi OE, Jordi R, Kaiser G, Kummer M, Rösslein R, Weibel M (1993) Congenital intracranial cysts: clinical findings, diagnosis,

treatment and follow-up. A multicenter, retrospective long-term evaluation of 72 children. Eur J Pediatr Surg 3: 323–334
Bleyer WA, Skinner AL (1976) Fatal neonatal hemorrhage after maternal anticonvulsant therapy. JAMA 235: 626–627
Bleyer WA, Hakami N, Shepard TH (1971) The development of hemostasis in the human fetus and newborn infant. J Pediatr 790: 838–853
Born D, Martinez EE, Almeida PAM, Santos DV, Carvalho ACC, Moron AF, Miyasaki CH, Moraes SD, Ambrose JA (1992) Pregnancy in patients with prosthetic heart valves: the effects of anticoagulation on mother, fetus, and neonate. Am Heart J 124: 413–417
Brambel CE, Fitzpatrick VdeP, Hunter RE (1951) Protracted anticoagulant therapy during the antenatal period. Trans N Engl Obstet Gynecol Soc 5: 27–45
Braun S, Roberts A, Chamorro G, Leontic E, Casanegra P, Albuquerque J (1984) Riesgo materno y fetal del embarazo en cardiopatas con y sin tratamiento anticoagulante. Rev Med Chil 112: 1009–1016
Brenner B, Tavori S, Zivelin A, Keller CB, Suttie JW, Tatarsky I, Seligsohn U (1990) Hereditary deficiency of all vitamin K-dependent procoagulants and anticoagulants. Br J Haematol 75: 537–542
Carson M, Reid M (1976) Warfarin and fetal abnormality. Lancert 1: 1356
Casanegra P, Aviles G, Maturana G, Dubernet J (1975) Cardiovascular management of pregnant women with a heart valve prosthesis. Am J Cardiol 36: 802–806
Chen WWC, Chan CS, Lee PK, Wang RY, Wong VCW (1982) Pregnancy in patients with prosthetic heart valves: an experience with 45 pregnancies. Q J M 203: 358–365
Chong MKB, Harvey D, DeSwiet M (1984) Follow-up study of children whose mothers were treated with warfarin during pregnancy. Br J Obstet Gynaecol 91: 1070–1073
Chung K-S, Bezeaud A, Goldsmith JC, McMillan CW, Ménaché D, Roberts HR (1979) Congenital deficiency of blood clotting factors II, VII, IX, and X. Blood 53: 776–787
Conard J, Horellou MH, VanDreden P, Lecompte T, Samama M (1990) Thrombosis and pregnancy in congenital deficiencies in ATIII, protein C or protein S: a study of 78 women. Thromb Haemost 63: 319–320
Cotrufo M, deLuca TSL, Calabró R, Mastrogiovanni G, Lama D (1991) Coumarin anticoagulation during pregnancy in patients with mechanical valve prostheses. Eur J Cardiothorac Surg 5: 300–305
Cox DR, Martin L, Hall BD (1977) Asplenia syndrome after fetal exposure to warfarin Lancet 2: 1134
Curry CHR, Magenis E, Brown M, Lanman JT, Tsai J, O'Lague P, Goodfellow P, Mohandas T, Bergner EA, Shapiro LJ (1984) Inherited chondrodysplasia punctata due to a deletion of the terminal short arm of an X chromosome. New Engl J Med 311: 1010–1015
Curtin T, Mulhern B (1980) Foetal warfarin syndrome. Ir Med J 73: 393–394
Dahlman TC (1993) Osteoporotic fractures and the recurrence of thromboembolism during pregnancy and the puerperium in 184 women undergoing thromboprophylaxis with heparin. Am J Obstet Gynecol 168: 1265–1270
Dahlman T, Lindvall N, Hellgren M (1990) Osteopenia in pregnancy during long-term heparin treatment: a radiological study post partum. Br J Obstet Gynaecol 97: 221–228
Dahlman T, Sjöberg HE, Hellgren M, Bucht E (1992) Calcium homeostasis in pregnancy during long-term heparin treatment. Br J Obstet Gynaecol 99: 412–416
Dahlman TC, Sjöberg HE, Ringertz H (1994) Bone mineral density during long-term prophylaxis with heparin in pregnancy. Am J Obstet Gynecol 170: 1315–1320
Dean H, Berliner S, Shoenfeld Y, Pinkhas J (1981) Warfarin treatment during pregnancy in patients with prosthetic mitral valves. Acta Haematol 66: 65–66

Deblay MF, Vert P, Andre M, Marchal F (1982) Transplacental vitamin K prevents haemorrhagic disease of infant of epileptic mother. Lancet 1: 1247

Demers C, Ginsberg JS (1992) Deep venous thrombosis and pulmonary embolism in pregnancy. Clin Chest Med 13: 645–656

deVries TW, vanderVeer E, Heijmans HSA (1993) Warfarin embryopathy: patient, possibility, pathogenesis and prognosis. Br J Obstet Gynaecol 100: 869–871

DiSaia PJ (1966) Pregnancy and delivery of a patient with a Starr-Edwards mitral valve prosthesis. Report of a case. Obstet Gynecol 28: 469–472

Dominguez R, Vila-Coro AA, Slopis JM, Bohan TP (1991) Brain and ocular abnormalities in infants with in utero exposure to cocaine and other street drugs. AJDC 145: 688–695

Fasco MJ, Hildebrandt EF, Suttie JW (1982) Evidence that warfarin anticoagulant action involves two distinct reductase activities. J Biol Chem 257: 11210–11212

Feteih R, Tassinari MS, Lian JB (1990) Effect of sodium warfarin on vitamin K-dependent proteins and skeletal development in the rat fetus. J Bone Miner Res 5: 885–894

Fischer M, Zweymüller E (1966) Kongenitaler kombinierter Mangel der Faktoren II, VII und X. Z Kinderheilkd 95: 309–323

Flessa HC, Kapstrom AB, Glueck HI, Will JJ (1965) Placental transport of heparin. Am J Obstet Gynecol 93: 570–573

Floyd RC, Gookin KS, Hess LW, Martin RW, Rawlinson KF, Moenning RK, Morrison JC (1991) Administration of heparin by subcutaneous infusion with a programmable pump. Am J Obstet Gynecol 165: 931–933

Fourie DT, Hay IT (1975) Warfarin as a possible teratogen. S Afr Med J 49: 2081–2083

Franco B, Meroni G, Parenti G, Levilliers J, Bernard L, Gebbia M, Cox L, Maroteaux P, Sheffield L, Rappold GA, Andria G, Petit C, Ballabio A (1995) A cluster of sulfatase genes on Xp22.3: mutations in chondrodysplasia punctata (CDPX) and implications for warfarin embryopathy. Cell 81: 15–25

Freude S, Pabinger-Fasching I, Kozel-Lachmann D, Braun F, Pollak A (1991) Warfarin-Embryopathie bei mütterlicher Coumarin-Therapie wegen Protein-S-Mangels. Pädiatr Padol 26: 239–241

Furie B, Furie BC (1990) Molecular basis of vitamin K-dependent γ-carboxylation. Blood 75: 1753–1762

Gallop PM, Lian JB, Hauschka PV (1980) Carboxylated calcium-binding proteins and vitamin K. N Engl J Med 302: 1460–1466

Gärtner BC, Seifert CB, Michalk DV, Roth B (1993) Phenprocoumon therapy during pregnancy: a case report and comparison of the teratogenic risk of different coumarin derivatives. Z Geburtshilfe Perinatol 197: 262–265

Ginsberg JS, Hirsh J (1988) Optimum use of anticoagulants in pregnancy. Drugs 36: 505–512

Ginsberg JS, Hirsh J (1989a) Use of anticoagulants during pregnancy. Chest 95: 156S-160S

Ginsberg JS, Hirsh J (1989b) Anticoagulants during pregnancy. Annu Rev Med 40: 79–86

Ginsberg JS, Hirsh J (1992) Use of antithrombotic agents during pregnancy. Chest 102: 385S–390S

Ginsberg JS, Hirsh J, Turner DC, Levine MN, Burrows R (1989a) Risks to the fetus of anticoagulant therapy during pregnancy. Thromb Haemost 61: 197–203

Ginsberg JS, Kowalchuk, G, Hirsh J, Brill-Edwards P, Burrows R (1989b) Heparin therapy during pregnancy. Risks to the fetus and mother. Arch Intern Med 149: 2233–2236

Goldsmith GH, Pence RE, Ratnoff OD, Adelstein DJ, Furie B (1982) Studies on a family with combined functional deficiencies of vitamin K-dependent coagulation factors. J Clin Invest 69: 1253–1260

Gooch WM, McLendon RE, Parvey LS, Wilroy RS (1978) Warfarin embryopathy: longitudinal and postmortem examination. Clin Res 26: 74A

Greaves M (1993) Anticoagulants in pregnancy. Pharmacol Ther 59: 311–327
Greb AE, Pauli RM, Kirby RS (1987) Accuracy of fetal death reports: comparison with data from an independent stillbirth assessment program. Am J Public Health 77: 1202–1206
Griffiths GC, Nicholas G, Asher JD, Flanagan B (1965) Heparin osteoporosis. J Am Med Assoc 193: 91–94
Griffiths HT, Liu DTY (1984) Severe heparin osteoporosis in pregnancy. Postgrad Med J 60: 424–425
Guidozzi F (1984) Pregnancy in patients with prosthetic cardiac valves. S Afr Med J 65: 961–963
Guillot M, Toubas PL, Mselati JC, Gamarra E, Moriette G, Relier JP (1979) Embryo-foetopathie coumarinique et maladie des epiphyses ponctuées. Arch Fr Pediatr 36: 63–66
Hale JE, Fraser JD, Price PA (1988) The identification of matrix gla protein in cartilage. J Biol Chem 263: 5820–5824
Hall BD (1989) Warfarin embryopathy and urinary tract anomalies: possible new association. Am J Med Genet 34: 292–293
Hall JG, Pauli RM, Wilson KM (1980) Maternal and fetal sequelae of anticoagulation during pregnancy. Am J Med 68: 122–140
Hanson JW, Smith DW (1975) The fetal hydantoin syndrome. J Pediatr 87: 285–290
Haram K, Hervig T, Thordarson H, Aksnes L (1993) Osteopenia caused by heparin treatment in pregnancy. Acta Obstet Gynecol Scand 72: 674–675
Harrod MJE, Sherrod PS (1981) Warfarin embryopathy in siblings. Obstet Gynecol 57: 673–676
Hauschka PV, Reid ML (1978) Timed appearance of a calcium-binding protein containing gamma-carboxyglutamic acid in developing chick bone. Dev Biol 65: 426–434
Hauschka PV, Lian JB, Gallop PM (1978) Vitamin K and mineralization. Trends Biochem Sci 3: 75–78
Hauschka PV, Lian JB, Cole DE, Gundberg CM (1989) Osteocalcin and matrix Gla protein: vitamin K-dependent proteins in bone. Physiol Rev 69: 990–1047
Hirsh J (1991) Heparin. N Engl J Med 324: 1565–1574
Hirsh J, Fuster V (1994) Guide to anticoagulant therapy. Part 2: oral anticoagulants. Circulation 89: 1469–1480
Hirsh J, Cade JF, Gallus AS (1970) Fetal effect of coumadin administered during pregnancy. Blood 36: 623–627
Holzgreve W, Carey JC, Hall BD (1976) Warfarin-induced fetal abnormalities. Lancet 2: 914–915
Hosenfeld D, Wiedemann HR (1989) Chondrodysplasia punctata in an adult recognized as vitamin K antagonist embryopathy. Clin Genet 35: 376–381
Howe AM, Webster WS (1990) Exposure of the pregnant rat to warfarin and vitamin K1: an animal model of intraventricular hemorrhage in the fetus. Teratology 42: 413–420
Howe AM, Webster WS (1992) The warfarin embryopathy: a rat model showing maxillonasal hypoplasia and other skeletal disturbances. Teratology 46: 379–390
Howe AM, Webster WS, Lipson AH, Halliday JL, Sheffield LJ (1992) Binder's syndrome due to prenatal vitamin K deficiency: a theory of pathogenesis. Aust Dent J 37: 453–460
Hutchins GM, Moore GW, Lipford EH, Haupt HM, Walker MC (1983) Asplenia and polysplenia malformation complexes explained by abnormal embryonic body curvature. Pathol Res Pract 177: 60–76
Iturbe-Alessio I, Fonseca M, Mutchinik O, Santos MA, Zajarias A, Salazar E (1986) Risks of anticoagulant therapy in pregnant women with artificial heart valves. N Engl J Med 315: 1390–1393
Javares T, Coto EO, Maiques V, Rincon A, Such M, Caffarena JM (1984) Pregnancy after heart valve replacement. Int J Cardiol 5: 731–743

Johnson CA, Chung KS, McGrath KM, Bean PE, Roberts HR (1980) Characterization of a variant prothrombin in a patient congenitally deficient in factors II, VII, IX and X. Br J Haematol 44: 461–469
Jones KL (1988) Recognizable patterns of human malformations, 4th edn. Saunders, Philadelphia, p 552
Jullian M, Youlton R, Rivera L (1983) Embriopatia por anticoagulante oral. Rev Med Chil 111: 1045–1048
Kandrotas RJ (1992) Heparin pharmacokinetics and pharmacodynamics. Clin Pharmacokinet 22: 359–374
Kaplan LC (1985) Congenital Dandy Walker malformation associated with first trimester warfarin: a case report and literature review. Teratology 32: 333–337
Kaplan LC, Anderson GG, Ring BA (1982) Congenital hydrocephalus and Dandy-Walker malformation associated with warfarin used during pregnancy. Birth Defects 18[3A]: 79–83
Kerber IJ, Warr OS, Richardson C (1968) Pregnancy in a patient with a prosthetic mitral valve. J Am Med Assoc 203: 223–225
Kort HI, Cassel GA (1981) An appraisal of warfarin therapy during pregnancy. S Afr Med J 60: 578–579
Kraus AP, Perlow A, Singer K (1949) Dangers of dicumarol treatment in pregnancy. J Am Med Assoc 139: 758–762
Kronick JN, Phelps NE, McCallion DJ, Hirsh J (1974) Effects of sodium warfarin administered during pregnancy in mice. Am J Obstet Gynecol 118: 819–823
Lamontagne JM, Leclerc JE, Carrier C, Bureau M (1984) Warfarin embryopathy – a case report. J Otolaryngol 13: 127–129
Lanfranchi C, Olivier C, Boudot del la Motte E, Bavoux F (1982) Anticoagulants et grossesse. Therapie 37: 493–495
LaPiedra OJ, Bernal JM, Ninot S, Gonzalez I, Pastor E, Miralles PJ (1986) Open heart surgery for thrombosis of a prosthetic mitral valve during pregnancy. Fetal hydrocephalus. J Cardiovasc Surg (Torino) 27: 217–220
Larrea JL, Núñez L, Reque JA, GilAquado M, Matarros R, Minguez JA (1983) Pregnancy and mechanical valve prostheses: a high-risk situation for the mother and the fetus. Ann Thorac Surg 36: 459–463
Lavelle PA, Lloyd QP, Gay CV, Leach RM (1994) Vitamin K deficiency does not functionally impair skeletal metabolism of laying hens and their progeny. J Nutr 124: 371–377
Lee C-N, Wu C-C, Lin P-Y, Hsieh F-J, Chen H-Y (1994) Pregnancy following cardiac prosthetic valve replacement. Obstet Gynecol 83: 353–356
Lee PK, Wang RYC, Chow JSF, Cheung KL, Wong VCW, Chan TK (1986) Combined use of warfarin and adjusted subcutaneous heparin during pregnancy in patients with an artificial heart valve. J Am Coll Cardiol 8: 221–224
Leicher-Düber A, Schuhmacher R, Spranger J (1990) Symptomatische Verkalkungen beim Neugeborenen. Fortschr Rontgenstrahlen Neuen Bildgeb Verfahr Erganzungsbd 152: 463–468
Leonard CO (1988) Vitamin responsive bleeding disorder: a genocopy of the warfarin embryopathy. Proc Greenwood Genet Cent 7: 165–166
Loeser RF, Wallin R (1991) Vitamin K-dependent carboxylation in articular chondrocytes. Connect Tissue Res 26: 135–144
Lutz DJ, Noller KL, Spittell JA, Danielson GK, Fish CR (1978) Pregnancy and its complications following cardiac valve prosthesis. Am J Obstet Gynecol 131: 460–468
Mandelbrot L, Guillaumont M, Leclercq M, Lefrere JJ, Gozin D, Daffos F, Forestier F (1988) Placental transfer of vitamin K1 and its implications in fetal hemostasis. Thromb Haemost 60: 39–43
Mason JDT, Jardine A, Gibbin KP (1992) Foetal warfarin syndrome – a complex airway problem. Case report and review of the literature. J Laryngol Otol 106: 1098–1099

Maternal and Neonatal Haemostasis Working Party of the Haemostasis and Thrombosis Task (1993) Guidelines on the prevention, investigation and management of thrombosis associated with pregnancy. J Clin Pathol 46: 489–496
Matorras R, Reque JA, Usandizaga JA, Minguez JA, Larrea JL, Núñez L (1985) Prosthetic heart valve and pregnancy. A study of 59 cases. Gynecol Obstet Invest 19: 21–31
Mätzsch T, Bergqvist D, Bergqvist A, Hodson S, Dawes J, Hedner U, Østergaard P (1991) No transplacental passage of standard heparin or an enzymatically depolymerized low molecular weight heparin. Blood Coagul Fibrinolysis 2: 273–278
McCallion D, Phelps NE, Hirsh J, Cade JF (1971) Effects of coumadin administered during pregnancy to rabbits and mice. Teratology 4: 235–236
McMillan CW, Roberts HR (1966) Congenital combined deficiency of coagulation factors II, VII, IX and X. New Engl J Med 274: 1313–1315
Moncada RM, Venta LA, Venta ER, Fareed J, Walenga JM, Messmore HL (1992) Tracheal and bronchial cartilaginous rings: warfarin sodium-induced calcification. Radiology 184: 437–439
Mountain KR, Hirsh J, Gallus AS (1970) Neonatal coagulation defect due to anticonvulsant drug treatment in pregnancy. Lancet 1: 265–268
Murray JC, Johnson JA, Bird TD (1985) Dandy-Walker malformation: etiologic heterogeneity and empiric recurrence risk. Clin Genet 28: 272–283
Neerhof MG, Krewson DP, Haut M, Librizzi RJ (1993) Heparin therapy for congenital antithrombin III deficiency in pregnancy. Am J Perinatol 10: 311–312
Newcomb T, Matter M, Conroy L, DeMarsh QB, Finch CA (1956) Congenital hemorrhagic diathesis of the prothrombin complex. Am J Med 20: 798–805
Normann EK, Stray-Pederson B (1989) Warfarin-induced fetal diaphragmatic hernia. Case report. Br J Obstet Gynaecol 96: 729–730
Olthof E, DeVries TW, Touwen BCL, Smrkovsky M, Geven-Boere LM, Heijmans HSA, Van der Veer E (1994) Late neurological, cognitive and behavioural sequelae of prenatal exposure to coumarins: a pilot study. Early Hum Dev 38: 97–109
O'Neill H, Blake S, Sugrue D, MacDonald D (1982) Problems in the management of patients with artificial valves during pregnancy. Br J Obstet Gynaecol 89: 940–943
Otawara Y, Price PA (1986) Developmental appearance of matrix Gla protein during calcification in the rat. J Biol Chem 261: 10828–10832
Pascual-Castroviejo I, Velez A, Pascual-Pascual SI, Roche MC, Villarejo F (1991) Dandy-Walker malformation: analysis of 38 cases. Childs Nerv Syst 7: 88–97
Pati S, Helmbrecht GD (1994) Congenital schizencephaly associated with in utero warfarin exposure. Reprod Toxicol 8: 115–120
Pauli RM (1988) Mechanism of bone and cartilage maldevelopment in the warfarin embryopathy. Pathol Immunopathol Res 7: 107–112
Pauli RM, Haun JM (1993) Intrauterine effects of coumarin derivatives. Dev Brain Dysfunct 6: 229–247
Pauli RM, Madden JD, Kranzler KJ, Culpepper W, Port R (1976) Warfarin therapy initiated during pregnancy and phenotypic chondrodysplasia punctata. J Pediatr 88: 506–508
Pauli RM, Lian JB, Mosher DF, Suttie JW (1987) Association of congenital deficiency of multiple vitamin K-dependent coagulation factors and the phenotype of the warfarin embryopathy: clues to the mechanism of teratogenicity of coumarin derivatives. Am J Hum Genet 41: 566–583
Pavankumar P, Venugopal P, Kaul U, Iyer KS, Das B, Sampathkumara A, Airon B, Rao IM, Sharma ML, Bhatta ML, Gopinath N (1988) Pregnancy in patients with prosthetic cardiac valve. Scand J Thorac Cardiovasc Surg 22: 19–22
Pawlow I, Pawlow V (1985) Kumarin-Embryopathie. Z Klin Med 40: 885–888
Pettifor JM, Benson R (1975) Congenital malformations associated with the administration of oral anticoagulants during pregnancy. J Pediatr 86: 459–462
Phoon CK, Neill CA (1994) Asplenia syndrome: insight into embryology through an analysis of cardiac and extracardiac anomalies. Am J Cardiol 73: 581–587

Price PA, Kaneda Y (1987) Vitamin K counteracts the effects of warfarin in liver but not in bone. Thromb Res 46: 121–131
Price PA, Lothringer JW, Nishimoto SK (1981) Absence of the vitamin K-dependent bone protein in fetal rat mineral. Evidence for another gamma-carboxyglutamic acid-containing component in bone. J Biol Chem 255: 2938–2942
Price PA, Williamson MK, Haba T, Dell RB, Jee WS (1982) Excessive mineralization with growth plate closure in rats on chronic warfarin treatment. Proc Natl Acad Sci USA 79: 7734–7738
Price PA, Urist MR, Otawara Y (1983) Matrix Gla protein: a new γ-carboxylglutamic acid-containing protein associated in organic matrix of bone. Biochem Biophys Res Commun 117: 765–771
Quarrell OWJ, Koch M, Hughes HE (1990) Maxillonasal dysplasia (Binder's syndrome). J Med Genet 27: 384–387
Quenneville G, Barton B, McDevitt E, Wright I (1959) The use of anticoagulants for thrombophlebitis during pregnancy. Am J Obstet Gynecol 77: 1135–1149
Raivio KO, Ikonen E, Saarikoski S (1977) Fetal risks due to warfarin therapy during pregnancy. Acta Paediatr Scand 66: 735–739
Richman EM, Lahman JE (1976) Fetal anomalies associated with warfarin therapy initiated shortly prior to conception. J Pediatr 88: 509–510
Robinson MJ, Pash J, Grimwade J, Campbell J (1978) Fetal warfarin syndrome. Med J Aust 1: 157
Roll R, Baer F (1967) Effects of coumarin (o-hydroxycinnamic acid lactone) on pregnant mice. Arzneimittelforschung 17: 97–100
Rutherford SE, Phelan JP (1991) Clinical management of thromboembolic disorders in pregnancy. Crit Care Clin 7: 809–828
Ruthnum P, Tolmie JL (1987) Atypical malformations in an infant exposed to warfarin during first trimester of pregnancy. Teratology 36: 299–301
Salazar E, Zajarias A, Gutierrez N, Iturbe I (1984) The problem of cardiac valve prostheses, anticoagulants, and pregnancy. Circulation 70 [Suppl 1]: 169–177
Sareli P, England MJ, Berk MR, Marcus RH, Epstein M, Driscoll J, Meyer T, McIntyre J, vanGelderen C (1989) Maternal fetal sequelae of anticoagulation during pregnancy in patients with mechanical heart valve prostheses. Am J Cardiol 63: 1462–1465
Sbarouni E, Oakley CM (1994) Outcome of pregnancy in women with valve prostheses. Br Heart J 71: 196–201
Shaul WL, Emery H, Hall JG (1975) Chondrodysplasia punctata and maternal warfarin use during pregnancy. Am J Dis Child 129: 360–362
Shearer MJ (1992) Vitamin K metabolism and nutriture. Blood Rev 6: 92–104
Sheffield LJ, Danks DM, Mayne V, Hutchinson AL (1976) Chondrodysplasia punctata – 23 cases of a mild and relatively common variety. J Pediatr 89: 916–923
Sheffield LJ, Halliday JL, Jensen FO, Danks DM (1990) To lump or split types of chondrodysplasia punctata? Proc Greenwood Genet Cent 10: 112–113
Sheffield LJ, Halliday JL, Jensen F (1991) Maxillonasal dysplasia (Binder's syndrome) and chondrodysplasia punctata. J Med Genet 28: 503–504
Sheikhzadeh A, Ghabusi P, Kakim SH, Wendler G, Sarram M (1983) Congestive heart failure in valvular heart disease in pregnancies with and without valvular prostheses and anticoagulant therapy. Clin Cardiol 6: 465–470
Sherman S, Hall BD (1976) Warfarin and fetal abnormality. Lancet 1: 692
Soff GA, Levin J (1981) Familial multiple coagulation factor deficiencies. I. Review of the literature: differentiation of single hereditary disorders associated with multiple factor deficiencies from coincidental concurrence of single factor deficiency states. Semin Thromb Hemost 7: 112–148
Solomon GE, Hilgartner MW, Kutt H (1972) Coagulation defects caused by diphenylhydantoin. Neurology 22: 1165–1171
Stenflo J, Suttie JW (1977) Vitamin K-dependent formation of gamma-carboxyglutamic acid. Ann Rev Biochem 46: 157–172

Stevenson RE, Burton OM, Ferlauto GJ, Taylor HA (1980) Hazards of oral anticoagulants during pregnancy. JAMA 243: 1549–1551

Stevenson RE, Kelly JC, Aylsworth AS, Phelan MC (1987) Vascular basis for neural tube defects: a hypothesis. Pediatrics 80: 102–106

Strüwe E, Reinwein H, Stier R (1984) Coumarin-Embryopathie. Radiologe 24: 68–71

Sturridge F, deSwiet M, Letsky E (1994) The use of low molecular weight heparin for thromboprophylaxis in pregnancy. Br J Obstet Gynaecol 101: 69–71

Sundaram KS, Lev M (1988) Warfarin administration reduces synthesis of sufatides and other sphingolipids in mouse brain. J Lipid Res 29: 1475–1479

Sundaram KS, Lev M (1990) Regulation of sulfotransferase activity by vitamin K in mouse brain. Arch Biochem Biophys 277: 109–113

Suttie JW (1980) The metabolic role of vitamin K. Fed Proc 39: 2730–2735

Suttie JW (1983) Current concepts of the mechanism of action of vitamin K and its antagonists. In: Lindenbaum J (ed) Nutrition in hematology. Churchill Livingstone, New York, p 245

Tamburrinni O, Bartolomeo-Deluri A, DiGuglielmo GI (1987) Chondrodysplasia punctata after warfarin. Case report with 18-month follow-up. Pediatr Radiol 17: 323–324

Tejani N (1973) Anticoagulant therapy with cardiac valve prosthesis during pregnancy. Obstet Gynecol 42: 785–793

Toriello HV, Lin A, Higgins JV (1990) Conditions with stippled epiphyses: two new syndromes/situations and a review of the literature. Proc Greenwood Genet Cent 10: 113–114

Uotila L (1990) The metabolic functions and mechanism of action of vitamin K. Scand J Clin Lab Invest 201: 109–117

Uszynski M (1992) Heparin neutralization by an extract of the human placenta: measurements and the concept of placental barrier to heparin. Gynecol Obstet Invest 33: 205–28

Vanlaeys R, Deroubaix P, Deroubaix G, LeLong M (1977) Les antivitamines K sont-elles teratogenes? Nouv Presse Med 6: 756

Ville Y, Jenkins E, Shearer MJ, Hemley H, Vasey DP, Layton M, Nicolaides KH (1993) Fetal intraventricular haemorrhage and maternal warfarin. Lancet 341: 1211

Vitali E, Donatelli F, Quaini E, Groppelli G, Pellegrini A (1986) Pregnancy in patients with mechanical prosthetic valves. Our experience regarding 98 pregnancies in 57 patients. J Cardiovasc Surg (Torino) 27: 221–227

vonSydow G (1947) Hypoprotrombinemi och hjärnskada hos barn till dikumarin-behandlad moder. Nord Med 34: 1171–1172

Warkany J (1975) A warfarin embryopathy? Am J Dis Child 129: 287–288

Warkany J, Bofinger M (1975) Le role de la coumadine dans les malformations congénitales. Med Hyg 33: 1454–1457

Weeninck GH, vanDijk-Wieda CA, Meyboom RHB, Koppe JG, Staalman CR, Treffers PE (1981) Teratogeen effect van coumarine-derivaten. Ned Tijdschr Geneeskd 125: 702–706

Whitfield MF (1980) Chondrodysplasia punctata after warfarin in early pregnancy. Case report and summary of the literature. Arch Dis Child 55: 139–142

Wong V, Cheng CH, Chan KC (1993) Fetal and neonatal outcome of exposure to anticoagulants during pregnancy. Am J Med Genet 47: 17–21

Wulfsberg EA, Curtis J, Jayne CH (1992) Chondrodysplasia punctata: a boy with X-linked recessive chondrodysplasia punctata due to an inherited X-Y translocation with a current classification of these disorders. Am J Med Genet 43: 823–828

Yokoyama T, Copeland NG, Jenkins NA, Montgomery CA, Elder FF, Overbeek PA (1993) Reversal of left-right asymmetry: a situs inversus mutation. Science 260: 679–682

Zakzouk MS (1986) The congenital warfarin syndrome. J Laryngol Otol 100: 215–219

Zipprich KW, Canzler E, Hundsdörfer S (1987) Zur teratogenen Wirkung von Kumarinen. Zentralbl Gynakol 1009: 364–368

CHAPTER 26

Antiviral Agents

R. STAHLMANN and S. KLUG

A. Introduction

As a result of extensive research activities during the last two decades, a considerable number of antiviral agents are now available for therapeutic use. As a matter of fact, we still face the unsatisfactory situation that none of the currently used antiviral drugs is effective in elimination of nonreplicating or latent viruses. In principle, the development of antiviral agents is more difficult than the development of drugs against other infectious agents, because viral replication depends crucially on the metabolic processes of the infected cell. However, an increasing need for antiviral agents arises from the increasing incidences of viral diseases such as genital herpes, cytomegalovirus (CMV) infections in immunocompromised patients, and last but not least the acquired immunodeficiency syndrome (AIDS) epidemic.

With respect to therapeutic use, antivirals are a rather heterogenous group of drugs (Table 1). Some are given only for very specific indications (e.g., foscarnet, ribavirin) or only for topical treatment (e.g., idoxuridine). Others, such as acyclovir, are now among the most often prescribed chemotherapeutic agents. Due to these differences in use, the data bases for the single antiviral drugs differ considerably. Whereas several studies in experimental animals as well as human data have been published with frequently used compounds, information on other compounds is very scarce. These differences should be considered when the reproductive toxicity of these drugs is evaluated. The lack of data for some compounds certainly does not mean that there might be no risks with these drugs.

In comparison to bacterial or mammalian cells, viruses differ significantly in their "morphology" as well as in their "reproduction". They possess only one type of nucleic acid (either DNA or RNA), and they have no anabolic system and no cellular organisation. The consequence is an obligatory intracellular parasitic existence. These specific properties and the coexistence with the infected cell make it difficult to develop antiviral agents with a high degree of selective toxicity for the infectious agent. Principally, agents that may inhibit the viruses are also likely to injure the host cells that harbor them.

It would be ideal to develop compounds which interact primarily and selectively with the replication of the virus or certain phases of the infection process without causing severe damage to human cells. So far, this has not

Table 1. The different antiviral compounds, their class, and their main target

Substance	Class	Main target
Acyclovir	Nucleoside analogue	DNA polymerase
Amantadine	Others	Membrane protein (uncoating)
Didanosine (ddI)	Nucleoside analogue	Viral reverse transcriptase
Dideoxyadenosine (ddA)	Nucleoside analogue	Viral reverse transcriptase
Foscarnet	Pyrophosphate analogue	DNA polymerase, reverse transcriptase
Ganciclovir	Nucleoside analogue	DNA polymerase
Idoxuridine	Nucleoside analogue	DNA polymerase
Ribavirin	Nucleoside analogue	RNA polymerase
Rimantadine	Others	Membrane protein (uncoating)
Vidarabine	Nucleoside analogue	DNA polymerase
Zalcitabine (ddC)	Nucleoside analogue	Reverse transcriptase
Zidovudine	Nucleoside analogue	Reverse transcriptase

been achieved successfully, but some efforts appear to be rather promising. Agents are now available that interfere with virus-specific events such as attachment to the cell, uncoating of the viral genome, or viral nucleic acid synthesis.

For example, amantadine delays or stops the uncoating process of influenza A viruses. Many antiviral drugs target viral nucleic acid synthesis. The replication of viral nucleic acids is catalyzed by virus-specific enzymes. These viral enzymes often catalyze the same reactions as corresponding cellular enzymes but differ in their structure or kinetics. This allows the synthesis of compounds which reveal a higher affinity to virus-specific enzymes than to the corresponding cellular enzyme. This is the principal mode of action of the nucleoside analogues. Foscarnet, an example of a drug with different structure and action, is an inorganic pyrophosphate analogue that directly inhibits viral polymerases or reverse transcriptases.

Most of the virustatic agents now used are nucleoside analogues. Although their exact mechanism of action has not been clarified, it is obvious that they alter the DNA metabolism in virus-infected cells. Because it has been well known for many years that compounds which can alter DNA metabolism often exhibit a pronounced prenatal toxicity, the reproductive toxicity of these drugs should be evaluated with the necessary care.

Several situations might occur in which a pregnant woman is exposed to antiviral agents. For example, when acyclovir is used for long-term suppressive therapy of recurrent genital herpes, a woman might become pregnant during the period of continuous drug intake. A woman might also be treated orally or intravenously for a primary genital herpes with acyclovir without knowledge of pregnancy. Last but not least, it is possible that a severe viral infection is diagnosed in a pregnant woman and that the patient must be treated. In all cases it is important to have as much information as possible on the reproductive toxicity of the drug in experimental animals, and a large epidemiological data base would also be desirable to advise these patients.

Zidovudine (AZT) is an antiviral agent with increasing use in AIDS patients. Besides prenatal drug exposure in the first trimester due to lack of knowledge of pregnancy, the drug will probably be taken by an increasing number of pregnant women because it was shown that this nucleoside analogue can be beneficial in pregnant AIDS patients (see Sect. B.I.8.b.γ). In a controlled study it has been demonstrated that transmission of HIV from mother to child can be reduced by pre-, peri-, and postnatal administration of zidovudine without significant harm to the newborn child (CONNOR et al. 1994). Other anti-HIV drugs might also be used for this indication in the future, and although the treatment will probably be restricted to the less critical second half of pregnancy, such drug uses require careful risk–benefit analysis.

B. Reproductive Toxicity of Antiviral Agents

I. Nucleoside Analogues

1. Acyclovir

a) Pharmacology and Clinical Use

Acyclovir is a synthetic nucleoside analogue and one of the most frequently used chemotherapeutic agents. The cyclic sugar in the molecule of the physiologic counterpart has been replaced by a linear side chain lacking the 3′-hydroxyl group (Fig. 1). Acyclovir inhibits the replication of herpesviruses, such as herpes simplex viruses (HSV)-1 and -2. The drug is preferentially phosphorylated to the monophosphate by viral thymidine kinases and subsequently converted by cellular enzymes to the triphosphate. This biologically active metabolite, which is present in 40- to 100-fold higher concentrations in HSV-infected than in uninfected cells, competitively inhibits DNA polymerases and can act as a chain terminator after incorporation into the DNA (FURMAN et al. 1979, 1981). The oral bioavailibility of the drug is poor (15%–21%). Up to 90% of an intravenously administered dose is excreted unchanged via the kidneys with a half-life of 2.5–3 h. Acyclovir has become the first-line drug for treatment of many types of infections by herpes simplex and

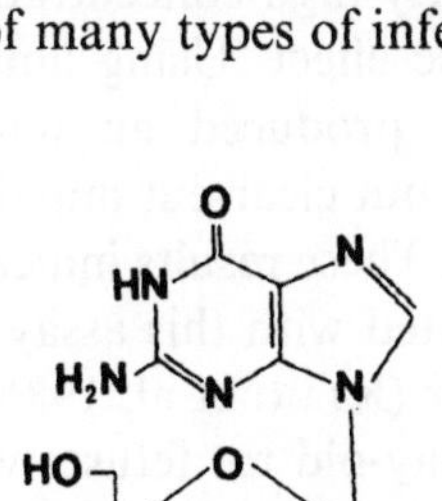

Fig. 1. Structure of acyclovir

varicella zoster viruses. The drug was superior to vidarabine in patients with HSV encephalitis and other diseases; the recommended therapeutic treatment schedule is 10 mg/kg i.v. three times daily for encephalitis and other severe infections, e.g., in immunocompromised patients. In immunocompetent patients, clinical efficacy, e.g., of a 5-day treatment with 800 mg five times daily for herpes zoster, is only demonstrable if treatment is initiated within 72 h of rash onset. Continuous administration of oral acyclovir was effective for suppression of recurrent genital herpes. Tolerance of acyclovir is generally good, though some patients show neurotoxic reactions (headache, nausea) or reversible renal dysfunction after intravenous treatment. Because the solubility of the drug is poor, it can crystallize in renal tubules after rapid injection; therefore, it should be given as a slow infusion over 1 h (LASKIN et al. 1982; WHITLEY and GNANN 1992; HAYDEN 1995).

b) Experimental Data

α) *In Vitro Data.* Data established with an in vitro method gave the first indication that acyclovir might interfere with prenatal development. Experiments with the rat whole embryo culture (KLUG et al. 1985a,b) showed that acyclovir disturbs normal development of rat embryos at concentrations which are in the same range as those achieved under therapeutic conditions. At the lowest concentration tested (10 μ*M* acyclovir), no adverse effects were observed. A retardation of the development of the ear anlagen – but no clear-cut abnormalities – was observed at concentrations of 25 μ*M*. At this concentration, some other routinely evaluated parameters such as crown–rump length, number of somites, protein content, or score were also significantly decreased. At concentrations of 50 μ*M* acyclovir or higher, additional disturbances of embryonic differentiation became obvious. Mainly affected was the telencephalon, and in some embryos the telencephalon anlage was completely missing. Histological examinations confirmed these findings. The ventricles of the brain and the central channel of the neural tube were dilated, and the neuroepithelium was monolayered instead of multilayered. The histological findings have been described in detail elsewhere (KLUG et al. 1985a).

Acyclovir was also tested in the limb bud culture system to investigate whether the drug interferes specifically with the morphogenetic differentiation of the limb anlagen. Only fairly high concentrations of the compound (more than 100 μ*M*) had an adverse effect. Using limb buds of 11-day-old mouse embryos, 200 μ*M* acyclovir produced an interference with development (mainly of the paw skeleton), but clear-cut impairment of limb differentiation was obtained only at 400 μ*M*. These results indicated that limbs at the stage of gestation which was investigated with this assay do not react very sensitively towards exposure to acyclovir (KLUG et al. 1985a; STAHLMANN et al. 1993).

The thymic lobes of 17-day-old rat fetuses were cultured for 7 days with the addition of acyclovir at concentrations of 3–300 μ*M*. The effect of the nucleoside analogue on lymphopoiesis in the organ culture was determined by isolation and quantification of the cultured thymocytes. A concentration of 10 μ*M* acyclovir was sufficient to cause a significant reduction of the thymocyte number. Increasing concentrations (30–300 μ*M*) of acyclovir led to a concentration-dependent inhibition of cell proliferation and to an apparent cytotoxicity. Light and electron microscopic investigations revealed a selective effect on lymphatic cells, whereas the morphology of the thymic epithelium was only slightly altered (FOERSTER et al. 1992a,b).

Using the chick embryo in ovo, another model for developmental toxicity independent of the maternal organism, the embryotoxic potential of the virustatic was also demonstrated. Growth retardation was noticed after intra-amniotic administration of a dose of 3 μg acyclovir or more. Impaired development (gross structural abnormalities) was observed in the concentration range of 6–30 μg/embryo (HEINRICH-HIRSCH and NEUBERT 1991).

β) In Vivo Data. Acyclovir has been studied following a segment II protocol (parenteral treatment on days 6–15 in rats). This study did not indicate any prenatal toxicity of the compound, but the highest dose used was only 25 mg/kg body weight (MOORE et al. 1983). Testing of higher doses was reported to be complicated by the fact that acyclovir crystallizes in renal tubules after treatment with doses above 25 mg/kg. Therefore, it was not possible to use higher doses for *multiple* daily treatment as required by the routinely used segment II protocol. Severe maternal nephrotoxicity must be expected if rats are dosed from day 6 to 15 of pregnancy with similar doses which had been shown to be teratogenic for related nucleoside analogues, e.g., vidarabine (200 mg/kg). It seems important to mention that (a) the highest doses of acyclovir tested in the routinely performed studies are close to the maximum recommended therapeutic ones (intravenous infusion of 10–15 mg/kg three times daily) and (b) nephrotoxic reactions have also been observed in patients after treatment with therapeutic doses (BRIDGEN et al. 1982).

From the results with acyclovir in the rat whole embryo culture system (see Sect. B.I.1.b.α), it was possible to deduce the period of high susceptibility and to initiate specific in vivo studies. Pregnant rats were treated on day 10 of gestation with subcutaneous injections of 100 mg/kg once, twice, or three times daily. To compare the results of this experiment directly with the outcome observed in the whole embryo culture, section was performed on day 11.5 of gestation and the embryos were evaluated with the same methods as those used for evaluation of embryos at the end of the whole embryo culture. A very similar morphological outcome was found in the embryos after in vitro and in vivo exposure to acyclovir. Macroscopically as well as histologically, the typical central nerve system (CNS) defects (telencephalon) were visible in embryos exposed to more than one injection of 100 mg aciclovir/kg on day 10 of gestation. All embryos exhibited abnormalities after treatment with three injections, and all other variables investigated were significantly altered in comparison to the controls. One single injection of 100 mg/kg on day 10 of gestation was sufficient to induce significant changes in the embryos (significant reduction of crown-rump length, number of somites, and protein content in comparison to control values), but only 1.5% of the embryos ($n=68$) were abnormal in this group (STAHLMANN et al. 1988a).

Subsequently, rat fetuses were evaluated on day 21 of gestation (skeletal system and visceral organs) after exposure to three doses of 100 mg acyclovir/kg on day 10 of gestation. These studies confirmed the pronounced prenatal toxicity of acyclovir at this dose level in rats. Mainly affected were the skull bones, and a typical finding was malformation of the os tympanicum. Multiple other malformations such as eye defects and tail anomalies were also seen in the majority of offspring after acyclovir exposure on day 10 of gestation (STAHLMANN et al. 1987; CHAHOUD et al. 1988).

The 1-day treatment caused slight and reversible nephropathia; however, it is extremely unlikely that the malformations observed were induced indirectly by the changes in the maternal organism, and since the morphological alteration of the 11.5-day-old embryos were very similar after in vitro and in vivo exposure, the results of the two experimental approaches can be regarded as confirming one another. In addition, evidence for a direct teratogenic action of the drug in rats comes from comparative studies with folic acid. This compound also crystallizes in renal tubules and causes nephropathia. Although similar changes in plasma urea and creatinine levels were seen in pregnant rats after treatment with folic acid or acyclovir, specific teratogenicity was only seen with acyclovir (STAHLMANN et al. 1988b).

Acyclovir exposure on day 10 of gestation (one or three injections of 100 mg/kg) affected the development of several organs of the immune system: thymus weights were decreased and spleen weights were increased in fetuses on day 21 as well as in 15-week-old offspring. Using an infectivity model with *Trichinella spiralis*, it was shown that in these offspring the function of the immune system was persistently impaired. Leukocyte counts in peripheral blood were low, and the CD4 to CD8 ratio was significantly increased in 9-month-old rats after prenatal exposure to acyclovir. Thus the prenatally induced alterations in the immune system are irreversible in rat offspring (STAHLMANN et al. 1991, 1992a, 1995).

The teratogenic potential of acyclovir in marmosets (*Callithrix jacchus*) after oral treatment at a dose level leading to plasma concentrations in the range of human therapeutic concentrations was studied by KLUG et al. (1992). Six animals were treated once daily with an oral dose of 200 mg/kg from day 50 or 51 to day 63 or 64 of gestation. Two animals were treated daily with the same dose twice a day from day 45 to 57. In this study, no gross structural anomalies (external or skeletal) were observed, but one abortion was noticed.

In a parallel study, STAHLMANN and coworkers (1992b) treated a group of 14 pregnant marmosets intravenously on different days of organogenesis (days 45–58) with single doses of 30, 60, or 90 mg acyclovir/kg. Five animals were treated with two to five doses of 30 mg/kg on consecutive days. The treatment led to a reversible increase of creatinine and urea concentrations in the blood plasma. Gross structural anomalies were not seen in any of the 19 animals, but there was a significant reduction of the fetal body weight after the 90-mg/kg dose, which might have been induced by maternal toxicity. In the whole group, three abortions were observed (not dose related).

γ) *Human Data.* To study the kinetics of acyclovir in pregnancy, the drug was given orally every 8 h from 38 weeks' gestation until delivery. Newborn plasma levels were similar to maternal ones, but the drug accumulated in amniotic fluid: in three patients (dosage, 200 mg t.i.d.), the maternal plasma concentrations at delivery ranged from 0.65 to 1.7 μmol/l, and the concentrations in amniotic fluid ranged from 1.87 to 6.06 μmol/l (FRENKEL et al. 1991).

Isolated case reports on fetal malformations associated with the use of acyclovir during pregnancy have been published, but of course in casuistics the cause–effect relationship remains obscure (GUBBELS et al. 1991).

There is as yet no indication from studies in humans that acyclovir or other virustatics present a risk for human embryos or fetuses. The data of several hundred women exposed to acyclovir during pregnancy have been collected in the Acyclovir in Pregnancy Registry, and no significant increase in the incidence of malformations or abortions was noticed, but the number is still too small to draw well-founded conclusions (ANDREWS et al. 1992; ELDRIGE et al. 1993, 1995). A report summarizing the progress and findings of the registry is prepared periodically and is available to health-care professionals from the following address: Acyclovir in Pregnancy Registry, Burroughs Wellcome Co., Research Triangle Park, NC 27709, USA.)

When evaluating the registry data, it should be borne in mind that most of the patients were exposed to relatively low oral doses of 200 mg five times daily, leading to average peak plasma concentrations of 0.4–0.8 mg/l. Although these data do not indicate a teratogenic effect under therapeutic conditions, a risk for humans might exist if higher doses are used (e.g., 4000 mg acyclovir daily for treatment of shingles) and/or with new derivatives with better bioavailability, such as valacyclovir (the *l*-valyl ester of acyclovir). The recommended daily dose for valacyclovir is 3000 mg, and the estimated bioavailability of acyclovir after valacyclovir administration is approximately four times greater than after oral acyclovir administration. Acyclovir peak plasma concentrations after valacyclovir treatment ranged from 2.8 to 16.1 mg/l (12.3–71.5 μM) in one large clinical trial with more than 750 patients (BEUTNER et al. 1995).

2. Didanosine and Dideoxyadenosine

a) Pharmacology and Clinical Use

Didanosine (2′,3′-dideoxyinosine, ddI) is active against HIV-1 and HIV-2 in vitro including zidovudine (AZT)-resistent isolates. It is converted intracellularly to 2′, 3′-dideoxyadenosine (ddA) triphosphate (ddATP; Figs. 2,3), which inhibits viral reverse transcriptase and acts as a chain terminator of viral DNA synthesis. The intracellular half-life of the triphosphate

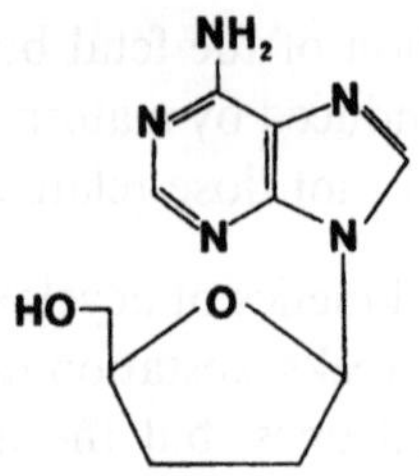

Fig. 2. Structure of 2′,3′-dideoxyadenosine

(ddATP) is reported to be 8–24 h. Oral bioavailability is approximately 35%–45%; the drug is excreted, with a half-life of approximately 0.6–1.5 h, predominantly unchanged via the kidneys. Pancreatitis is an important side effect of the drug. ddI is used for patients at advanced stages of HIV infection (FAULDS and BROGDEN 1992; HAYDEN 1995; KHOO and WILKINS 1995).

b) *Experimental Data*

α) *In Vitro data.* The effect of ddI on the in vitro development of thymic lobes from 17-day-old rat fetuses was studied in an organ culture system. At concentrations of 30 μ*M* ddI and higher, a significant reduction in the number of thymocytes was observed (FOERSTER et al. 1992b).

In the rat whole embryo culture system, no effect on normal development of 9.5-day-old rat embryos at concentrations up to 200 μ*M* ddA was seen. Histological examination showed that the compound affected the development of the neural epithelium, the brain vesicles, and the neural tube at concentrations of 500 and 1000 μ*M* (KLUG et al. 1991).

β) *In Vivo Data.* Groups of pregnant mice (five to 15) were treated with 10, 30, 100, or 300 mg ddI/kg per day subcutaneously for the whole period of pregnancy. No adverse effects were noted (SIEH et al. 1992).

No adverse effects were seen in routinely performed fertility, teratology, or peri- and postnatal studies with ddI. The only adverse effect noticed was a decreased food intake and decreased body weight gain in dams and pups at 1000 mg/kg during mid and late gestation in the rat fertility study (cited after MCLAREN et al. 1991; more detailed data have not been published).

Three subcutaneous injections of 200 mg ddA/kg on day 10 of pregnancy induced adverse effects on normal embryonic development in rats. When the embryos were evaluated with regard to growth and differentiation on day 11.5 of gestation, they showed significantly decreased values for the end points crown–rump length, number of somites, protein content, and the developmental score, but no dysmorphogenesis was observed (KLUG et al. 1991).

γ) *Human Data.* Fetoplacental passage of ddI was studied in two pregnant women during the second trimester of pregnancy. Concentrations in amniotic

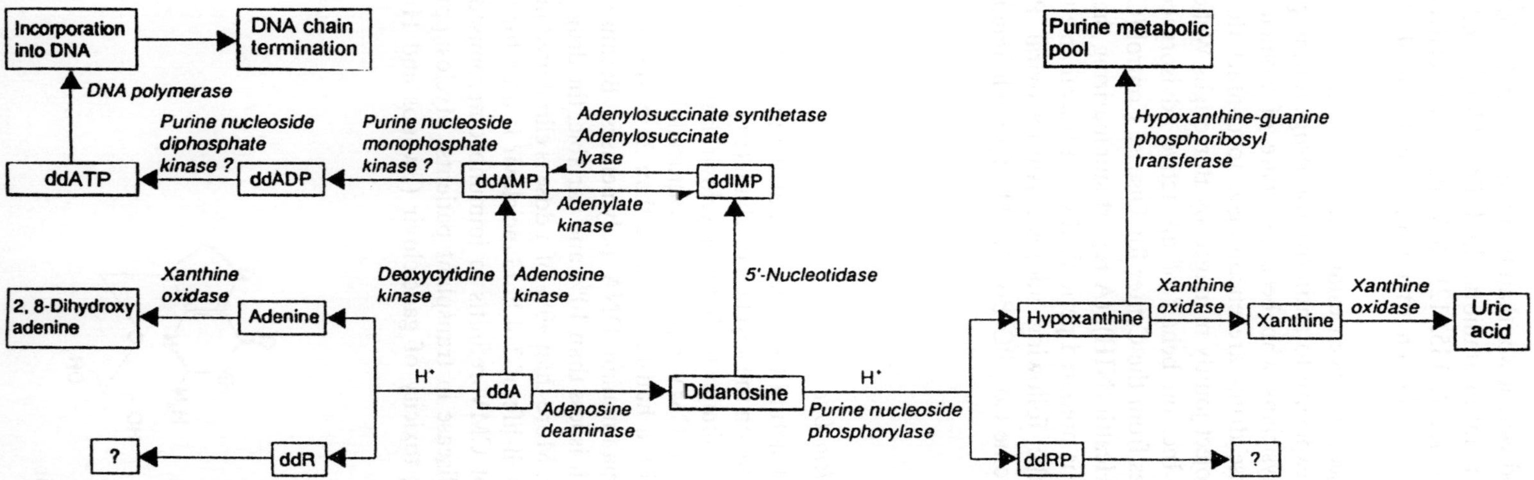

Fig. 3. Activation and metabolic pathways of didanosine (ddI) and 2′,3′-dideoxyadenosine (ddA). Didanosine undergoes extensive metabolism with formation of either 2′,3′-dideoxyadenosine -5′-triphosphate (ddATP) or uric acid or entry into the purine metabolic pool. Acid hydrolysis results in the conversion of didanosine to hypoxanthine. ddADP 2′,3′-dideoxyadenosine-5′ -diphosphate; ddAMP 2′,3′ -dideoxyadenosine-5′-monophosphate; ddIMP, 2′,3′-dideoxyinosine-5′-monophosphate; ddR, 2′,3′-dideoxyribose; ddRP, 2′,3′-dideoxyribose-1-phosphate; H^+ acid hydrolysis. The DNA polymerase (*top left*) is retroviral DNA polymerase (reverse transcriptase) or cellular DNA polymerase. (After HARTMAN et al. 1990; FAULDS and BROGDEN 1992).

fluid as well as in fetal blood were lower than those in maternal blood (patient no 1, less than 5, 42, 295 ng/ml in amniotic fluid, fetal blood, and maternal blood, respectively; patient no. 2, 135, 121, 629 ng/ml, respectively). Concentrations were measured 65 or 78 min after a single oral dose of 375 mg ddI (PONS et al. 1991).

An epidemiologic project has been established to collect observational, non-experimental data on exposure to antiretroviral drugs during pregnancy (ELDRIGE et al. 1993). Physicians who become aware of women receiving treatment with one of these drugs are encouraged to contact the registry, which is a collaborative project jointly managed by Burroughs Wellcome Co. and Hoffmann-La Roche Inc. on behalf of an active advisory committee composed of representatives from the Center for Disease Control (CDC) and the National Institutes of Health (NIH). (A report summarizing the progress and findings of the registry is prepared periodically and is available to health-care professionals from the following address: Antiretroviral Pregnancy Registry, Burroughs Wellcome Co., PO Box 12700, Research Triangle Park, NC 27709–2700, USA.)

3. Ganciclovir

a) Pharmacology and Clinical Use

Ganciclovir has a similar structure to acyclovir, but the molecule has an additional hydroxymethyl group on the acyclic side chain (Fig. 4). The inhibitory action against most viruses is similar to that of acyclovir, but it is considerably more potent for inhibition of CMV replication. Intracellular ganciclovir triphosphate levels decline with a half-life of more than 24 h; this active metabolite inhibits viral and mammalian DNA polymerases. Because the oral bioavailability of ganciclovir is less than 10% and variable, the drug should be administered intravenously. More than 90% of a dose is eliminated unchanged by renal excretion with a half-life of 2–4 h. Ganciclovir is used for treatment and chronic suppression of CMV retinitis in immunocompromised patients and prevention of CMV disease in transplant patients. Myelosuppression is the principal dose-limiting toxicity of ganciclovir (FAULDS and HEEL 1990; HAYDEN 1995).

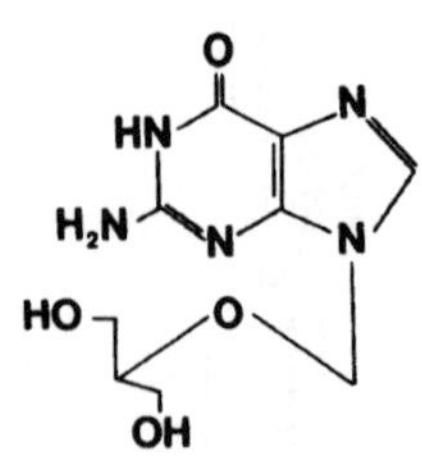

Fig. 4. Structure of ganciclovir

b) Experimental Data

α) In Vitro Data. Ganciclovir was tested in the whole embryo culture system with 9.5-day-old rat embryos at concentrations between 25 and 500 μ*M*. Similiar to the effects observed with other virustatics in this system, ganciclovir induced specific alterations of the head (100–500 μ*M*). Targets of the compound were the neural epithelium of the brain vesicles and the neural tube. At concentrations of 50 μ*M* and higher, the embryos showed significantly reduced values for the end points crown–rump length, number of somites, protein content, and the developmental score. At the highest concentration tested, 5% of the exposed embryos showed a failure in rotation. It is of interest that ganciclovir was the only virustatic from a series of related compounds which slightly influenced the development of the heart; it caused a dilated pericard in some of the exposed embryos (KLUG et al. 1991).

At a concentration of 10 μ*M*, ganciclovir caused a significant reduction in the number of thymocytes in cultured thymic lobes from 17-day-old rat fetuses (FOERSTER et al. 1992b).

Hepatocytes from rat fetuses (day 19 of gestation) were exposed to ganciclovir for 24 h using concentrations ranging from 0.5 mg/l (approximately 2 μ*M*) to 30 mg/l. The growth rate of hepatocytes was not altered by the drug. At a concentration of 6 mg/l medium, ganciclovir did not induce significant leakage of lactate dehydrogenase, but ^{51}Cr release was slightly greater than in control hepatocytes (HENDERSON et al. 1993).

Using a single, isolated perfused human placental cotyledon system and human placental vesicles, the placental transfer characteristics of the drug were studied. Ganciclovir was concentrated initially at the maternal placental surface and then crossed passively into the fetal compartment (HENDERSON et al. 1993).

β) In Vivo Data. Very little data on the reproductive toxicity of ganciclovir have been published. According to the manufacturer the result of the routinely performed studies are as follows:

> Female mice exhibited decreased fertility, decreased mating behavior, and increased embryo death after daily intravenous doses of 90 mg/kg. Daily intravenous doses of up to 20 mg/kg and daily oral doses of up to 1000 mg/kg did not impair female fertility. In male mice, fertility was reversibly decreased after daily intravenous doses of 2 mg/kg and daily oral doses of 10 mg/kg. At higher doses these effects were irreversible (SYNTEX 1988).
> In mice, doses of 108 mg ganciclovir/kg on days 7–16 of gestation, which were maternally toxic, caused an increased resorption rate and decreased fetal body weight. No teratogenic changes were seen (SYNTEX 1988).
> In rabbits, daily doses of 20 and 60 mg ganciclovir/kg during gestation caused fetal growth retardation, embryo death, and teratogenicity (SYNTEX 1988).

Three subcutaneous injections of 200 mg ganciclovir/kg on day 10 of pregnancy induced abnormal embryonic development in rats; 67% of the embryos evaluated on day 11.5 of gestation were malformed (KLUG et al. 1991).

Testicular hypoplasia was reported in rat offspring after intrauterine exposure to ganciclovir. Dams had been treated with three subcutaneous injections of 100 mg ganciclovir/kg on day 10 of gestation (HARTMANN et al. 1991).

4. Idoxuridine

a) Pharmacology and Clinical Use

Idoxuridine is an iodinated thymidine analogue that inhibits replication of several DNA viruses (Fig. 5). It can be used for topical treatment of HSV keratitis, herpes labialis, genital herpes, and herpes zoster, but alternative drugs exhibit a higher efficacy and/or better tolerance (PRUSOFF 1988; HAYDEN 1995).

b) Experimental Data

Pregnant mice were treated intraperitoneally on single days of gestation (7, 8, 9, 10, or 11) with 100, 300, or 500 mg idoxuridine/kg. The two highest doses were embryolethal. After treatment with 100 mg idoxuridine/kg, malformations such as exencephaly, polydactyly, and cleft palate were observed at high incidences (SKALKO and PACKARD 1973).

Pregnant mice and rats were treated during late gestation for three consecutive days either subcutaneously (100, 200, or 400 mg/kg in mice) or orally (200 or 400 mg/kg in rats). The first treatment day in mice was day 16 of gestation (in rats, day 18 of gestation). Cerebellum, eye, and kidneys of the offspring were examined histologically on day 10 or 20 postnatally. The highest doses induced postnatal mortality in both species; lower doses caused minimal to moderate lesions in the cerebellum or kidneys (PERCY 1975).

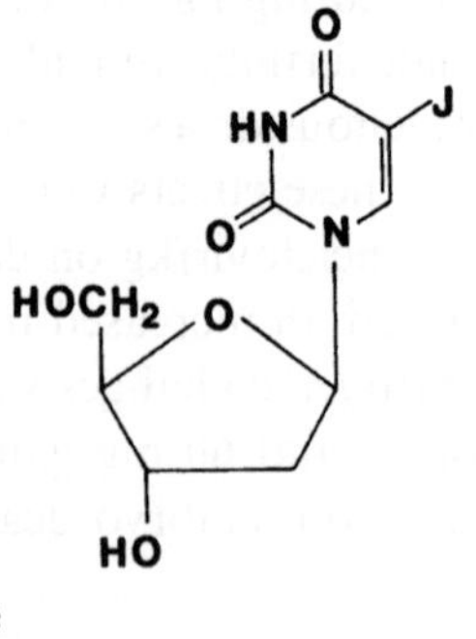

Fig. 5. Structure of idoxuridine

In newborn rats, the drug caused retinal dysplasia, cerebellar hypoplasia, and focal renal cortical dysplasisa when administered subcutaneously at doses of 200 and 400 mg/kg on postnatal days 1 – 6 (PERCY et al. 1973).

Idoxuridine was teratogenic in rabbits after local administration. Dams received idoxuridine eye drops several times daily (a 0.1% solution) between the sixth and the 18th day of gestation. Among the 121 fetuses evaluated, 28 (23%) showed malformations such as exophthalmus and clubbed forelimb (ITOI et al. 1975).

5. Ribavirin

a) *Pharmacology and Clinical Use*

In contrast to most other nucleoside analogues currently used as antiviral agents, the sugar moiety in the ribavirin molecule is ribose and thus a physiological one. However, instead of a regular purine base, this drug contains a nonphysiological heterocyclus (Fig. 6). It has multiple sites of action, eg., inhibition of guanine nucleotides and inhibition of viral RNA polymerase. The compound inhibits influenza and respiratory syncytial virus at concentrations less than 10 mg/l. Due to its toxicity (e.g., hematotoxicity, neurotoxicity) when administered orally or intravenously, ribavirin is only applied by aerosol using a special nebulizer. By this route of administration, it shortens the duration of shedding of virus and improves clinical symptoms in children with pneumonia and bronchiolitis due to respiratory syncytial virus (HAYDEN 1995).

b) *Experimental Data*

α) *In Vitro Data.* The effects of ribavirin on liver cells from 13 to 14-day-old mouse fetuses were studied in vitro. The drug inhibited the erythroid colony formation with an IC_{50} of 0.62 μ*M*. These cells reacted more sensitively towards the action of the nucleoside analogue than bone marrow cells from adult mice (GOGU et al. 1989).

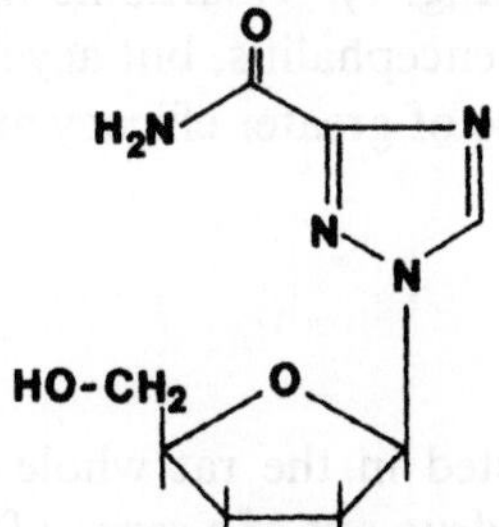

Fig. 6. Structure of ribavirin

The effects of ribavirin were studied in a limb bud culture system. Fore- and hindlimb buds from stage-18 mouse embryos (day 12) were cultured for 6 days in the presence of 1, 5, 10, 25, 50, or 100 mg ribavirin/l medium. At 10 mg/l, the drug caused a slight growth inhibition. Higher concentrations induced abnormal development (KOCHHAR et al. 1980; KOCHHAR 1982).

β) In Vivo Data. Pregnant mice were injected at seven different time points between day 10 and day 13 of gestation (embryonic stages 14–20) with single intraperitoneal injections of ribavirin at doses between 10 and 200 mg/kg. Doses in excess of 25 mg/kg were teratogenic. Defects of the orofacial bones occurred dose and phase dependently. For example, treatment on day 10.5 resulted in shortened maxilla, while treatment on either day 11 or 11.5 resulted in a reduction of the length of the jaws; treatment on day 12 caused deformation of the mandible (KOCHHAR et al. 1980).

Ribavirin was teratogenic in hamsters (single i.v., i.p., and p.o. doses of 2.5–5 mg/kg between day 7 and day 9 of gestation), causing abnormalities of the limbs, eyes, and the brain. In rats, higher doses were necessary (single i.p. and p.o. doses of 25–50 mg/kg on day 9 of gestation) to induce malformations, which were generally restricted to the head region (FERM et al. 1978).

In rabbits, doses as low as 1 mg/kg caused embryolethality, but no teratogenic effects were noted in baboons that received 120 mg ribavirin/kg during critical periods of gestation (HILLYARD 1980, cited after KACMAREK 1991).

γ) Human Data. The description of the teratogenic potential of ribavirin in several animal species has raised concern, as the drug is applied in the form of an aerosol and exposure of care-givers occurs. Although there is no evidence that such an exposure presents a definite hazard, it has been proposed that maximum efforts should be made to minimize environmental exposure of caregivers (KACMAREK 1991; KOREN and ITO 1993).

6. Vidarabine

a) Pharmacology and Clinical Use

Vidarabine triphosphate competitively inhibits viral and, to a lesser extent, cellular DNA polymerases (Fig. 7). Vidarabine has proven to be effective in treatment of herpes simplex encephalitis, but acyclovir has replaced it for this and other indications because of greater efficacy and/or safety (WHITLEY et al. 1980; HAYDEN 1995).

b) Experimental Data

α) In Vitro Data. When tested in the rat whole embryo culture system, vidarabine was the most toxic drug out of a series of related antivirals, with 50% of the embryos affected at concentrations of 3–10 μM. After exposure to the adenine derivative, all dysmorphogenic embryos revealed alterations of the

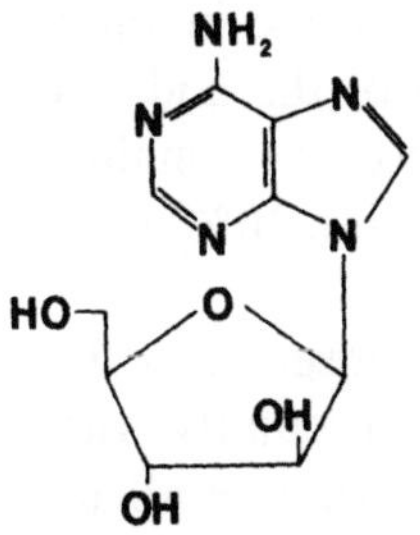

Fig. 7. Structure of vidarabine

head region exclusively. Histologically, the substance severely affected the development of the neural epithelium of the brain vesicles and of the neural tube. Physiologically multilayered at the end of the culture period, the neuroepithelium of the treated embryos in the middle and higher concentrations changed to being monolayered and the cells were less densely packed. In a few of the exposed embryos, exencephaly, covered by epidermal epithelium, could be demonstrated (KLUG et al. 1991).

Vidarabine at a concentration of 10 μ*M* induced a reduction of thymocytes in cultured thymic lobes from 17-day-old rat fetuses. A concentration of 300 μ*M* caused morphological alterations of the lymphatic cells (FOERSTER et al. 1992b).

β) In Vivo Data. The prenatal toxicity of vidarabine has been studied in several animal species using multiple routes of application. In addition to review articles (KURTZ 1975; KURTZ et al. 1977), one paper is available in which the main studies are described in detail (SCHARDEIN et al. 1977). In a comparative study with a series of antivirals, the monophosphate of the drug was studied in rats (KLUG et al. 1991).

Rather low incidences of malformations (1.1%–2.7% of fetuses) were seen in rat fetuses on day 21 after intramuscular treatment on days 6–15 of pregnancy with doses of 30, 100, or 150 mg vidarabine/kg. Daily dosing with 200 or 250 mg/kg induced malformations at higher incidences (23% or 77% of the fetuses), but simultaneously induced signs of general toxicity in the pregnant animals. When applied intravaginally to pregnant rats on days 15–21 of gestation, no adverse effects on the offspring were observed (SCHARDEIN et al. 1977).

Three subcutaneous injections of 200 mg vidarabine phosphate/kg on day 10 of pregnancy induced abnormal embryonic development in all 47 embryos evaluated on day 11.5 of gestation (KLUG et al. 1991).

In rabbits, malformations were found at low incidences of 3% and 11% after daily treatment on days 6–18 of gestation with 5 or 25 mg vidarabine/kg, respectively. In contrast, the 5′-monophosphate of the drug was not teratogenic under these conditions, which might be explained by differences in pharmacokinetics (KURTZ et al. 1977; SCHARDEIN et al. 1977).

Vidarabine was also teratogenic in rabbits when applied topically as a 10% ointment to 5% or 10% of the body surface on days 6–18 of gestation once daily. Five out of 30 fetuses were malformed (16.7%; in controls, 1.2%) when 6 g vidarabine was applied to 10% of the body surface (Schardein et al. 1977).

No teratogenic effect was observed with vidarabine in a limited study in rhesus monkeys treated intramuscularly with 15 or 25 mg vidarabine/kg on days 16–30 of gestation (Schardein et al. 1977).

7. Zalcitabine

a) *Pharmacology and Clinical Use*

Zalcitabine (2,3-deoxycytidine, ddc; Fig. 8) is active against HIV-1 and HIV-2, including strains resistant to zidovudine. It is phosphorylated in infected cells to the active metabolite dideoxycytidine 5′-triphosphate (ddCTP), which inhibits reverse transcriptase competitively and probably causes chain termination of viral DNA elongation. The intracellular half-life of ddCTP is approximately 3 h. The drug is well absorbed after oral administration with a bioavailability of more than 80% in adults. The primary route of elimination of zalcitabine is renal excretion; the plasma half-life ranges from 1 to 3 h. Zalcitabine is used for treatment of adults with advanced HIV infection. Painful sensimotor peripheral neuropathy is the major dose-limiting side effect. It develops in up to 30% of patients; the risk of this adverse reaction increases with doses above 0.03 mg/kg per day (Whittington and Brogden 1992; Hayden 1995).

b) *Experimental Data*

α) *In Vitro Data.* The effects of zalcitabine on liver cells from 13- to 14-day-old mouse fetuses were studied in vitro. The drug inhibited the erythroid colony formation with an IC_{50} of 8.0 μM. These cells reacted more sensitively towards the action of the nucleoside analogue than bone marrow cells from adult mice (Gogu et al. 1989).

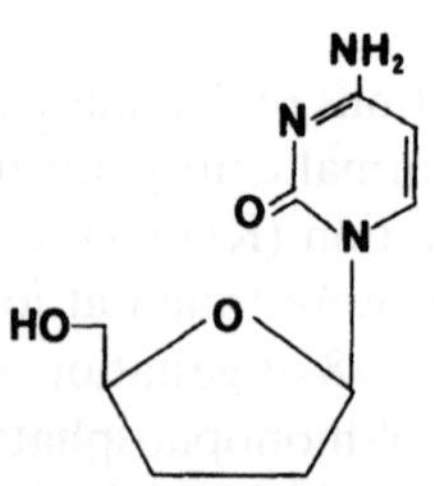

Fig. 8. Structure of 2′,3′-dideoxycytidine (zalcitabine)

Rat embryos (day 11.5) showed alterations of the head region after in vitro exposure to ddC for 48 h and additionally abnormal development of other organs. In 17% of all embryos with abnormalities, the neural tube was irregularly shaped and incompletely fused. Furthermore, the development of the ear anlage was impaired in 3% of all embryos, the ear vesicle resembling a bubble. The embryonic flexure was disturbed in 5% of the embryos at the highest concentration tested (500 μ*M*; KLUG et al. 1991).

Zalcitabine at a concentration of 10 μ*M* caused a reduction of thymocytes in cultured thymic lobes from 17-day-old mouse fetuses. A concentration of 100 μ*M* induced morphological alterations of the lymphatic cells (FOERSTER et al. 1992b).

β) *In Vivo Data.* Prenatal toxicity of zalcitabine was studied in mice after oral treatment with doses of 0, 100, 200, 500, or 1000 mg zalcitabine/kg. Doses were administered via gavage twice daily on gestational days 6–15. Weight gain of zalcitabine-treated mice, when corrected for uterine weight, did not differ from the controls. The percentage of resorptions per litter and mean litter size increased significantly in the highest dose group. Average fetal body weight per litter decreased significantly, and the percentage of malformed fetuses per litter increased significantly in the two highest dose groups. Only two of 95 fetuses showed skeletal malformations in the group receiving 100 mg/kg twice daily, compared to 114 of 115 fetuses in the group receiving 200 mg/kg twice daily (LINDSTRÖM et al. 1990).

Three subcutaneous injections of 200 mg zalcitabine/kg on day 10 of pregnancy induced only slight impairment of embryonic development (e.g., reduced protein concentration) when evaluated on day 11.5 of gestation (KLUG et al. 1991).

γ) *Human Data.* Sufficient data regarding the safety of zalcitabine in human pregnancy are not available. An epidemiologic project has been established to collect observational, nonexperimental data on exposure to ddI, zalcitabine, or zidovudine during pregnancy (see Sect. B.I.2.b.*γ*).

8. Zidovudine

a) Pharmacology and Clinical Use

Zidovudine (Fig. 9) is active against HIV-1, HIV-2, and other retroviruses. It is phosphorylated in infected cells to the active triphosphate, which inhibits viral RNA-dependent DNA polymerase (reverse transcriptase) and causes chain termination of viral DNA elongation. The intracellular half-life of the triphosphate is approximately 3–4 h. The drug is well absorbed after oral administration, with a bioavailability of approximately 60%–70%. In patients receiving 100 mg every 4 h mean peak and trough plasma levels are 0.4–0.5 and 0.1 mg/l. Zidovudine is rapidly converted to its 5′-*O*-glucuronide; the plasma half-life ranges from 1 to 1.5 h. Zidovudine is the initial agent of choice for treatment of HIV infection in patients with less than 500 CD4 cells/μl

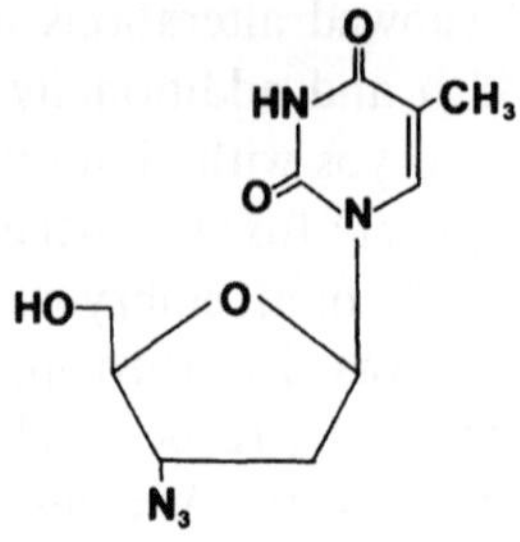

Fig. 9. Structure of zidovudine

blood; however, many open questions exist in the treatment of HIV-positive patients, and the optimal therapeutic strategies are still a matter of controversy. Granulocytopenia and anemia are the most common adverse reactions seen in patients treated with zidovuidine (YARCHOAN et al. 1989; KHOO and WILKINS 1995; HAYDEN 1995).

b) Experimental Data

α) *In Vitro Data.* In comparison to other, closely related virustatics, zidovudine showed very low activity in the rat whole embryo culture system. Only the highest concentration tested (300 μ*M*) induced abnormalities of the head region in 20% of the cultured embryos. Dysmorphogenesis differed significantly from that induced by the other virustatics tested; it led only to minor protrusions in the prosencephalon, while the other brain vesicles differentiated normally. The extremely low potential of zidovudine to interfere with development of mammalian embryos at this developmental stage was one of the most surprising results of the comparative in vitro experiments with several antiviral drugs. There is as yet no satisfying explanation for the low potential of this nucleoside analogue in prenatal toxicity (KLUG et al. 1991).

Zidovudine at a concentration of 30 μ*M* caused only a slight reduction in the number of thymocytes in cultured thymic lobes from 17-day-old rat fetuses. Even at a tenfold higher concentration, thymocytes were still frequent among the cytoplasmatic ramifications of epithelial cells in the reticulum (FOERSTER et al. 1992b).

The effects of zidovudine on liver cells from 13- to 14-day-old mouse fetuses were studied in vitro. The drug inhibited the erythroid colony formation with an IC_{50} of 2.1 μ*M*. These cells reacted more sensitively towards the action of the nucleoside analogue than bone marrow cells from adult mice (GOGU et al. 1989).

β) *In Vivo Data.* Administration of zidovudine at 1.0 mg/ml in drinking water to pregnant mice on gestational days 1–13 reduced the number of fetuses by 60% as evaluated on day 13. Half of the dose, which is equivalent to a daily dose of approximately 70 mg zidovudine/kg, still induced significant decreases in litter size (12 ± 1.5 vs. 16.5 ± 2.2 fetuses) and fetal size (8.5 ± 0.6 vs. 10.5 ± 0.8 mm) in comparison to a control group ($n = 10$ per group). The hematocrit in zidovudine-treated pregnant animals dropped from a control value of 42.6 ± 2.5 to 33.5 ± 1.7. Concomitant administration of erythropoetin, vitamin E, or interleukin-3 to the pregnant mice caused significant reversal of the zidovudine-induced maternal and fetal toxicity. In an attempt to elucidate the mechanism of the prenatal toxicity observed in this study, the authors determined the number of colony-forming units of erythroid progenitor cells both in bone marrow cells obtained from the pregnant mice and in the fetal hepatic cells isolated from the hepatic tissue of the fetuses on day 13 of gestation. From the results obtained, it is likely that the protective effects on fetal development observed by the three agents may be partially related to the reversal of the maternal anemia, but direct effects on fetal hepatic cells might also play a role (GOGU et al. 1992).

Administration of zidovudine in drinking water (0.25 mg/ml) to female mice for 6 weeks before mating, during the mating period, and for an additional 7–10 days afterwards led to a considerable number of pregnancy failures (males were untreated). The number of fetuses per litter was significantly decreased, and the resorption rate was significantly increased in comparison to a control group (TOLTZIS et al. 1991).

In vivo exposure of the fertilized oocyte of mice through the first 3 days postcoitus resulted in significant inhibition of in vitro blastocyst hatching and outgrowth development when the embryos were harvested immediately prior to implantation. Similarly, when two-cell embryos harvested from unexposed females were exposed to zidovudine at a concentration of 1 μM zidovudine in vitro over 24 h, development beyond the blastocyte stage was inhibited. In contrast, drug exposure during in vitro blastocyst and postblastocyst development had little or no toxic effect (TOLTZIS et al. 1993).

Groups of pregnant mice (five to 15) were treated with 10, 30, 100, or 300 mg zidovudine/kg per day orally for the whole period of pregnancy. No adverse effects were noted in this study (SIEH et al. 1992).

Alterations of several parameters of the hematopoetic and reproductive system of young male rats have been described after exposure for 4 weeks via drinking water at two dose levels (approximately 15 and 150 mg/kg, as estimated from the average drinking water intake). However, the results of this study should be interpreted with caution, because even in the high-dose group zidovudine was not detectable by high-performance liquid chromatography (HPLC) in serum samples (SIKKA et al. 1991).

When zidovudine was administered orally to pregnant Sprague-Dawley rats at doses of 125–500 mg/kg per day on days 6–15 of gestation (segment II protocol), no embryotoxicity was observed (AYERS 1988).

In another study in pregnant rats, zidovudine was injected three times on day 10 of gestation at a dose of 200 mg/kg. Embryos were evaluated on day 11.5 of gestation. A slight but significant reduction in the number of somites and the crown-rump length of the embryos was noticed, but no abnormalities were identified (KLUG et al. 1991).

When pregnant Wistar rats were treated with three oral doses of 100 mg zidovudine/kg on day 10 of gestation, no adverse effects were noted on maternal body weight, reproductive capacity, or hematology. In addition, no effects on growth or survival of the offspring were noted (GREENE et al. 1991)

These results differ from earlier findings with prenatally administered zidovudine in the same rat strain, when in two independent experimental series a high postnatal mortality was found under very similar experimental conditions (STAHLMANN et al. 1988c). The reason for the difference in the results from these two studies is unknown.

In two short reports, the effects of zidovudine in pregnant nonhuman primates (*Macaca nemestrina*) are described. Zidovudine was administered to macaques (1.5 mg/kg every 4 h) via a gastric catheter for at least 10 days before conception occurred, and treatment continued throughout pregnancy. It took significantly more matings to achieve the first six zidovudine pregnancies than the six control pregnancies. Fifteen pregnancies (nine zidovudine-treated animals, six vehicle controls) were brought to term. One fetus from each group died at full-term delivery. Three additional pregnancies resulted in first-trimester loss (two in the zidovudine group, one control; cause unknown). Birth weights were similar in the two groups, but weight gain in zidovudine-exposed offspring was slower. The data presented are preliminary, and no clear-cut conclusions can be drawn from these papers until the study is finished (HA et al. 1994; NOSBISCH et al. 1994).

γ) Human Data. Zidovudine pharmacokinetics was determined in three HIV-positive women receiving zidovudine (200 mg orally every 4 h) in weeks 19–39 of pregnancy and postpartum. Cord serum levels were slightly higher (113% and 129% of simultaneous maternal concentrations in two patients) and amniotic fluid were 4.7 and 2.5 times higher (3.3 and 6.5 μmol/l) than the corresponding cord serum levels (WATTS et al. 1991).

Sufficient data on human exposure during first trimester of pregnancy are not available. However, several clinical trials have been published to evaluate the efficacy and tolerance of zidovudine in later stages of pregnancy and to elucidate the possibility of treating pregnant women with zidovudine to reduce the risk of vertical HIV transmission.

A retrospective survey was performed in 43 pregnant women taking zidovudine at a dose of 300–1200 mg a day; 24 of these women took the drug for at least two trimesters. In two cases, the dose had to be reduced due to maternal toxic reactions (gastrointestinal, hematological). All infants, including two sets of twins, were born alive. No gross structural abnormalities were observed in 12 infants with first-trimester exposure. Two cases of intrauterine growth retardation were reported among the infants delivered at term. Three of seven newborns with hemoglobin values of less than 13.5 g/l were born prematurely, effects which could, in part, have resulted from their mother's treatment with zidovudine (SPERLING et al. 1992).

In a large double-blind, placebo-controlled trial, the efficacy and safety of zidovudine in reducing the risk of maternal–infant HIV transmission was studied in a total of 409 women giving birth to 415 infants. Only HIV-infected pregnant women (14–34 weeks of gestation) with CD4 cell counts above 200/μl blood who had not received antiretroviral therapy during the current pregnancy were enrolled. Zidovudine was administered antepartum, intrapartum, and postnatally for 6 weeks to the newborns. Women who received zidovudine had a transmission rate of 8.3%, as compared with 25.5% among those who received placebo, a reduction in the risk of transmission of 67.5%. This significant result indicates that it is possible to prevent a devastating disease in children by treating pregnant women with zidovudine. Only minimal adverse reactions were observed: the level of hemoglobin at birth in the zidovudine-exposed infants was significantly lower than in the placebo group. By 12 weeks of age, hemoglobin values in the two groups were similar. Although the results of this clinical trial are impressive, many questions remain unanswered. Most importantly, it must be remembered that more than 99% of HIV-infected children are born in developing countries where due to social and financial obstacles a corresponding zidovudine therapy is not available (CONNOR et al. 1994; ROGERS and JAFFE 1994; BAYER 1994).

Sufficient data regarding the safety of zidovudine in human pregnancy are not available. An epidemiologic project has been established to collect observational, nonexperimental data on exposure to ddI, zalcitabine, or zidovudine during pregnancy (see Sect. B.I.2.b.γ).

9. Comparative Studies of Several Nucleoside Analogues

Only a few comparative studies have been performed with regard to the effects of nucleoside analogues on prenatal development. However, especially for a series of structurally related compounds, in vitro data can provide valuable information. The effects of several nucleoside analogues studied comparatively in embryo culture systems have been reviewed elsewhere and are discussed here only briefly (STAHLMANN et al. 1993).

a) In Vitro Data

The effects of antiretroviral nucleosides on in vitro murine embryonic development was tested by exposing two-cell embryos to four virustatics at concentrations of 1, 10, and 100 μ*M*. The effects of stavudine (d4T), zalcitabine, and ddI on in vitro development of early murine embryos were markedly different from that of zidovudine. Exposure of two-cell embryos to zidovudine was consistently associated with significant inhibition of blastocyst formation at 1 μ*M*. In contrast, the effects of d4T and ddC were not detectable at concentrations under 100 μ*M*, and no toxicity was observed after ddI exposure (TOLTZIS et al. 1994). Corresponding findings were reported in another paper with zidovudine and ddI (SIEH et al. 1992).

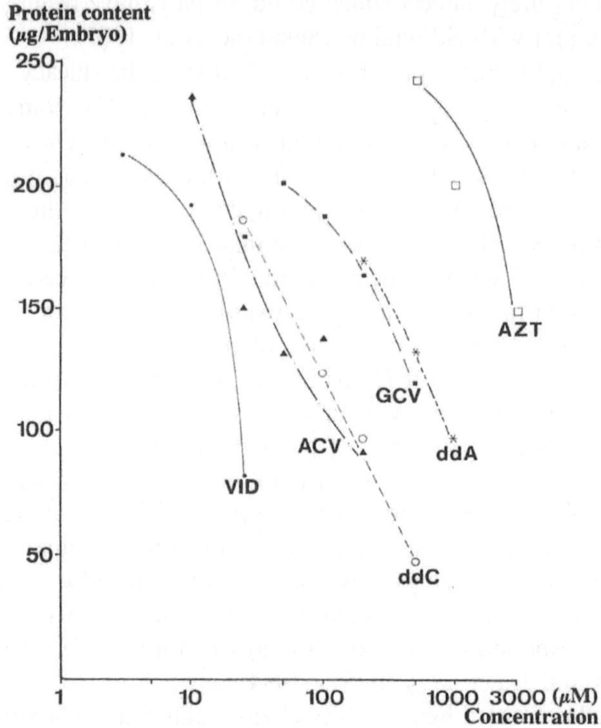

Fig. 10. Established concentration–response relationships of six nucleoside analogues with respect to protein content of the embryos at the end of the culture period. Protein content is the most sensitive variable to react to exposure to the nucleoside analogues. The result closely resembles the outcome observed with the scores shown in Fig. 11. *VID*,vidarabine; *ACV*, acyclovir; *ddC*, 2,3-dideoxycytidine (zalcitabine); *GCV*, ganciclovir; *ddA*, 2′3′-dideoxyadenosine; *AZT*, zidovudine

In Figs. 10 and 11, data from comparative in vitro experiments are compiled demonstrating significant differences in the potency of the drugs to induce adverse effects on embryonic development (rat whole embryo culture). With increasing concentrations of the drugs, the values for protein content and the score data of the embryos declined. As far as the potency of the various substances is concerned, the following rank order of decreasing toxic potency was established: vidarabine phosphate > acyclovir = zalcitabine > ganciclovir > ddA > zidovudine. The most toxic compound was vidarabine phosphate, with 50% of the embryos affected at concentrations between 3 and 10 µ*M*. The corresponding concentrations for acyclovir and zalcitabine were between 50 and 100 µ*M*, for ganciclovir between 100 and 200 µ*M*, for ddA between 500 and 1000 µ*M*, and for zidovudine greater than 3000 µ*M* (KLUG et al. 1985a, 1991).

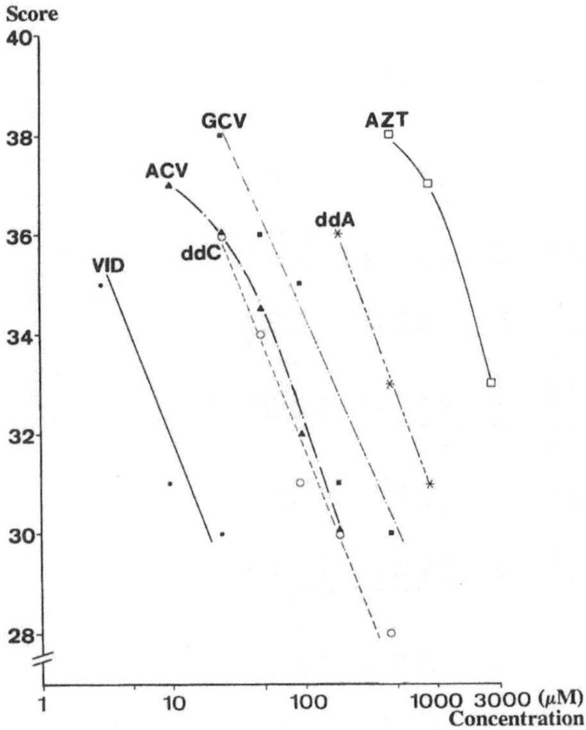

Fig. 11. Established concentration–response relationships of six nucleoside analogues with respect to the score of the embryos at the end of the culture period. The score reflects the disturbance of regular differentiation in vitro. Vidarabine (VID) obviously interferes most and zidovudine (*AZT*) least with the development. All other drugs show activity between these two antivirals. *ddC*, 2,3-dideoxycytidine; *ACV*, acyclovir; *GCV*, ganciclovir; *ddA*, 2′3′-dideoxyadenosine

Table 2 shows the data of the evaluation of the embryos after acyclovir and vidarabine exposure at concentrations between 10 and 100 μ*M* and between 3 and 25 μ*M*, respectively.

The effects of several virustatic agents on liver cells from 13- to 14-day-old mouse fetuses and on bone marrow cells were studied in vitro. Zidovudine, zalcitabine; d4T and ribavirin inhibited the erythroid colony formation with IC_{50} values of 2.1, 8.0, 9.0, and 0.62 μ*M*, respectively. These cells reacted more sensitively towards the action of the nucleoside analogue than bone marrow cells from adult mice (Gogu et al. 1989).

When tested in an organ culture system of thymic lobes from 17-day-old rat fetuses, all of the tested nucleoside analogues led to a reduction in the number of lymphatic cells as compared to controls. However, clear-cut differences in the potency to cause this effect were observed. Zidovudine and ddI caused a significant reduction of thymocytes at a concentration of 30 μ*M*,

Table 2. Evaluation of rat embryos (day 11.5) after a 48-h exposure to acyclovir or vidarabine in the culture medium (whole embryo culture)

	Concentration	Embryos (n)	Crown–rump length (mm)			Somites (n)			Protein (µg/embryo)			Score			Dysmorphogenic embryos
	(µM)		Q_1	Med	Q_3	Q_1	Med	Q_3	Q_1	Med	Q_3	Q_1	Med	Q_3	(%)
Control	–	44	3.60	3.36	3.18	27	26	25	284	235	176	38	37	35	0
Acyclovir	10	19	3.60	3.30	3.12	27	26	25	264	237	209	37	37	35	0
	25	27	3.30	3.06^{**}	2.88	26	25^{**}	23	196	150^{**}	124	36	36^{**}	34	0
	50	18	3.36	3.15^{**}	2.88	27	26	24	175	130^{**}	110	38	35^{**}	33	28
	100	19	3.36	3.00^{**}	2.82	27	26^{**}	25	175	138^{**}	122	33	32^{**}	31	95
Vidarabine	3	14	3.66	3.45	3.18	26	25	25	269	213	203	36	35	35	0
	10	19	3.36	3.18^{*}	3.00	25	24^{**}	20	207	193^{*}	141	33	31^{**}	30	79
	25	16	2.67	2.46^{**}	2.33	22	20^{**}	18	97	82^{**}	62	30	30^{**}	30	100

Med, Median; Q_1, first quartile; Q_3, third quartile.
$^{**}p<0.01$.

whereas acyclovir, ganciclovir, vidarabine, and zalcitabine caused a reduction of cell numbers at a concentration of 10 μ*M*. Zalcitabine clearly was the most and zidovudine the least potent compound tested (FOERSTER et al. 1992b).

b) In Vivo Data

In Table 3, data obtained after treatment with one to three injections of 100 mg acyclovir/kg on day 10 of gestation and after treatment with vidarabine (three injections of 100, 200, or 300 mg/kg) are compiled. Obviously, under these conditions acyclovir exhibits a higher potential for prenatal toxicity than vidarabine. After treatment with three doses of 100 mg/kg no abnormalities were observed with vidarabine, whereas all embryos exposed to acyclovir were abnormal. The higher in vivo activity of acyclovir is in contrast to the results obtained with the whole embryo culture and might be explained by the fact that acyclovir is metabolized less than vidarabine. Corresponding experiments were performed with several other virustatic agents. Pregnant rats were treated on day 10 of gestation with three subcutaneous injections of vidarabine, ganciclovir, ddA, zalcitabine, and zidovudine (200 mg/kg). The embryos were evaluated on day 11.5 of gestation. Under these conditions, significant alterations of the score or crown–rump length were seen with all compounds. Abnormalities were seen only with vidarabine and ganciclovir (KLUG et al. 1991).

Taking all the evidence from these comparative studies together, it is obvious that the rank order established with the rat whole embryo culture system was confirmed under in vivo conditions. One exception is the position of acyclovir: under in vivo conditions acyclovir possesses a higher potential for prenatal toxicity in rats than other structurally related virustatics.

II. Other Virustatics (Non-nucleoside Analogues)

1. Amantadine and Rimantadine

a) Pharmacology and Clinical Use

Amantadine (Fig. 12) specifically inhibits the replication of influenza A viruses. Rimantadine is a closely related derivative that is up to tenfold more active in certain in vitro assays. Amantadine interacts with the M2 protein of influenza viruses (an integral membrane protein) and with agglutinin, thus stabilizing the virus capsid (LUBECK et al. 1978; HAY et al. 1985). The efficacy of these drugs differs significantly depending on the virus strain. Furthermore, the degree of selectivity is rather low. After oral administration, both drugs are well absorbed and excreted either unmetabolized in the urine (amantadine) or after extensive metabolism (rimantadine). Their main use is the prevention of influenza A viral infection. If drug intake is started within 1–2 days after the onset of symptoms of an influenza A infection, these drugs also show a certain therapeutic effect (HIRSCH and SWARTZ 1980; DOUGLAS 1990; HAYDEN 1995).

Table 3. Evaluation of rat embryos (day 11.5) after s.c. treatment of rats on day 10 of gestation with acyclovir or vidarabine

	Dose (mg/kg)	Embryos (*n*)	Crown–rump length (mm)			Somites (*n*)			Protein (μg/embryo)			Score			Dysmorphogenic embryos %
			Q_1	Med	Q_3	Q_1	Med	Q_3	Q_1	Med	Q_3	Q_1	Med	Q_3	
Control	Not treated	49	4.08	3.96	3.72	29	28	28	418	322	279	410	40	40	0
Acyclovir	1 × 100	68	3.72	3.54**	3.36	26	25*	24	272	226**	203	40	40	38	2
	2 × 100	95	3.66	3.48**	3.24	26	25**	24	244	212**	180	37	37**	35	75
	3 × 100	31	3.18	3.12**	2.88	25	24**	20	166	140**	86	35	34**	32	100
Vidarabine	3 × 100	17	3.78	3.72**	3.63	26	26	26	277	260	241	40	40	39	0
	3 × 200	47	3.18	3.00**	2.76	24	24**	23	151	122**	92	31	31	30	100
	3 × 300	31	2.86	2.76**	2.57	24	23**	22	99	90**	72	32	30**	31	100

Med, Median; Q_1, first quartile; Q_3, third quartile.
** $p < 0.01$.
* $0.01 < p < 0.05$.

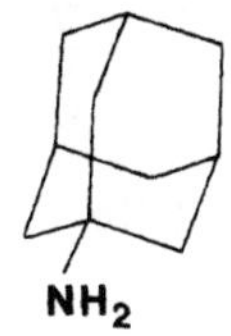

Fig. 12. Structure of amantadine

b) Experimental Data

α) In Vivo Data. In mice receiving 10 or 40 mg amantadine/kg orally on days 7–12 of gestation, fetal mortality was increased. Body weights of dams and the surviving offspring were reduced after the high dose only (ANONYMOUS 1985, cited after LEVY et al. 1991).

Amantadine was administered to rats in a three-generation study via diet providing a dose of 10 mg amantadine/kg daily. The rats remained on this drug dosage continuously through the experiment, except that the dosage was raised to 32 mg amantadine/kg at 3 weeks before the third mating (week 30 of test). The treatment resulted in no observed abnormalities except for a decreased fertility and lactation index for the third litter when the higher dose was applied (VERNIER et al. 1969).

LAMAR and coworkers treated rats and rabbits orally on days 7–14 of gestation with doses of up to 100 mg amantadine/kg. Teratogenic effects were seen with higher doses in rats but not in rabbits. The defects included malrotated hindlimbs and skeletal defects such as absence of ribs of the lumbar and sacral portion of the spinal column (LAMAR et al. 1970).

β) Human Data. Isolated case reports have been published on women exposed during the first trimester. In two reports, amantadine exposure was associated with teratogenicity (NORA et al. 1975; QAMAR et al. 1993), but these case reports do not allow us to draw any conclusions with respect to risks in humans. LEVY and coworkers report the case of a woman who used amantadine throughout two of her pregnancies and subsequently delivered tow normal infants (LEVY et al. 1991).

2. Foscarnet

a) Pharmacology and Clinical Use

Foscarnet (trisodium phosphonoformate; Fig. 13) is very similar in its structure to pyrophosphate and interacts with the binding sites of the viral DNA polymerase and reverse transcriptase. The inhibition is not competitive and seems to be different in comparison to the nucleoside analogues. Virus-induced polymerases reveal a higher sensitivity to foscarnet than corresponding cellular enzymes, thus causing a relative selectivity. Foscarnet inhibits all herpesviruses and HIV, including most strains of CMV and HSV that are

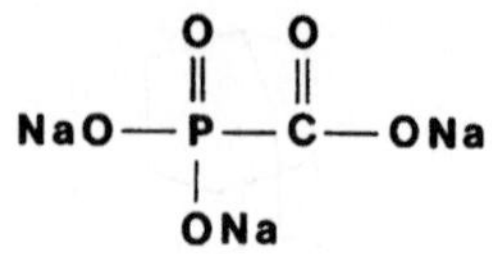

Fig. 13. Structure of foscarnet

resistant to nucleoside analogues. Because oral bioavailability is less than 20%, the drug is administered by intravenous infusion (ÖBERG 1989; WAGSTAFF and BRYSON 1994; HAYDEN 1995).

b) Experimental Data

α) In Vivo Data. Very little information on the reproductive toxicology of foscarnet is available. The studies performed routinely have not been published in detail, but a summary of the findings is given in a review (ÖBERG 1989):

> A fertility study in rats, teratogenicity studies in rats and rabbits, and a peri- and postnatal development study in rats have been carried out according to current guidelines. Rats were given 12–150 mg foscarnet/kg body wt and rabbits 12–75 mg foscarnet/kg body wt subcutaneously. Neither fertility and general reproductive performance, nor parturition and postnatal development were adversely affected under these experimental conditions. No adverse effects of foscarnet on the dams were seen in any group in these studies. Furthermore, foscarnet does not adversely affect litter size, fetal loss, litter and mean pup masses or frequency of malformations.

Administration of foscarnet to newborn rats at a dose of 10 mg/kg resulted in hypoplasia of dental enamel (CARACATSANIS et al. 1989). It is unknown whether the enamel organs of teeth that develop before birth are targets of prenatal toxicity of the drug or not.

C. Summary

Most of the virustatic agents used today are nucleoside analogues. Although their exact mechanism of action is not clarified, it is obvious that they alter the DNA metabolism in virus-infected cells.

It has been shown that small amounts of the biologically active derivatives are also formed in non-infected cells. Because it has been well known for many years that compounds which can alter DNA metabolism often exhibit a pronounced prenatal toxicity the reproductive toxicity of these drugs should be evaluated with the necessary carefullness.

Aciclovir and zidovudine are the two most widely used virustatics today. For these nucleoside analogs a considerable number of reproductive toxicity studies are available, the data base for the other antivirals is much smaller and for several drugs of this class routine studies - although performed by the manufacturer - have not been published.

Aciclovir - and valaciclovir, a prodrug with better bioavailability - are widely used for treatment of herpes virus infections, including zoster. Data established with the rat whole-embryo-culture gave the first indication that aciclovir disturbs normal development of rat embryos at concentrations which are in the same range as those achieved under therapeutic conditions (i.e. intravenous treatment with aciclovir or oral valaciclovir). For example, a retardation of the development of the ear anlagen in rat embryos was observed with aciclovir in vitro at concentrations of 25 μM.

When aciclovir was studied in rats following a segment-II-protocol (parenteral treatment from day 6 to 15; highest dose 25 mg/kg) no indication of prenatal toxicity was observed, however, after exposure on day 10 of gestation at one or multiple s.c.-injections of 100 mg/kg prenatal toxicity was obvious in embryos (day 11.5), fetuses (day 21) or offspring. Mainly affected were skull bones and typical findings were malformations of the os tympanicum. After treatment on day 10 the drug also affected the development of the rat immune system: thymus weights were decreased, spleen weights increased and in an infectivity model it was shown that the function of the immune system was persistently impaired.

Results from comparative studies with several nucleoside compounds in pregnant rats showed that obviously aciclovir possesses a higher potential for prenatal toxicity in rats than other structurally related virustatics. Zidovudine is the initial drug of choice for treatment of HIV-infection in patients with less than 500 $CD4^+$ cells/μl blood. In several in vitro systems (e.g. the rat whole-embryo-culture) zidovudine showed very low toxicity. The low potential of zidovudine to interfere with development of mammalian embryos during organogenesis was confirmed by studies in rats. In studies with mice it was shown that very early stages of embryonic development (before implantation) are impaired by zidovudine at low concentrations. These results are also in aggreement with in vitro data.

Transmission rate of human immunodeficiency virus (HIV) was significantly reduced, when women were treated with zidovudine during gestation and labour and the newborn for six weeks postnatally. Because no significant adverse reactions were found in the infants, the risk-benefit assessment seems to enfavour the treatment of HIV-infected, pregnant women with zidovudine.

Up to now, there is no indication from epidemiological data that aciclovir or other virustitics exhibit a risk for human embryos or fetuses. However, it should be considered that only a very large epidemiological data base allows a scientifically sound evaluation of the situation.

Drug registries have been established by the manufacturers which collect cases with exposure to often used virustatics such as aciclovir or antiretroviral drugs during pregnancy. If the data from the aciclovir registry are evaluated it should be considered, that most of the cases documented so far were exposed to relatively low oral doses (as recommended initially when the drug was marketed). Although these data indicate no risk of prenatal toxicity, a risk for humans might exist if higher doses are used and/or with new derivatives with better bioavailability - such as valaciclovir (the *l*-valyl ester of aciclovir).

Furthermore, in the context of the registry only gross-structural malformations are registered and it can not be excluded that there might be functional defects in the infants after prenatal exposure to aciclovir or other antivirals. Only a very restrictive use of the drugs in pregnant women (and non-pregnant women at child-bearing age, as long as it is not definitely excluded that they are pregnant) can help to minimize the potential risk of prenatal toxicity from these drugs. The only exception is the clinical use of antivirals for those indications, which derive from large clinical trials with unequivocal results indicating a clear-cut benefit and low risk.

References

Andrews EB, Yankaskas BC, Cordero JF, Schoeffler K, Hampp S (1992) Acyclovir in pregnancy registry: six years' experience. Obstet Gynecol 79: 7–13
Ayers KM (1988) Preclinical toxicology of zidovudine. An overview. Am J Med 85: 186–188
Bayer R (1994) Ethical challenges posed by zidovudine treatment to reduce vertical transmission of HIV. N Engl J Med 331: 1223–1225
Beutner KR, Friedman DJ, Forszpaniak C, Andersen PL, Wood MJ (1995) Valaciclovir compared with acyclovir for improved therapy for herpes zoster in immunocompetent adults. Antimicrob Agents Chemother 39: 1546–1553
Bridgen D, Rosling AE, Woods NC (1982) Renal function after aciclovir intravenous injection. Am J Med 73 Suppl: 182
Caracatsanis M, Fouda N, Hammsarstrom L (1989) Effects of phosphonoformic and phosphonoacetic acids on developing enamel of rat molars. Scand J Dent Res 97: 139–149
Chahoud I, Stahlmann R, Bochert G, Dillmann I, Neubert D (1988) Gross-structural defects in rats after aciclovir application on day 10 of gestation. Arch Toxicol 62: 8–14
Connor EM, Sperling RS, Gelber R, Kiselev P, Scott G, O'Sullivan MJ, VanDyke R, Bey M, Shearer W, Jacobson RL, Jimenez E, O' Neill E, Bazin B, Delfraissy J-F, Culnane M, Coombs R, Elkins M, Moye J, Stratton P, Balsley J (1994) Reduction of maternal-infant transmission of human immunodeficiency virus type 1 with zidovudine treatment. N Engl J Med 331: 1173–1180
Douglas RG (1990) Prophylaxis and treatment of influenza. N Engl J Med 322: 443–450
Eldrige RR, White AD, Tennis PS, Andrews EB, Dai WS, Cordero JF (1993) Three prospective registries to monitor prenatal maternal exposures: the acyclovir in pregnancy, the antiretroviral pregnancy and the lamotrigine in pregnancy registries. Reprod Toxicol 7: 637
Eldrige RR, White AD, Andrews EB (1995) Monitoring birth outcomes in the aciclovir pregnancy registry (Abstr k97). Interscience Conference on Antimicrobial Agents and Chemotherapy, American Society for Microbiology, San Francisco
Faulds D, Brogden RN (1992) Didanosine. A review of its antiviral activity, pharmacokinetic properties and therapeutic potential in human immunodeficiency virus infection. Drugs 44: 94–116
Faulds D, Heel RC (1990) Ganciclovir. A review of its antiviral activity, pharmacokinetic properties and therapeutic efficacy in cytomegalovirus infections. Drugs 39: 597–638
Ferm VH, Willhite C, Kilham L (1978) Teratogenic effects of ribavirin on hamster and rat embryos. Teratology 17: 93–102

Foerster M, Kastner U, Neubert R (1992a) Effect of six virustatic nucleoside analogues on the development of fetal rat thymus in organ culture. Arch Toxicol 66: 688–699
Foerster M, Merker H-J, Stahlmann R, Neubert D (1992b) In vitro effect of aciclovir on lymphopoiesis in fetal rat thymus. Toxicol In Vitro 6: 207–217
Frenkel LM, Brown ZA, Bryson YJ, Corey L, Unadket JD, Hensleigh PA, Arvin AM, Prober CG, Connor JD (1991) Pharmacokinetics of acyclovir in the term human pregnancy and neonate. Am J Obstet Gynecol 164: 569–576
Furman PA, St Clair MH, Fyfe JA, Rideout JL, Keller PM, Elion GB (1979) Inhibition of herpes simplex virus-induced DNA polymerase activity and viral DNA replication by 9-(2-hydroxyethoxymethyl) guanine and its triphosphate. J Virol 32: 72–77
Furman PA, de Miranda P, St Clair MH, Elion GB (1981) Metabolism of acyclovir in virus-infected and uninfected cells. Antimicrob Agents Chemother 20: 518–524
Gogu SR, Beckman BS, Agrawal KC (1989) Anti HIV-drugs: comparative toxicities in murine fetal liver and bone marrow erythroid progenitor cells. Life Sci 45: iii–vii
Gogu SR, Beckman BS, Agrawal KC (1992) Amelioration of zidovudine-induced fetal toxicity in pregnant mice. Antimicob Agents Chemother 36: 2370–2374
Greene JA, Ayers KM, DeMiranda P, Tucker WE Jr (1991) Postnatal survival in Wistar rats following oral dosage with zidovudine on gestation day 10. Fundam Appl Toxicol 15: 201–206
Gubbels JL, Gold WR, Bauserman S (1991) Prenatal diagnosis of fetal diastematomyelia in a pregnancy exposed to acyclovir. Reprod Toxicol 5: 517–520
Ha JC, Nosbisch C, Conrad SH, Ruppenthal GC, Sackett GP, Abkowitz J, Unadkat JD (1994) Fetal toxicity of zidovudine (azidothymidine) In *Macaca nemestrina*: preliminary observations. J AIDS 7: 154–157
Hartman NR, Yarchoan R, Pluda JM, Thomas RV, Marczyk KS et al (1990) Pharmacokinetics of 2′, 3′ dideoxyadenosine and 2′, 3′ dideoxyinosine in patients with severe human immunodeficiency virus infection. Clin Pharmacol Ther 47: 647–654
Hartmann J, Chahoud I, Buerkle B, Neubert D (1991) Testis hypoplasia in rat offspring prenatally exposed to ganciclovir (Abstr p 21) Teratology 44:25A
Hay AJ, Wolstenholme AJ, Skehel JJ, Smith MH (1985) The molecular basis of the specific anti-influenza action of amantadine. EMBO J 4: 3021–3024
Hayden FG (1995) Antiviral agents. In: Mandell GL, Bennett JE, Dolin R (eds) Mandell, Douglas and Bennett's principles and practice of infectious diseases, 4th edn. Churchill Livingstone, Edinburgh, pp 411–450
Heinrich-Hirsch B, Neubert D (1991) Effect of aciclovir on the development of the chick embryo in ovo. Arch Toxicol 65: 402–408
Henderson GI, Hu ZQ, Yang Y, Perez TB, Devi BG, Frosto TA, Schenker S (1993) Ganciclovir transfer by human placenta and its effects on rat fetal cells. Am J Med Sci 306: 151–156
Hirsch MS, Swartz MN (1980) Antiviral agents. N Engl J Med 302: 903–907
Itoi M, Gefter JW, Kaneko N, Ishii Y, Ramer RM, Gasset AR (1975) Teratogenicities of ophthalmic drugs. I. Antiviral ophthalmic drugs. Arch Ophthalmol 93: 46–51
Kacmarek RM (1991) Care-giver protection from exposure to aerolized pharmacologic agents – is it necessary? Chest 100: 1104–1105
Khoo SH, Wilkins EGL (1995) Review: controversies in anti-retroviral therapy of adults. J Antimicrob Chemother 35: 245–262
Klug S, Lewandowski C, Blankenburg G, Merker HJ, Neubert D (1985a) Effect of acyclovir on mammalian embryonic development in culture. Arch Toxicol 58: 89–96
Klug S, Lewandowski C, Neubert D (1985b) Modification and standardization of the culture of early postimplantation embryos for toxicological studies. Arch Toxicol 58: 84–88
Klug S, Lewandowski C, Merker H-J, Stahlmann R, Wildi L, Neubert D (1991) In vitro and in vivo studies on the prenatal toxicity of five virustatic nucleoside

analogues in comparison to aciclovir. Arch Toxicol 65: 283–291
Klug S, Stahlmann R, Golor G, Foerster M, Felies A, Bochert G, Rahm U, Neubert D (1992) Aciclovir in pregnant marmoset monkeys – oral treatment (Abstr). Teratology 45: A64
Kochhar DM (1982) Embryonic limb bud organ culture in assessment of teratogenicity of environmental agents. Teratogen Carcinogen Mutagen 2: 303–312
Kochhar DM, Penner JD, Knudsen TB (1980) Embryotoxic, teratogenic, and metabolic effects of ribavirin in mice. Toxicol Appl Pharmacol 52: 99–112
Koren G, Ito S (1993) Exposures of pregnant health care workers to ribavirin through inhalation of contaminated air. Reprod Toxicol 7: 158
Kurtz SM (1975) Toxicology of adenine arabinoside. In: Pavan-Langston D, Buchanan RA, Alford CA (eds) Adenine arabinoside: an antiviral agent. Raven, New York, pp 145–157
Kurtz SM, Fitzgerald JE, Schardein JL (1977) Comparative animal toxicology of vidarabine and its 5′ monophosphate. Ann NY Acad Sci 284: 6–8
Lamar JK, Calhoun FJ, Darr AG (1970) Effects of amantadine hydrochloride on cleavage and embryonic development in the rat and rabbit. Toxicol Appl Pharmacol 17: 272 (abstr)
Laskin OL, Longstreth JA, Saral R, de Miranda P, Keeney R, Lietman PS (1982) Pharmacokinetics and tolerance of acyclovir, a new anti-herpesvirus agent in humans. Antimicrob Agents Chemother 21: 393–398
Levy M, Pastuszak A, Koren G (1991) Fetal outcome following intrauterine amantadine exposure. Reprod Toxicol 5: 79–81
Lindström P, Harris M, Hoberman AM, Dunnick JK, Morrissey RE (1990) Developmental toxicity of orally administered 2′,3′-dideoxycytidine in mice. Teratology 42: 131–136
Lubeck MD, Schulman JL, Palase P (1978) Susceptibility of influenza A virus to amantadine is influenced by the gene coding for M protein. J Virol 28: 710–716
McLaren C, Datema R, Knupp CA, Buroker RA (1991) Review: didanosine. Antiviral Chem Chemother 2: 321–328
Moore HL Jr, Szczech GM, Rodwell DE, Kapp RW Jr, de Miranda P, Tucker WE Jr (1983) Preclinical toxicology studies with aciclovir: teratologic, reproductive and neonatal tests. Fundam Appl Toxicol 3: 560–568
Nora JJ, Nora AH, Way GL (1975) Cardiovascular maldevelopment associated with maternal exposure to amantadine. Lancet 2: 607
Nosbisch C, Ha JC, Sackett GP, Conrad SH, Ruppenthal GC, Unadkat JD (1994) Fetal and infant toxicity of zidovudine in Macaca nemestrina (Abstr p 115). Teratology 49: 415 (abstr)
Öberg B (1989) Antiviral effects of phosphonoformate (PFA, foscarnet sodium). Pharmacol Ther 40: 213–285
Percy DH (1975) Teratogenic effects of the pyrimidine analogues 5-iododeoxyuridine and cytosine arabinoside in late fetal mice and rats. Teratology 11: 103–111
Percy DH, Albert DM, Amemiya T (1973) Ocular defects in newborn rats treated with 5-iododeoxyuridine. Proc Soc Exp Biol Med 142: 1272–1276
Pons JC, Boubon MC, Taburet AM, Singlas E, Chambrin V, Frydman R, Papiernik E, Delfraissy JF (1991) Fetoplacental passage of 2,3-dideoxyinosine. Lancet 337: 732
Prusoff WH (1988) Idoxuridine or how it all began. In: DeClerq E (ed) Clinical use of antiviral drugs. Nijhoff, Norwell, pp 15–24
Qamar IU, Pandit PB, Jefferies AL, Landes A, Koren G, Chitayat D (1993) Tibial hemimelia and tetralogy of Fallot associated with first trimester exposure to amantadine. Teratology 47: 345–381
Rogers MF, Jaffe HW (1994) Reducing the risk of maternal-infant transmission of HIV: a door is opened. N Engl J Med 331: 1222–1223
Schardein JL, Hentz DL, Petrere JA, Fitzgerald JE, Kurtz SM (1977) The effect of vidarabine on the development of the offspring of rats, rabbits and monkeys.

Teratology 15: 231–242
Sieh E, Coluzzi ML, Cusella de Angelis MG, Mezzogiorno A, Floridia M, Canipari R, Cossu G, Vella S (1992) The effects of AZT and DDI on pre- and postimplantation mammalian embryos: an in vivo and in vitro study. AIDS Res Hum Retroviruses 8: 639–649
Sikka SC, Gogu SR, Agrawal KC (1991) Effect of zidovudine (AZT) on reproductive and hematopoietic systems in the male rat. Biochem Pharmacol 42: 1293–1297
Skalko RG, Packard DS (1973) The teratogenic response of the mouse embryo to 5-iododeoxyuridine. Experientia 29: 198–200
Sperling RS, Stratton P, O'Sullivan MJ, Boyer P, Watts DH, Lambert JS, Hammill H, Livingston EG, Gloeb DJ, Minkoff H, Fox HE (1992) A survey of zidovudine use in pregnant women with human immunodeficiency virus infection. N Engl J Med 326: 857–861
Stahlmann R, Klug S, Lewandowski C, Chahoud I, Bochert G, Merker H-J, Neubert D (1987) Teratogenicity of acyclovir in rats. Infection 15: 261–262
Stahlmann R, Klug S, Lewandowski C, Bochert G, Chahoud I, Rahm U, Merker H-J, Neubert D (1988a) Prenatal toxicity of acyclovir in rats. Arch Toxicol 61: 468–479
Stahlmann R, Franz R, Merker H-J (1988b) Embryotoxicity of aciclovir in rats after single application. Naunyn Schmiedebergs Arch Pharmacol 337: R35 (abstr)
Stahlmann R, Chahoud I, Bochert G, Klug S, Korte M, Neubert D (1988c) Prenatal toxicity of zidovudine in rats. Teratology 38: 28A (abstr)
Stahlmann R, Chahoud I, Korte M, Thiel R, van Loveren H, Vos JG, Foerster M, Merker H-J, Neubert D (1991) Structural anomalies of thymus and other organs and impaired resistance to *Trichinella spiralis* infection in rats after prenatal exposure to aciclovir. In: (eds) Harwood Thymus update vol 4. Kendall MD Ritter MA (eds) Harwood, New York, pp 129–155
Stahlmann R, Korte M, Van Loveren H, Vos JG, Thiel R, Neubert D (1992a) Abnormal thymus development and impaired function of the immune system in rats after prenatal exposure to aciclovir. Arch Toxicol 66: 551–559
Stahlmann R, Golor G, Klug S, Renschler C, Stürje H, Merker H-J, Neubert D (1992b) Aciclovir in pregnant marmoset monkeys – intravenous treatment. Teratology 45: A4 (abstr)
Stahlmann R, Klug S, Foerster M, Neubert D (1993) Significance of embryo culture methods for studying the prenatal toxicity of virustatic agents. Reprod Toxicol 7: 129–143
Stahlmann R, Golor G, Thiel R, Chahoud I, Kuschel S, Baumann-Wilschke I (1995) Persistent changes of components of the immune system in rats after exposure to aciclovir during organogenesis. Teratology 51: 168–169 (abstr)
Syntex (1988) CYMEVENE and cytomegalovirus infections. Syntex Pharmaceuticals
Toltzis P, Marx CM, Kleinman N, Levine EM, Schmidt EV (1991) Zidovudine-associated embryonic toxicity in mice. J Infect Dis 163: 1212–1218
Toltzis P, Mourton T, Magnuson T (1993) Effect of zidovudine on preimplantation murine embryos. Antimicrob Agents Chemother 37: 1610–1613
Toltzis O, Mourton T, Magnuson T (1994) Comparative embryonic cytotoxicity of antiretroviral nucleosides. J Infect Dis 169: 1100–1102
Vernier VG, Harmon JB, Stump JM, Lynes TE, Marvel JP, Smith DH (1969) The toxicologic and pharmacologic properties of amantadine hydrochloride. Toxicol Appl Pharmacol 15: 642–665
Wagstaff AJ, Bryson HM (1994) Foscarnet. A reappraisal of its antiviral activity, pharmacokinetic properties and therapeutic use in immunocompromised patients with viral infections. Drugs 48: 199–226
Watts DH, Brown ZA, Tartaglione T, Burchett SK, Opheim K, Coombs R, Corey L (1991) Pharmacokinetic disposition of zidovudine during pregnancy. J Infect Dis 163: 226–232
Whitley R, Alford C, Hess F, Buchanan R (1980) Vidarabine: a preliminary review of its pharmacological properties and therapeutic use. Drugs 20: 267–282

Whitley RJ, Gnann JW (1992) Acyclovir: a decade later. N Engl J Med 327: 782–789
Whittington R, Brogden RN (1992) Zalcitabine. A review of its pharmacology and clinical potential in acquired immunodeficiency syndrome (AIDS). Drugs 44: 656–683
Yarachoan R, Mitsuya H, Myers CE, Broder S (1989) Clinical pharmacology of 3′-azido-2′, 3′-dideoxythymidine (zidovudine) and related dideoxynucleosides. N Engl J Med 321: 726–739

CHAPTER 27

Angiotensin-Converting Enzyme Inhibitor Fetopathy

M. BARR

A. Introduction

Angiotensin-converting enzyme (ACE) inhibitors (ACEI), increasingly popular antihypertension agents, have several effects other than control of blood pressure. One of these effects, fetal damage, is the focus of this chapter.

The observation that the venom of a Brazilian snake contained a bradykinin-potentiating factor that caused vasodilation (FERREIRA 1965) was followed by the demonstration that a peptide mixture from the venom inhibited ACE (BAKLHE 1968). Captopril, the first clinically usable ACEI, was synthesized in 1975 by CUSHMAN and ONDETTI (1991), and thereafter there was rapid development of additional drugs of this class. In the United States, captopril (Capoten), was approved for clinical use in 1981, followed by enalapril (Vasotec) in 1985, lisinopril (Prinivil, Zestril) in 1987, benazepril (Lotensin), fosinopril (Monopril), and ramipril (Altace) in 1991, and quinapril (Accupril) in 1992. Many other ACEI have been discussed in the literature, but are not currently marketed in the United States; these include abutapril, alacepril, ceronapril, cilazepril, delapril, idrapril, imidapril, moexipril, pentopril, perindopril, rentiapril, spirapril, temocapril, trandolapril, zabicipril, and zofenapril, Since their introduction, the ACEI have gained wide acceptance as effective antihypertensive agents.

Two of the currently marketed ACEI, captopril and lisinopril, are in the active form; the others are proactive drugs that must be deesterified in the liver to the active form (designated by the suffix -prat). The prodrugs were developed to enhance absorption, prolong activity, or both. Within the class of ACEI, there is a considerable range both in extent and timing of absorption. Once absorbed, the drugs vary in their degree of protein binding, from none for lisinopril to 97% for quinapril. Captopril binds the enzyme by means of a sulfhydryl, fosinopril by a phosphodyl, and the others by a carboxyl group (MATERSON and PRESTON 1994). In general, their pharmacological activity persists beyond measurable plasma concentrations, presumably reflecting tissue binding of the drug. ACEI are excreted principally through the kidney, although there is up to 50% fecal excretion of ramipril and fosinopril (WILLIAMS 1988; manufacturers' literature).

Placental passage of the various ACEI appears to be both species and drug dependent. Some ACEI cross the placenta of common laboratory animals only in small amounts or not at all (Endo et al. 1992). However, from measurements in exposed human newborns, it has been established that enalapril, captopril, and lisinopril do cross the human placenta in pharmacologically significant amounts, and it is assumed that the other available ACEI will also do so (Schubiger et al. 1988; Pryde et al. 1993). Once in the fetus, it is presumed that at least a portion of an ACEI will be excreted via the fetal urine (provided there is urine production) largely as the active form of the drug. This excreted fraction in the amniotic fluid may then be swallowed and recirculated. What significance such drug recirculation might have on fetal physiology has not been assessed.

B. Review of the Renin–Angiotensin System

ACE, a zinc metalloprotease, is the controlling factor in two parallel systems, the renin–angiotensin (RAS) and bradykinin systems (Fig. 1). It catalyzes the hydrolysis of carboxy-terminal dipeptides from several oligopeptide substrates, most importantly the decapeptide angiotensin I and the nonapeptide bradykinin (Ehlers and Riordan 1989). Angiotensin I, which has no known biological action in humans, is formed from the action of renin on the precursor angiotensinogen. Angiotensin I is rapidly converted by ACE to the biologically active octapeptide angiotensin II (Fig. 2). A parallel system involves the generation of bradykinin and its inactivation by kininase II, which is identical with ACE. There are two catalytic sites for ACE, one near the carboxyl terminus and one near the amino acid terminus. While these sites may have some substrate specificity, angiotensin I and bradykinin appear to be catalyzed equally at both sites (Johnston et al. 1993). Angiotensin II is a potent vasoconstrictor that stimulates aldosterone secretion from the adrenal cortex and suppresses renin release by increasing sodium retention. Angio-

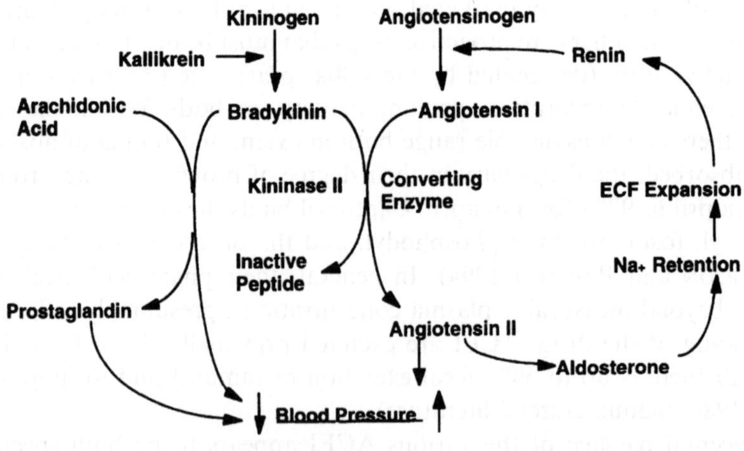

Fig. 1. The renin–angiotensin and bradykinin systems. ECF, extracellular fluid

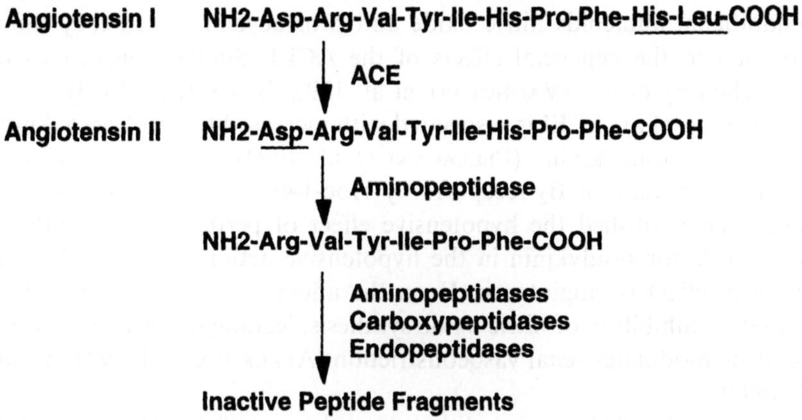

Fig. 2. The conversion of angiotensin I and degradation of angiotensin II

tensin II acts to raise the blood pressure, while bradykinin, if not inactivated, lowers the blood pressure.

The RAS is both a circulating and a tissue hormonal system, and, of the two, tissue systems may be the more important (Ehlers and Riordan 1989; Johnston et al. 1992; Ganong 1994). ACE is widely distributed in the body; it is found not only in soluble form in body fluids, but also in membrane-bound form in such diverse cells as arterial endothelium, epithelia with brush borders (placenta, kidney, intestine, and choroid plexus), neuroepithelium, and in the male genital tract (testis, prostate, and epididymis) (Erdos and Skidgel 1987). In some tissues it appears that angiotensinogen is produced by some cells and transported in paracrine fashion to other renin-containing cells, where it is converted to angiotensin II (Ganong 1994). Evidence is emerging that plasma ACE inhibition and hemodynamic responses are separable and subject to local regulatory factors (Lees et al. 1992). For example, prolonged treatment with an ACEI results in increased plasma ACE but decreased ACE in the renal cortex and renal tubular brush border (Michel et al. 1993). The renal proximal convoluted tubule contains angiotensinogen, ACE, and angiotensin II receptors (Moe et al. 1993; Jackson et al. 1991). Renin could not be found in tubules from untreated rats, but after enalapril administration, renin was found, indicating the existence of a tissue RAS that generates angiotensin II to regulate sodium absorption locally (Moe et al. 1993). The principal functions of angiotensin II in the kidney are inhibition of renin release by the juxtaglomerular cells, regulation of sodium reabsorption by the distal convoluted tubules, biosynthesis of prostaglandins by the cortical tubules, and vasoconstriction of the efferent and afferent arterioles (Degasparo and Levens 1994; Kang et al. 1994).

ACEI may also have a locally acting, prostaglandin-dependent component to their hypotensive action. Linz et al. (1993) reported that bradykinin degradation inhibition by ramipril was associated with enhanced formation of the endothelial autacoids nitric oxide and prostacyclin, which they thought contributed to the beneficial effects of the ACEI. Similar conclusions have been reached by others (VANHOUTTE et al. 1993; BUSSE et al. 1993). Chronic administration of an ACEI is associated with increased levels of bradykinin B_2 receptor-stimulating kinins (PELLACANI et al. 1994). BOUAZIZ et al. (1994) noted that blockade of B_2 receptors by Hoe-140 prior to administration of perindoprilat abolished the hypotensive effect of perindoprilat in rats, suggesting a role for bradykinin in the hypotensive action of ACEI. The vasoconstrictive effect of angiotensin II on the afferent glomerular arteriole was enhanced by inhibition of nitric oxide synthesis, leading to the conclusion that nitric oxide modulates renal vasoconstriction (ALBEROLA et al. 1994; YOSHIDA et al. 1994).

ARIMA et al. (1994) suggested that the resistance of the efferent arteriole in rats is regulated by prostaglandins released from the upstream glomerulus, while prostaglandins synthesized in the afferent arteriole modulate angiotensin II action locally. From experimentation in dogs, BUGGE and STOKKE (1994) found that an angiotensin II-induced reduction of renal blood flow of 20% was not accompanied by a reduction in glomerular filtration rate (GFR). However, when renal blood flow was decreased by 30% there was an 18% reduction in GFR. When prostaglandin synthesis was blocked, the vasoconstrictive action of angiotensin II was doubled, and all decreases in renal blood flow were paralleled by decreases in GFR. Until renal blood flow was decreased by 25%–30%, angiotensin II increased afferent and efferent arteriolar resistance equally, but with greater decrease in renal blood flow afferent resistance increased more than efferent resistance, and prostaglandin blockade had no effect on this pattern. BUGGE and STOKKE (1994) thought that inhibition of prostaglandin synthesis causes a shift in angiotensin II's effect mainly on renal blood flow to a combined effect on both renal blood flow and GFR. They concluded that renal prostaglandins attenuate the effect of angiotensin II on the glomerulus rather than on the afferent arteriole.

In contrast, GERBER et al. (1993) found that in humans, while the prostaglandin inhibitor indomethacin reduced the urinary excretion of prostacyclin metabolite by more than 50%, it had no effect on the hypotensive effect of either captopril or enalapril. They concluded that neither ACEI has a significant prostacyclin-dependent component to its hypotensive action. GANSEVOORT et al. (1994) similarly concluded that the antiproteinuric and renal hemodynamic effects of ACEI in humans are due to blockade in the RAS, rather than in the bradykinin system. The relation of the clinical effects of the ACEI to bradykinin degradation and prostaglandin synthesis and the species and agent specificity of such effects are currently unresolved. These two hormones may be acting at the tissue level without correlated change in blood or excretion levels. Involvement of the prostaglandin system in ACEI fetopathy seems possible, particularly given a number of reports of renal tubular dysgenesis and fetal–neonatal anuria associated with nonsteroidal anti-inflammatory drug exposure (see below).

During normal human pregnancy, plasma levels of renin, angiotensinogen, angiotensin II, and aldosterone increase compared with the nonpregnant state. In pregnancies complicated by hypertension and proteinuria, maternal plasma angiotensin II is reduced to nonpregnant levels (WEIR et al. 1973). In pregnancy-induced hypertension, mean arterial blood pressure is correlated with serum ACE activity (LI et al. 1992) and is characterized by enhanced angiotensin II sensitivity due to an increase in angiotensin II receptor number (GRAVES et al. 1992). Bradykinin infused in vitro through the fetal vessels of human placentas was inactivated, but prior administration of captopril blocked this inactivation, demonstating that ACE occurs in fetal placental vessels (DE MOURA and VALE 1986). YAGAMI et al. (1994) noted that the amount of ACE mRNA in the human placenta increases over the course of pregnancy but decreases near term, while placental ACE activity continues to increase from the first trimester with no decrease in late gestation. They concluded that the placenta contributes to fetal plasma angiotensin II. Of course, the story is even more complicated, for as HAGEMANN et al. (1994) have noted, angiotensin II is only one factor in the complex regulation of uteroplacental blood flow. Not only are there species differences in fetal RAS activation, but species differences in the expression of the uteroplacental RAS as well.

The RAS becomes crucially important under conditions of low renal perfusion pressure (HALL et al. 1977; BLYTHE 1983). In these conditions, which pertain to the fetus, angiotensin II-mediated efferent arteriolar resistance is essential to the maintenance of the GFR and production of urine (RUDOLPH et al. 1971; GUIGNARD 1982). Activation of bradykinin by suppression of angiotensin II can reduce glomerular filtration pressure by virtue of bradykinin's dilation of the efferent arteriole (KON et al. 1993). In the adult, ACEI increase renal blood flow by also dilating the afferent arteriole. Usually this occurs without significant alteration of the GFR (WILLIAMS 1988). However, in cases of renal artery stenosis, bilateral or unilateral in a solitary kidney, glomerular filtration pressure is reduced by the ACEI, because the ability to increase renal blood flow is insufficient to compensate for the efferent arteriolar dilation (HRICIK et al. 1983).

The RAS is certainly active in fetal life, when it functions to maintain GFR under conditions of low renal perfusion pressure (JELINEK et al. 1986). At least in the lamb, the RAS appears to be more active in the fetus than in the neonate (BINDER and ANDERSON 1992). However, the relative gestational timing of activation of the RAS varies among species, with late activation (day 19) in the rat and presumably much earlier activation in the human. For this reason, caution is needed in the extrapolation of experimental results from another species to the human.

The ACE gene is developmentally regulated in a tissue-specific manner and plays a role in the regulation of both renal function and growth (YOSIPIV et al. 1994). In the fetal period, tissue kinin generation and degradation are coordinately regulated, while circulating angiotensin II and ACE activity change reciprocally (YOSIPIV et al. 1994). Plasma ACE levels are higher in children than in adults. In rats, plasma ACE rises markedly a few days after birth and then declines to adult levels by 2 weeks of age. When various tissues were studied for their ACE mRNA activity, it was found that gut, kidney, and testis showed different patterns of ontogenesis, indicating that ACE regulation is organ specific (COSTEROUSSE et al. 1994).

C. Genetic Diversity and Receptor Types

At least in humans, the ACE gene is polymorphic, with either an insertion or deletion of a 250-bp fragment in intron 16 of the gene (RIGAT et al. 1990). Homozygosity for the deletion (DD) is associated with the highest serum levels of ACE activity, while homozygosity for the insertion (II) correlates with the lowest levels. The prevalence of the gene types vary among different populations (LEE 1994; BARLEY et al. 1994). It has been observed that the homozygous insertion genotype is less common in diabetic patients with nephropathy compared with those without nephropathy (MARRE et al. 1994). Similarly, the homozygous deletion genotype has been suggested as a marker for increased risk of developing left ventricular hypertrophy (IWAI et al. 1994; SCHUNKERT et al. 1994). The human angiotensinogen gene is also polymorphic, but there is no linkage with the ACE genotype (RUTLEDGE et al. 1994). Whether or not the maternal or fetal ACE genotype is correlated with the occurrence of adverse effects from fetal ACEI exposure is unknown.

Angiotensin II receptors have two principal types. Type 1 (AT_1) is present in a wide variety of tissues in the adult and seems to mediate all the known effects on angiotensin II related to the RAS. Type 2 (AT_2), which in the adult is apparently not involved in blood pressure control, is found in the mature renal capsule (DOUGLAS and HOPFER 1994; KANG et al. 1994). Notably, AT_2 predominates in the fetus, where it presumably plays a role in nephric development (GRADY et al. 1991; GRONE et al. 1992; DEGASPARO and LEVENS 1994). During maturation in the rat, the mRNA of AT_1 changes from wide distribution in the nephrogenic cortex to localization in the glomeruli, arteries, and vasa recta, suggesting a role for the receptor in renal growth and development (TUFRO-MCREDDIE et al. 1993). In the rat kidney, AT_1 receptor mRNA is developmentally regulated, and its expression in the cortex antedates AT_1 receptor ligand binding (AGUILERA et al. 1994). Inhibition of AT_1 results both in restriction of kidney and somatic growth and in increased renin mRNA levels (GOMEZ et al. 1993). It has been shown that the AT_1 receptor mediates renal and somatic angiotensin II-induced growth by regulating cell hyperplasia (TUFRO-MCREDDIE et al. 1994). In the mouse, both AT_1 and AT_2 are present by gestation day (GD) 14. AT_1 is found in the mature glomeruli and primitive nephrons; by GD 16 it appears also in the proximal and distal convoluted tubules. In contrast, AT_2 is found in the renal mesenchyme, and its expression subsides within 3 weeks after birth. The temporal pattern of expression of binding for both receptor types suggests that, whereas AT_2 receptors are involved in growth and differentiation of the nephron, AT_1 receptors have dual functions, at first in nephron development and later in renal function. In some species intrarenal AT_2 receptors disappear shortly after birth, while in others these receptors persist into adulthood, where their function has not been clearly defined (DEGASPARO and LEVENS 1994).

In rodents, the AT_1 receptor has two subtypes, AT_{1A} and AT_{1B}, the genes for which are located on different chromosomes. These subtypes were initially said to show few pharmacological differences and were also presumed to function as growth factors. Humans on the other hand have only a single AT_1 gene (YOSHIDA et al. 1992). According to BURSON et al. (1994), expression of AT_{1A} in adult mice is found in kidney, liver, adrenal, ovary, brain, testis, adipose tissue, lung, and heart, while AT_{1B} is absent from most of these tissues but is detectable in brain, testis and adrenal. BURSON et al. (1994) reported that expression of AT_{1A}, but not AT_{1B}, was found in placenta, fetal kidney, liver, and heart in 16.5- and 18.5-GD fetuses. However, in other studies of the mouse kidney, while AT_{1A} mRNA is the predominant isoform in mesangial and juxtaglomerular cells, proximal tubules, vasa recta, and interstitial cells of the kidney, AT_{1B} was detected in mesangial and glomerular cells and in the renal pelvis (GASC et al. 1994). In fetal rats, kidney (glomeruli) and adrenal have also been found to express both AT_{1A} and AT_{1B} mRNA as early as GD 10, while only AT_{1A} was found in liver, lung, heart, large blood vessels, and pitutary (SHANMUGAM et al. 1994a,b). LLORENS-CORTES et al. (1994) noted that, in the face of sodium depletion, rat adrenal AT_{1A} and AT_{1B} receptor mRNA levels increased 60% and 110%, respectively. In contrast, in renovascular hypertension, rat adrenal AT_{1B} receptor mRNA levels decreased 50%, while there was no change in AT_{1A} mRNA. These findings suggest that, despite a high degree of homology between the coding sequences of the AT_{1A} and AT_{1B} genes, they exhibit disparate tissue-specific expression profiles and that the receptor subtypes may be involved in the mediation of different biological effects of angiotensin II.

Angiotensin II exerts an effect on the growth of its target tissues. For example, adrenal cortical cells are induced to divide in the presence of angiotensin II. The mechanism by which angiotensin II induces hyperplasia of its target tissues is largely unknown, but may involve a direct action on protooncogene synthesis or an indirect action on growth factor secretion (CLAUSER et al. 1992). As previously noted, the subtypes of angiotensin II receptors appear to function differentially as fetal growth factors (TUFRO-MCREDDIE et al. 1993; FLUCK and RAINE 1994; SHANMUGAM et al. 1994a,b), suggesting that in particular the ACEI may inhibit cell proliferation of mesangial cells and hypertrophy of the renal tubular cells (FLUCK and RAINE 1994). While the roles of angiotensin receptors in fetal growth and development are becoming clearer, they are still not well understood and are the subject of continuing investigation. The effects of specific receptor blockade on form and function of the fetal kidneys of appropriate species could be quite helpful in understanding the mechanisms underlying ACEI fetopathy.

While the different ACEI are broadly similar in their actions, they are not clinically interchangeable for all indications. It has been noted that affinities of the various ACEI for tissue ACE vary in different organs, and thus responses in therapeutic situations may vary by the particular ACEI used (Ruzicka and Leenen 1995).

D. Clinical Uses of Angiotensin-Converting Enzyme Inhibitors

Since their introduction, ACEI have come to be regarded as a major advance in the treatment of hypertension (Thurston 1992; Rodicio and Ruilope 1993; Materson and Preston 1994). Guthrie (1993) noted that prescriptions for ACEI increased nearly 250% between 1986–1990, testifying to their popularity in clinical use, and many articles have appeared extolling their use as first-line agents in the treatment of hypertension. It has been observed that most types of hypertension respond favorably to ACEI, either as monotherapy or with the addition of diuretics (Sassano et al. 1987; Garay et al. 1994).

Antihypertensives such as β-adrenergic receptor blockers and diuretics tend to increase peripheral resistance and may have undesirable side effects on the metabolism of electrolytes, glucose, and lipids. In contrast, ACEI decrease vascular resistance, improve glucose handling, control left ventricular mass, and offer a degree of myocardial protection (Williams 1988; Gavras 1988; Materson and Preston 1994). In the absence of heart failure, ACEI produce little change in heart rate, cardiac output, or pulmonary wedge pressure in normal or hypertensive people (Vidt et al. 1982; Todd and Heel 1986; Gomez et al. 1987). A principal use of the ACEI is for myocardial protection (Williams 1988; Gavras 1988; Materson and Preston 1994; Ruzicka and Leenen 1995). Angiotensin II-induced myocardial hypertrophy is mediated, at least in part, through AT_1 receptors (Miyata and Haneda 1994). Both short-term and long-term outcome after myocardial infarction may be improved by institution of ACEI therapy soon after the event (Ambrosioni et al. 1995). Chronically high levels of plasma ACE have been associated with thickening of arterial walls (Bonithon-Koop et al. 1994); presumably the ACEI will inhibit this response. It has also been asserted that they contribute to an improved quality of life for the hypertensive patient, particularly when compared with beta blockers (Materson and Preston 1994).

Another use for ACEI that has aroused considerable clinical interest stems from the demonstration that they slow progression of the proteinuria and nephropathy associated with both type 1 and 2 diabetes mellitus (Ravid et al. 1993; Bakris 1993; Chagnac et al. 1994; Lebovitz et al. 1994; Mulec et al. 1994; Viberti et al. 1994). Defronzo and Ferrannini (1991) suggested that hypertension may be an effect of insulin resistance. The occurrence of insulin resistance has also been linked to hypertension in the gestational diabetic (Bevier et al. 1994). The ACEI reduce microalbuminuria, preserve renal function, and increase insulin sensitivity; these effects are believed to be independent of the RAS-mediated, antihypertensive action of these drugs (RAVID et al. 1993; REMUZZI et al. 1994). The ACEI have also been reported to retard the progression of nondiabetic renal failure; however, the many variables involved in this diagnosis make assessment of this mode of therapy difficult (BECKER et al. 1994).

E. Adverse Effects of Angiotensin-Converting Enzyme Inhibitors

In the enthusiasm to use ACEI as first-line agents for treatment of hypertension and to retard development of diabetic nephropathy, their safety has sometimes been overstated. Although they are clinically very useful drugs, ACEI are not devoid of adverse effects. Aside from exaggerated hypotensive responses to therapy, and rare occurrences of angioedema, neutropenia, agranulocytosis, and hepatic failure, several adverse effects stand out. Because they suppress aldosterone secretion, ACEI are liable to produce hyperkalemia, particularly in the face of sodium restriction, heart failure, diabetes, or coadministration of potassium-sparing diuretics or nonsteroidal anti-inflammatory drugs (WILLIAMS 1988; SCHLUETER et al. 1994). A not uncommon (2%–34%) side effect of the ACEI is a dry, irritating cough, probably due to inhibition of bradykinin degradation (FLETCHER et al. 1994; KANG et al. 1994). Ironically, this symptom has generated more literature than has the possibility of fetal damage.

The use of ACEI in certain circumstances, such as bilateral renal artery stenosis, hypertensive nephrosclerosis, autosomal dominant polycystic kidney disease, advanced diabetic nephropathy, and chronic congestive heart failure, can contribute to, not ameliorate, renal dysfunction (COOKE and DEBESSE 1994; TOTO 1994). Because glomerular filtration pressure is reduced by the ACEI, and the ability to increase renal blood flow through a stenotic arterial supply is limited, there may be insufficient compensation for the efferent arteriolar dilation (HRICIK et al. 1983; FLUCK and RAINE 1994). In some cases the renal dysfunction may be reversible on discontinuation of the drug (WOOD et al. 1991), and in others irreversible (DEVOY et al. 1992). Renal scanning after administration of an ACEI has been used to detect and, more reliably, to predict the outcome of therapy of renal artery stenosis (MCGRATH et al. 1983; DAVIDSON and WILCOX 1991; CANZANELLO and TEXTOR 1994; DERKX and SCHALEKAMP 1994). In the presence of sodium volume depletion, the effects of ACEI on the kidney are potentiated and may lead to decreased renal function (MANDAL et al. 1994; TOTO 1994). This may be reversed or prevented by sodium replacement. Among children with congenital and acquired heart disease, ACEI-associated renal failure has been related to young age, low body weight, and left-to-right shunting (LEVERSHA et al. 1994).

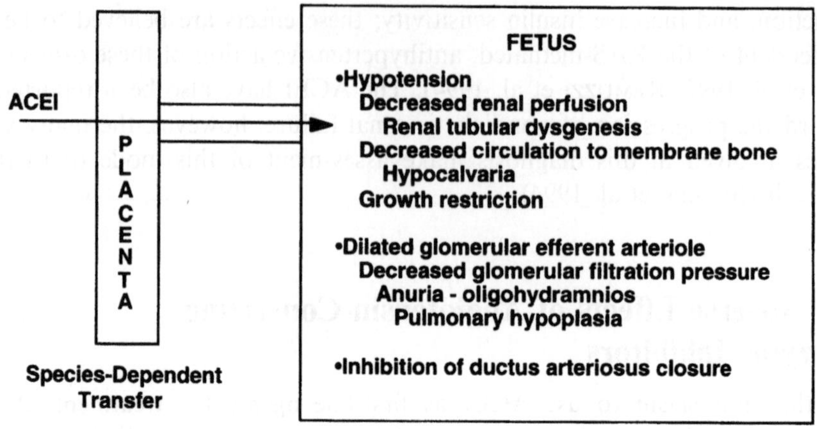

Fig. 3. Angiotensin-converting enzyme inhibitor (ACEI) fetopathy

A particularly troubling adverse effect may occur when the human fetus is exposed to an ACEI (Fig. 3). The abnormal outcome most commonly reported is second- to third-trimester onset of oligohydramnios and growth restriction, followed by delivery of an infant whose neonatal course is complicated by prolonged and often profound hypotension and anuria (ROSA et al. 1989). Affected infants commonly have the oligohydramnios deformation sequence and its lethal component, pulmonary hypoplasia (GUIGNARD et al. 1981; MEHTA and MODI 1989; CUNNIFF et al. 1990; PRYDE et al. 1993). In the few cases that have had microscopy of the kidneys, renal tubular dysgenesis has been a consistent finding (CUNNIFF et al. 1990; PRYDE et al. 1993; M. BARR, unpublished observations). Delayed development of calvarial bone has been noted in enough cases to consider it a part of the syndrome (BARR and COHEN 1991). Postnatal persistence of a patent ductus arteriosus (PDA) may also be part of the syndrome.

F. Animal Developmental Toxicity Studies

Studies of the ACEI given during the period of organogenesis to rats and rabbits have failed to show malformations, even from high doses (FUJII et al. 1985; FUJII and NAKATSUKA 1985; ROBERTSON et al. 1986; AL-SHABANAH et al. 1991; AL-HARBI et al. 1992). However, with more prolonged exposure and at the highest doses tested, fetal growth restriction and delayed ossification have been noted in conjunction with evidence of maternal toxicity (AL-HARBI et al. 1992; VALDES et al. 1992). When ACEI are administered to rats and rabbits in late gestation and during lactation, in doses comparable to those used in human therapy, there has been significant fetal and neonatal death (manufacturers' literature; BROUGHTON PIPKIN et al. 1982; MINSKER et al. 1990). Nonetheless, developmental toxicity studies in animals do shed some light on the mechanisms underlying this human ACEI fetopathy.

I. Rabbit

The adult rabbit is particularly sensitive to ACEI, showing a greater antihypertensive response than does the rat or dog. As determined by inhibition of the acute pressor response to angiotensin I, enalaprilat and captopril are equally potent in pregnant and nonpregnant rabbits (J.M. MANSON, personal communication). In rabbits, 2.5–5.0 mg/kg per d of captopril resulted in pregnancy loss rates ranging from 37% to 92% (BROUGHTON PIPKIN et al. 1980, 1982; FERRIS and WEIR 1982). Although no gross abnormalities were found at autopsy of the stillborn kits, microscopic examination of their kidneys was not reported (BROUGHTON PIPKIN et al. 1982). The ED_{50} values for blockade of the pressor response to angiotensin I by enalaprilat in the rabbit, rat, and dog were 2.8, 5.1, and 6.4 mg/kg, respectively. Comparable ED_{50} values for captopril in these three species were 2.88, 26.1, and 80 mg/kg, respectively (J.M. MANSON, personal communication). The basis for this marked sensitivity in rabbits is not known.

In late gestation, the rabbit fetus is highly sensitive to the fetotoxic effects of ACEI. When administered in the human therapeutic range, ACEI produce fetal deaths in mid- to late gestation, with a peak effect on GD 26 (KEITH et al. 1982; BROUGHTON PIPKIN et al. 1982; MINSKER et al. 1990). Elevation of maternal blood urea nitrogen (BUN) and creatinine, and occasionally death, markers of maternal toxicity, occur at the same treatment dosages as those causing fetal deaths. Fetal deaths in rabbits associated with ACEI treatment may result from diminished placental perfusion secondary to maternal systemic hypotension, rather than from a direct effect on fetal renal perfusion (FERRIS and WEIR 1982; BINDER et al. 1989; BINDER and FABER 1992). Captopril studies in the rabbit have shown an effect neither on amniotic fluid volume and electrolyte composition nor on fetal plasma urea nitrogen, creatinine, and calcium levels (J.M. MANSON, personal communication). When maternal plasma volume of the rabbit was expanded by saline administration, maternal toxicity from enalapril was reduced, but fetal deaths still occurred, particularly at higher dosage levels (30 mg/kg per day) (MINSKER et al. 1990).

II. Rat

The maturation of the RAS in the fetal and neonatal rat kidney has been studied (ICE et al. 1988; AGUILERA et al. 1994). The RAS develops late in gestation, beyond the time when treatment typically is administered in a segment II rat developmental toxicity study (GD 6–17). Therefore, it should not be surprising that such studies have failed to show adverse fetal effects, beyond decreases in fetal body weight and delayed ossification at doses that also produce evidence of maternal toxicity.

Expression of the angiotensin II receptor gene was not detected in newborn rat liver, but in the kidney gene expression was 2.5 times greater than in the adult (IWAI et al. 1991). Decreasing concentrations of plasma renin from fetal, to newborn, to neonatal rats are indicative of increased activity of the RAS in immature animals (JELINEK et al. 1986). Renin is first detected by immunostaining in the kidney on GD 19, when it is located in the arcuate and interlobular arteries. On GD 20, renin also appears in the afferent glomerular arteriole. As age increases, detectable renin disappears progressively from the arcuate and interlobular arteries until it is restricted primarily to the afferent arteriole in the 20-day-old postnatal pup and finally only to the juxtaglomerular apparatus in the adult kidney. A similar, but more rapid progression of maturation has been observed in the mouse (reviewed by ICE et al. 1988). Studies of the synthesis of renin mRNA in the rat have shown the same pattern and timing (GOMEZ et al. 1989).

Interestingly, female fertility and late gestation/lactation studies, in which ACEI exposure occurred during the critical period of fetal RAS development (GD 19), have shown adverse effects on neonates (ROBERTSON et al. 1986). At high doses (30–300 mg/kg per day for enalapril and lisinopril), pup deaths occurred during the lactation period. At lower treatment level (10 mg/kg per day), decreased pre- and postweaning body weight gain was observed. Similar results have been obtained with other ACEI, quinapril (DOSTAL et al. 1991) and rentiapril (COZENS et al. 1987).

Given the late development of the RAS relative to birth, late gestation/lactation studies (GD 15 to the end of lactation) would seem to provide the most sensitive appraisal of ACEI effects on rat pups. Kidneys from weanlings whose mothers received quinapril during late gestation (GD 15 onward) and/or lactation had juxtaglomerular cell hypertrophy (DOSTAL et al. 1991; GRAZIANO et al. 1993), although the latter authors reported finding no morphological changes in the renal tubules and no adverse effects on renal function. In contrast, FRIBERG et al. (1994) reported that, in the spontaneously hypertensive rat and the normotensive strains WKY and WR, early postnatal treatment with the ACEI enalapril and captopril and the AT_1 blocker losartan, but not the AT_2 blocker PD-123319, produced persistent and irreversible histopathological renal changes in adult life, long after the cessation of treatment. These abnormalities were associated with impaired urine-concentrating ability. For the most part, microscopic examination of such kidneys has either not been done or not been reported in the literature.

All in all, it does not appear that the rat is a suitable model for predicting or understanding ACEI fetopathy in humans. The RAS develops much later in gestation in rats than in humans. In rats, ACEI toxicity is manifested in neonates and weanlings, while in humans the major effects occur prenatally and consequently affect adaptation to extrauterine life. From the available data, it also appears that in rats the placenta may serve as more of a barrier to ACEI penetration into the fetal compartment than it does in humans, although more data are needed on this point (ENDO et al. 1992).

III. Sheep

The chronically catheterized sheep is an excellent preparation for study of maternal and fetal cardiovascular and renal changes that occur during pregnancy (MAGNESS et al. 1985; HILL and LUMBERS 1988). Captopril crosses the sheep placenta and blocks the fetal RAS. When administered in late pregnancy (GD 119–133; term, 147 days) at 2.8–3.5 mg/kg IV to the ewe, following preadministration of angiotensin I, maternal blood pressure was transiently reduced but returned to normal within 2 h. In contrast, fetal blood pressure was reduced and remained so for up to 2 days, and seven of eight captopril-exposed lambs were stillborn (BROUGHTON PIPKIN et al. 1982). Although no gross abnormalities were found at autopsy, microscopic examination of the fetal kidneys was not reported. In a similar model, fetal renal function was studied in sheep given captopril (GD 123–133) at an infusion rate of 15 mg IV for 4 days (LUMBERS et al. 1992, 1993). Fetal GFR decreased from an initial 4.2 ml/min to 2.7 ml/min after maternal captopril administration, and 3 days later it was still lower at 1.5 ml/min. When these fetuses were given angiotensin II intravenously (6μg/kg per h), fetal GFR returned to a mean of 3.4 ml/min. It was concluded that a small decrease in fetal arterial pressure partly contributed to the lowering of fetal GFR, but efferent arteriolar tone also fell, so that glomerular filtration pressure decreased further. Thus maintenance of fetal renal function depends on the integrity of the fetal RAS and its control of glomerular efferent arteriolar tone.

The sheep model has also been used for studies of enalapril (BROUGHTON PIPKIN and WALLACE 1986). At a median gestational age of 128 days, a pressor response to exogenous angiotensin I (5 μg IV) was first established in the maternal compartment. Maternally administered angiotensin I had no effect on fetal blood pressure. Following the administration of enalapril (1 mg/kg), maternal blood pressure was decreased for 30 min, but no effect on fetal heart rate, plasma renin, or ACE activity was found. With a higher dose of enalapril (2 mg/kg), there was a transient fall in fetal blood pressure, considered to be due to changes in uteroplacental blood flow rather than to a direct effect of the drug on the fetus. Only trace amounts of enalapril were found in fetal plasma, and levels of its active metabolite, enalaprilat, were undetectable in three of four fetuses. One unexplained fetal death occurred in the 2-mg/kg group. It was concluded that there was minimal transfer of enalapril into the fetal compartment (BROUGHTON PIPKIN and WALLACE 1986). Thus the sheep model, while informative about the physiological effects of captopril in gestation, is not applicable to the investigation of human fetal effects from exposure to all other ACEI, for enalapril does cross the human placenta and evidently acts directly on the fetus.

IV. Need for a Better Animal Model

Similar to the experience in humans, the adverse effects reported in laboratory animals show fetotoxicity but not teratogenicity of the ACEI. To date, the sheep model has provided data most applicable to the human fetopathy, provided placental transfer of the drug occurs. However, as discussed above, laboratory animals and sheep, while providing valuable information about the physiology of the RAS and ACEI, do not answer all the questions raised by the features of the human fetopathy. So far, little exploration of ACEI effects on maternal and fetal physiology seems to have been done in nonhuman primates. A recent study, using a sequential crossover design, in baboons treated with placebo versus a therapeutic dose of enalapril did demonstrate a high incidence of adverse effects (Harewood et al. 1994). Four of nine liveborn infants exposed to enalapril had growth restriction, and there were four fetal deaths, including one set of twins. This was compared with no adverse outcomes among 11 infants in the placebo arm of the study. No gross abnormalities were noted at autopsy of the dead fetuses, but no growth analysis or renal histology of these fetuses were reported. The results of this study do suggest that further investigations in nonhuman primates might be helpful in elucidating the mechanisms of the fetopathy seen in humans.

G. Human Angiotensin-Converting Enzyme Inhibitor Fetopathy

As early as 1980, reports of fetal wastage in ACEI-exposed experimental animals were published (Broughton Pipkin et al. 1980, 1982; Ferris and Weir 1982). The first adverse outcome in a human pregnancy was reported in 1981 (Duminy and Burger 1981). This was followed in short order by a number of other cases implicating both captopril and enalapril as potential fetotoxicants (Guignard 1982; Boutroy et al. 1984; Caraman et al. 1984; Rothberg and Lorenz 1984). Specific warnings about the use of ACEI in human pregnancy appeared in the literature as early as 1985 (Lindheimer and Katz 1985). The original U.S. Food and Drug Administration use-in-pregnancy classification of all ACEI was category C. When attention was finally directed to contemporary and published adverse fetal outcomes, the classification was changed, in early 1992, to category D (for second- and third-trimester use). At this time, a boxed warning and extended discussion were added to the manufacturers' literature, and a "Dear Doctor" letter was sent out. Despite this record, several recent reviews of the ACEI still include no mention of the possibility of adverse fetal effects (Kang et al. 1994; Kellow 1994; Materson and Preston 1994; Pinkney and Yudkin 1994; Wilson 1994), while a few do warn against use of these drugs in pregnancy (Mogensen 1994; Molitch 1994; Robertson 1994).

While it appears that the majority of ACEI fetopathy cases have ensued after enalapril exposure, cases associated with captopril and lisinopril exposure are recorded (HANSSENS et al. 1991; PRYDE et al. 1993). As pointed out by BRENT and BECKMAN (1991), there is no reason to assume that any of the other ACEI will not produce the same result. Similarly, many of the reported cases have involved coexposure to other antihypertensive agents, and, although combination therapy may possibly increase the risk to the fetus, monotherapy with an ACEI has produced the full-blown fetopathy.

The use of ACEI in women for myocardial protection probably poses little concern to the teratologist, since the vast majority of such use will be in patients beyond the child-bearing years. However, the use of ACEI for mild or essential hypertension, and in such conditions as the collagen vascular diseases and diabetes mellitus, will involve a considerable number of women in the child-bearing years, some of whom will become pregnant. The use of ACEI for the management of pregnancy-induced hypertension and pre-eclampsia would mean certain fetal exposure.

The frequency of ACEI use in pregnancy is largely unknown. PIPER et al. (1992) reported that among 106 813 Tennessee Medicaid patients who delivered a live or stillborn infant during the period 1983–1988, 19 were exposed to an ACEI. Given the expansion of use of these agents to the current time, the number of fetal exposures could increase, unless there is a wider appreciation of the danger to the fetus from their use.

The true rates of adverse fetal effects from ACEI use in human pregnancy cannot be determined from available information. PRYDE et al. (1993) summarized 29 affected cases from the literature and added three more. Two more cases have been reported since the review by PRYDE et al. (PIPER et al. 1992; THORPE-BEESTON et al. 1993), and I have received at least partial reports of five others from clinicians and attorneys. To be sure, a number of exposed pregnancies have resulted in no detectable adverse effect (KREFT-JAIS et al. 1988; PIPER et al. 1992). In the largest published collection there were 31 pregnancies exposed either to captopril (n = 22) or enalapril (n = 9) (KREFT-JAIS et al. 1988). The adverse outcomes included nine cases of intrauterine growth restriction (IUGR), three intrauterine deaths, and two infants with PDA, of which one ultimately required ligation. Twenty-two cases in this report were presumably unaffected. MITCHELL et al. (1993) reported a case of a fetus exposed to enalapril (5 mg/day) from conception to 23 weeks of gestation who was delivered as a healthy baby at 38 weeks. MILLAR et al. (1983) reported the case of an unaffected infant delivered at 29 weeks, after 2 week exposure to captopril. I have personally examined two fetuses exposed in the second trimester, neither of whom showed evidence of growth restriction or histological abnormality of the kidneys (unpublished observation).

The adverse developmental effects of ACEI are purposely designated ACEI *fetopathy*, because to date there is no convincing evidence that there is harm from exposure in the first trimester of human gestation or its equivalent in laboratory animals. In humans, in the case reported by THORPE-BEESTON et al. (1993) renal cystic dysplasia was found, a disorder distinct from the renal tubular dysgenesis found in other cases. The renal dysplasia in this case may be entirely coincidental to enalapril exposure but, even if related, it is still likely to be of fetal rather than embryonic origin. In the captopril-exposed case reported by DUMINY and BURGER (1981), there was a terminal transverse deficiency of a limb, again a defect likely to originate in the fetal period. KALER et al. (1987) reported a singular case of an infant exposed throughout gestation to captopril, propranolol, furosemide, and minoxidil. The infant had marked hypertrichosis, presumably associated with the minoxidil exposure, omphalocele, ventricular septal defect, and several minor anomalies of the face and hands. Hypotension was noted for the first 24 h of life, but thereafter resolved, and renal function was normal. There are no features of this case that are clearly indicative of a fetal ACEI effect. The second case mentioned by PIPER et al. (1992) was remarkable for microcephaly and encephalocele, the latter almost certainly a problem of first-trimester origin. I believe that this infant was misclassified as having hypocalvaria. I am aware of a single enalapril exposure-associated case of anencephaly. These few disparate anomalies associated with ACEI exposure do not constitute sufficient evidence of true teratogenicity, although admittedly as yet there is insufficient evidence to be truly confident of safety in the first trimester.

I. Renal Tubular Dysgenesis

Although there are few reports of the histology of the kidneys in ACEI-exposed fetuses and infants, they are consistent in their demonstration of renal tubular dysgenesis (RTD). RTD is characterized by diminished to absent differentiation of proximal convoluted tubules (SWINFORD et al. 1989), which have been shown to be shorter and straighter than normal (VOLAND et al. 1985). There may also be increased cortical and medullary mesenchyme (and later fibrosis) and dilation of Bowman's spaces and tubules. The histological changes in the kidney strongly suggest ischemic injury, with the added component of deficient tubular differentiation most easily confirmed by failure to demonstrate the brush borders of proximal convoluted tubules with periodic acid-Schiff stain. RTD was fully described in four cases of ACEI fetopathy (CUNNIFF et al. 1990; PRYDE et al. 1993), it was probably the lesion in the case reported by KNOTT et al. (1989), and it has been observed in a number of unreported cases. It has been my experience that, unless specifically looked for, the histopathological changes of RTD in the fetal/neonatal kidney can be and have been missed.

RTD, with lack of proximal tubular differentiation and clinical neonatal anuria or oliguria, has also been reported as a congenital autosomal recessive disease, with features of oligohydramniotic deformation and pulmonary hypoplasia (VOLAND et al. 1985; BERNSTEIN 1988; SWINFORD et al. 1989; ALLANSON et al. 1992). BERNSTEIN and BARAJAS (1994) demonstrated increased renin in preglomerular arterioles, glomerular hila, and mesangial areas in this form of RTD. They hypothesized that the increased renin accumulation reflected strong local vasoconstriction, which caused reduced glomerular perfusion. The renin accumulation suggested faulty feedback control of renin secretion in the genetic variety of RTD, which would link this disorder to the RTD seen in ACEI fetopathy.

RTD has also been reported in association with exposure to nonsteroidal anti-inflammatory drugs, particularly indomethacin (SIMEONI et al. 1989; RESTAINO et al. 1991; BAVOUX 1992; BUDERUS et al. 1993; VOYER et al. 1994). The GFR of prenatally indomethacin-exposed infants was significantly reduced when measured on day 3 of life (VAN DEN ANKER et al. 1994). Long-term indomethacin treatment during pregnancy may cause fetal renal failure and irreversible renal damage characterized by cystic dilation of superficial nephrons, ischemic changes of the deep cortex, and increased intrarenal renin content (VAN DER HEIJDEN et al. 1994). It is tempting to relate these findings to ACEI-associated RTD, but WALKER et al. (1994), using indomethacin and an arginine vasopressin V2 receptor antagonist in sheep, concluded that the fetal oliguria associated with indomethacin is mediated through the stimulation of the renal arginine vasopressin V2 receptor and that inhibition of prostaglandin synthesis by indomethacin may affect renal tubular sodium handling. The arginine vasopressin V2 receptor appears to ameliorate the fetal hypertensive response to indomethacin. They suggested that fetal oliguria and hypertension resulted from indomethacin stimulation of circulating arginine vasopressin and enhancement of peripheral arginine vasopressin effects in the fetus. By way of contrast, ACEI-induced anuria is accompanied by hypotension, while indomethacin-exposed infants have had normal blood pressures (VAN DER HEIJDEN et al. 1994).

The question arises as to whether the RTD seen in association with ACEI exposure is due to a primary interference with tubular differentiation or whether it is a manifestation of a derangement of fetal physiology. To examine the specificity of this set of histopathological findings for ACEI exposure, MARTIN et al. (1992) studied the fetuses of nine women who were chronically hypertensive. Three of these mothers used antihypertensive agents throughout pregnancy, including one who used an ACEI. The tubular morphology of the kidneys was compared with the renal tubules of 20 normal controls, 13 fetuses with various multiple malformation syndromes, and six cases of the twin–twin transfusion syndrome. Features of RTD were identified in the ACEI-exposed case, one methyldopa-exposed case, and two cases of twin–twin transfusion syndrome (MARTIN et al. 1992). In a study of 24 cases of the twin–twin transfusion syndrome, RTD was found in the donor twin in 11 instances, compared with no cases among their recipient twins or 22 sets of dizygous twins (M. BARR, unpublished observation). To tie these findings and evidence from the literature together, it was hypothesized that the primary mechanism by which ACEI affect development of the fetal kidney is through decreased renal blood flow (MARTIN et al. 1992).

Possible support for the renal ischemia hypothesis comes from LANDING et al. (1994), who noted that RTD has been seen as a unilateral lesion in young infants with renal artery stenosis due to arteritis or medial arterial calcinosis and that it resembles the renal tubular atrophy of a variety of end-stage kidney diseases, such as glomerulonephritis, tubulointerstitial kidney disease, obstructive uropathy and pyelonephritis, graft rejection of transplanted kidneys, and the renal parenchymal changes associated with protracted dialysis therapy. LANDING et al. (1994) reported that labeled lectins that differentially mark proximally convoluted, distally convoluted, connecting, and collecting tubules showed no distinctive differences in the staining patterns of the hypoplastic renal tubules in RTD compared with kidneys affected by postnatal renal artery obstruction or various types of end-stage renal disease. These findings suggest that the renal tubular changes in the conditions cited, including ACEI-associated RTD, result from renal ischemia. However, the issue of mechanism is far from settled and appears to be more complicated than renal ischemia leading directly to RTD. In the fetal sheep, when renal oxygen delivery was limited by a prolonged reduction in hematocrit, sodium and water excretion increased and resulted in an *increase* in amniotic fluid volume (DAVIS et al. 1994), in contrast to the oligohydramnios seen in ACEI fetopathy. Thus it would appear that, in the case of ACEI exposure, the initiating event for RTD is a derangement of fetal physiology and function of the RAS by the drug, which secondarily interferes with the differentiation of proximally convoluted tubules. While RTD is not specific for ACEI exposure, ACEI are rather proficient at creating the conditions that lead to RTD.

II. Hypocalvaria

Six reported, ACEI-exposed cases have had hypoplasia of the membrane bones of the skull or hypocalvaria (DUMINY and BURGER 1981; ROTHBERG and LORENZ 1984; MEHTA and MODI 1989; PRYDE et al. 1993). In this condition, the calvarial bones are found to be normal in position, shape, and histologic appearance, but greatly reduced in size. As a result, the sutures and fontanelles are described as symmetrically enlarged, and in severe cases the normally developing brain may be essentially unprotected by skull and liable to trauma during labor and delivery (BARR and COHEN 1991). Lacking age-specific morphometrics for the size of the calvarial bones, which would be more relevant than fontanelle measurement, a diagnosis of hypocalvaria in milder cases is admittedly a subjective assessment. Although there are now only six published cases in which this skull lesion is specifically described, I think it is more common but unrecognized. Among several unpublished cases that I have reviewed, mention was made of enlarged fontanelles with split sutures in instances in which hydrocephaly or macroencephaly was not present. From a single case experience, it has been noted that growth of the calvaria apparently resumes postnatally, and eventually the skull is indistinguishable from normal (PRYDE et al. 1993).

The cause of the hypoplastic calvaria found with ACEI exposure is unknown. A possibility is that inhibition of angiotensin II may concomitantly inhibit one of the variety of growth factors involved in calvarial bone development (BARR and COHEN 1991; BERNSTEIN and BARAJAS 1994). Endochondral bone and membrane bone grow and develop in different ways. Since nutrition takes place by diffusion through the cartilaginous epiphyses, long bones normally develop in a low-oxygen environment. Membrane bones, on the other hand, have a high degree of vascularity, and presumably a high oxygen tension is required for their normal growth (SCHUMACHER 1985). It has been hypothesized that presumed fetal hypotension produced by ACEI exposure results in relative hypoxia and thereby affects calvarial growth (BARR and COHEN 1991).

The foregoing is not meant to imply that endochondral bone growth is unaffected in ACEI fetopathy; it may be affected. In a single case (PRYDE et al. 1993, case no. 1), limb lengths were disproportionately short by measurement, although not to a degree that a diagnosis of short-limbed dwarfism would be made. No data on limb lengths are given for other cases. From the finding of decreased limb lengths, disproportionate to crown–rump length, in other fetuses with presumed hypotension (e.g., hydrops fetalis of various etiologies), it is suggested that relative growth lag of the limbs is a marker of an adaptive shift of circulatory pattern in the compromised fetus, in which circulation is diminished to the periphery to preserve central perfusion (unpublished observations). Thus it could be predicted that an additional feature of ACEI fetopathy would be some shortening of the limbs.

III. Intrauterine Growth Restriction

IUGR (birth weight centile < 10%) has been observed in 21 of 27 cases of ACEI fetopathy for which birth weight information was given (PRYDE et al. 1993). Thus IUGR is one of the cardinal features of ACEI fetopathy, although its basis is still largely unresolved. The occurrence of IUGR is common in pregnancies complicated by maternal hypertension. Whether or not such growth restriction is worsened by the addition of ACEI exposure is uncertain.

The status of the fetal RAS in pregnancies complicated by severe generic IUGR, and its possible relationship to elevated fetoplacental vascular resistance, was explored by KINGDOM et al. (1993). In an IUGR group, cord venous angiotensin II was markedly elevated compared with controls, but there was no change in angiotensin receptor concentration or angiotensin receptor affinity. However, KNOCK et al. (1994) state that the capacity and affinity of angiotensin II-binding sites are significantly lower in placentas associated with pre-eclampsia and IUGR compared with normal term placentas. They stated that this was not due to prior binding of the receptors by endogenous ligand, but appeared to represent downregulation of the receptors in these conditions.

The fetal RAS is augmented in at least some forms of IUGR, and responsiveness of the fetoplacental vasculature to angiotensin II is not diminished as might be expected from the elevated plasma angiotensin II levels. Angiotensin II may contribute to the increased fetoplacental vascular resistance observed in this disorder. However, BOURA et al. (1994) noted that fetal vessels of the placenta constrict intensely on exposure to thromboxane A_2, prostaglandin $F_2\alpha$, and prostaglandin E_2 but show only a weak response to angiotensin II. This would indicate that the prostaglandin system is more significant than the angiotensin system in the determination of IUGR. Further research in the human placenta in vitro and in nonhuman primates could help assess the effect of ACE inhibition on placental function and its potential impact on IUGR.

It has been suggested that IUGR itself is associated with retarded renal development. Among six severely affected IUGR stillbirths, five had nephron number estimates below a matched control group's 5% prediction limit, and nephron numbers were significantly reduced in a group of infants with IUGR who died within 1 year of birth (HINCHLIFFE et al. 1992). Thus, while there appears to be a direct adverse effect of ACEI on the developing kidney, the co-occurrence of IUGR may further restrict renal development.

IV. Patent Ductus Arteriosus

A final observation that appears on reviewing reports of neonatal outcomes in pregnancies exposed to ACEI is the persistence of a PDA. While this may be associated with the increased incidence of prematurity among the cases reported, there may also be an increased need for surgical ductal ligation. Of seven PDA cases reported in association with ACEI exposure, three required surgical intervention (BOUTROY et al. 1984; CARAMAN et al. 1984; PLOUIN and TCHOBROUTSKY 1985; KREFT-JAIS et al. 1988). Given the possible effect of the ACEI on the fetal bradykinin–prostaglandin system, which would be expected to increase prostaglandin E and prostacyclin systemically and perhaps locally, it is possible that prenatal ACEI may inhibit ductal closure. Certainly, more data are needed before any firm conclusion can be reached on this subject.

V. Could It Be the Maternal Disease and Not the Drug?

ACEI fetopathy infants have been born to mothers whose hypertension was of widely varying etiology, including lupus erythematosus, renovascular hypertension, nephrotic syndrome, glomerulonephritis, renal transplantation, pre-eclampsia, and essential hypertension (PRYDE et al. 1993). Fetal distress, IUGR, and oligohydramnios are not uncommon complications of hypertensive pregnancies. However, there is evidence that the oligohydramnios, hypotension, and anuria of ACEI fetopathy are truly drug-related rather than attributable to the underlying maternal disease process. The evolution of oligohydramnios followed by the delivery of a neonate with prolonged hypotension and anuria is not included in the well-described list of complications of maternal hypertension and its traditional therapy, although one case of RTD has been reported under these circumstances (MARTIN et al. 1992). In the case of ACEI exposure, it is highly likely that the oligohydramnios is in fact due to a drug-related fetal hypotension and renal failure, because the hypotension and renal failure persist in the neonate. Evidence in support of this assertion is the occurrence of profound hypotension and anuria in some neonates given low doses of ACEI postnatally for treatment of hypertension (TACK and PERLMAN 1988; PERLMAN and VOLPE 1989; WELLS et al. 1990). Further, when ACE activity has been measured in prenatally ACEI-exposed hypotensive neonates, it has been found to be profoundly blunted and normalizes only after dialysis removes the otherwise renally excreted drug (SCHUBIGER et al. 1988; PRYDE et al. 1993). There are reports that the onset of oligohydramnios was temporally related to the initiation of maternal ACEI therapy (GUIGNARD et al. 1981; SCHUBIGER et al. 1988), and in one case the amniotic fluid volume increased toward normal after the ACEI was discontinued, although that fetus did not survive for long after birth (BROUGHTON PIPKIN et al. 1989).

VI. Caution About Angiotensin Receptor Antagonists

Although beyond the scope of this chapter, it should be noted that many angiotensin II receptor antagonists have been developed and are very likely to gain wide clinical use. It is noted that the reproductive and developmental toxicity of the AT_1-selective receptor antagonist losartan in the rat has a great many similarities to that of the ACEI (FRIBERG et al. 1994). Because of this, it would be prudent to assume that receptor antagonists will have fetopathic effects similar to the ACEI in humans and to restrict their use in pregnant women.

H. Summary

Occasionally there is a drug whose use in pregnancy is so frequently associated with adverse outcome of so specific a pattern that it becomes clear that its use must be restricted before its harmfulness can be validated by epidemiological studies. I believe this to be the case with the drug class of angiotensin-converting enzyme inhibitors. There are mammalian models suggesting substantial fetotoxicity in a dose-related fashion. There is a strong and consistent pattern to the reported cases of ACEI-related adverse outcomes: the syndrome of oligohydramnios-anuria, neonatal hypotension, renal tubular dysgenesis and hypocalvaria is too specific in association with the use of these drugs to be ignored. There is a very plausible biologic mechanism to explain the relationship. The features of ACEI fetopathy suggest that the underlying pathogenetic mechanism is fetal hypotension, which may also result from other exposures and disorders. Thus, while the fetopathy may not be caused only by ACEIs, they are particularly liable to produce adverse fetal effects.

I. Recommendations

It is advised that ACEI not be prescribed to pregnant women. Based on current information, exposure to ACEI in the first trimester should not be considered an indication for terminating a pregnancy. In such cases, an alternative antihypertensive regimen should be substituted for the drug prior to entering the second trimester. These drugs should not be used in the management of pregnancy-induced hypertension or pre-eclampsia (BROUGHTON PIPKIN and RUBIN 1994). If a woman is inadvertently on ACEI therapy in the second or third trimesters of pregnancy, she should be monitored for signs of fetal toxicity, including oligohydramnios, growth restriction, or fetal distress. Although oligohydramnios was observed to reverse in a single case when the ACEI was discontinued, a cautionary note must be sounded. The signs of fetal toxicity may not be detected until after irreversible damage to the fetus has occurred.

In the case of an ACEI-exposed fetus, at the time of delivery or before, the pediatricians should be notified of the potential for neonatal hypotension and anuria and the possible need to attempt early dialysis to remove the otherwise renally excreted drug. However, the combination of hypotension and anuria, often complicated by prematurity and growth restriction, makes both peritoneal dialysis and hemodialysis extremely difficult, and the mortality rate is exceptionally high (SEDMAN et al. 1995) Because the damage that can occur from these drugs is so severe, it is urged that their prescription to women of child-bearing potential be limited and, if they must be used, that the women be warned about the risk to the fetus and be monitored very closely for the occurrence of pregnancy.

References

Aguilera G, Kapur S, Feuillan P, Sunarakbasak B, Bathia AJ (1994) Developmental changes in angiotensin II receptor subtypes and AT(1) receptor mRNA in rat kidney. Kidney Int 46: 973–979

Alberola AM, Salazar FJ, Nakamura T, Granger JP (1994) Interaction between angiotensin II and nitric oxide in control of renal hemodynamics in conscious dogs. Am J Physiol 267: R 1472–R1478

Al-Harbi MM, Al-Shabanah OA, Al-Gharably NMA, Islam MW (1992) The effect of maternal administration of enalapril on fetal development in the rat. Res Commun Chem Pathol Pharmacol 77: 347–385

Al-Shabanah OA, Al-Harbi MM, Al-Gharably NMA, Islam MW (1991) The effect of maternal administration of captopril on fetal development in rat. Res Commun Chem Pathol Pharmacol 73: 221–230

Allanson JE, Hunter AGW, Mettler GS, Jimenez C (1992) Renal tubular dysgenesis – a not uncommon autosomal recessive syndrome. Am J Med Genet 43: 811–814

Ambrosioni E, Borghi C, Magnani B (1995) The effect of the angiotensin-converting enzyme inhibitor zofenopril on mortality and morbidity after anterior myocardial infarction. N Engl J Med 332: 80–85

Arima S, Ren Y, Juncos LA, Carretero OA, Ito S (1994) Glomerular prostaglandins modulate vascular reactivity of the downstream efferent arterioles. Kidney Int 45: 650–658

Bakhle YS (1968) Conversion of angiotensin I to angiotensin II by cell-free extracts of dog lung. Nature 220: 919–921
Bakris GL (1993) Angiotensin-converting enzyme inhibitors and progression of diabetic nephropathy. Ann Intern Med 118: 643–644
Barley J, Blackwood A, Carter ND et al. (1994) Angiotensin converting enzyme insertion/deletion polymorphism: association with ethnic origin. J Hypertens 12: 955–957
Barr M, Cohen MM (1991) ACE inhibitor fetopathy and hypocalvaria: the kidney-skull connection. Teratology 44: 485–495
Bavoux F (1992) Toxicite foetal des anti-inflammatoires non steroidiens. Presse Med 21: 1909–1912
Becker GJ, Whitworth JA, Ihle BU, Shahinfar S, Kincaid-Smith PS (1994) Prevention of progression in non-diabetic chronic renal failure. Kidney Int 45 [Suppl 45]: S167–S170
Bernstein J (1988) Renal tubular dysgenesis. Pediatr Pathol 8: 453–456
Bernstein J, Barajas L (1994) Renal tubular dysgenesis: evidence of abnormality in the renin-angiotensin system. J Am Soc Nephrol 5: 224–227
Bevier WC, Jovanovic-Peterson L, Burns A, Peterson CM (1994) Blood pressure predicts insulin requirement and exogenous insulin is associated with increased blood pressure in women with gestational diabetes mellitus. Am J Perinatol 11: 369–373
Binder ND, Anderson DF (1992) Plasma renin activity responses to graded decreases in renal perfusion pressure in fetal and newborn lambs. Am J Physiol 262: R524–R529
Binder ND, Faber JJ (1992) Effects of captopril on blood pressure, placental blood flow and uterine oxygen consumption in pregnant rabbits. J Pharmacol Exp Ther 260: 294–299
Binder N, Faber J, Anderson D (1989) The effect of captopril on blood pressure, uterine blood flow and oxygen consumption in the pregnant rabbit. Clin Res 37: 6A
Blythe WB (1983) Captopril and renal autoregulation. N Engl J Med 308: 390–391
Bonithon-Koop C, Ducimetiere P, Touboul PJ et al. (1994) Plasma angiotensin-converting enzyme activity and carotid wall thickening. Circulation 89: 952–954
Bouaziz H, Joulin Y, Safar M, Benetos A (1994) Effects of bradykinin B-2 receptor antagonism on the hypotensive effects of ACE inhibition. Br J Pharmacol 113: 717–722
Boura ALA, Walters WAW, Read MA, Leitch IM (1994) Autacoids and control of human placental blood flow. Clin Exp Pharmacol Physiol 21: 737–748
Boutroy MJ, Vent P, Hunault de Ligny B, Milton A (1984) Captopril administration in pregnancy impairs fetal angiotensin converting enzyme activity and neonatal adaptation. Lancet 2: 935–936
Brent RL, Beckman DA (1991) Angiotensin-converting enzyme inhibitors, an embryopathic class of drugs with unique properties: information for clinical teratology counselors. Teratology 43: 543–546
Broughton Pipkin F, Wallace CP (1986) The effect of enalapril (MK-421), an angiotensin-converting enzyme inhibitor, on the conscious pregnant ewe and her foetus. Br J Pharmacol 87: 533–542
Broughton Pipkin F, Rubin PC (1994) Pre-eclampsia – the 'disease of theories.' Br Med Bull 50: 381–396
Broughton Pipkin F, Turner SR, Symonds EM (1980) Possible risk with captopril during pregnancy: some animal data. Lancet 2: 1256
Broughton Pipkin F, Symonds EM, Turner SR (1982) The effect of captopril (SQ 14,225) upon mother and fetus in the chronically cannulated ewe and in the pregnant rabbit. J Physiol 323: 415–422
Broughton Pipkin F, Baker PN, Symonds EM (1989) Angiotensin-converting enzyme inhibitors in pregnancy. Lancet 2: 96–97

Buderus S, Thomas B, Fahnenstich H, Kowalewski S (1993) Renal failure in two preterm infants: toxic effect of prenatal maternal indomethacin treatment. Br J Obstet Gynaecol 100: 97–98
Bugge JF, Stokke ES (1994) Angiotensin II and renal prostaglandin release in the dog – interactions in controlling renal blood flow and glomerular filtration rate. Acta Physiol Scand 150: 431–440
Burson JM, Aquilera G, Gross KW, Sigmund CD (1994) Differential expression of angiotensin receptor 1A and 1B in mouse. Am J Physiol 267: E260–E267
Busse R, Fleming I, Hecker M (1993) Endothelium-derived bradykinin: implications for angiotensin-converting enzyme inhibitor therapy. J Cardiovasc Pharamacol 22 [Suppl 5]: S31–S36
Canzanello VJ, Textor SC (1994) Noninvasive diagnosis of renovascular disease. Mayo Clin Proc 69: 1172–1181
Caraman PL, Miton A, Hurault de Ligny B et al. (1984) Grossesses sous captopril. Therapie 39: 59–62
Chagnac A, Korzets A, Zevin D et al. (1994) Effect of enalapril on the microvascular albumin leakage in patients with diabetic microangiopathy and normal or mildly elevated blood pressure. Clin Nephrol 41: 144–148
Clauser E, Bihoreau C, Monnot C et al. (1992) L'angiotensine II est-elle un facteur de croissance? Diabete Metab 18: 129–136
Cooke HM, Debesse A (1994) Angiotensin-converting enzyme inhibitor-induced renal dysfunction: recommendations for prevention. Int J Clin Pharmacol Ther 32: 65–70
Costerousse O, Allegrini J, Huang HM, Bounhik J, Alhenc-Gelas F (1994) Regulation of ACE gene expression and plasma levels during rat postnatal development. Am J Physiol 267: E745–E753
Cozens DD, Barton SJ, Clark R et al. (1987) Reproductive toxicity studies of rentiapril. Arzneimittelforschung 37: 164–169
Cunniff C, Jones KL, Phillipson K, Short S, Wujek J (1990) Oligohydramnios and renal tubular malformation associated with maternal enalapril use. Am J Obstet Gynecol 162: 187–189
Cushman DW, Ondetti MA (1991) History of the design of captopril and related inhibitors of angiotensin-converting enzyme. Hypertension 17: 589–592
Davidson R, Wilcox CS (1991) Diagnostic usefulness of renal scanning after angiotensin converting enzyme inhibitors. Hypertension 18: 299–303
Davis LE, Hohimer AR, Woods LL (1994) Renal function during chronic anemia in the ovine fetus. Am J Physiol 266: R1759–R1764
Defronzo RA, Ferrannini E (1991) Insulin resistance. A multifaceted syndrome responsible for NIDDM, obesity, hypertension, dyslipidemia, and atherosclerotic cardiovascular disease. Diabetes Care 14: 173–194
Degasparo M, Levens NR (1994) Pharmacology of angiotensin II receptors in the kidney. Kidney Int 46: 1486–1491
de Moura RS, Vale NS (1986) Effect of captopril on bradykinin inactivation by human foetal placental circulation. Br J Clin Pharmacol 21: 143–148
Derkx FHM, Schalekamp MADH (1994) Renal artery stenosis and hypertension. Lancet 344: 237–239
Devoy MAB, Tomson CRV, Edmunds ME, Feehally J, Walls J (1992) Deterioration in renal function associated with angiotensin-converting enzyme inhibitor therapy is not always reversible. J Intern Med 232: 493–498
Dostal LA, Kim SN, Schardein JL, Anderson JA (1991) Fertility and perinatal/postnatal studies in rats with the angiotensin-converting enzyme inhibitor, quinapril. Fundam Appl Toxicol 17: 684–695
Douglas JG, Hopfer U (1994) Novel aspect of angiotensin receptors and signal transduction in the kidney. Annu Rev Physiol 56: 649–669
Duminy PC, Burger PduT (1981) Fetal abnormality associated with the use of captopril during pregnancy. S Afr Med J 60: 805

Ehlers MW, Riordan JF (1989) Angiotensin-converting enzyme: new concepts concerning its biological role. Biochemistry 28: 5311–5316

Endo M, Yamada Y, Kohno M et al. (1992) Metabolic fate of the new angiotensin-converting enzyme inhibitor imidapril in animals 4. Placental transfer and secretion into milk in rats. Arzneimittelforschung 42: 483–489

Erdos EG, Skidgel RA (1987) The angiotensin I-converting enzyme. Lab Invest 56: 345–348

Ferreira SH (1965) A bradykinin-potentiating factor (BPF) present in the venom of Bothrops jararaca. Br J Pharmacol 24: 163–169

Ferris TF, Weir EK (1982) Effect of captopril on uterine blood flow and prostaglandin E synthesis in the pregnant rabbit. J Clin Invest 71: 809–815

Fletcher AE, Palmer AJ, Bulpitt CJ (1994) Cough with angiotensin converting enzyme inhibitors: how much of a problem? J Hypertens 12 [Suppl 2]: S43–S47

Fluck RJ, Raine AEG (1994) ACE inhibitors in non-diabetic renal disease. Br Heart J 72 [Suppl 3]: S46–S51

Friberg P, Sundelin B, Bohman SO et al. (1994) Renin-angiotensin system in neonatal rats: induction of a renal abnormality in response to ACE inhibition or angiotensin-II antagonism. Kidney Int 45: 485–492

Fujii T, Nakatsuka T (1985) Enalapril (MK-421) oral teratogenicity in the rat. Yakuri to Chiryo 13: 529–548

Fujii T, Nakatsuka T, Hanada S, Komatsu T, Okiyama H (1985) Enalapril (MK-421) oral teratogenicity study in the rat. Yakuri to Chiryo 13: 519–528

Ganong WF (1994) Origin of the angiotensin II secreted by cells. Proc Soc Exp Biol Med 205: 213–219

Gansevoort RT, de Zeeuw D, de Jong PE (1994) The antiproteinuric effect of ACE inhibition mediated by interference in the renin-angiotensin system. Kidney Int 45: 861–867

Garay R, Senn N, Ollivier JP (1994) Erythrocyte ion transport as indicator of sensitivity to antihypertensive drugs. Am J Med Sci 307 [Suppl 1]: S120–S125

Gasc JM, Shanmugam S, Sibony M, Corvol P (1994) Tissue-specific expression of type 1 angiotensin II receptor subtypes – an in situ hybridization study. Hypertension 24: 531–537

Gavras H (1988) The place of angiotensin-converting enzyme inhibitors in the treatment of cardiovascular diseases. N Engl J Med 319: 1541–1543

Gerber JG, Franca G, Byyny RL, Loverde M, Nies AS (1993) The hypotensive action of captopril and enalapril is not prostacyclin dependent. Clin Pharmacol Ther 54: 523–532

Gomez HJ, Cirillo VJ, Moncloa F (1987) The clinical pharmacology of lisinopril. J Cardiovasc Pharmacol 9 [Suppl 3]: S27–S34

Gomez R, Lunch K, Sturgill B et al. (1989) Distribution of renin mRNA and its protein in the developing kidney. Am J Physiol 257: F850–F858

Gomez RA, Tufro-McReddie A, Everett AD, Pentz ES (1993) Ontogeny of renin and AT1 receptor in the rat. Pediatr Nephrol 7: 635–638

Grady EF, Sechi LA, Griffin CA, Schambelan M, Kalinyak JE (1991) Expression of AT2 receptors in the developing rat fetus. J Clin Invest 88: 921–933

Graves SW, Moore TJ, Seely EW (1992) Increased platelet angiotensin-II receptor number in pregnancy-induced hypertension. Hypertension 20: 627–632

Graziano MJ, Dethloff LA, Griffin HE, Pegg DG (1993) Effects of the angiotensin-converting enzyme inhibitor quinapril on renal function in rats. Arch Intern Pharmacodyn Ther 324: 87–104

Grone HJ, Simon M, Fuchs E (1992) Autoradiographic characterization of angiotensin receptor subtypes in fetal and adult human kidney. Am J Physiol 262: F326–F331

Guignard JP (1982) Renal function in the newborn infant. Pediatr Clin North Am 29: 777–790

Guignard JP, Burgener F, Calaine A (1981) Persistent anuria in a neonate: a side effect of captopril? Intern J Pediatr Nephrol 2: 133

Guthrie R (1993) Fosinopril: an overview. Am J Cardiol 72: H22–H24
Hagemann A, Nielsen AH, Poulsen K (1994) The uteroplacental renin-angiotensin system: a review. Exp Clin Endocrinol 102: 252–261
Hall JE, Guyton AC, Jackson TE et al (1977) Control of glomerular filtration rate by renin-angiotensin system. Am J Physiol 233: F366–F372
Hanssens M, Keirse MJNC, Vankelecom F, Van Assch FA (1991) Fetal and neonatal effects of treatment with angiotensin-converting enzyme inhibitors in pregnancy. Obstet Gynecol 78: 128–135
Harewood WJ, Phippard AF, Duggin GG, Horvath JS, Tiller DJ (1994) Fetotoxicity of angiotensin-converting enzyme inhibition in primate pregnancy: a prospective, placebo-controlled study in baboons (Papio hamadryas). Am J Obstet Gynecol 171: 633–642
Hill K, Lumbers E (1988) Renal function in adult and fetal sheep. J Dev Physiol 10: 149–159
Hinchliffe SA, Lynch MRJ, Sargent PH, Howard CV, van Velzen D (1992) The effect of intrauterine growth retardation on the development of renal nephrons. Br J Obstet Gynaecol 99: 296–301
Hricik DE, Browing PJ, Kopelman R et al. (1983) Captopril induced functional renal insufficiency in patients with bilateral renal artery stenosis or renal artery stenosis in a solitary kidney. N Engl J Med 308: 373–376
Ice KS, Geary KM, Gomez RA et al. (1988) Cell and molecular studies of renin secretion. Clin Exp Hypertens Theor Pract A10: 1169–1187
Iwai N, Yamano Y, Chaki S et al (1991) Rat angiotensin II receptor: cDNA sequence and regulation of the gene expression. Biochem Biophys Res Commun 177: 299–304
Iwai N, Ohmichi N, Nakamura Y, Kinoshita M (1994) DD genotype of the angiotensin-converting enzyme gene is a risk factor for left ventricular hypertrophy. Circulation 90: 2622–2628
Jackson B, Mendelsohn FAO, Johnston CI (1991) Angiotensin-converting enzyme inhibition: prospects for the future. J Cardiovasc Pharmacol 18: S4–S8
Jelinek J, Hackenthal R, Hilgenfeldt U, Schaechtelin G, Hackenthal E (1986) The renin-angitensin system in the perinatal period in rats. J Dev Physiol 8: 33–41
Johnston CI, Burrell LM, Perich R, Jandeleit K, Jackson B (1992) The tissue renin-angiotensin system and its functional role. Clin Exp Pharmacol Physiol [Suppl 19]: 1–5
Johnston CI, Fabris B, Yoshida K (1993) The cardiac renin-angiotensin system in heart failure. Am Heart J 126: 756–760
Kakuchi J, Ichiki T, Kiyama S et al. (1995) Developmental expression of renal angiotensin II receptor genes in the mouse. Kidney Int 47: 140–147
Kaler SG, Patrinos ME, Lambert GH et al. (1987) Hypertrichosis and congenital anomalies associated with maternal use of minoxidil. Pediatrics 79: 434–436
Kang PM, Landau AJ, Eberhardt RT, Frishman WH (1994) Angiotensin II receptor antagonists: a new approach to blockade of the renin-angiotensin system. Am Heart J 127: 1388–1401
Keith IM, Will JA, Weir EK (1982) Captopril: association with fetal death and pulmonary vascular changes in the rabbit. Proc Soc Exp Biol Med 170: 378–383
Kellow NH (1994) The renin-angiotensin system and angiotensin converting enzyme (ACE) inhibitors. Aaesthesia 49: 613–622
Kingdom JCP, McQueen J, Connell JMC, Whittle MJ (1993) Fetal angiotensin-II levels and vascular (type-I) angiotensin receptors in pregnancies complicated by intrauterine growth retardation. Br J Obstet Gynaecol 100: 476–482
Knock GA, Sullivan MHF, McCarthy A et al. (1994) Angiotensin II (AT(1)) vascular binding sites in human placentae from normal-term, preeclamptic and growth retarded pregnancies. J Pharmacol Exp Ther 271: 1007–1015
Knott PD, Thorp SS, Lamont CAR (1989) Congenital renal dysgenesis possibly due to captopril. Lancet 1: 451

Kon V, Fogo A, Ichikawa I, Hellings SE, Bills T (1993) Bradykinin causes selective efferent arteriolar dilation during angiotensin-I converting enzyme inhibition. Kidney Int 44: 545–550

Kreft-Jais C, Plouin C, Tchobroutsky C, Boutroy MJ (1988) Angiotensin-converting enzyme inhibitors during pregnancy: a survey of 22 patients given captopril and nine given enalapril. Br J Obstet Gynaecol 95: 420–422

Landing BH, Ang SM, Herta N, Larson EF, Turner M (1994) Labeled lectin studies of renal tubular dysgenesis and renal tubular atrophy of postnatal renal ischemia and end-stage kidney disease. Pediatr Pathol 14: 87–99

Lebovitz HE, Wiegmann TB, Cnaan A et al. (1994) Renal protective effects of enalapril in hypertensive NIDDM – role of baseline albuminuria. Kidney Int 45 [Suppl 45]: S150–S155

Lee EJD (1994) Population genetics of the angiotensin-converting enzyme in Chinese. Br J Clin Pharmacol 37: 212–214

Lees KR, Squire IB, Reid JL (1992) The clinical pharmacology of ACE inhibitors: evidence for clinically relevant differences. Clin Exp Pharmacol Physiol [Suppl 19]: 49–53

Leversha AM, Wilson NJ, Clarkson PM et al. (1994) Efficacy and dosage of enalapril in congenital and acquired heart disease. Arch Dis Child 70: 35–39

Li J, Hu HY, Zhao YN (1992) Serum angiotensin-converting enzyme activity in pregnancy-induced hypertension. Gynecol Obstet Invest 33: 138–141

Lindheimer MD, Katz AI (1985) Hypertension in pregnancy (current concepts). N Engl J Med 313: 675–680

Linz W, Wiemer G, Scholkens BA (1993) Contribution of bradykinin to the cardiovascular effects of ramipril. J Cardiovasc Pharmacol 22 [Suppl 9]: S1–S8

Llorens-Cortes C, Greenberg B, Huang HM, Corvol P (1994) Tissular expression and regulation of type 1 angiotensin II receptor subtypes by quantitative reverse transcriptase polymerase chain reaction analysis. Hypertension 24: 538–548

Lumbers ER, Kingsford NM, Menzies RI, Stevens AD (1992) Acute effects of captopril, an angiotensin-converting enzyme inhibitor, on the pregnant ewe and fetus. Am J Physiol 262: R754–R760

Lumbers ER, Burrell JH, Menzies RI, Stevens AD (1993) The effects of a converting enzyme inhibitor (captopril) and angiotensin-II on fetal renal function. Br J Pharmacol 110: 821–827

Magness R, Osei-Boaten K, Mitchell M, Rosenfeld C (1985) In vitro prostacyclin production by ovine uterine and systemic arteries: effects of angiotensin II. J Clin Invest 76: 2206–2212

Mandal AK, Markert RJ, Saklayen MG, Mankus RA, Yokokawa K (1994) Diuretics potentiate angiotensin converting enzyme inhibitor-induced acute renal failure. Clin Nephrol 42: 170–174

Marre M, Bernadet P, Gallois Y et al. (1994) Relationships between angiotensin I converting enzyme gene polymorphism, plasma levels, and diabetic retinal and renal complications. Diabetes 43: 384–388

Martin RA, Jones KL, Mendoza A, Barr M, Benirschke K (1992) Effect of ACE inhibition on the fetal kidney: decreased renal blood flow. Teratology 46: 317–321

Materson BJ, Preston RA (1994) Angiotensin-converting enzyme inhibitors in hypertension. Arch Intern Med 154: 513–523

McGrath BP, Matthews PG, Johnston CI (1983) Use of captopril in the diagnosis of renal hypertension. Aust N Z J Med 11: 359–363

Mehta N, Modi N (1989) ACE inhibitors in pregnancy. Lancet 2: 96

Michel B, Stephan D, Grima M, Barthelmebs M, Imbs JL (1993) Effects of one-hour and one-week treatment with ramipril on plasma and renal brush border angiotensin-converting enzyme in the rat. Eur J Pharmacol 242: 237–243

Millar JA, Wilson PD, Morrison M (1983) Management of severe hypertension in pregnancy by a combined drug regimen including captopril: case report. N Z Med J 96: 796–798

Minsker DH, Bagdon WJ, MacDonald JS, Robertson RT, Bokelman DL (1990) Maternotoxicity and fetotoxicity of an angiotensin-converting enzyme inhibitor, enalapril, in rabbits. Fundam Appl Toxicol 14: 461–470
Mitchell TH, Stafford M, Jenkins DM (1993) Case report-1 – Enalapril in a pregnant, diabetic woman. J Drug Dev 6: 21–22
Miyata S, Haneda T (1994) Hypertrophic growth of cultured neonatal rat heart cells mediated by type 1 angiotensin II receptor. Am J Physiol 266: H2443–H2451
Moe OW, Ujiie K, Star RA et al (1993) Renin expression in renal proximal tubule. J Clin Invest 91: 774–779
Mogensen CE (1994) Renoprotective role of ACE inhibitors in diabetic nephropathy. Br Heart J 72 [Suppl 3]: S38–S45
Molitch ME (1994) ACE inhibitors and diabetic nephropathy. Diabetes Care 17: 756–760
Mulec H, Johnsen SA, Bjorck S (1994) Long-term enalapril treatment in diabetic nephropathy. Kidney Int 45 [Suppl 45]: S141–S144
Pellacani A, Brunner HR, Nussberger J (1994) Plasma kinins increase after angiotensin-converting enzyme inhibition in human subjects. Clin Sci (Colch) 87: 567–574
Perlman JM, Volpe JJ (1989) Neurologic complications of captopril treatment of neonatal hypertension. Pediatrics 83: 47–52
Pinkney JH, Yudkin JS (1994) Antihypertensive drugs: issues beyond blood pressure control. Prog Cardiovasc Dis 36: 397–415
Piper JM, Ray WA, Rosa FW (1992) Pregnancy outcome following exposure to angiotensin-converting enzyme inhibitors. Obstet Gynecol 80: 429–432
Plouin PF, Tchobroutsky C (1985) Inhibition de l'enzyme de conversion de l'angiotensine au cours de la grossesse humaine: quinze cas. Presse Med 14: 2175–2178
Pryde PG, Sedman AB, Nugent CE, Barr M (1993) Angiotensin-converting enzyme inhibitor fetopathy. J Am Soc Nephrol 3: 1575–1582
Ravid M, Savin H, Jutrin I et al. (1993) Long-term stabilizing effect of angiotensin-converting enzyme inhibition on plasma creatinine and on proteinuria in normotensive type-II diabetic patients. Ann Intern Med 118: 577–581
Remuzzi A, Imberti O, Puntorieri S et al. (1994) Dissociation between antiproteinuric and antihypertensive effect of angiotensin converting enzyme inhibitors in rats. Am J Physiol 267: F1034–F1044
Restaino I, Kaplan BS, Kaplan P et al. (1991) Renal dysgenesis in a monozygotic twin: association with in utero exposure to indomethacin. Am J Med Genet 39: 252–257
Rigat B, Hubert C, Alhenc-Gelas F et al. (1990) An insertion deletion polymorphism in angiotensin I converting enzyme gene accounting for half the variance of serum enzyme levels. J Clin Invest 86: 1343–1346
Robertson JIS (1994) Role of ACE inhibitors in uncomplicated essential hypertension. Br Heart J 72 [Suppl 3]: S15–S23
Robertson R, Minsker DH, Bokelman DL (1986) MK-421 (enalapril maleate): late gestation and lactation study in rats. Jpn Pharmacol Ther 14: 43–55
Robillard JE, Schutte BC, Page WV et al. (1994) Ontogenic changes and regulation of renal angiotensin II type 1 receptor gene expression during fetal and newborn life. Pediatr Res 36: 755–762
Rodicio JL, Ruilope LM (1993) Angiotensin-converting enzyme inhibitors in the treatment of mild arterial hypertension. Clin Exp Hypertens 15: 1277–1289
Rosa FW, Bosco LA, Fossum-Graham C et al. (1989) Neonatal anuria with maternal angiotensin-converting enzyme inhibition. Obstet Gynecol 74: 371–374
Rothberg AD, Lorenz R (1984) Can captopril cause fetal and neonatal renal failure? Pediatr Pharmacol 4: 189–192
Rudolph AM, Heyman MA, Teramo KAW, Barrett CT, Raiha NCR (1971) Study on the circulation of the previable human fetus. Pediatr Res 5: 452–465
Rutledge DR, Browe CS, Ross EA (1994) Frequencies of the angiotensinogen gene and angiotensin I converting enzyme (ACE) gene polymorphisms in African Amer-

icans. Biochem Mol Biol Int 34: 1271–1275
Ruzicka M, Leenen FHH (1995) Relevance of blockade of cardiac and circulatory angiotensin-converting enzyme for the prevention of volume overload-induced cardiac hypertrophy. Circulation 91: 16–19
Sassano P, Chatellier G, Billaud E et al. (1987) Treatment of mild to moderate hypertension with or without the converting enzyme inhibitor enalapril: results of a six-month double-blind trial. Am J Med 83: 227–235
Schlueter W, Keilani T, Batlle DC (1994) Tissue renin angiotensin systems: theoretical implications for the development of hyperkalemia using angiotensin-converting enzyme inhibitors. Am J Med Sci 307 [Suppl 1]: S81–S86
Schubiger G, Flury G, Nussberger J (1988) Enalapril for pregnancy-induced hypertension: acute renal failure in the neonate. Ann Intern Med 108: 215–216
Schumacher GH (1985) Factors influencing craniofacial growth. In: Dixon AD, Sarnat BG (eds) Normal and abnormal bone growth. Basic and clinical research. Liss, New York, pp 3–22
Schunkert H, Hense HW, Holmer SR et al. (1994) Association between a deletion polymorphism of the angiotensin-converting-enzyme gene and left ventricular hypertrophy. N Engl J Med 330: 1634–1638
Sedman AB, Kershaw DB, Bunchman TE (1995) Recognition and management of angiotensin converting enzyme inhibitor fetopathy. Pediatr Nephrol (to be published)
Shanmugam S, Corvol P, Gasc JM (1994a) Ontogeny of the two angiotensin II type 1 receptor subtypes in rats. Am J Physiol 267: E828–E836
Shanmugam S, Monnot C, Corvol P, Gasc JM (1994b) Distribution of type 1 angiotensin II receptor subtype messenger RNAs in the rat fetus. Hypertension 23: 137–141
Simeoni U, Messer J, Weisburd P, Haddad J, Willard D (1989) Neonatal renal dysfunction and intrauterine exposure to prostaglandin synthesis inhibitors. Eur J Pediatr 148: 371–373
Swinford AE, Bernstein J, Toriello HV, Higgins JV (1989) Renal tubular dysgenesis: delayed onset of oligohydramnios. Am J Med Genet 32: 127–132
Tack ED, Perlman JM (1988) Renal failure in sick hypertensive premature infants receiving captopril therapy. Pediatrics 112: 805–810
Thorpe-Beeston JG, Armar NA, Dancy M et al. (1993) Pregnancy and ACE inhibitors. Br J Obstet Gynaecol 100: 692–693
Thurston H (1992) Angiotensin-converting enzyme inhibition as 1st-line treatment for hypertension. Clin Exp Pharmacol Physiol [Suppl 19]: 67–71
Todd PA, Heel RC (1986) Enalapril: a review of its pharmacodynamic properties, and therapeutic use in hypertension and congestive heart failure. Drugs 31: 198–248
Toto RD (1994) Renal insufficiency due to angiotensin-converting enzyme inhibitors. Miner Electrolyte Metab 20: 193–200
Tufro-McReddie A, Harrison JK, Everett AD, Gomez RA (1993) Ontogeny of type 1 angiotensin II receptor gene expression in the rat. J Clin Invest 91: 530–537
Tufro-McReddie A, Johns DW, Geary KM et al. (1994) Angiotensin II type 1 receptor: role in renal growth and gene expression during normal development. Am J Physiol 266: F911–F918
Valdes G, Marinovic D, Falcon C, Chuaqui R, Duarte I (1992) Placental alterations, intrauterine growth retardation and teratogenicity associated with enalapril use in pregnant rats. Biol Neonate 61: 124–130
Van den Anker JN, Hop WCJ, de Groot R et al. (1994) Effects of prenatal exposure to betamethasone and indomethacin on the glomerular filtration rate in the preterm infant. Pediatr Res 36: 578–581
van der Heijden BJ, Carlus C, Narcy F et al. (1994) Persistent anuria, neonatal death, and renal microcystic lesions after prenatal exposure to indomethacin. Am J Obstet Gynecol 171: 617–623

Vanhoutte PM, Boulanger CM, Illiano SC et al. (1993) Endothelium-dependent effects of converting-enzyme inhibitors. J Cardiovasc Pharmacol 22 [Suppl 5]: S10–S16

Viberti G, Mogensen CE, Groop LC et al. (1994) Effect of captopril on progression to clinical proteinuria in patients with insulin-dependent diabetes-mellitus and microalbuminuria. JAMA 271: 275–279

Vidt DG, Bravo EL, Fouad FM (1982) Captopril. N Engl J Med 306: 214–219

Voland JR, Hawkins EP, Wells TR, et al. (1985) Congenital hypernephronic nephromegaly with tubular dysgenesis: a distinctive inherited renal anomaly Pediatr Pathol 4: 341–345

Voyer LE, Drut R, Mendez JH (1994) Fetal renal maldevelopment with oligohydramnios following maternal use of piroxicam. Pediatr Nephrol 8: 592–594

Walker MPR, Moore TR, Brace RA (1994) Indomethacin and arginine vasopressin interaction in the fetal kidney: a mechanism of oliguria. Am J Obstet Gynecol 171: 1234–1241

Weir RJ, Fraser R, Lever AF et al. (1973) Plasma renin, renin substrate, angiotensin II, and aldosterone in hypertensive disease of pregnancy. Lancet 1: 291–294

Wells TG, Bunchman TE, Kearns GL (1990) Treatment of neonatal hypertension with enalapril. J Pediatr 117: 664–667

Williams GH (1988) Converting enzyme inhibitors in the treatment of hypertension. N Engl J Med 319: 1517–1524

Wilson PAP (1994) When to start an ACE inhibitor and in whom. Cardiovasc Drugs Ther 8: 111–114

Wood EG, Bunchman TE, Lynch RE (1991) Captopril-induced reversible acute renal failure in an infant with coarctation of the aorta. Pediatrics 88: 816–818

Yagami H, Kurauchi O, Murata Y et al. (1994) Expression of angiotensin-converting enzyme in the human placenta and its physiologic role in the fetal circulation. Obstet Gynecol 84: 453–457

Yoshida H, Kakuchi J, Guo DF et al. (1992) Analysis of the evolution of the angiotensin II type 1 receptor gene in mammals (mouse, rat, bovine, human). Biochem Biophys Res Commun 186: 1042–1049

Yoshida H, Tamaki T, Aki Y et al. (1994) Effects of angiotensin II on isolated rabbit afferent arterioles. Jpn J Pharmacol 66: 457–464

Yosipiv IV, Dipp S, El-Dahr SS (1994) Ontogeny of somatic angiotensin-converting enzyme. Hypertension 23: 369–374

CHAPTER 28

Anesthetics

M. Fujinaga

A. Introduction

In general, anesthetics are drugs that are used to provide anesthesia for surgical procedures. Anesthesia is divided into two major classes, general and regional anesthesia. General anesthesia usually requires the following three elements: amnesia, analgesia, and muscle relaxation. Practically speaking, various kinds of drugs are administered to patients during the course of general anesthesia to meet all these requirements (Table 1). For example, in a typical case, anesthesia is induced by a bolus injection of intravenous anesthetic, and an endotracheal tube is placed into the patient's trachea after administration of a muscle relaxant. Anesthesia is then maintained with either inhalational anesthetics administered as gases through the lungs or with various drugs administered intravenously. Regional anesthesia, including infiltration and peripheral and central nerve blocks (Table 2), is produced by a class of drugs known as local anesthetics, although sometimes these are supplemented with opioids when spinal (intrathecal) or epidural anesthesia is involved. In addition, patients who have regional anesthesia often also receive other kinds of drugs, such as intravenous sedatives and opioids. Furthermore, regional anesthesia is sometime used with general anesthesia to reduce drug requirements for the latter.

In addition to anesthesia during surgery, pre- and postoperative care is an important part of the clinical practice of anesthesia. It often involves administering additional medication, including sedatives, analgesics, antiemetics, and drugs for prophylaxis of pulmonary aspiration (Table 1). Furthermore, many other kinds of drugs are used during anesthesia to control physiologic functions of patients, including vasoactive agents, vasodilators, diuretics and various kinds of intravenous fluids for controlling the cardiovascular system. Thus not only those drugs actually called anesthetics but also many other kinds of drugs are administered during anesthesia. This makes the effects of individual drugs used during anesthesia difficult to examine, since no single drug is used alone. Furthermore, it is very difficult to separate the effects of anesthesia from surgery or from the original condition that necessitated it. Despite these difficulties, this chapter will review what is known about developmental toxicity of anesthetics. First of all, anesthesia for in vitro fertilization procedures is considered, focusing on the effects of anesthetics on

Table 1. Inclusive list of drugs that are now commonly used for "general anesthesia" in the United States

Drug type	Drug
Premedication	
Sedatives	
Barbiturates	Pentobarbital (D), secobarbital (D)
Nonbarbiturate sedatives	Droperidol (C), hydroxyzine (C), promethazine (C)
Benzopiazepines	Diazepam (D), lorazepam (D), midazolam (D)
Opioids	Meperidine (B), morphine (C)
α_2-Adrenergic agonists	Clonidine (C)
Anticholinergics	Atropine (C), glycopyrrolate (B), scopolamine (C)
Prophylaxis for pulmonary aspiration	Cimetidine (B), metoclopramide (C), nonparticulate Antacids, ranitidine (B)
Induction of anesthesia	
Intravenous anesthetics	
Barbiturates	Methohexital (B), thiamylal (?), thiopental (?)
Benzodiapepines	Midazolam (D)
Steroids	Propofol (B)
Benzylimidazoles	Etomidate (C)
Dissociative anesthetics	Ketamine (?)
Opioids[a]	Alfentanil (C), fentanyl (C), sufentanil (C)
Muscle relaxants	
Depolarizing agents	Succinylcholine (C)
Nondepolarizing agents	Atracurium (C), doxacurium (C), metocurine (C) mivacurium (C), pancuronium (C), pipecuronium (C), rocuronium (B), d-tubocurarine (C), vecuronium (C)
Maintenance of anesthesia	
Inhalational anesthetics	
Volatile anesthetics	Desflurane (B), enflurane (?), halothane (C), isoflurane (?), sevoflurane (?)
Gaseous anesthetics	Nitrous oxide (?)
Intravenous anesthetics	
Barbiturates	Methohexital (B), thiamylal (?), thiopental(?)
Benzodiapepines	Diazepam (D), midazolam (D)
Benzylimidazoles	Etomidate (C)
Neuroleptics	Droperidol (C)
Steroids	Propofol (B)
Opioids	Alfentanil (C), fentanyl (C), mepperidine (B), morphine (C), sufentanil (C)
Muscle relaxants	
Nondepolarizing agents	Atracurium (C), doxacurium (C), metocurine (C), mivacurium (C), pancuronium (C), pipecuronium (C), rocuronium (B), d-tubocurarine (C), vecuronium (C)
Reversal of anesthesia	
Agents to revere neuromuscular blockage	
Anticholinesterases	Edrophonium (C), neostigmine (C), pyridostigmine (C)
Anticholinergic drugs	Atropine (C), glycopyrrolate (B)
Opioid antagonists	Naloxone (B)
Benzodiapemine antagonists	Flumazenil (C)

Table 1. (*Contd.*)

Drug type	Drug
Postoperative medication	
Sedatives	Midazolam (D), haloperidol (C)
Analgesics	
Opioids	Fentanyl (C), butorphanol (C), meperidine (B), morphine (C)
Nonopioids	Acetaminophen (C), ketorolac (C)
Antiemetics	Droperidol (C), haloperidol (C), metoclopramide (C), ondansetron (B), prochlorperazine (C), promethazine (C)

Letters in parentheses indicate the Food and Drug Administration's Pregnancy Categories (U.S.A): A, controlled studies show no risk; B, no evidence of risk in humans; C, risk cannot be ruled out; D, positive evidence of risk; X, contraindicated in pregnancy.
[a]High-dose opioids are sometimes used for induction of patients for cardiac surgery.

Table 2. Classificaiton of regional anesthesia and inclusive list of drugs that are now commonly used in the United States

Infiltration	
Local infiltration	
Local anesthetics	Bupivacaine (C), lidocaine (B), procaine (?)
Intravenous regional	
Local anesthetics	Lidocaine (B), prilocaine (B)
Peripheral nerve blockade	
Nerve block	
Local anesthetics	Bupivacaine (C), chloroprocaine (C), etidocaine (B), procaine (?), lidocaine (B), tetracaine (C)
Plexus block	
Local anesthetics	Bupivacaine (C), chloroprocaine (C), etidocaine (B), procaine (?), lidocaine (B), tetracaine (C)
Central nerve blockade	
Epidural anesthesia	
Local anesthetics	Bupivacaine (C), chloroprocaine (C), etidocaine (B), procaine (?), lidocaine (B), tetracaine (C)
Opioids	Dilaudid (C), fentanyl (C), morphine (C)
Spinal anesthesia	
Local anesthetics	Bupivacaine (C), lidocaine (B), procaine (?), tetracaine (C)
Opioids[a]	Morphine (C), sufentanil (C)

The choice of local anesthetics depends on the duration of desired anesthesia, e.g., chloroprocaine for short duration, lidocaine or mepivacaine for intermediate duration, and bupivacaine or etidocaine for long duration. Letters in parentheses indicate the Food and Drug Administrations's Pregnancy Categories (U.S.A): A, controlled studies show no risk; B, no evidence of risk in humans; C, risk cannot be ruled out; D, positive evidence of risk; X, contraindicated in pregnancy.
[a]Used for intrathecal injection.

fertilization and preimplantational development. This is followed by a discussion of anesthesia for nonobstetrical surgery during pregnancy, focusing on the effects of anesthetics on postimplantational development including organogenesis. Two other areas, anesthesia for fetal therapy/surgery, a new area of medicine that is usually performed during the late gestational period, and anesthesia for delivery, focusing on the effects of anesthesia for delivery on postnatal development, are then examined. Finally, we look at waste inhalational anesthetics in the work place, discussing a potential hazard that has long been debated.

B. In Vitro Fertilization Procedures

In vitro fertilization in humans began in the late 1970s. For many years, laparoscopy was the standard method for oocyte retrieval and is still used. It is usually performed under general anesthesia, although spinal or epidural anesthesia is also used in some facilities. However, transvaginal ultrasound-guided oocyte retrieval is rapidly replacing laparoscopy, and is usually performed under regional anesthesia involving infiltration of local anesthetic into the vaginal wall.

I. Human Studies

The major concern during in vitro fertilization procedures is follicular fluid concentrations of drugs used during the procedures (Table 3). During local infiltration anesthesia, the peak concentration of lidocaine, a local anesthetic, is usually less than 1 μg/ml. However, it may occasionally reach as high as 100 μg/ml, probably when the aspiration needle for oocyte retrieval passes through the infiltrated area (BAILEY-PRIDHAM et al. 1990). During general anesthesia, follicular fluid concentrations of intravenously administered drugs appear to be negligible (Table 3).

Several early studies on laparoscopic oocyte retrieval under general anesthesia reported that fertilization or cleavage rates of the oocytes that were retrieved at the end of the procedures were lower than those of the oocytes retrieved at the beginning (BOYERS et al. 1987; ENDLER et al. 1987: HAYES et al. 1987). These results led investigators to speculate that anesthesia causes deleterious effects on oocytes. However, many later studies have shown that no anesthetic regimen for oocyte retrieval results in lower fertilization, cleavage, or pregnancy rates (Table 4). A preliminary report by PALOT et al. (1990) showed a lower cleavage rate when nitrous oxide (N_2O) was used with continuous propofol infusion for laparoscopic oocyte retrieval. The same investigators and others (ROSEN et al. 1987) reported that coadministration of N_2O to either halothane or isoflurane did not decrease the cleavage rate, in contrast to when halothane or isoflurane was used alone. VINCENT et al. (1995) also reported that a combination of N_2O and propofol used for laparoscopic

Table 3. Follicular fluid concentrations of the drugs used during oocytes retrieval

Drugs	Doses	Follicular fluid concentrations	References
Regional anesthesia			
Lidocaine[a]	50 mg, s.m.	Mean = 0.36 ± 1.1 μg/ml (n = 46, S.D.)	Wikland et al. 1990
Lidocaine	100 mg, s.m.	Peak = < 1 μg/ml (n = 5) except for three patients with 4.6, 12.2, and 118.0 μg/ml	Bailey-Pridham et al. 1990
Lidocaine[a]	25 mg, s.m.	Mean = 0.4 μg/ml (n = 9; range, 0.03–1.2 μg/ml)	Wikland et al. 1988
General anesthesia			
Fentanyl[b]	3.7 ± 1 μg/kg (S.D), i.v.	Not detected (n = 12), except for in one patient (0.21 ng/ml after 6 μg/kg, i.v.)	Schoeffler et al. 1988
Midazolam[b]	0.1 mg/kg, i.v.	Not detected (n = 19)	Chopineau et al. 1993
Propofol[b]	2.5 mg/kg, i.v.	0.1–0.2 fg/ml (n = 85)	Palot et al. 1988
Thiamylal[b]	5 mg/kg, i.v.	Mean = 1.67 ± 0.83 μg/ml (n = 9, S.D.)	Endler et al. 1987
Thiopental[b]	5 mg/kg, i.v.	Mean = 1.62 ± 0.61 μg/ml (n = 15, S.D.)	Endler et al. 1987

i.v., intravenous injection; s.m., submucosal injection for paracervical block.
[a]Pethidine was given intramuscularly for premedication.
[b]Several other drugs were used to induce and maintain general anesthesia.

Table 4. Summary of human studies on the effects of anesthesia for in vitro fertilization procedures

References	Experiments	Findings
Oocyte retrieval		
Sia-kho et al. 1993	General anesthesia for transvaginal oocyte retrieval. Groups: (1) propofol/opioid ($n=283$), (2) propofol/midazolam/opioid ($n=274$), (3) midazolam/opioid ($n=101$)	No difference in PR
Palot et al. 1990	General anesthesia for laparoscopy. Groups: (1) continuous propofol infusion (CPI) ($n=50$), (2) CPI + 50% N_2O ($n=10$), (3) CPI + halothane ($n=50$), (4) 50% N_2O + halothane ($n=50$)	Lower C.R. in group 2
Rosen et al. 1987	General anesthesia for laparoscopy. Groups: (1) 0.7% isoflurane + 60% N_2O in O_2 ($n=47$), (2) 1.4% isoflurane in O_2 ($n=51$)	No difference in FR and PR
Belaisch-Allart et al. 1985	Groups: (1) laparoscopy under general anesthesia ($n=225$), (2) laparoscopy under regional anesthesia supplemented with intravenous anesthetics ($n=97$), (3) sonography with regional anesthesia ($n=85$)	No difference in CR and PR
Gamete intrafallopian transfer with laparoscopy		
Pierce et al. 1992	Induction of general anesthesia[a]. Groups: (1) thiopental ($n=126$), (2) propofol ($n=151$)	No difference in PR
Van der Ven et al. 1988	Groups: (1) without anesthesia ($n=86$), (2) with general anesthesia (thiopental and alfentanil; $n=131$)	No difference in PR
Placement of conceptuses after in vitro fertilization		
Fishel et al. 1987[b]	Groups: (1) 2.3% enflurane + 70% N_2O/30% O_2 ($n=269$), (2) 1%–1.5% halothane + 70% N_2O/30% O_2 ($n=87$)	Lower IR in group 2 (17% vs. 34%, group 1)
Laparoscopic pronuclear stage transfer		
Vincent et al. 1995	Maintenance of general anesthesia[a]. Groups: (1) propofol/N_2O ($n=56$), (2) isoflurane/N_2O ($n=50$)	Lower PR in group 1 (30% vs. 54%, group 2)

The following references were not included in the table because of unclearness of the experimental designs or the number of patients are too small: Critchlow et al. 1991; Hood et al. 1988; Lefebvre et al. 1988. CR, cleavage rate; FR, fertilization rate; IR, implantation rate; PR, pregnancy rate.

[a] Several other drugs were used to induce, maintain, and reverse general anesthesia.

[b] General anesthesia with enflurane/N_2O/O_2 was used for oocyte retrieval for all patients.

pronuclear stage transfer resulted in a lower pregnancy rate compared to a combination of N_2O and isoflurane. Although further work is needed to clarify the issue, these results argue against the use of a combination of N_2O and propofol, but not volatile anesthetics for in vitro fertilization procedures.

Although no substantial study has been reported, in vitro fertilization procedures are currently believed not to be associated with increased incidences of spontaneous abortion or morphological abnormalities once pregnancy is established.

II. Animal Studies

The mouse in vitro fertilization system is now widely used as an animal model to examine the effects of drugs on fertilization rate and blastocyst development. Most drugs, including N_2O, appear to have no effects at clinically relevant concentrations (Table 5). Nevertheless, at least two drugs need to be further investigated. One of these is lidocaine. SCHNELL et al. (1992) reported that lidocaine exposure to oocytes at more than 1 μg/ml for 30 min, clinically relevant conditions in some patients (Table 2), resulted in decreased fertililzation rate, and produced a deleterious effect on blastocyst. In contrast, WIKLAND et al. (1988) found no effects on blastocyst development at 0.1–100 μg/ml, although the duration of lidocaine exposure was not reported. McFARLAND et al. (1989) also reported preliminary data that lidocaine at 0.1–100 μg/ml for 4 h did not cause adverse effects on blastocyst development. Another drug of concern is isoflurane. CHETKOWSKI and NASS (1988) reported that administration of isoflurane for 30 min during the two-cell stage mouse either with or without N_2O resulted in a lower rate of blastocyst development. WARREN et al. (1992) also reported similar adverse effects at higher concentrations (3% and 5%) and longer exposure period (2 h). Further investigation is needed to clarify these issues, particularly the effects of lidocaine, since this is now the most commonly used local anesthetic for ultrasound-guided oocyte retrieval.

To date, no study has been conducted to examine the effects of drug exposure to oocytes during in vitro fertilization procedures on later morphological development.

C. Nonobstetrical Surgery During Pregnancy

All women of child-bearing age scheduled for surgery should be carefully questioned regarding the possibility of pregnancy. Nevertheless, the most frequent and serious error is unnecessary delay of an urgently required surgery. The reason is that delay in such treatment may lead to an increase in both maternal and fetal morbidity and mortality that far outweighs any potential risk to the fetus. The common reasons for delay are innate fear of anesthesia based on old assumptions, failure to perform indicated diagnostic

Table 5. Summary of the effects of anesthetics in animal studies using mouse in vitro fertilization system

Species	Experiments	Findings	References
Fertilization[a]			
Bupivacaine	0.01–100 μg/ml for 30 min	No effects up to 100 μg/ml	SCHNELL et al. 1992
Chloroprocaine	0.01–100 μg/ml for 30 min	No effects at 0.01 μg/ml Lower FR at >0.1 μg/ml	SCHNELL et al. 1992
Lidocaine	0.01–100 μg/ml for 30 min	No effects at 0.01 and 0.1 μg/ml Lower FR. at >1 μg/ml	SCHNELL et al. 1992
Lidocaine	0.01–100 μg/ml for 4 h	No effects	MCFARLAND et al. 1989
Blastocyst development			
Fentanyl[b]	1.5 ng/ml for 30 min	No effects	CHETKOWSKI and NASS 1988
Isoflurane	1.5%–5% for 2 h	No effects at 1.5%; low BDR at 3% and 5%	WARREN et al. 1992
Isoflurane[b]	1.5% for 30 min	Lower BDR: 43.9% ($n=82$) vs. 79.4% (control, $n=131$)	CHETKOWSKI and NASS 1988
Lidocaine[a]	0.01–100 μg/ml for 30 min	No effects at 0.01 and 0.1 μg/ml Lower FR at >1 μg/ml	SCHNELL et al. 1992
Lidocaine[a]	0.01–100 μg/ml for 4 h	No effects	MCFARLAND et al. 1989
Lidocaine	0.01–100 μg/ml	No effects	WIKLAND et al. 1988
Meperidine[b]	250 ng/ml for 30 min	No effects	CHETKOWSKI and NASS 1988
Midazolam[b]	5 ng-50 μg/ml for 72 h	No effects up to 12.5 μg/ml, 0% at > 25 μg/ml	SWANSON and LEAVITT 1992
N_2O	60% for 30 min	No effects except for one specific conditions[c]	WARREN et al. 1990
N_2O[b]	60% for 30 min	No effects	CHETKOWSKI and NASS 1988
N_2O + isoflurane[b]	60% N_2O and 0.75% isoflurane for 30 min	Lower BDR: 64.9% ($n=94$) vs. 79.4% (control, $n=131$)	CHETKOWSKI and NASS 1988

The following reference was not included in the table because of unclearness of the experimental design: MATT et al. 1991.
BDR, blastocyst development rate; FR Fertility rate.
[a]Oocytes were exposed to test agents and examined after 72 h.
[b]Test chemicals were administered at the two-cell stage and the effects were examined after 72 h.
[c]When N_2O was administered at the two-cell stage within 4 h of expected cleavage.

procedures such as X-rays for fear of affecting the fetus, and difficulty in arriving at a correct diagnosis (SLATER and AUFSES 1991). In general, if the surgical problem is not an acute one, surgery should be postponed until after delivery. If that is not possible, the surgery may be performed at any time during pregnancy based on epidemiological findings, as discussed later. However, the patient and her family should be informed that currently available data indicate that there is an increased risk of low birth weight, premature delivery, and postnatal death after surgery at any stage of pregnancy, no increased risk of congenital defects even after surgery during the first trimester, and unknown risks for abortion and behavioral deficiency.

To date, there is no evidence that any anesthetic drug or technique is safer than another. Thus the type of anesthesia should be that considered appropriate for the particular case and should be one that minimizes changes in maternal physiology. Similarly, appropriate premedication of any kind should be given if necessary. Generally, regional anesthesia is preferred to general anesthesia when both are applicable, because the embryo/fetus is exposed to the minimum number of drugs and lowest possible drug mass (COHEN 1994). Some have recommended that N_2O not be administered to pregnant women or, if it is used, that it be given with folinic acid to bypass the metabolic block created by methionine synthase inactivation (MARX 1985). However, the scientific evidence at the present time does not support this recommendation (discussed later). Based on current information, many recommend that N_2O be used if it is necessary for the appropriate conduction of anesthesia, especially as techniques that compensate for an anesthetic regimen without N_2O have not been shown to be safer.

Choosing the appropriate dosages of anesthetic agents requires extra caution, because sensitivity to many drugs is increased from as early as the first trimester. During pregnancy, the dose requirement for many anesthetic agents is known to decrease by as much as 30%–40% in humans (GIN and CHAN 1994) and animals (PALAHNIUK et al. 1974; STROUT and NAHRWOLD 1981). This is probably because of the sedative effects of progesterone (DATTA et al. 1989) and/or the increased levels of endogenous opiates (GINTZLER 1980; SANDER et al. 1989). Inhalational anesthesia is rapidly induced in the pregnant patient, because decreased functional residual capacity results in less dilution of inspired gases and quicker achievement of desired alveolar concentrations. Prevention of pulmonary aspiration also needs special attention during pregnancy because of decreased gastric motility, increased gastric content, and difficulty of endotracheal intubation because of weight gain. Local anesthetic requirement also decreases by as much as 30% from the first trimester (FAGRAEUS et al. 1983). Again, progesterone has been implicated as the underlying cause. Animal studies also have shown that there is a greater sensitivity of nerve fibers to local anesthetics during pregnancy (DATTA et al. 1983). Epidural venous engorgement reduces the volume of both the cerebrospinal fluid and the epidural space and leads to additional decreases of drug requirement for spinal and epidural anesthesia as pregnancy advances.

After the first trimester, the enlarging uterus may compress the aorta and inferior vena cava and thereby decrease venous return and cardiac output. A substantial decrease in uteroplacental perfusion may follow even if overt maternal hypotension is absent. Thus great care should be taken to provide adequate uterine displacement during both regional and general anesthesia if progressive fetal deterioration is to be avoided. Maternal blood pressure is often high during either regional or general anesthesia in the supine position because of elevated catecholamine levels secondary to light anesthesia. Again, one cannot assume that placental perfusion is adequate. Hypoxia and hypercapnia occur more rapidly in the pregnant patient than in the nonpregnant patient because of decreased functional residual capacity, which results in reduced O_2 storage in the lungs, increased O_2 consumption, early airway closure, and decreased cardiac output in the supine position. Thus preoxygenation is recommended before endotracheal intubation and supplemental O_2 during regional anesthesia.

I. Human Studies

Epidemiological studies on developmental toxicity of anesthetics are extremely difficult to perform for several reasons. First, it is difficult to separate the effects of anesthetic drugs from the effects of surgery or from the original condition that necessitated it. Second, most patients receive many drugs (Table 1), and thus it is difficult to separate out the effects of the drug of interest. Finally, the rate of anesthesia for nonobstetric reasons during pregnancy is not very high, perhaps 1%–2%. Despite these difficulties, several reports have been published on fetal outcome following anesthesia for nonobstetrical surgery (Table 6); reports on obstetrical operations, mainly Shirodkar's (cerclage of the cervix), were excluded because the risk of a poor outcome is so high following these procedures (CRAWFORD and LEWIS 1986; SHNIDER and WEBSTER 1965; ALDRIDGE and TUNSTALL 1986)

The following conclusions can be drawn from these epidemiological studies with some degree of certainty (Table 7). First, morphological abnormalities, so-called "birth defects," are an unlikely consequence of anesthesia and surgery no matter when in pregnancy the surgery is performed. Second, the incidence of low birth weight and premature delivery is increased if surgery is performed at any time during pregnancy (MAZZE and KÄLLÉN 1989). Finally, the incidence of postnatal death is increased if surgery is performed during the second and third trimester. Much less certain is whether anesthesia and surgery increase the incidence of spontaneous abortion. Studies by DUNCAN et al. (1986) and BRODSKY et al. (1980) suggest that there is an increased incidence of spontaneous abortion after surgery. However, studies of spontaneous abortion are difficult to interpret, as is commonly known. One difficulty in their results is a reported background incidence of spontaneous abortion below 10%. Such a low background incidence emphasizes the dangers of responder bias and failure to medically confirm spontaneous abortion. In addition, no

Table 6. Summary of epidemiological studies on fetal outcome after nonobstetrical surgery

	Mazze and Källén (1989)[a]	Duncan et al. (1986)	Crawford and Lewis (1986)[b]	Brodsky et al. (1980)	Shnider and Webster (1965)[b]	Smith (1963)
Source	Swedish records	Manitoba records[c]	Hospitals records	Dental personnel[d]	Hospital records	Hospital records
Period	1973–1981	1973–1981	1968–1984	1968–1978	1959–1964	1957–1961
Pregnancies examined (n)	720 000	n.d.	n.d.	12929	9073	18248
Operations (n)	5405[e]	2565	59	287[f]	129[g]	67[h]
During first trimester	2252	n.d.	35	187	45	10
During second trimester	1881	n.d.	n.d.	100	42	45
During third trimester	1272	n.d.	n.d.	0	42	12
Categories of anesthesia						
General anesthesia	2 929	911	35	n.d.	45	24
Regional anesthesia	369	46	0	n.d.	15	43
Local anesthesia	375	337	0	n.d.	50	0
No anesthesia	0	503	0	n.d.	19	0
Unknown	1732	768	0	n.d.	0	0
Manifestation of developmental toxicity (first/second/third trimester)						
Death						
Spontaneous abortion	Not examined	–[i]	n.a. (0/35)	+/+/ n.d.	n.d.	n.d.
Stillbirth	–/–/–	n.d.	n.d.	n.d.	n.d.	n.a. (9/67)
Postnatal death	–/+/+	n.d.	n.d.	n.d.	n.a. (5/129)	n.d.
Morphological abnormality	–/–/–	–	n.a. (1/35)	–/–/n.d.	n.a. (8/129)	n.a. (0/67)
Growth retardation						
Very low birth weight (< 500 g)	+/+/+	n.d.	n.d.	n.d.	n.d.	n.d.
Low birth weight (< 2500 g)	+/+/+	n.d.	n.a. (0/35)	n.d.	n.a. (15/129)	n.d.
Premature delivery	+/+/+	n.d.	n.d.	n.d.	n.a. (8/129)	n.d.
Functional deficiency	Not examined	Not examined	Not examined	Not examined	Not examined	Not examined

+, manifestation of developmental toxicity found; –, manifestation not found; n.d., no data presented or not examined; n.a., no statistical analysis was made. [a]See Källén and Mazze (1990) for a subsequent report. [b]Data from patients having cervical cerclage were included in the original reports but have been omitted from this table. [c]Entire population of Manitoba, Canada, of approximately 1 million. [d]Postal survey of dentists, dental assistants, and their spouses in the United States. Response rate was 70%. [e]0.75% of total pregnancies examined. [f]2.22% of total pregnancies examined. [g]1.42% of total pregnancies examined. [h]0.36% of total pregnancies examined. [i]There was no difference between control and operated groups. However, there was a significant increase for the subgroup having general anesthesia.

Table 7. Conclusions from epidemiological studies on fetal outcome following anesthesia for nonobstetrical surgery

1. Morphological abnormalities are an unlikely consequence no matter when surgery/anesthesia occurs during pregnancy
2. Incidences of low birth weight and premature delivery increases when surgery/anesthesia occurs at any time during pregnancy
3. Incidence of postnatal death increases when surgery/anesthesia occurs during the second and third trimester
4. Whether incidence of spontaneous abortion is increased is uncertain
5. No study has been conducted for the effects of surgery/anesthesia on postnatal function

conclusions about postnatal functional deficiency after surgery can be made, since relevant studies have not been performed.

II. Animal Studies

Animal studies on developmental toxicity of anesthetics are also difficult to perform, because the situation with many anesthetic drugs is quite the reverse of normal toxicological screening when concentrations of drugs far above those used clinically are tested. Many drugs used in anesthesia have such potent physiologic effects that they are often not tested in animals at the same high dosages used clinically. This is because a large number of animals must be used in reproductive studies, and it is not practical to control the many physiologic changes that occur in each animal. In contrast, anesthesiologists carefully control such changes in each patient to prevent morbidity. Even at low concentrations, physiologic effects produced by the drug under test conditions must be taken into account when interpreting results.

1. Anesthetic Agents

a) Inhalational Anesthetics

A number of studies have been conducted at anesthetic concentrations of inhalational anesthetics using various species of animals (Table 8). Among currently used inhalational anesthetics, N_2O is the only one that has been consistently shown to cause developmental toxicity in mammals (discussed later). A number of investigators have examined neurobehavioral effects of volatile anesthetics (BOWMAN and SMITH 1977; CHALON et al. 1981; KOËTER and RODIER 1986; RICE 1986; RODIER and KOËTER 1986; SMITH et al. 1978) and N_2O (HOLSON et al. 1989; KOËTER and RODIER 1986; MULLENIX et al. 1986; RICE 1990; RODIER and KOËTER 1986; TASSINARI et al. 1986). However, their results are not consistent and much more work is needed to clarify the issue.

Table 8. Summary of animal studies on the effects of anesthetic concentrations of volatile anesthetics

Species	Experiments	Findings	References
Desflurane			
Rat/rabbit	FDA reproductive testing segments (10 and 13 cumulative MAC-h during organogenesis)	No effects	ANONYMOUS 1995
Enflurane			
Rat	1.65% for 6 h daily on GD 8–10, 11–13, or 14–16	No effects[a]	MAZZE et al. 1986
Rat	1.25% for 1 h daily during organogenesis	No effects	SAITO et al. 1974[b]
Mouse	0.75% for 1 h daily during organogenesis	No effects	SAITO et al. 1974[b]
Halothane			
Rat	0.27% for 24 h on GD 8	No effects	MAZZE et al. 1988
Rat	0.8% for 6 h daily on GD 8–10, 11–13, or 14–16	No effects[a]	MAZZE et al. 1986
Rat	1.35%–1.43% for 1 h daily on GD 1–5, 6–10, or 11–15	No effects	KENNEDY et al. 1976
Rabbit	2.16%–2.3% for 1 h daily on GD 6–9, 10–14, or 15–18	No effects	KENNEDY et al. 1976
Rat	0.8% for 12 h on GD 6, 7, 8, 9, or 10	High incidence of SDV on GD 8, 9, or 10	BASFORD and FINK 1968
Isoflurane			
Rat	0.35% for 24 h on GD 8	No effects	FUJINAGA et al. 1987
Rat	1.05% for 6 h daily on GD 8–10, 11–13, or 14–16	No effects[a]	MAZZE et al. 1986
Mouse	0.6% for 4 h daily on GD 6–15	High incidence of SDV and VA, including cleft palate	MAZZE et al. 1985
Rat	1.6% for 1 h daily for 5 days from GD 1, 6, or 11	No effects	KENNEDY et al. 1977
Rabbit	2.3% for 1 h daily for 24 days from GD 6, 10, or 15	No effects	KENNEDY et al. 1977
Sevoflurane			
	No available data		

FDA, Food and Drug Administration; GD, gestational day (GD 0=plug positive day, when defined); MAC, minimum alveolar concentration; SDV, skeletal developmental variants; VA, visceral abnormalities.
[a]Decreased fetal weight was observed only those exposure occurred on GD 8–10 and 14–16, but not 11–13.
[b]Cited from SHEPARD (1992).

b) Combination of N_2O and Other Agents

In clinical practice, N_2O is seldom administered alone, but rather is combined with other volatile or intravenous anesthetics. To date, only a few studies have been performed to examine the effects of such combinations (Table 9). Addition of opioid to N_2O does not significantly add to developmental toxicity of N_2O alone in rats (Mazze et al. 1987). Although the mechanism remains unclear (discussed later), addition of a volatile anesthetic has been reported to prevent N_2O-induced developmental toxicity (Fujinaga et al. 1987; Mazze et al. 1988). Interestingly, similar unexplained inhibitory effects of volatile anesthetic have been reported for the effects of N_2O on cleavage rate (Palot et al. 1990) and neurofunctional development (Koëter 1990; Smits-Van Prooije et al. 1989).

c) Local Anesthetics

Commonly used local anesthetics have not been found to cause developmental toxicity at clinically relevant concentrations in in vivo models (Table 10), unless cocaine is classified as a local anesthetic. Nevertheless, lidocaine has been reported to cause neural tube defects in embryos of mice in vitro (O'Shea and Kaufman 1980). A similar report has been made concerning lidocaine and other local anesthetics in early chick embryos (Lee and Nagele 1985; Lee et al. 1988). However, lidocaine does not produce similar lesions in rat embryos in vitro (Fujinaga and Baden 1993), suggesting that the effect may be species specific. Two studies have reported behavioral deficits following lidocaine in rats (Smith et al. 1986; Teiling et al 1987), but the results are inconsistent and further work is necessary to clarify the issue.

d) Intravenous Anesthetics

None of the intravenous anesthetics has been shown to cause developmental toxicity at clinically relevant concentrations (Table 11). Although barbiturates and benzodiazepines have been reported to cause developmental toxicity at high doses, such drug exposure is not relevant to clinical anesthesia. They are commonly used as anitconvulsant agents and are discussed in Chap. 29.

e) Opioids

There are numerous reports that prenatal opioid exposure causes developmental toxicity, including morphological abnormalities and neurobehavioral toxicity. However, in most studies, the opioids were injected to animals as a bolus, and from the doses and side effects that were reported it is likely that test animals treated with high doses experienced severe respiratory depression that was the probable cause of toxicity. Impaired nutrition and opioid withdrawal are also confounding factors of those experimental results (Lichtblau and Sparber 1984; Raye et al. 1977). In contrast, several well-designed studies aimed at minimizing respiratory depression indicate that currently used

Table 9. Summary of animal studies on the effects of combination of N_2O and other agents

Species	Experiments	Findings	References
N_2O + Fentanyl			
Rat	35% and 50% N_2O for 24 h on GD 8 and fentanyl, 500 μg/kg per day, s.c. (pump), on GD 7–21	No additive effects	Mazze et al. 1987
N_2O + Halothane			
Hamster	60% N_2O + 0.6% halothane for 3 h on GD 8, 9, or 10	Increased RR on GD 10 Increased GR on GD 10 and 11	Bussard et al. 1974[a]
Rat	50%–75% N_2O + 0.27% halothane for 24 h on GD 8	Adverse effects[b] of N_2O were inhibited	Mazze et al. 1988
N_2O + Isoflurane			
Rat	50% N_2O + 0.35% isoflurane for 24 h on GD 8	Adverse effects[b] of N_2O were inhibited	Fujinaga et al. 1987

GD, gestational day (GD 0 = plug positive day, when defined); GR, growth retardation; pump, Alzet osmotic minipump (Alza Co., Palo Alto, CA); RR, resorption rate; s.c., subcutaneous.
[a]Teratological examination was not performed.
[b]Reproductive toxicity and teratogenicity.

Table 10. Summary of animal studies on the effects of local anesthetics

Species	Experiments	Findings	References
Bupivacaine			
Rat/rabbit	FDA reproductive testing segments	Decreased postnatal survival at nine times the human dose (rat); embryotoxic effects at five times the human dose (rabbit)	Anonymous 1995
Chloroprocaine			
	No available data		
Etidocaine			
Rat/rabbit	FDA reproductive testing segments	No effects at one to seven times the human dose	Anonymous 1995
Lidocaine			
Rat	9, 18, 36, 72 mg/kg per day (with 1:100 000 epinephrine), i.m. on GD 6–15	No effects (maternal toxicity at 72 mg/kg)	Laborde et al. 1988
Rat	100, 250, 500 mg/kg per day, s.c. (pump), 2 weeks before mating – GD 21 (100, 250) or GD 3–17 (500)	No effects except for decreased fetal weight at 500 mg/kg per day (time of delivery was delayed in this group)	Fujinaga and Mazze 1986
Rat	55 mg/kg per day, i.p., on GD 5–7, 8–10, 12–14, 15–17	No effects	Ramazzotto et al. 1985
Prilocaine			
Rat	FDA reproductive testing segments	No effects at 300 mg/kg, i.m.	Anonymous 1995
Procaine			
Rat	FDA reproductive testing segments	MA and RR at 500 mg/kg in diet and 20–30 mg/kg per day by gavage	Anonymous 1995
Tetracaine			
	No available data		

The following reference was not included in the table because of unclearness of experimental designs: Martin and Jurand 1992.
FDA, Food and Drug Administration; GD, gestational day (GD 0 = plug positive day, when defined); i.m., intramuscular injection; i.p., intraperitoneal injection; MA, morphological abnormalities; pump, Alzet osmotic minipump (Alza Co., Palo Alto, CA); RI, reproductive indices; RR, resorption rate; s.c., subcutaneous implantation.

Table 11. Summary of animal studies on the effects of intravenous anesthetics

Species	Experiments	Findings	References
Droperidol			
Rat	FDA reproductive testing segments	Decrease in postnatal survival at 4.4 times the human dose, i.v.	ANONYMOUS 1995
Etomidate			
Rat	FDA reproductive testing segments	No teratogenic effects but some reproductive effects at one and four times the human dose	ANONYMOUS 1995
Ketamine			
Rat	120 mg/kg, i.m., on GD 9–13	No effects	EL-KARIM and BENNY 1976[a]
Methohexital			
Rat/rabbit	FDA reproductive testing segments	No reproductive effects at four and seven times the human dose	ANONYMOUS 1995
Midazolam			
Rat/rabbit	FDA reproductive testing segments	No reproductive and teratogenic effects at five and ten times the human dose	ANONYMOUS 1995
Propofol			
Rat	FDA reproductive testing segments	No reproductive effects at 15 mg/kg per day	ANONYMOUS 1995
Thiamylal			
Mouse	20–140 mg/kg, i.p., once or twice a day, either on GD 7–14	High incidence of foot joint malformation at >60 mg/kg on GD 10	TANIMURA 1965[a]
Thiopental			
Mouse	50, 100 mg/kg, i.p.?, on GD 11	No teratogenic effects, but GR	TANIMURA et al. 1967[a]
Rat	50, 100 mg, i.p.?, on GD 4	No effects	PERSAUD 1965[a]

FDA, Food and Drug Administration. GD, gestational day (GD 0 = plug positive day, when defined); GR, growth retardation; i.p., intraperitoneal injection; i.m., intramuscular injection; i.v., intravenous injection.
[a]Cited from SHEPARD (1992).

opioids do not cause developmental toxicity at clinically relevant concentrations (Table 12).

f) Muscle Relaxants

Muscle relaxants are especially difficult to test for developmental toxicity in standard in vivo animal models because of the respiratory depression that they cause in the mother. It is difficult to compensate for this effect with artificial ventilation because of technical difficulties and the large number of animals required for most toxicity studies. The few in vivo studies performed in rodents have indicated that none of the clinically used muscle relaxants causes developmental toxicity (Table 13). A rat whole embryo culture study also indicates that commonly used nondepolarizing muscle relaxants do not show any adverse effects at clinically relevant concentrations (Fujinaga et al. 1992a)

Although there is no evidence that muscle relaxants produce developmental toxicity during organogenesis, there is some evidence that they do so later in gestation. For example, prolonged disruption of muscle activity induced by various cholinergic agents, including muscle relaxants, causes axial deformities, peripheral skeletal malformations, and limb deformities in the chick (Bueker and Platner 1956; Drachman and Coulombre 1962; Landauer 1960, 1977; Meiniel 1981; Upshall et al. 1968). However, these effects are seldom seen in mammals, possibly because the placental barrier acts to reduce the concentrations of muscle relaxants presented to the fetus (Meiniel 1981). One would assume that they would be seen in humans only under exceptional circumstances. In fact, there is a report of such a case, an infant born with arthrogryposis to a mother who was treated with *d*-tubocurarine for tetanus for 19 days starting on about the 55th day of gestation (Jago 1970). The author speculated that drug-induced immobilization of the fetus by *d*-tubocurarine at the time of or shortly after the development of the joint cavities was the most probable cause. However, this is uncertain, because the patient suffered several episodes of severe hypoxia, bronchopneumonia, and myocarditis and was treated with various other drugs including diazepam, chlorpromazine, digoxin, hydrochlorothiazide, and antibiotics.

g) Others

According to available data, including those in Anonymous (1995), none of the other drugs listed in Table 1 appear to cause developmental toxicity in animals at clinically relevant concentrations used for anesthesia.

2. Abnormal Physiological Conditions

If anesthesia eventually proves to be a contributing factor to poor fetal outcome after surgery, changes in maternal physiology are more likely to be an etiology than direct toxicity to anesthetic drugs (see Chap. 20 for details). The ultimate result of most physiologic insults is fetal hypoxia, which can certainly

Table 12. Summary of animal studies on the effects of opioids

Species	Experiments	Findings	References
Alfentanil			
Rat	8 mg/kg per day, s.c. (pump), GD 5–20	No effects	FUJINAGA et al. 1988a
Butorphanol			
Rat	FDA reproductive testing segments	Increased stillbirths at 1 mg/kg, s.c.	ANONYMOUS 1995
Rabbit	FDA reproductive testing segments	Increased post-IL at 60 mg/kg, p.o.	ANONYMOUS 1995
Fentanyl			
Rat	500 μg/kg per day, s.c. (pump), GD 5–20	No effects	FUJINAGA and MAZZE 1988
Rat	500 μg/kg per day, s.c.(pump), 2	No effects	MAZZE et al. 1987
Rat	10, 100, 500 μg/kg per day, s.c.(pump), 2 weeks before mating – GD 20	No effects	FUJINAGA et al. 1986
Meperidine			
	No available data		
Morphine			
Rat	10, 35, 70 mg/kg per day, s.c.(pump), GD 5–20	No effects at 10 mg/kg per day; increased Pre-IL at 35, 70 mg/kg per day; increased PND at 35 mg/kg per day	FUJINAGA and MAZZE 1988
Sufentanil			
Rat	10, 50, 100 μg/kg per day, s.c. (pump), GD 5–20	No effects	FUJINAGA et al. 1988a

The following references were not included in the table because opioid-induced respiratory depression is not considered in the experimental designs: teratological studies (ARACURI and GAUTIERI 1973; CIOCIOLA and GAUTIERI 1983; GEBER and SCHRAMM 1975; HARPEL and GAUTIERI 1968; IULIUCCI and GAUTIERI 1971; MARTIN and JURAND 1992; ZAGON and McLAUGHLIN 1977b), neurobehavioral studies (DAVIS and LIN 1972; FRIEDLER and COCHIN 1972; JOHANNESSON and BECKER 1972; O'CALLANGHAN and HOLTZMAN 1976; SMITH and JOFFE 1975; SOBRIAN 1977; STEELE and JOHANNESSON 1975; ZAGON and McLAUGHLIN 1977a; ZIMMERBERG et al. 1974).

FDA, Food and Drug Administration; GD, gestational day (GD 0 = plug positive day, when defined); IL, preimplantation loss; PND, postnatal death; p.o., per os; pump, Alzet osmotic minipump (Alza Co., Palo Alto, CA); s.c., subcutaneous implantation.

Table 13. Summary of animal studies on the effects of muscle relaxants

Species	Experiments	Findings	References
Atracurium			
Rabbit	0.15 mg/kg daily or 0.1 mg/kg twice daily s.c., GD 6–18	No effects	SKARPA et al. 1983
Rat (WEC)	25–125 μg/ml	No effects at < 25 μg/ml	FUJINAGA et al. 1992a
Doxacurium			
Rat/mouse	FDA reproductive testing segments	No effects at MSD, s.c.	ANONYMOUS 1995
Metocurine			
	No available data		
Mivacuronium			
Rat/mouse	FDA reproductive testing segments	No effects at MSD, s.c.	ANONYMOUS 1995
Pancuronium			
Rabbit	8 μg/kg, i.v., GD 8–16	No effects	SUZUKI et al. 1970
Rat (WEC)	1–40 μg/ml	No effects at < 1 μg/ml	FUJINAGA et al. 1992a
Pipecuronium			
	No available data		
Rocuronium			
Rat	FDA reproductive testing segments	No effects at 0.3 mg/kg	ANONYMOUS 1995
Succinylchorine			
	No available data		
d-Tubocurarine			
Mouse	0.6, 5 mg/kg, i.m., GD 13.5	No effects on palate formation	JACOBS 1971
Rat (WEC)	3–150 μg/ml	No effects at < 3 μg/ml	FUJINAGA et al. 1992a
Vecuronium			
Rat (WEC)	50–300 μg/ml	No effects at < 50 μg/ml	FUJINAGA et al. 1992a

GD, gestational day (GD 0 = plug positive day, when defined); i.m., intramuscular injection; i.v., intravenous injection; MSD, maximum subparalyzing dose; s.c., subcutaneous injection; WEC, whole embryo cluture; FDA, Food and Drug Administration.

cause developmental toxicity to various degrees. However, designing well-controlled, in vivo experiments to study developmental toxicity of a specific physiologic change is very difficult, because most physiologic parameters are interdependent and change continually and concurrently. Although numerous studies of such topics as hypoxia and acidosis during pregnancy have been reported, almost no conclusions have been drawn. For example, most experiments of hypoxia have been performed by exposing animals to low O_2 concentrations for various durations but since arterial O_2 tensions have not been reported interpretation of results is almost impossible. In contrast to the uncertain results from in vivo studies, results from many in vitro studies have shown developmental toxicity of hypoxia under relatively well defined conditions. Nevertheless, it is difficult to extrapolate such results to clinical situations.

Increasing concern has been expressed in recent years that changes in body temperature can cause developmental toxicity (EDWARDS 1988). Although hyperthermia rarely occurs in modern anesthetic practice, cases of malignant hyperpyrexia do occur and probably are associated with adverse fetal outcome. Hypothermia is also known to cause developmental toxicity in animals (SMOAK and SADLER 1991), although there are many case reports of uneventful birth after hypothermia used during cardiac surgery (reviewed by STRICKLAND et al. 1991).

3. Mechanisms of N_2O-Induced Developmental Toxicity

N_2O is the only inhalational anesthetic that has definitely been shown to cause developmental toxicity in animals. Although several studies reported embryonic toxicity of N_2O in chick (MOBBS et al. 1966; PARBROOK et al. 1965; RECTOR and EASTWOOD 1964; SMITH et al. 1965; SNERGIREFF et al. 1968), FINK et al. (1967) were the first to report it in mammals. They exposed pregnant rats continuously to 50% N_2O for 2, 4 or 6 days starting on gestational day (GD) 8 (GD 0 being the day when copulatory plug was observed), and found increased incidences of resorptions and skeletal abnormalities. Subsequently, they established that exposure to 70% N_2O for 24 h during the organogenesis period consistently caused developmental toxicity (SHEPARD and FINK 1968). Since then, this model has been used extensively to investigate the mechanisms of N_2O-induced developmental toxicity (FUJINAGA et al. 1987, 1989, 1990, 1991; KEELING et al. 1986; LANE et al. 1980; MAZZE et al. 1984, 1987, 1988). Although the threshold concentration for the effects was found to be about 50% (MAZZE et al. 1987), the threshold exposure time for the effects is yet to be accurately determined. Under similar condition, N_2O has been shown to cause developmental toxicity in hamsters (SHAH et al. 1979), but no investigators have reported similar effects in mice.

The developmental abnormalities caused by N_2O exposure in rats are now clearly established. They are resorptions, growth retardation, skeletal abnormalities including major and minor rib and vertebral defects, and visceral abnormalities including situs inversus (FUJINAGA et al. 1989, 1990; SHEPARD and FINK 1968). The effects of N_2O on postnatal function have not been studied using the standard model. The days of gestation when embryos are most susceptible to the effects of N_2O are as follows: GD 8 (resorption, situs inversus, minor skeletal anomalies such as extra cervical rib), GD 9 (major skeletal malformations), and GD 11 (resorptions) (FUJINAGA et al. 1989). Although the etiology of the resorptions is unclear, it has been demonstrated that, after exposure on GD 8, the highest incidence of fetal death occurs between GD 13 and 14 (FUJINAGA et al. 1990). This is the time when the liver takes over from the yolk sac as the dominant organ for hematopoiesis; thus it has been suggested that N_2O exposure on GD 8 might damage primordial liver cells. The etiology of the resoptions after exposure on GD 11 has been suggested to be due to the failure of the embryo to switch from dependence on the yolk sac to dependence on the chorioallantoic placenta for vital nutritional and metabolic functions, a switch which normally occurs on this day (FUJINAGA et al. 1989).

Until recently, N_2O-induced developmental toxicity was thought to be caused solely by the oxidation of vitamin B_{12}, which cannot then function as a coenzyme for methionine synthase. This enzyme catalyzes the transmethylation from methyltetrahydrofolate and homocysteine to produce tetrahydrofolate and methionine (Fig. 1). The expected result of its inhibition is decreased tetrahydrofolate, which may lead to impaired DNA synthesis, and decreased methionine, which may lead to impaired methylation reactions (reviewed by NUNN and CHANARIN 1985). Inactivation of methionine synthase by N_2O is known to cause a pernicious anemia-like syndrome consisting of subacute combined degeneration of the spinal cord (SCOTT et al. 1981), megaloblastic anemia, and pancytopenia in humans (CHANARIN 1980). Because the hematologic changes in humans are prevented by folinic acid (5-formyl tetrahydrofolate) administered with N_2O (O'SULLIVAN et al. 1981), presumably because DNA synthesis is restored to normal, impaired DNA synthesis was proposed to account for N_2O's developmental toxicity (NUNN 1987). However, the following evidence indicates that lack of tetrahydrofolate is not the main cause of N_2O-induced developmental toxicity. First, maximum reduction of methionine synthase activity and decrease in DNA synthesis occur at dosages of N_2O that are well below those that cause developmental toxicity (BADEN et al. 1984). Second, supplementation with folinic acid, which should restore DNA synthesis to normal, only partially reduces the high incidence of only one type of malformation, minor skeletal anomalies (KEELING et al. 1986; MAZZE et al. 1988). Third, supplementation with methionine, but not with folinic acid, almost completely prevents N_2O-induced growth retardation and all malformations other than situs inversus in an in vitro whole embryo culture system (FUJINAGA and BADEN 1994).

It is not clear at this time why decreased tetrahydrofolate plays almost no role in N_2O-induced developmental toxicity. Certainly, N_2O's effect on DNA

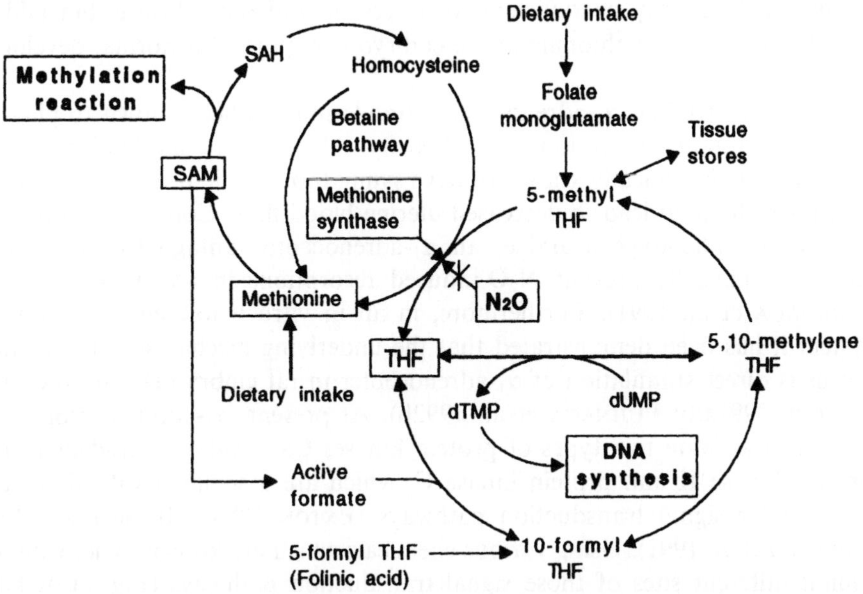

Fig. 1. Metabolic pathways depicting the inhibition of methionine synthase by N_2O and the consequences of interference with this reaction. N_2O inactivates vitamin B_{12} by oxidizing its cobalt, leading to inhibition of methionine synthase activity, which requires vitamin B_{12} in the fully reduced state for normal activity. As a consequence of the inhibition of methionine synthase activity, homocysteine is not converted to methionine, nor is 5-methyl tetrahydrofolate (THF) converted to THF. Decreased availability of THF is presumed to result in decreased levels of 5, 10-methylene THF, which is the one-carbon donor for the conversion of deoxyuridine monophosphate (*dUMP*) to triphosphate (*dTMP*), one of the four essential DNA bases. Decreased methionine is presumed to lead to decreased SAM, thus to impaired methylation reactions. *SAH*, *S*-adenosyl homocysteine; *SAM*, *S*-adenosyl methionine.

synthesis is only partial (BADEN et al. 1983, 1984). Furthermore, it is likely that the salvage pathway for thymidylate, which is known to exist in many mammalian tissues, becomes more active in the embryo to compensate for the decreased de novo synthesis of thymidylate. The mechanism for the preventive effects of methionine is also unclear. It has been reported that rat embryos grown in methionine-deficient culture medium develop abnormalities, including neural tube closure defects (COELHO and KLEIN 1990; FLYNN et al. 1987; KLUG et al. 1990). However, the types of abnormalities caused by methionine-deficient culture medium are not the same as those caused by N_2O (BADEN and FUJINAGA 1991; FUJINAGA et al. 1988b). For example, N_2O does not cause neural tube closure defects, suggesting that either N_2O-induced toxicity is not simply due to decreased methionine synthesis or that the experiments designed to examine the effects of methionine deficiency in the culture medium are accompanied by another condition which leads to neural tube closure defects. It is also possible that preexisting methionine in the culture medium may be sufficient to process neural tube closure, but additional synthesis of methionine in the embryo is required for normal development.

In recent studies, α_1-adrenoceptor stimulation has been shown to play a role in N_2O-induced developmental toxicity. Although the mechanism is yet unclear, N_2O is clinically known to have sympathomimetic properties (EISELE 1985), which could lead to decreased uterine blood flow. Consistent with this speculation, phenoxybenzamine, an α_1-adrenoceptor antagonist, has been shown to partially prevent N_2O-induced resorptions in an in vivo model (FUJINAGA et al. 1991). Furthermore, in an in vitro whole embryo culture system it has been demonstrated that the underlying mechanism of situs inversus is direct stimulation of α_1-adrenocepter in rat embryo (FUJINAGA and BANDEN 1991a,b; FUJINAGA et al. 1992b). At present, α_1-adrenoceptors are known to activate two types of protein kinases Ca^{2+}/calmodulin-dependent protein kinase II and protein kinase C, which are associated with different intracellular signal transduction pathways (EXTON 1988; MINNEMAN 1988; RUFFOLO et al. 1991). Using various chemicals which are known to activate or inhibit different sites of those signal transduction pathways (Fig. 2), it has been demonstrated that the effect is mediated by α_{1A}- but not α_{1B}-adrenoceptor subtype, and by Ca^{2+}/calmodulin-dependent protein kinase II, but not protein kinase C (FUJINAGA et al. 1994, 1995b). Most recently, there is evidence that phenylephrine, an α_1-adrenoceptor agonist, increases the gene expression of c-myc, a transcriptional factor, in the embryo, suggesting that c-myc may be involved in causing situs inversus by α_1-adrenoceptor stimulation (FUJINAGA et al. 1995a). In addition, because α_1-adrenocepter antagonist alone does not cause situs inversus (FUJINAGA et al. 1992b), it has been suggested that α_1-adrenoceptor stimulation interferes with whatever signal transduction pathway is actually involved in the normal development of the left–right body axis, leading to situs inversus.

Clearly, N_2O-induced developmental toxicity is multifactorial, and much further work is needed to fully understand it. Figure 3 shows hypothetical mechanisms of N_2O-induced developmental toxicity based on current knowledge and some speculation. Interestingly, this schema provides a possible answer for the preventive effects of volatile anesthetic on N_2O-induced developmental toxicity without recovering methionine synthase activity (FUJINAGA et al. 1987; MAZZE et al. 1988). In other words the volatile anesthetic inhibits N_2O-induced catecholamine release, thereby maintaining uterine blood flow, which provides nutritional supply to the embryo including methionine. However, whether sympathomimetic effects of N_2O are due to catecholamine release, and whether volatile anesthetics actually block such effects, remains to be demonstrated. In addition, there may be yet unknown nutrition that might be involved in N_2O-induced developmental toxicity under in vivo situation, if decreased uterine blood flow plays a role.

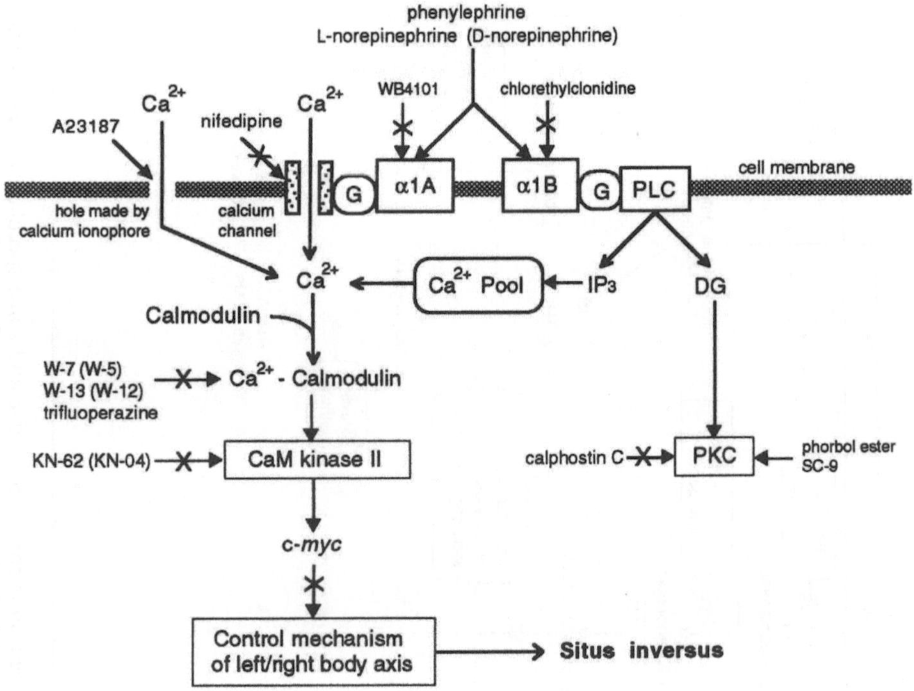

Fig. 2. Signal transduction pathways related with α_1-adrenoceptors and the chemicals that are known to activate or inhibit different sites of pathways; the chemicals in *parenthesis* are negative controls. α_1-Adrenoceptors are known to activate two types of protein kinases, Ca^{2+}/calmodulin-dependent protein kinase II (*CaM kinase II*) and protein kinase C (*PKC*), which are associated with different intracellular signal transduction pathways. Stimulation of α_1-adrenoceptor has been shown to cause situs inversus in rat embryo mediated by stimulation of α_{1A}- but not α_{1B}-adrenoceptor subtype, and by CaM kinase II but not PKC. Most recently, evidence has been provided that c-myc, a transcriptional factor, may be involved in this effect. *α1A*, α_{1A}-adrenoceptor subtype; α1B, α_{1B}-adrenoceptor subtype; *DG*, diacylglycerol; *G*, guanosine nucleotide-binding protein; *IP₃*, inositol triphosphate; *PLC*, phospholipase C

D. Fetal Therapy/Surgery

Fetal therapy/surgery is a new area of medicine (reviewed by ROSEN 1991). Some procedures, such as intrauterine blood transfusion for erythroblastosis fetalis and placement of vesicoamniotic shunt catheter, are considered minor since they can be performed percutaneously. Others, such as repair of diaphragmatic hernia and excision of cystic adenomatoid malformation, are more major and are performed via a hysterectomy. Local anesthetic infiltration of the maternal abdomen is usually sufficient for percutaneous placement of needles and catheters. General anesthesia is usually used for procedures involving hysterectomy. Placement of a lumbar epidural catheter for postoperative analgesia with opioids is often performed before induction of

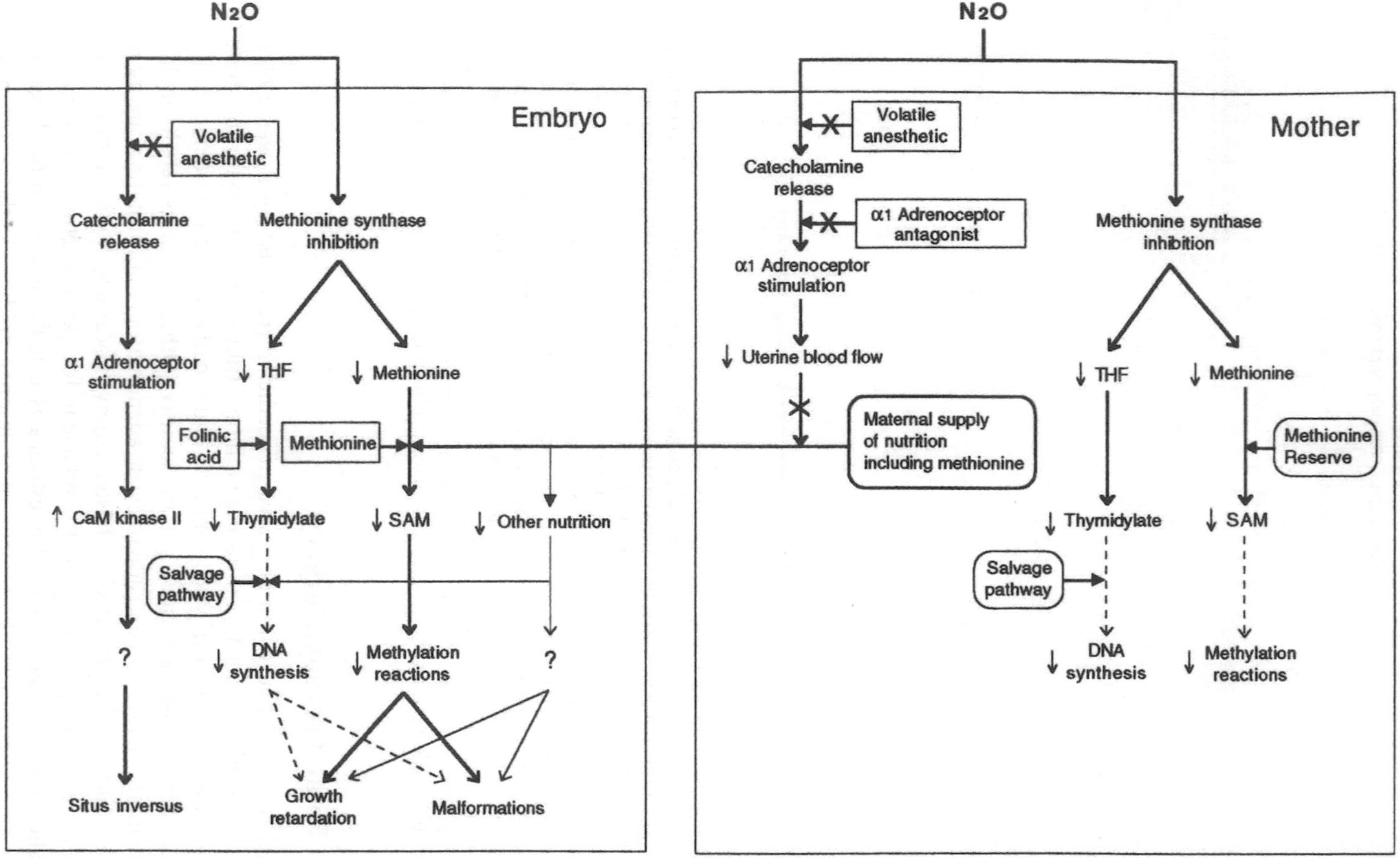

Fig. 3. Hypothetical mechanisms of N_2O-induced developmental toxicity based on current knowledge and some speculation. *THF*, tetrahydrofolate; *CaM kinase II*, Ca^{2+}/calmodulin-dependent protein kinase II; *SAM*, *S*-adenosyl methionine. (See text for details)

anesthesia. Neuromuscular blocking agents may be needed for fetal immobilization, but because the placental transfer of these agents is limited, they should preferably be injected directly into the fetus either intramuscularly or into the umbilical vein under ultrasonic guidance.

I. Human Studies

The major concern after major fetal therapy/surgery is a high incidence of premature labor resulting in preterm delivery (LONGAKER et al. 1991). Thus tocolytic agents including ritodrine, terbutaline, or magnesium sulfate are routinely administered postoperatively (ROSEN 1991). The cause is thought to be uncontrollable uterine contraction resulting from hysterectomy. To date, there are no reports on the effects of anesthesia during fetal surgery/therapy on fetal outcome. Compared with the original conditions that necessitated fetal therapy/surgery and the severity of surgical procedures on the fetuses, the effects of anesthetics, if any, would be expected to be minimal.

II. Animal Studies

No relevant study has been conducted in animals to examine the effects of anesthetics during fetal therapy/surgery. Again, any anesthetic effects would probably be minimal.

E. Delivery

The choice of anesthesia for delivery depends on obstetric requirements and patient's desire. For vaginal delivery, regional anesthesia, particularly epidural anesthesia, is now most commonly used. For cesarean section, both regional and general anesthesia are commonly used. Each has its advantages and disadvantages for the mother as well as for the fetus, as described elsewhere in textbooks of anesthesiology.

I. Human Studies

Effects of epidural anesthesia for delivery on postnatal behavior have been examined in many studies (ABBOUD et al. 1982; KANGAS-SAARELA et al. 1987; SCANLON et al. 1976; SEPKOSKI et al. 1992; THALME et al. 1974; TISON et al. 1994). Although some claim that babies whose mothers received epidural anesthesia during delivery show neurobehavioral deficits, those arguments are extremely controversial, and to date there is no strong evidence against the use of epidural anesthesia for delivery. On the other hand, several studies have shown that epidural anesthesia decreases maternal plasma catecholamines and probably improves uterine blood flow and fetal well-being (JOUPPILA et al. 1978; SHNIDER et al. 1983).

II. Animal Studies

No relevant study of the effects of anesthetics during delivery on fetal outcome has been conducted in animals.

F. Waste Inhalational Anesthetics in the Workplace

In the United States alone, more than 200 000 people are estimated to be exposed to waste inhalational anesthetics, including physicians, dentists, veterinarians, nurses, and technicians (WHITCHER 1985); see COHEN 1980 for further reading. Among inhalational anesthetics, N_2O is the major concern, because its clinical concentration reaches as high as almost 75% of inspired gas, whereas those of other agents are much less. The NATIONAL INSTITUTE FOR OCCUPATIONAL SAFETY AND HEALTH (NIOSH) has suggested that no worker should be exposed to time weighted (average) N_2O concentrations exceeding 25 ppm in the operating room or 50 ppm in the dental office (1977). By using adequate scavenging systems, these standards can be easily achieved (WHITCHER 1985).

I. Human Studies

Many epidemiological studies have been conducted to evaluate the reproductive performance of personnel who work in operating rooms or dental offices and their spouses. At present, most investigators agree with the report from the epidemiologists who were commissioned by the American Society of Anesthesiologists to conduct an independent review of studies (BURING et al. 1985). They found that there were small increases in the rate of spontaneous abortion and congenital defects of offspring among women directly exposed to waste anesthetic gases that were within the range that normally occurs with studies of the type reviewed and that are often due to various biases and uncontrolled confounding factors. No consistent increase of adverse reproductive effects were found for the partners of exposed men. The epidemiologists pointed out that, even if the increases were real, many factors such as X-rays and viruses could have accounted for them. Based on this review and subsequent registry-based studies that are generally regarded as more credible than the original studies (ERICSON and KÄLLÉN 1985; HEMMINKI et al. 1985), hazards from low levels of waste anesthetic gases appear to be minimal or absent. Nevertheless, this issue is still under active investigation, and some investigators believe that hazards may exist under certain circumstances. For example ROWLAND et al. (1992) recently reported that high levels of N_2O exposure in dental assistance adversely affect a woman's fertility, although the study was based on questionnaires, which always contain a risk of responder bias.

II. Animal Studies

Most studies to date have shown that no adverse reproductive effects are caused by chronic exposure to trace concentrations of volatile anesthetics (GREEN et al. 1982; HALSEY et al. 1981; LANSDOWN et al. 1976; MAZZE 1985; PETERS and HUDSON 1982; POPE and PERSAUD 1978; POPE et al. 1975, 1978; WHARTON et al. 1978, 1979) and N_2O (MAZZE et al. 1982; POPE et al. 1978; RAMAZZOTTO et al. 1979), although several studies reported adverse effects on reproductive indices (COATE et al. 1979a,b; CORBETT et al. 1973; POPOVA et al. 1979; RICE et al. 1985; VIEIRA 1979; VIEIRA et al. 1980, 1983) and neurobehavioral functions (LEVIN et al. 1990; QUIMBY et al. 1974, 1975). There are also some reports that chronic exposure to anesthetic or subanesthetic concentrations of volatile anesthetics caused developmental toxicity (MAZZE et al. 1985; VIEIRA 1979; WHARTON et al. 1981). However, as discussed by those investigators, the effects seem to be due to maternal effects rather than direct effects of the drugs.

References

Abboud TK, Khoo SS, Miller F, Doan T, Henriksen EH (1982) Maternal, fetal and neonatal responses after epidural anesthesia with bupivacaine, 2-chloroprocaine, or lidocaine. Anesth Analg 61: 638–644

Aldridge LM, Tunstall ME (1986) Nitrous oxide and the fetus. A review and the results of a retrospective study of 175 cases of anaesthesia for insertion of Shirodkar suture. Br J Anaesth 59: 1348–1356

Anonymous (1995) Physicians' desk reference, 49th edn. Medical Economics, Montrale

Arcuri PA, Gautieri RF (1973) Morphine-induced fetal malformations. III. Possible mechanisms of action. J Pharm Sci 62: 1626–1634

Baden JM, Fujinaga M (1991) Effects of nitrous oxide on day 9 rat embryos grown in culture. Br J Anaesth 66: 500–503

Baden JM, Rice SA, Serra M, Kelley M, Mazze RI (1983) Thymidine and methionine syntheses in rats exposed to nitrous oxide. Anesth Analg 62: 738–741

Baden JM, Serra M, Mazze RI (1984) Inhibition of fetal methionine synthetase by nitrous oxide. Br J Anaesth 56: 523–526

Bailey-Pridham DD, Cook CL, Reshef E, Hurst HE, Drury K, Yussman MA (1990) Follicular fluid lidocaine levels during transvaginal oocyte retrieval. Fertil Steril 53: 171–173

Basford AB, Fink BR (1968) The teratogenicity of halothane in the rat. Anesthesiology 29: 1167–1173

Belaisch-Allart JC, Hazout A, Guillet-Rosso F, Glissant M, Testart J, Frydman R (1985) Various techniques for oocyte recovery in an in vitro fertilization and embryo transfer program. J In Vitro Fertil Embryo Trans 2: 99–104

Bowman RE, Smith RF (1977) Behavioral and neurochemical effects of prenatal halothane. Environ Health Perspect 21: 189–193

Boyers SP, Lavy G, Russell JB, DeCherney AH (1987) A paired analysis of in vitro fertilization and cleavage rates of first-versus last-recovered preovulatory human oocytes exposed to varying intervals of 100% CO_2 pneumoperitoneum and general anesthesia. Fertil Steril 48: 969–974

Brodsky JB, Cohen EN, Brown BW, Wu ML, Whitcher C (1980) Surgery during pregnancy and fetal outcome. Am J Obstet Gynecol 138: 1165–1167

Bueker ED, Platner WS (1956) Effect of cholinergic drugs on development of chick embryo. Proc Soc Exp Biol Med 91: 539–543

Buring JE, Hennekens CH, Mayrent SL, Rosner B, Greenberg ER, Colton T (1985) Health experiences of operating room personnel. Anesthesiology 62: 325–330
Bussard DA, Stoelting RK, Peterson C, Ishaq M (1974) Fetal changes in hamsters anesthetized with nitrous oxide and halothane. Anesthesiology 41: 275–278
Chalon J, Tang C-K, Ramanathan S, Eisner M, Katz R, Turndorf H (1981) Exposure to halothane and enflurane affects learning function of murine progeny. Anesth Analg 60: 794–797
Chanarin I (1980) Cobalamins and nitrous oxide: a review. J Clin Pathol 33: 909–916
Chetkowski RJ, Nass TH (1988) Isoflurane inhibits early mouse embryo development in vitro. Fertil Steril 49: 171–173
Chopineau J, Bazin J-E, Terrisse M-P, Sautou V, Janny L, Schoeffler P, Bastide P (1993) Assay for midazolam in liquor folliculi during in vitro fertilization under anesthesia. Clin Pharmacy 12: 770–773
Ciociola AA, Gautieri RF (1983) Evaluation of the teratogenicity of morphine sulfate administered via a miniature implantable pump. J Pharm Sci 72: 742–745
Coate WB, Kapp RW, Lewis TR (1979a) Chronic exposure to low concentrations of halothane-nitrous oxide: reproductive and cytogenetic effects in the rat. Anesthesiology 50: 310–318
Coate WB, Kapp RW Jr, Ulland BM (1979b) Toxicity of low concentration long-term exposure to an airborne mixture of nitrous oxide and halothane. J Environ Path Toxicol 2: 209–231
Coelho CND, Klein NW (1990) Methionine and neural tube closure in cultured rat embryos: morphological and biochemical analyses. Teratology 42: 437–451
Cohen EN (1980) Anesthetic exposure in the workplace. PSG, Littleton
Cohen SE (1994) Non-obstetric surgery during pregnancy. In: 45th Annual refresher course lectures and clinical update program. American Society of Anesthesiologists, Park Ridge, p 272
Corbett TH, Cornell RG, Endres JL, Millard RI (1973) Effects of low concentrations of nitrous oxide on rat pregnancy. Anesthesiology 39: 299–301
Crawford JS, Lewis M (1986) Nitrous oxide in early human pregnancy. Anaesthesia 41: 900–905
Critchlow BM, Ibrahim Z, Pollard BJ (1991) General anaesthesia for gamete intrafallopian transfer. Eur J Anaesthesiol 8: 381–384
Datta S, Lambert DH, Gregus J, Gissen AJ, Covino BG (1983) Differential sensitivities of mammalian nerve fibers during pregnancy. Anesth Analg 62: 1070–1072
Datta S, Migliozzi RP, Flanagan HL, Krieger NR (1989) Chronically administered progesterone decreases halothane requirements in rabbits. Anesth Analg 68: 46–50
Davis WM, Lin CH (1972) Prenatal morphine effects on survival and behavior of rat offspring. Res Commun Chem Pathol Pharmacol 3: 205–214
Drachman DB, Coulombre AJ (1962) Experimental club foot and arthrogryposis multiplex congenita. Lancet 2: 523–526
Duncan PG, Pope WDB, Cohen MM, Greer N (1986) The safety of anesthesia and surgery during pregnancy. Anesthesiology 64: 790–794
Edwards MJ (1988) Hyperthermia as a teratogen: a review of experimental studies and their clinical significance. Teratog Carcinog Mutagen 6: 563–582
Eisele JHJr (1985) Cardiovascular effects of nitrous oxide. In: Eger EI II (ed) Nitrous oxide/N_2O. Elsevier, New York, p 125
El-Karim AHB, Benny R (1976) Embryotoxic and teratogenic action of ketamine hydrochloride in rats. Ain Shavis Med J 27: 459–463
Endler GC, Hayes MF, Stout M, Moghissi KS, Magyar DM, Sacco AG (1987) Follicular fluid concentrations of thiopental and thiamylal during laparoscopy for oocyte retrieval. Fertil Steril 48: 828–833
Ericson HA, Källén AJB (1985) Hospitalization for miscarriage and delivery outcome among Swedish nurses working in operating rooms 1973–1978. Anesth Analg 64: 981–988

Exton JH (1988) The roles of calcium and phosphoinositides in the mechanisms of α1-adrenergic and other agonists. Rev Physiol Biochem Pharmacol 111: 117–231
Fagraeus L, Urban BJ, Bromage PR (1983) Spread of epidural analgesia in early pregnancy. Anesthesiology 58: 184–187.
Fink BR, Shepard TH, Blandau RJ (1967) Teratogenic activity of nitrous oxide. Nature 214: 146–148
Fishel S, Webster J, Faratian B, Jackson P (1987) General anesthesia for intrauterine placement of human conceptuses after in vitro fertilization. J In Vitro Fertil Embryo Trans 4: 260–264
Flynn TJ, Friedman L, Black TN, Klein NW (1987) Methionine and iron as growth for rat embryos cultured in canine serum. J Exp Zool 244: 319–324
Friedler G, Cochin J (1972) Growth retardation in offspring of female rats treated with morphine prior to conception. Science 175: 654–656
Fujinaga M, Baden JM (1991a) Evidence for an adrenergic mechanism in the control of body asymmetry. Dev Biol 143: 203–205
Fujinaga M, Baden JM (1991b) Critical period of rat development when sidedness of asymmetric body structures is determined. Teratology 44: 453–462
Fujinaga M, Baden JM (1993) Effects of lidocaine on rat embryos grown in culture during the early postimplantation period. Teratology 47: 418 (abstr)
Fujinaga M, Baden JM (1994) Methionine prevents nitrous oxide-induced teratogenicity in rat embryos grown in culture. Anesthesiology 81: 184–189
Fujinaga M, Mazze RI (1986) Reproductive and teratogenic effects of lidocaine in Sprague-Dawley rats. Anesthesiology 65: 626–632
Fujinaga M, Mazze RI (1988) Teratogenic and postnatal developmental studies of morphine in Sprague-Dawley rats. Teratology 38: 401–410
Fujinaga M, Stevenson JB, Mazze RI (1986) Reproductive and teratogenic effects of fentanyl in Sprague-Dawley rats. Teratology 34: 51–57
Fujinaga M, Baden JM, Yhap EO, Mazze RI (1987) Reproductive and teratogenic effects of nitrous oxide, isoflurane, and their combination in Sprague-Dawley rats. Anesthesiology 67: 960–964
Fujinaga M, Mazze RI, Jackson EC, Baden JM (1988a) Reproductive and teratogenic effects of sufentanil and alfentanil in Sprague-Dawley rats. Anesth Analg 67: 166–169
Fujinaga M, Mazze RI, Baden JM, Fantel AG, Shepard TH (1988b) Rat whole embryo culture: an in vitro model for testing nitrous oxide teratogenicity. Anesthesiology 69: 401–404
Fujinaga M, Baden JM, Mazze RI (1989) Susceptible period of nitrous oxide teratogenicity in Sprague-Dawley rats. Teratology 40: 439–444
Fujinaga M, Baden JM, Shepard TH, Mazze RI (1990) Nitrous oxide alters body laterality in rats. Teratology 41: 131–135
Fujinaga M, Baden JM, Suto A, Myatt JK, Mazze RI (1991) Preventive effects of phenoxybenzamine on nitrous oxide induced reproductive toxicity in Sprague-Dawley rats. Teratology 43: 151–157
Fujinaga M, Baden JM, Mazze RI (1992a) Developmental toxicity of nondepolarizing muscle relaxants in cultured rat embryos. Anesthesiology 76: 999–1003
Fujinaga M, Maze M, Hoffman BB, Baden JM (1992b) Activation of α-1 adrenergic receptors modulates the control of left/right sidedness in rat embryos. Dev Biol 150: 419–421
Fujinaga M, Hoffman BB, Baden JM (1994) Receptor subtype and intracellular signal transduction pathway associated with situs inversus induced by α1 adrenergic stimulation in rat embryos. Dev Biol 162: 558–567
Fujinaga M, Okazaki M, Baden JM, Hoffman BB (1995a) Gene expression of transcriptional factor induced by α1 adrenoceptor stimulation in rat embryos at the beginning of neurulation. Teratology 51: 184 (abstr)

Fujinaga M, Park HW, Shepard TH, Mirkes PE, Baden JM (1995b) Staurosporine does not block adrenergic-induced situs inversus, but causes a unique syndrome of defects in rat embryos grown in culture. Teratology 50: 261–274
Geber WF, Schramm LC (1975) Congenital malformations of the central nervous system produced by narcotics and analgesics in the hamster. Am J Obstet Gynecol 123: 705–713
Gin T, Chan MTV (1994) Decreased minimum alveolar concentration of isoflurane in pregnant humans. Anesthesiology 81: 829–832
Gintzler AR (1980) Endorphin-mediated increases in pain threshold during pregnancy. Science 210: 193–195
Green CJ, Monk SJ, Knight JF, Doré C, Luff NP, Halsey MJ (1982) Chronic exposure of rats to enflurane 200 p.p.m.: no evidence of toxicity or teratogenicity. Br J Anaesth 54: 1097–1104
Halsey MJ, Green CJ, Monk SJ, Doné C, Knight JF, Luff NP (1981) Maternal and paternal chronic exposure to enflurane and halothanc: fetal and histological changes in the rat. Br J Anaesh 53: 203–215
Harpel HS, Gautieri RF (1968) Morphine-induced fetal malformations. I. Exencephaly and axial skeletal fusions. J Pharm Sci 57: 1590–1597
Hayes MF, Magyar DM, Sacco AG, Endler GC, Savoy-Moore RT, Moghissi KS (1987) Effect of general anesthesia on fertilization and cleavage of human oocytes in vitro. Fertil Steril 48: 975–981
Hemminki K, Kyyrönen P, Lindbohm ML (1985) Spontaneous abortions and malformations in the offspring of nurses exposed to anesthetic gases, cytostatic drugs, and other potential hazards in hospitals, based in registered information of outcome. J Epidemiol Community Health 39: 141–147
Holson RR, Hansen DK, Laborde J, Bate H (1989) Chronic exposure of pregnant rats to trace levels of nitrous oxide (N_2O) does not alter performance of offspring on a behavioral test battery. Teratology 39: 508 (abstract)
Hood A, Brown J, Haddy S, Marrs R (1988) The effect of anesthesia on outcome following gamete intra-fallopian transfer (GIFT). Anesthesiology 69: A662 (abstr)
Iuliucci JD, Gautieri RF (1971) Morphine-induced fetal malformations. II. Influence of histamine and diphenhydramine. J Pharm Sci 60: 420–424
Jacobs RM (1971) Failure of muscle relaxants to produce cleft palate in mice. Teratology 4: 25–30
Jago RH (1970) Arthrogryposis following treatment of maternal tetanus with muscle relaxants. Arch Dis Child 45: 227–279
Johannesson T, Becker BA (1972) The effects of maternally-administered morphine on rat foetal development and resultant tolerance to the analgesic effect of morphine. Acta Pharmacol Toxicol 31: 305–313
Jouppila R, Jouppila P, Hollmen A, Kuikka J (1978) Effect of segmental extradural analgesia on placental blood flow during normal labour. Br J Anaesth 50: 563–567
Källén B, Mazze RI (1990) Neural tube defects and first trimester operations. Teratology 41: 717–720
Kangas-Saarela T, Jouppila R, Jouppila P, Hollman A, Puukka M, Juujarvi K (1987) The effect of segmental epidural analgesia on the neurobehavioural response of newborn infants. Acta Anaesthesiol Scand 31: 347–351
Keeling PA, Rocke DA, Nunn JF, Monk SJ, Lumb MJ, Halsey MJ (1986) Folinic acid protection against nitrous oxide teratogenicity in the rat. Br J Anaesth 58: 528–534
Kennedy GL Jr, Smith SH, Keplinger ML, Calandra JC (1976) Reproductive and teratologic studies with halothane. Toxicol Appl Pharma 35: 467–474
Kennedy GL Jr, Smith SH, Keplinger ML, Calandra JC (1977) Reproductive and teratologic studies with isoflurane. Drug Chem Toxicol 1: 75–88
Klug S, Lewandowski C, Wildi L, Neubert D (1990) Bovine serum: an alternative to rat serum as a culture medium for the rat whole embryo culture. Toxicol In Vitro 4: 598–601

Koëter HBWM (1990) Adverse effects of halothane N_2O-mixtures on postnatal development of rats. Teratology 41: 572 (abstr)
Koëter HBWM, Rodier PM (1986) Behavioral effects in mice exposed to nitrous oxide or halothane: prenatal vs. postnatal exposure. Neurobehav Toxicol Teratol 8: 189–194
Laborde JB, Holson RR, Bates HK (1988) Developmental toxicity evaluation of lidocaine in CD rats. Teratology 37: 472 (abstr)
Landauer W (1960) Nicotine-induced malformations of chick embryos and their bearing on the phenocopy problem. J Exp Zool 143: 107–122
Landauer W (1977) Cholinomimetic teratogens: studies with chick embryos. Teratology 12: 125–140
Lane GA, Nahraold ML, Tait AR, Taylor-Busch M, Cohen PJ, Beaudoin AR (1980) Anesthetics as teratogens: nitrous oxide is fetotoxic, xenon is not. Science 210: 899–901
Lansdown ABG, Pope WDB, Halsey MJ, Bateman PE (1976) Analysis of fetal development in rats following maternal exposure to subanesthetic concentrations of halothane. Teratology 13: 299–304
Lee H, Nagele RG (1985) Neural tube defects caused by local anesthetics in early chick embryos. Teratology 31: 119–127
Lee H, Bush KT, Nagele RG (1988) Time-lapse photographic study of neural tube closure defects caused by xylocaine in the chick. Teratology 37: 263–269
Lefebvre G, Vauthier D, Seebacher J, Henry M, Thormann F, Darbois Y (1988) In vitro fertilization: a comparative study of cleavage rates under epidural and general anesthesia – interest for gamete intrafallopian transfer. J In Vitro Fertil Embryo Trans 5: 305–306
Levin ED, DeLuna R, Uemura E, Bowman RE (1990) long-term effects of developmental halothane exposure on radial arm maze performance in rats. Behav Brain Res 36: 147–154
Lichtblau L, Sparber SB (1984) Opioids and development: a prospective on experimental models and methods. Neurobehav Toxicol Teratol 6: 3–8
Longaker MT, Golbus MS, Filly RA, Rosen MS, Chang SW, Harrison MR (1991) Maternal outcome after open fetal surgery. A review of the first 17 human cases. JAMA 265: 737–741
Martin LVH, Jurand A (1992) The absence of teratogenic effects of some analgesics used in anaesthesia Additional evidence from a mouse model. Anaesthesia 47: 473–476
Marx GF (1985) The N_2O dilemma. Obst Anesth Digest 5: 126–128
Matt DW, Steingold KA, Dastvan CM, James CA, Dunwiddie W (1991) Effects of sera from patients given various anesthetics on preimplantation mouse embryo development in vitro. J In Vitro Fertil Embryo Trans 8: 191–197
Mazze RI (1985) Fertility, reproduction, and postnatal survival in mice chronically exposed to isoflurane. Anesthesiology 63: 663–667
Mazze RI, Källén B (1989) Reproductive outcome after anesthesia and operation during pregnancy: a registry study of 5405 cases. Am J Obstet Gynecol 161: 1178–1185
Mazze RI, Wilson AI, Rice SA, Baden JM (1982) Reproduction and fetal development in mice chronically exposed to nitrous oxide. Teratology 26: 11–16
Mazze RI, Wilson AI, Rice SA, Baden JM (1984) Reproduction and fetal development in rats exposed to nitrous oxide. Teratology 30: 259–265
Mazze RI, Wilson AI, Rice SA, Baden JM (1985) Fetal development in mice exposed to isoflurane. Teratology 32: 339–345
Mazze RI, Fujinaga M, Rice SA, Harris SB, Baden JM (1986) Reproductive and teratogenic effects of nitrous oxide, halothane, isoflurane and enflurane in Sprague-Dawley rats. Anesthesiology 64: 339–344

Mazze RI, Fujinaga M, Baden JM (1987) Reproductive and teratogenic effects of nitrous oxide, fentanyl and their combination in Sprague-Dawley rats. Br J Anaesth 59: 1291–1297
Mazze RI, Fujinaga M, Baden JM (1988) Halothane prevents nitrous oxide teratogenicity in Sprague-Dawley rats; folinic acid does not. Teratology 38: 121–127
Meiniel R (1981) Neuromuscular blocking agents and axial teratogenesis in the avian embryo: can axial morphogenetic disorders be explained by pharmacological action upon muscle tissue? Teratology 23: 259–271
Minneman KP (1988) α1-Adrenergic receptor subtypes, inositol phosphates, and source of cell Ca^{2+}. Pharmacol Rev 40: 87–119
Mobbs IG, Parbrook GD, McKenzie (1966) The effect of various concentrations of nitrous oxide on the 24-hour explanted chick embryo. Br J Anaesth 38: 866–870
Mullenix PJ, Moore PA, Tassinari MS (1986) Behavioral toxicity of nitrous oxide in rats following prenatal exposure. Toxicol Ind Health 2: 261–271
McFarland CW, Witt BR, Wheeler CA, Thorneycroft IH (1989) Effects of short term exposure to lidocaine on mouse embryo development in vitro. Abstract book for 45th Annual Meeting of the American Fertility Society, S146 (abstr P-212)
National Institute for Occupational Safety and Health, DHEW (NIOSH) (1977) Criteria for a recommended standard. Occupational exposure to waste anesthetic gases and vapors. Publication No. 77–140. United States Department of Health, Education and Welfare, Public Health Service, Center for Disease control, Cincinnati
Nunn JF (1987) Clinical aspects of the interaction between nitrous oxide and vitamin B_{12}. Br J Anaesth 59: 3–13
Nunn JF, Chanarin I (1985) Nitrous oxide inactivates methionine synthase. In: Eger EI II (ed) Nitrous ;oxide/N_2O. Elsevier, New York, p 211
O'Callanghan JP, Holtzman SG (1976) Prenatal administration of morphine to the rat: tolerance to the analgesic effect of morphine in the offspring. J Pharmacol Exp Ther 197: 533–544
O'Shea KS, Kaufman MH (1980) Neural tube closure defects following in vitro exposure of mouse embryos to xylocaine. J Exp Zool 214: 235–238
O'Sullivan H, Jannings F, Ward K, McCann S, Scott JM, Weir DG (1981) Human bone marrow biochemical function and megaloblastic hematopoiesis after nitrous oxide anesthesia. Anesthesiology 55: 645–649
Palahniuk RJ, Shnider SM, Eger EI II (1974) Pregnancy decreases the requirement for inhaled anesthetic agents. Anesthesiology 41: 82–83
Palot M, Harika G, Pigeon F, Lamiable D, Rendoing J (1988) Propofol in general anesthesia for IVF (by vaginal and transurethral route); follicular fluid concentration and cleavage rate. Anesthesiology 69: A573 (abstract)
Palot M, Visseaux H, Harika G, Carré-Pigeon F, Rendoing J (1990) Effects of nitrous oxide and/or halothane on cleavage rate during general anesthesia for oocyte retrieval. Anesthesiology 73: A930 (abstr)
Parbrook GD, Mobbs I, Mackinzie J (1965) Effects of nitrous oxide on the early chick embryo. Br J Anaesth 37: 990–991
Persaud TVN (1965) Tierexperimentelle Untersuchungen zur Frage der teratogenen Wirkung von Barbituraten. Acta Biol Med Ger 14: 89–90
Peters MA, Hudson PM (1982) Postnatal development and behavior in offspring of enflurane-exposed pregnant rats. Arch Int Pharmacodyn Ther 256: 134–144
Pierce ET, Smalky M, Alper MM, Hunter JA, Armhein RL, Pierce EC (1992) Comparison of pregnancy rates following gamete intrafallopian transfer (GIFT) under general anesthesia with thiopental sodium or propofol. J Clin Anesth 4: 394–398
Pope WDB, Persaud TVN (1978) Foetal growth retardation in the rat following chronic exposure to the inhalational anaesthetic enflurane. Experientia 34: 1332–1333
Pope WDB, Halsey MJ, Lansdown ABG, Bateman PE (1975) Lack of teratogenic dangers with halothane. Acta Anaesthesiol Belg 26 [Suppl]: 169–173

Pope WDB, Halsey MJ, Lansdown ABG, Simmonds A, Bateman PE (1978) Fetotoxicity in rats following chronic exposure to halothane, nitrous oxide, or methoxyflurane. Anesthesiology 48: 11–16
Popova S, Virgieva T, Atanasova J, Atanasov A, Sahatchiev B (1979) Embryotoxicity and fertility study with halothane subanesthetic concentration in rats. Acta Anaesthesiol Scand 23: 505–512
Quimby KL, Aschkenase LJ, Bowman RE (1974) Enduring learning deficits and cerebral synaptic malformation from exposure to 10 parts of halothane per million. Science 185: 625–627
Quimby KL, Katz J, Bowman RE (1975) Behavioral consequences in rats from chronic exposure to 10 ppm halothane during early development. Anesth Analg 54: 628–633
Ramazzotto LJ, Carlin RD, Warchalowski GA (1979) Effects of nitrous oxide during organogenesis in the rat. J Dent Res 58: 1940–1943
Ramazzotto LJ, Paterson J, Tanner P, Coleman M (1985) Lidocaine administered during pregnancy. J Dent Res 64: 1214–1217
Raye JR, Dubin JW, Blechner JN (1977) Fetal growth retardation following maternal morphine administration: nutritional or drug effect? Biol Neonate 32: 222–228
Rector GHM, Eastwood DW (1964) The effects of an atmosphere of nitrous oxide and oxygen on the incubating chick. Anesthesiology 25: 109
Rice SA (1986) Behavioral effects of in utero isoflurane exposure in young SW mice. Teratology 33: 100C (abstr)
Rice SA (1990) Effect of prenatal N_2O exposure on startle reflex reactivity. Teratology 42: 373–381
Rice SA, Mazze RI, Baden JM (1985) Effects of subchronic intermittent exposure to nitrous oxide in Swiss Webster mice. J Environ Pathol Toxicol Oncol 6: 271–281
Rodier PM, Koëter HBWM (1986) General activity from weaning to maturity in mice exposed to halothane or nitrous oxide. Neurobehav Toxicol Teratol 8: 195–199
Rosen MA (1991) Anesthesia for procedures involving the fetus. Semin Perinatol 15: 410–417
Rosen MA, Roizen MF, Eger EI II, Glass RH, Martin M, Dandekar PV, Dailey PA, Litt L (1987) The effect of nitrous oxide on in vitro fertilization success rate. Anesthesiology 67: 42–44
Rowland AS, Baird DD, Weinberg CR, Shore DL, Shy CM, Wilcox AJ (1992) Reduced fertility among women employed as dental assistants exposed to high levels of nitrous oxide. N Engl J Med 327: 993–997
Ruffolo RR, Nichols AJ, Oriowo MA (1991) Interaction of vascular alpha-1 adrenoceptors with multiple signal transduction pathways. Blood Vessels 28: 122–128
Saito N, Urakawa M, Ito R (1974) Influence of enflurane on fetus and growth after birth in mice and rats. Oyo Yakuri 8: 1269–1276
Sander HW, Kream RM, Gintzler AR (1989) Spinal dynorphin involvement in the analgesia of pregnancy. Eur J Pharmacol 159: 205–209
Scanlon JW, Ostheimer GW, Lurie AO, Brown WU, Weiss JB, Alper MH (1976) Neurobehavioral responses and drug concentrations in newborns after maternal epidural anesthesia with bupivacaine. Anesthesiology 45: 400–405
Schnell VL, Sacco AG, Savoy-Moore RT, Ataya KM, Moghissi KS (1992) Effects of oocyte exposure to local anesthetics on in vitro fertilization and embryo development in the mouse. Reprod Toxicol 6: 323–327
Schoeffler PF, Levron JC, Jany L, Brenas FJ, Pouly JL (1988) Follicular concentration of fentanyl during laparoscopy for oocyte retrieval. Correlation with in vitro fertilization results. Anesthesiology 69: A663 (abstr)
Scott JM, Wilson P, Dinn JJ, Weir DG (1981) Pathogenesis of subacute combined degeneration: a result of methyl group deficiency. Lancet 8: 334–340
Sepkoski CM, Lester BM, Ostheimer GW, Brazelton TB (1992) The effects of maternal epidural anesthesia on neonatal behavior during the first month. Dev Med Child Neuro 34: 1072–1080

Shah RM, Burdett DN, Donaldson D (1979) The effects of nitrous oxide on the developing hamster embryos. Can J Physiol Pharmacol 57: 1229–1232
Shepard TH (1992) Catalog of teratogenic agents, 7th edn. John Hopkins University Press, Baltimore
Shepard TH, Fink BR (1968) Teratogenic activity of nitrous oxide in rats. In: Fink BR (ed) Toxicity of anesthetics. Williams and Wilkins, Baltimore, p 308
Shnider SM, Webster GM (1965) Maternal and fetal hazards of surgery during pregnancy. Am J Obstet Gynecol 92: 891–900
Shnider SM, Abboud TK, Artal R, Henriksen EH, Stefani SJ, Levinson G (1983) Maternal catecholamines decrease during labor after lumbar epidural anesthesia. Am J Obstet Gynecol 147: 13–15
Sia-Kho E, Grifo J, Liermann A, Mills A, Rosenwaks Z (1993) Comparison of pregnancy rates between propofol and midazolam in IVF-vaginal retrieval of oocytes. Anesthesiology 79: A1012 (abstr)
Skarpa M, Dayan AD, Follenfant M, James DA, Moore WB, Thompson PM, Lucke JN, Morgan M, Lovell R, Medd R (1983) Toxicity testing of atracurium. Br J Anaesth 55: 27S–29S
Slater GI, Aufses AH (1991) Complications of pregnancy: surgical aspects. In: Cherry SH, Merkatz IR (eds) Complications of pregnancy: medical, surgical, gynecologic, psychosocial, and perinatal, 4th edn. Williams and Wilkins, Baltimore, p 722
Smith BE (1963) Fetal prognosis after anesthesia during gestation. Anesth Analg 42: 521–526
Smith BE, Gaub ML, Moya F (1965) Teratogenic effects of anesthetic agents: nitrous oxide. Anesth Analg 44: 726–732
Smith DJ, Joffe JM (1975) Increased neonatal mortality in offspring of male rats treated with methadone or morphine before mating. Nature 253: 202–203
Smith RF, Bowman RE, Katz J (1978) Behavioral effects of exposure to halothane during early development in the rat: sensitive period during pregnancy. Anesthesiology 49: 319–323
Smith RF, Wharton GG, Kurtz SL, Mattran KM, Hollenbeck AR (1986) Behavioral effects of mid-pregnancy administration of lidocaine and mepivacaine in the rat. Neurobehav Toxicol Teratol 8: 61–68
Smits-van Prooije AE, Waalkens-Berendsen IDH, Koëter HBWM (1989) Adverse effects of chemical exposure on pre- and postnatal development of rats. Toxicology Tribune no. 7, TNO-CIVO Toxicology and Nutrition Institute, Aj Zeist, Netherlands
Smoak IW, Sadler TW (1991) Hypothermia: teratogenic and protective effects on the development of mouse embryos in vitro. Teratology 43: 635–641
Snergireff SL, Cox JR, Eastwood DW (1968) The effect of nitrous oxide, cyclopropane or halothane on neural tube mitotic index, weight, mortality and gross anomaly rate in the developing chick embryo. In: Fink BR (ed) Toxicity of anesthetics. Williams and Wilkins, Baltimore, p 279
Sobrian SK (1977) Prenatal morphine administration alters behavioral development in the rat. Pharmacol Biochem Behav 7: 285–288
Steele WJ, Johannesson T (1975) Effects of prenatally-administered morphine on brain development and resultant tolerance to the analgesic effect of morphine in offspring of morphine treated rats. Acta Pharmacol Toxicol 36: 243–256
Strickland RA, Oliver WC, Chantigian RC, Ney JA, Danielson GK (1991) Anesthesia, cardiopulmonary bypass, and the pregnant patient. Mayo Clin Proc 66: 411–429
Strout CD, Nahrwold ML (1981) Halothane requirement during pregnancy and lactation in rats. Anesthesiology 55: 322–323
Suzuki Y, Masuda H, Tanase H, Okonogi T (1970) The safety test of pancuronium bromide. Ann Sankyo Res Lab 22: 187–208
Swanson RJ, Leavitt MG (1992) Fertilization and mouse embryo development in the presence of midazolam. Anesth Analg 75: 549–554

Tanimura T (1965) Effect of administration thiamylal sodium to pregnant mice upon the development of their offspring. Kaibogaku Zasshi 40: 323–328
Tanimura T, Owaki Y, Nishimura H (1967) Effect of administration thiopental sodium to pregnant mice upon the development of their offspring. Okajimas Folia Anat Jpn 43: 219–226
Tassinari M, Mullenix P, Moore P (1986) The effects of nutrous oxide after exposure during middle and late gestation. Toxicol Ind Health 2: 261–271
Teiling AKY, Mohammed AK, Minor BG, Jarbe TUC, Hiltunen AJ, Archer T (1987) Lack of effects of prenatal exposure to lidocaine on development of behavior in rats. Anesth Analg 66: 533–541
Thalme B, Belfrage P, Raabe N (1974) Acid-base balance and clinical condition of mother, fetus and newborn child. Acta Obstet Gynecol Scand 53: 27–35
Tison CA, Reynolds F, Cabrol D (1994) The effects of maternal epidural anaesthesia on neonatal behaviour during the first month. Dev Med Child Neuro 367: 91–92
Upshall DG, Roger JC, Casida JE (1968) Biochemical studies on the teratogenic action of bidrin and other neuroactive agents in developing hen eggs. Biochem Pharmacol 17: 1529–1542
Van der Ven H, Diedrich K, Al-Hasani S, Pless V, Krebs D (1988) The effect of general anaesthesia on the success of embryo transfer following human in-vitro fertilization. Hum Reprod 3 [Suppl 2]: 81–83
Vieira E (1979) Effect of the chronic administration of nitrous oxide 0.5% to gravid rats. Br J Anaesth 51: 283–287
Vieira E, Cleaton-Jones P, Austin JC, Moyes DG, Shaw R (1980) Effects of low concentrations of nitrous oxide on rat fetuses. Anesth Analg 59: 175–177
Vieira E, Cleaton-Jones P, Moyes D (1983) Effects of low intermittent concentrations of nitrous oxide on the developing rat fetus. Br J Anaesth 55: 67–69
Vincent RD Jr, Syrop CH, Van Voorhis BJ, Chestnut DH, Sparks AET, McGrath JM, Choi WW, Bates JN, Penning DH (1995) An evaluation of the effect of anesthetic technique on reproductive success after laparoscopic pronuclear stage transfer. Anesthesiology 82: 352–358
Warren JR, Shaw B, Steinkampf MP (1990) Effects of nitrous oxide on preimplantation mouse embryo cleavage and development. Biol Reprod 43: 158–161
Warren JR, Shaw B, Steinkampf MP (1992) Inhibition of preimplantation mouse embryo development by isoflurane. Am J Obstet Gynecol 166: 693–698
Wharton RS, Mazze RI, Baden JM, Hitt BA, Dooley JR (1978) Fertility, reproduction and postnatal survival in mice chronically exposed to halothane. Anesthesiology 48: 167–174
Wharton RS, Wilson AI, Mazze RI, Baden JM, Rice SA (1979) Fetal morphology in mice exposed to halothane. Anesthesiology 51: 532–537
Wharton RS, Mazze RI, Wilson AI (1981) Reproduction and fetal development in mice chronically exposed to enflurane. Anesthesiology 54: 505–510
Whitcher C (1985) Controlling occupational exposure to nitrous oxide. In: Eger EI II (ed) Nitrous oxide/N_2O. Elsevier, New York, p 313
Wikland M, Enk L, Hammarberg K, Nilsson L, Hamberger L (1988) Oocyte retrieval under the guidance of a vaginal transducer. Ann NY Acad Sci 541: 103–110
Wikland M, Evers H, Jakobsson A-H, Sandqvist U, Sjöblom P (1990) The concentration of lidocaine in follicular fluid when used for paracervical block in a human IVF-ET programme. Hum Reprod 5: 920–923
Zagon IS, McLaughlin PJ (1977a) Morphine and brain growth retardation in the rat. Pharmacology 15: 276–282
Zagon IS, McLaughlin PJ (1977b) Effects of chronic morphine administration on pregnant rats and their offspring. Pharmacology 15: 302–310
Zimmerberg B, Charap AD, Glick SD (1974) Behavioral effects of in utero administration of morphine. Nature 247: 376–377

CHAPTER 29

Alcohols: Ethanol and Methanol

J.M. Rogers and G.P. Daston

A. Introduction

Ethanol is a human teratogen with a long history, and experimental studies of its teratogenicity extend back to the turn of the century (Feré 1895; Stockard 1910; Pearl 1916). Methanol is also a well-known human toxicant, producing ocular toxicity and death following acute exposure by a number of routes (Tephly and McMartin 1984; Kavet and Nauss 1990), but human developmental toxicity has not been reported. Methanol has recently been identified as a rodent teratogen in studies undertaken because of the potential for occupational exposures or increased use of this alcohol in vehicle fuels (Nelson et al. 1985; Rogers et al. 1993; Bolon et al. 1993). This chapter describes the dysmorphogenic effects of prenatal exposure to ethanol in humans and experimental animals and reviews the literature germane to understanding the pathogenesis and underlying mechanisms of ethanol- and methanol-induced birth defects. These alcohols have common pathways of metabolism which must be considered when evaluating putative mechanisms by which these alcohols might produce dysmorphogenesis following maternal exposure; aspects of the pharmacokinetics of ethanol and methanol relevant to evaluating animal models and mechanisms of action are presented. In this analysis, we also discuss in vitro approaches, principally whole embryo culture, which have been important in elucidating the developmental toxicity of ethanol, methanol, and their metabolites, as well as for furthering our understanding of mechanisms of action.

B. Human Toxicity

I. Ethanol: Fetal Alcohol Syndrome

Despite recurrent historical references to ethanol-induced developmental toxicity, clinical reports establishing a link between prenatal ethanol exposure and birth defects were fairly recent, first in France (Lemoine et al. 1968), and then in the United States by Jones and coworkers, who described the fetal alcohol syndrome (FAS) (Jones and Smith 1973; Jones et al. 1973). Since the coining of this term, there have been hundreds of clinical, epidemiological, and experimental studies of the effects of ethanol exposure during gestation.

FAS comprises abnormalities in three categories:

1. Prenatal and/or postnatal growth retardation
2. Central nervous system involvement such as mental retardation or other neurological abnormalities
3. Characteristic craniofacial dysmorphism (microcephaly, microphthalmia and/or short palpebral fissures, poorly developed philtrum, thin upper lip, and flattening of the maxillary area)

An individual must exhibit an abnormality in each of the three categories for a diagnosis of FAS to be made (ABEL 1990). These are minimal criteria, as numerous other alcohol-related birth defects (ARBDs), including cardiac, urogenital, and skeletal abnormalities, have been associated with FAS (Table 1). FROSTER and BAIRD (1992) recently reported an association between transverse limb defects and maternal ethanol abuse.

FAS has only been observed in children born to alcoholic mothers. There are numerous methodological difficulties involved in estimating the level of maternal ethanol consumption associated with FAS, but estimates of a minimum of 3–4 oz ethanol per day have been made (CLARREN et al. 1987a; ERNHART et al. 1987). ERNHART and coworkers (1987) based their estimate of 3 oz ethanol (six drinks) on a study of 359 newborns whose mothers' drinking histories were assessed prospectively during antenatal care visits. The Michigan Alcoholism Screening Test (MAST) was used, and women testing positive for alcoholism were compared to women testing negative, after matching for race, smoking, parity, prepregnancy weight, drug abuse, and other variables. The estimated threshold for blacks in this study was significantly lower, four drinks per day, than for whites (six drinks per day).

Although this chapter deals principally with congenital malformations, other forms of developmental toxicity also result from ethanol exposure during gestation. For example, heavy drinking episodes by women during the first trimester have been associated with an increase in spontaneous abortions (SOKOL et al. 1980). Cognitive deficits in FAS children are devastating. The average intelligence quotient (IQ) of FAS children has been reported to be 67-68 (ABEL 1990; STREISSGUTH et al. 1991a) and changes little over time (STREISSGUTH et al. 1991b). Effects of maternal alcohol consumption during pregnancy on attention, short-term memory, and performance on standardized tests have been noted in a longitudinal, prospective study of 462 children (STREISSGUTH et al. 1994a,b). A number of measures of alcohol intake were related to these effects, but the number of drinks per drinking occasion was the strongest predictor.

The most common effect of gestational alcohol exposure is prenatal growth retardation (ABEL 1982, 1990). In a study by WEGMAN (1987), the incidence of live-born FAS infants weighing less than 2500 g at birth was 77%, compared to an overall incidence of 6.8% in the United States. Effects of alcohol on weight are clear at birth, and deficits in weight and height can persist through the preschool years (STREISSGUTH et al. 1985). However, an-

Table 1. Abnormalities associated with fetal alcohol syndrome. (Modified from ABEL 1990)

Type	Abnormality
Cardiac	Ventricular septal defect
	Atrial septal defect
	Pulmonary stenosis
	Patent ductus arteriosus
	Tetralogy of Fallot
	Aberrant great vessels
Liver	Fibrosis
Kidney/urinary	Hydronephrosis
	Hypoplastic, dysplastic, or aplastic kidney
	Obstruction of uteropelvic function
Genital	Clitoromegaly
	Hypoplastic labia
	Undescended testicle
	Hypospadias
Cutaneous	Hemangiomas
	Hirsutism
	Abnormal palmar creases
Musculature	Hernias (diaphragm, umbilicus)
Skeletal	Hypoplastic nails
	Clinodactyly
	Camptodactyly
	Radioulnar synostosis
	Scoliosis
	Shortened fifth digit
	Hip dislocation
	Pectus excavatum and carinatum
	Fusion of vertebrae
	Transverse limb defects
Neural tube defects	Meningomyelocele
	Myelomeningocele
	Anencephaly
	Hydrocephaly
	Porencephaly
	Aberrant neuronal/glial migration
Tumors	Adrenal carcinoma
	Ganglioneuroblastoma
	Hepatoblastoma
	Neuroblastoma
	Nephroblastoma
	Rhabdomyosarcoma
	Wilm's tumor
	Medulloblastoma
	Sacrococcygeal teratoma
	Hodgkin's disease
	Leukemia

other report indicates that effects of ethanol were not evident based on weight between 8 months and 14 years of age (STREISSGUTH et al. 1994c). Some studies have found that alcohol consumption can affect birth weight in a dose-related fashion even if the mother is not alcoholic. LITTLE (1977) studied 800 women prospectively to evaluate the effects of drinking on birth weight. After adjusting for smoking, gestational age, maternal height, age, and parity, and sex of the child, it was found that for each ounce of absolute ethanol (the equivalent of two drinks) consumed per day during late pregnancy there was a 160-g decrease in birth weight. ROSETT et al. (1983) reported a 700-g decrease in birth weight for women consuming five or more drinks per occasion or more than 45 drinks per month during pregnancy. However, nearly all of this decrease in weight was attributed to shortened gestation.

Isolated or more subtle expressions of the toxicity of prenatal ethanol exposure have been termed fetal alcohol effects (FAE) (ROSETT 1980), although it has recently been proposed that the clinical use of this term be abandoned because it implies causation based on rather nonspecific diagnostic features (AASE et al. 1995). The difficulty for clinicians is that no single feature of FAS is pathognomonic for fetal alcohol exposure, and many are within the normal range of human variability or may be associated with factors other than alcohol.

Data from the national Birth Defects Monitoring Program indicate that FAS identified in coded hospital discharge diagnoses among newborns in the United States increased from 1 per 10 000 births in 1979 to 6.7 per 10 000 (0.67 per 1000) births in 1993 (Fig. 1; CENTERS FOR DISEASE CONTROL AND PREVENTION 1995). This may indicate a true increase in FAS, but there is likely a contribution from increased awareness and diagnosis of FAS by clinicians. ABEL (1995) estimated the U.S. incidence at 1.95 per 1000, but stressed that there are large differences in incidence among different racial and socioeconomic groups. The incidence at sites serving a primarily Caucasian/middle socioeconomic status (SES) population was 0.26 per 1000, almost ten times lower than the 2.29 per 1000 rate found where the population is mostly African American/low SES. Worldwide incidence was estimated at 0.97 per 1000 live births. FAS only occurs among alcoholics, and the rates of FAS among identified populations of alcohol-abusing mothers are much higher. Estimates of the incidence of FAS in children of women characterized as "heavy drinkers" (defined as consuming an average of two or more drinks per day, *or* five to six drinks per occasion, *or* a positive MAST score, *or* clinical diagnosis for alcohol abuse) range from about 22 per 1000 to more than 300 per 1000, with an overall rate of 43.1 per 1000 (ABEL 1995).

It is clear that a relatively small percentage of alcoholic mothers give birth to FAS children. SOKOL and coworkers (1980) reported that only 2.5% of 204 pregnant women identified as abusive drinkers gave birth to children with FAS. The biology underlying this relatively low incidence of FAS among the offspring of alcohol-abusing mothers has not been elucidated, but a number of maternal risk factors have been identified. Alcoholic women that have one

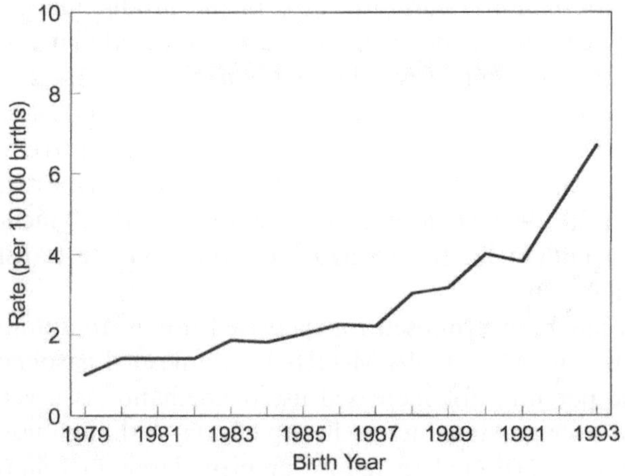

Fig. 1. Reported rate (per 10 000 births) of fetal alcohol syndrome by birth year identified in coded hospital discharge diagnoses (International Classification of Diseases, ninth revision clinical modification, code 760.71: "Noxious influences affecting fetus or newborn via placenta or breast milk, specifically alcohol; including fetal alcohol syndrome") in the United States from 1979 to 1993. Data from the National Birth Defects Monitoring Program. (Modified from CENTERS FOR DISEASE CONTROL AND PREVENTION 1995, with permission)

FAS child are at much greater risk of giving birth to another child with FAS than are alcoholics that have not already had an FAS child. SOKOL et al. (1986) found that women giving birth to FAS children were more likely to be black, multiparous, older, and heavier drinkers. Characteristics of high-risk pregnancies include chronic drinking, consumption of more than ten drinks per day, multiparity, age 30 years or older, low prepregnancy weight, history of spontaneous abortions, low SES, and general poor health (ABEL 1990). Mothers of FAS children also tend to gain little weight during pregnancy (HINGSON et al. 1982). ABEL and HANNIGAN (1995) propose that cultural risk factors such as low SES are "permissive" for FAS because of related conditions such as poor diet (especially for antioxidants), smoking, increased parity, stress, and increased exposure to environmental pollutants. These conditions combine with alcohol intake patterns and blood alcohol levels to produce FAS, with maternal/fetal hypoxia and free radical damage playing key roles (see Fig. 7 and below). The cellular mechanisms of ethanol teratogenesis will be discussed later.

II. Methanol: Adult Human Toxicity

Methanol has been widely used as a solvent for paint and industrial cleaning, as well as a chemical feedstock, since purification and deodorizing processes were developed in the 1890s. It has also been used as an adulterant in alcoholic beverages, resulting in recurrent episodes of large-scale poisonings (KAVET and

NAUSS 1990). Methanol is currently used in the production of methyl *tert*-butyl ether (MTBE), formaldehyde, acetic acid, and methyl methacrylate. It is a component of a number of widely used consumer products, including antifreeze, windshield washer fluids, and solvents for duplicating machines (CONIBEAR 1988). The UNITED STATES ENVIRONMENTAL PROTECTION AGENCY (1994) reported that methanol was the chemical with highest release to the environment in the 1992 Toxic Release Inventory of 23 630 facilities, totaling more than 214 million lb. It also had the largest off-site transfers, totaling almost 114 million lb.

Methanol has been proposed for increased use as an automobile fuel in gasoline blends, e.g., M-85 (85% MeOH, 15% unleaded gasoline), or as pure methanol. The potential for increased use of methanol as a vehicle fuel has raised concern over possible human health effects. Although use of methanol as a vehicle fuel is expected to reduce ambient levels of criteria pollutants including ozone and nitrogen oxides, there may be increases in airborne levels of methanol as well as the combustion product formaldehyde.

Although developmental toxicity of methanol in humans has not been reported, the ocular and systemic toxicity of methanol in humans has been recognized since the turn of the twentieth century. By 1904, BULLER and WOOD had collected 235 cases of blindness or death among individuals exposed to methanol orally, percutaneously, or by inhalation (BULLER and WOOD 1904; WOOD and BULLER 1904; cited in TEPHLY and MCMARTIN 1984). In 1987, 1601 methanol poisonings were reported to the American Association of Poison Control Centers, and the total incidence of methanol poisoning in the United States was estimated to be about 6400 cases (LITOVITZ 1988; cited in KAVET and NAUSS 1990).

Typically, victims of acute methanol poisoning experience a central nervous system depression similar to that produced by ethanol, but to a lesser degree. An asymptomatic latent period lasting from several hours to as much as 1–2 days follows. Subsequently, visual disturbances, intense abdominal pain, headache, dizziness, weakness, and nausea occur. Breathing problems associated with metabolic acidosis occur in severe cases, and convulsions and coma can ensue prior to death. The cause of death is usually respiratory failure. Blindness is a characteristic feature in survivors of acute methanol poisoning, usually resulting from severe edema of the optic disk.

Metabolic acidosis was first reported in a methanol poisoning victim by HARROP and BENEDICT (1920). Largely because it was not observed in lower animals, metabolic acidosis was not generally accepted as a major feature of methanol poisoning until CHEW et al. (1946) showed that alkali therapy benefited victims of methanol poisoning. As discussed in the next section, metabolic acidosis is due to the buildup of formate resulting from methanol metabolism.

There is wide variation among individuals in the amount of methanol required to elicit toxicity. For example, BENNETT and coworkers (1953) reported that as little as 15 ml of 40% methanol produced death in some victims, while in others as much as 500 ml did not induce permanent damage.

It should be emphasized that, although the known acute toxic effects of methanol in humans have been attributed to formate buildup, the consequences of exposure to lower levels of methanol not producing a formate overload are unknown. Thus, in understanding the developmental toxicity of methanol, it is important to elucidate the toxicity of the parent compound as well as its metabolites, formaldehyde and formate.

C. Pharmacokinetics and Metabolism

I. Absorption and Distribution

Ingested ethanol and methanol are absorbed from the entire length of the gastrointestinal tract, most rapidly in the small intestine. Absorption of methanol is slower than absorption of ethanol (TEPHLY and MCMARTIN 1984). The presence of food, factors that retard gastric emptying, or high concentrations of alcohol that inhibit gastric motility can slow absorption. Absorption of orally administered ethanol can also be affected by nutritional status, total body water content, and volume and rate of ethanol consumption. In rats and mice, absorption of methanol from the gastrointestinal tract appears to have both a fast and a slow component (WARD et al. 1995). Following an oral dose of 2.5 g methanol/kg, a significant portion of the dose was absorbed very rapidly, with the remainder absorbed for up to 5 h postgavage in mice and up to 10 h postgavage in rats. Both ethanol and methanol may also be absorbed via inhalation. Methanol is known to be absorbed through the skin, whereas percutaneous absorption of ethanol appears to be negligible (GUMMER and MAIBACH 1986). Once absorbed, methanol and ethanol rapidly distribute to the total body water and readily cross cell membranes by diffusion.

The pharmacokinetics of ethanol during pregnancy have been studied in several species under different experimental conditions. Ethanol passes relatively freely between the maternal and embryofetal compartments (DILTS 1970; KESÄNIEMI and SIPPEL 1975; BLAKLEY and SCOTT 1984a). Administration of single low doses of ethanol to pregnant women at term (IDANPAAN-HEIKKILA et al. 1972) and to third-trimester pregnant ewes (AYROMLOOI et al. 1979; BRIEN et al. 1985; COOK et al. 1981; CUMMING et al. 1984; NG et al. 1982; URFER et al. 1984), rhesus and cynomolgus monkeys (HILL et al. 1983), and guinea pigs (CLARKE et al. 1985) demonstrated bidirectional flux of ethanol across the placenta and similar elimination kinetics in the mother and fetus. In second-trimester pregnant women (BRIEN et al. 1983) and third-trimester pregnant ewes (BRIEN et al. 1985; NG et al. 1982) and guinea pigs (CLARKE et al. 1985), transfer of ethanol into amniotic fluid showed an initial lag followed by a higher ethanol concentration in the amniotic fluid compared to maternal blood during elimination. CLARKE et al. (1986) administered four doses of 1 g ethanol/kg at 1-h intervals to third-trimester guinea pigs and followed the pharmacokinetics of ethanol and acetaldehyde. The maternal and fetal blood profiles for ethanol were very similar, as were blood and brain concentrations in both the mother and the fetus. Higher concentrations of ethanol in the amniotic fluid were again observed during elimination, indicating that this compartment may serve as a reservoir for ethanol. Acetaldehyde concentrations in tissues were variable but at least 1000-fold lower than corresponding ethanol concentrations.

WARD and POLLACK (1996a) found that, in both rats and mice, methanol rapidly diffused from the dam to the conceptus after an i.v. or oral bolus dose, with distribution equilibrium typically reached within 30 min. At equilibrium and during elimination, conceptal methanol concentrations tended to be slightly higher than those in the dam. The rate of methanol diffusion to the fetus was dose -dependent, with proportionally less methanol reaching the fetus as the maternal dosage increased. Subsequent studies using the conceptal uptake of ^3H_2O as a marker of uteroplacental perfusion indicated that methanol may interfere with regional blood flow in a concentration-dependent fashion in both rats and mice (WARD and POLLACK 1996b).

HORTON et al. (1992) constructed a physiologically based pharmacokinetic model for inhaled methanol in rats, monkeys, and humans. Fischer-344 rats and rhesus monkeys were exposed to methanol via inhalation at 50–2000 ppm for 6 h. These exposures did not produce an elevation in blood formate in either species. Metabolism accounted for almost 97% of excretion of an i.v. methanol dose in rats. The models were used to predict the exposure range over which methanol pharmacokinetics would be similar in the laboratory species and humans. The authors found that similar end-of-exposure blood methanol concentrations would be expected in humans and experimental animals at exposure concentrations below about 1200 ppm, but that at higher concentrations there would be significant species differences.

PERKINS et al. (1995a) point out that blood methanol concentrations may be expected to vary widely between species and that species differences may not necessarily be due to readily determined toxicokinetic parameters. These authors reported two- to three fold higher blood methanol levels in mice than in rats after similar inhalation exposures (Fig. 2), despite a two fold higher V_{max} for elimination in mice. Higher blood levels in mice account in part for the greater sensitivity of mice than rats to the developmental toxicity of inhaled methanol (see below). Further, PERKINS et al. (1995b) developed a pharmacokinetic model of inhaled methanol in humans and predicted that, following an 8-h exposure to 5000 ppm methanol vapor, blood methanol concentrations in humans would be five fold lower than in rats and 13- and 18-fold lower than in the mouse (the most sensitive species to date for developmental toxicity; ROGERS et al. 1993a). These finding clearly demonstrate that species differences in absorption kinetics for inhaled methanol are a crucial consideration for interspecies extrapolation and human risk assessment.

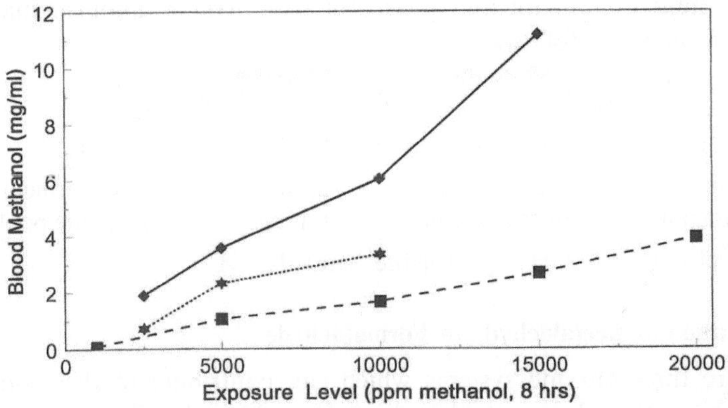

Fig. 2. Blood methanol concentrations in rats and mice following 8-h exposures to various concentrations of methanol. Different levels of activity for individually versus group-housed mice affected ventilation rate and blood methanol accumulation, so separate lines are shown for these two conditions. *Squares*, rats, *Stars*, group-housed mice; *diamonds*,individually-housed mice. (Data from PERKINS et al. 1995a)

Therefore, to the extent possible, any human risk assessment of inhaled methanol should be based on blood levels rather than exposure concentrations.

II. Metabolism

The major route of elimination for both ethanol and methanol is via oxidative metabolism, although excretion in expired air and urine may each account for 1%–5% of total ethanol excretion (BATT 1989). Likewise, following low dosages of methanol to rhesus monkeys and rats, as much as 90% is metabolized (OPPERMAN 1984).

Once a sufficient interval after dosing has passed so that the kinetic effects of absorption and distribution are small, blood ethanol clearance curves approximate linearity. The maximal elimination rate (V_{max}) for ethanol is quite variable among individuals, but is approximately 8.5g/h. End-phase elimination concentrations drop below the K_m values of the enzymes involved (probably below 1 mM), at which point metabolism becomes first order (VON WARTBURG 1989). It is important to note that blood ethanol levels attained after drinking alcoholic beverages greatly exceed the K_m for ethanol elimination.

A similar situation exists for methanol clearance. Below saturation, in vivo metabolism of methanol to CO_2 in Holtzman rats and rhesus monkeys obeys first-order (Michaelis-Menten) kinetics, and methanol metabolism in the two species have a similar V_{max} and K_m (TEPHLY et al. 1964; MAKAR et al. 1968). Above saturation, methanol metabolism exhibits zero-order kinetics. In cynomolgus monkeys, NOKER et al. (1980) showed a transition from zero-order to first-order kinetics at around 10 mM methanol in the blood.

The steps of ethanol and methanol oxidative metabolism may be represented simply as follows:

	Step 1		*Step 2*		*Step 3*	
$R\text{-}CH_2OH$	------>	R-CHO	------>	$R\text{-}COO^-$	------>	$CO_2 + H_2O$
ethanol		acetaldehyde		acetate		(and intermediary
methanol		formaldehyde		formate		metabolic pools)

For ethanol $R = CH_3$ and for methanol $R = H$.

1. Oxidation to Acetaldehyde or Formaldehyde

There are three enzyme systems which can contribute to the oxidation of ethanol or methanol to acetaldehyde or formaldehyde. These are the alcohol dehydrogenases (ADH), catalase, and the microsomal ethanol oxidizing system (MEOS):

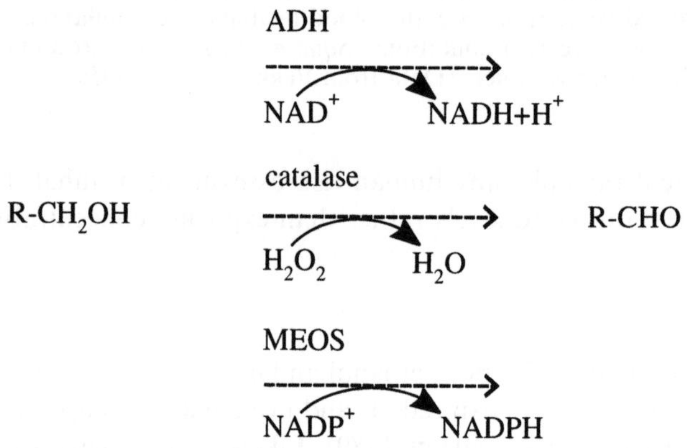

ADH is a well-characterized enzyme family present in all animals analyzed. In vertebrates, ADH has been demonstrated to catalyze numerous alcohol–aldehyde interconversions, including ethanol (KREBS and PERKINS 1970; LI 1977), methanol (VALLEE and BAZZONE 1983) and retinol (MEZEY and HOLT 1971). Human liver ADH is a zinc metalloenzyme with five classes of multiple molecular forms produced by the association of eight different types of subunits (LIEBER 1994). Five gene loci are involved, ADH1–5 (BOSRON et al. 1993). Both class I and class II ADH isozymes oxidize ethanol in the human liver, but class I isozymes have a much lower K_m (0.2–2 mM compared to 34 mM for class II); class II isozymes do not oxidize ethanol in the liver because of their very low affinity for that substrate (LIEBER 1994). Class IV ADH has been purified from human stomach (YIN et al. 1990; MORENO and PARÉS 1991), and another new form (class V) in liver and stomach has been reported (PARÉS et al. 1990; YASUNAMI et al 1991). First-pass metabolism by the gastric forms of ADH decreases the availability of ethanol and can be a "protective barrier" against systemic effects when ethanol is consumed in small amounts; however, gastric ADH activity is decreased in alcoholics (LEIBER 1994). ADH is the primary route of ethanol metabolism in both primates and rodents.

Primates also metabolize methanol principally via ADH, although only the class I isozymes appear to be involved (VALLEE and BAZZONE 1983).

Stage- and tissue-specific expression of the gene *Adh-1* (the only murine class I ADH) had been examined in the mouse embryo (VONESCH et al. 1994). This gene, which also oxidizes retinol to retinaldehyde, was first detected at gestation day 10.5 in mesenchyme of the mesonephros, and expression extended to other tissues as development progressed. It has been proposed that ethanol may exert its developmental toxicity at least in part by inhibiting retinol oxidation (DUESTER 1994a). The putative role of retinol metabolism in FAS will be discussed below.

The catalase pathway is the primary route of methanol oxidation to formaldehyde in rodents (TEPHLY and MCMARTIN 1984). Both ethanol and methanol can be oxidized to the corresponding aldehydes via catalase, but these reactions require a source of hydrogen peroxide (H_2O_2). The oxidation of fatty acids, urate, and glycolate in peroxisomes generates H_2O_2, and this process is the rate-limiting step in the oxidation of ethanol and methanol via catalase (THURMAN et al. 1989). Species differences in the activity of the catalase pathway, especially for methanol, are likely due to the absent to very low activity of hepatic urate oxidase, glycolate oxidase, xanthine oxidase, and other peroxide-generating systems in monkeys and humans compared to rodents (TEPHLY and MCMARTIN 1984). Despite the qualitative differences between rodents and primates in the oxidation of methanol to formaldehyde, this metabolic step proceeds at similar rates in nonhuman primates and rats (TEPHLY et al. 1964; MAKAR et al. 1968).

In rodents, unlike primates, catalase appears to be a major pathway of ethanol metabolism in addition to being the primary pathway for methanol metabolism. The deer mouse (*Peromyscus maniculatus*) mutant lacking hepatic class I ADH (FELDER et al. 1983; POSCH et al. 1989) has been used for studies of ethanol elimination by non-ADH systems. In this mutant, catalase predominates in the oxidation of ethanol. Recently, 4-methylpyrazole, a known ADH inhibitor, was found to reduce ethanol metabolism in class I ADH-deficient mice. Decreased ethanol metabolism was due to inhibition of fatty acylcoenzyme A (acyl-CoA) synthetase, resulting in decreased H_2O_2 production from peroxisomal fatty acid oxidation (BRADFORD et al. 1993a). This finding suggests that previous estimates of ethanol oxidation by ADH based on 4-methylpyrazole inhibition may be too high. Indeed, subsequent studies in ADH-positive deer mice demonstrated that the contribution of catalase was about 50% at low doses of ethanol (i.e., those producing blood levels of around 50 mg/dl) and approached 100% as the blood alcohol concentration was elevated (BRADFORD et al. 1993b).

The MEOS system comprises both enzymic and nonenzymic components (KOOP 1989). Isozymes of the cytochromes p-450 that can oxidize ethanol or methanol have been identified, and cytochrome p-450IIE1 (CYP2E1) is known to be inducible by ethanol (KOOP and TIERNEY 1990) or methanol (ALLIS et al. 1992). The MEOS system has a relatively high K_m for ethanol (8–10 mM, well within the range of heavy drinkers) and may therefore account for a larger proportion of ethanol oxidation at high ethanol concentrations in primates, but probably not in rodents (CROW and HARDMAN 1989). Nonenzymatic microsomal oxidation of ethanol and methanol may be catalyzed by physiologically reduced iron chelates, dependent on the presence of H_2O_2. The contribution of this nonenzymic pathway to the oxidation of ethanol or methanol in vivo has not been well characterized.

2. Oxidation to Acetate or Formate

Liver aldehyde dehydrogenase (AlDH) has broad specificity and in the presence of NAD^+ can oxidize acetaldehyde or formaldehyde to acetic or formic acid. All mammalian species studied to date have at least one hepatic form of AlDH with a low K_m for acetaldehyde (less than 10 μM), as well as higher K_m forms (CROW and HARDMAN 1989). In addition, there is a specific formaldehyde dehydrogenase (FDH) present in both rat and human liver which requires both NAD^+ and reduced glutathione (GSH) for activity (GOODMAN and TEPHLY 1970).

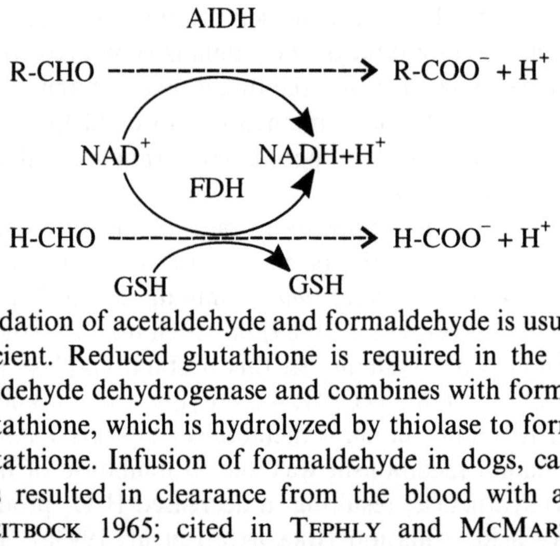

Oxidation of acetaldehyde and formaldehyde is usually rapid, and clearance is efficient. Reduced glutathione is required in the reaction catalyzed by formaldehyde dehydrogenase and combines with formaldehyde to form *s*-formyl glutathione, which is hydrolyzed by thiolase to form formic acid and reduced glutathione. Infusion of formaldehyde in dogs, cats, rabbits, guinea pigs, or rats resulted in clearance from the blood with a half-life of about 1 min (REITBOCK 1965; cited in TEPHLY and MCMARTIN 1984). However, formaldehyde is highly reactive with proteins, nucleic acids, and other endogenous compounds. Thus, intracellular formation of formaldehyde at the target site could play an important role in methanol toxicity.

3. Conversion to CO_2 and H_2O

a) Ethanol

Much of the acetate formed by the oxidation of acetaldehyde is exported from the liver and enters normal extrahepatic intermediary metabolic pools; it is ultimately converted to CO_2 and H_2O or incorporated into lipids after conversion to acetyl-CoA (CROW and GREENWAY 1989). Within 25 min of an acute ethanol dose, hepatic venous acetate concentration rises about ten fold and remains elevated until the ethanol has been metabolized.

b) Methanol

Differences in rates of metabolism of formate derived from methanol are a key factor underlying species differences in the toxicity of methanol. All of the known acute toxic effects of methanol in humans have been ascribed to a buildup of formate. Rodents can metabolize formate to CO_2 via catalase, but this appears to be a minor pathway. In both rodents and primates, formate is oxidized through the folate biochemical pathway after combining with tetrahydrofolate (THF), catalyzed by formyl-THF synthetase (f-THF-S), to produce 10-formyl-THF. The formyl group is then oxidized to CO_2 by 10-formyl-THF oxidoreductase (10-f-THF-O) (TEPHLY and MCMARTIN 1984).

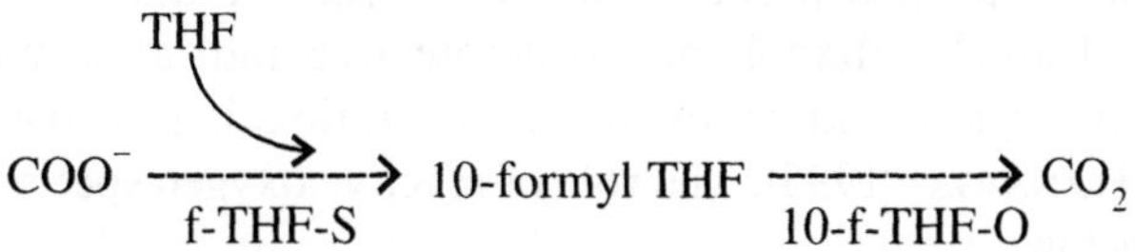

Although the folate pathway is utilized in both rodents and primates, in rodents it is more efficient. Both formate clearance from the blood and production of CO_2 proceed at least twice as fast in rats than in nonhuman primates (KAVET and NAUSS 1990). While rats show no significant increases in blood formate at any dosage of methanol, high dosages of methanol in primates result in accumulations of formate that can produce toxicity (KAVET and NAUSS 1990). Thus, at high methanol dosages, the maximal rate of formate oxidation in rodents exceeds the rate at which upstream metabolism generates formate, while in primates the rate of formate oxidation via folate is slower than the upstream supply. Despite this species difference, it is important to note that the accumulation of formate following methanol exposure in primates is a high-dose phenomenon and that at low doses no formate buildup would be expected in either primates or rodents.

Among various animal species, there appears to be a good correlation between the hepatic concentration of THF and the rate of formate oxidation. Folate deficiency can exacerbate the toxicity of methanol in monkeys (MCMARTIN et al. 1977), while monkeys treated with folate or 5-formyl-THF accumulate less blood formate than controls following acute methanol administration. Folate deficient rats exhibit slowed oxidation of formate (PALESE and TEPHLY 1975), which accumulates in these rats following a high dose of methanol (MAKAR and TEPHLY 1976). Because this accumulation of formate following methanol administration has some similarity to that observed in primates, the folate-deficient rat has been used as an animal model of human methanol poisoning. The oxidation of formate in rats can also be slowed by treatment with nitrous oxide, which blocks methionine synthetase, which in turn catalyzes the conversion of 5-methyl-THF to THF (EELLS et al. 1981, 1982).

4. Free Radicals

Acute administration of alcohols results in metabolic disturbances of both nicotinamide adenine dinucleotides (NAD^+/NADH) and NAD phosphate ($NADP^+$/NADPH) in favor of the reduced forms. The resultant reducing tissue environment can supply electrons to appropriate receptors such as oxygen, primarily via NADH-linked enzymes. Among a number of subsequent redox disturbances (see LIEBER 1994), this can result in an increased production of free radical intermediates, including those derived from the alcohol itself as well as secondary products such as hydrogen peroxide (SLATER 1988). The significance of alcohol-inducible CYP2E1 may include not only the oxidation of ethanol and methanol, but also the capacity of CYP2E1 to generate reactive oxygen intermediates such as superoxide radicals (DAI et al. 1993; CASTILLO et al. 1992). Acetaldehyde itself is capable of causing lipid peroxidation in isolated perfused rat liver (MÜLLER and SIES 1982). In addition, ethanol and methanol may promote free radical damage by causing ischemia/reperfusion and/or by depleting antioxidant defense mechanisms (ABEL and HANNIGAN 1995). The role of reactive oxygen species in FAS will be discussed below.

D. Animal Models of Fetal Alcohol Syndrome

The animal models discussed in this section pertain principally to the gross structural aspects of FAS. There is an abundant literature on postnatal behavioral effects of ethanol (RILEY and BARRON 1989; RILEY 1990; DRISCOLL et al. 1990), which will not be reviewed here. Additionally, the effects of maternal ethanol exposure specifically during preimplantation development will not be reviewed here, but have been extensively studied in mice and rats by SANDOR, CHECIU, and coworkers (SANDOR 1979, 1988; SANDOR et al. 1980; CHECIU and SANDOR 1981, 1982, 1983, 1986, 1987; CHECIU et al. 1984; CHECIU 1993; FAZAKAS-TODEA et al. 1985; FAZAKAS-TODEA 1993a,b).

Generally, two approaches have been taken in developing whole animal models of FAS. The first, which is probably a better model of FAS pregnancy, entails chronic exposure to ethanol, usually by the oral route in a liquid diet (LIEBER and DECARLI 1963, 1982; YEH and CERKLEWSKI 1984) or drinking water. Such studies have been carried out in a variety of species, including primates. The second approach has been developed to study the pathogenesis and mechanisms underlying the dysmorphogenesis produced by acute ethanol exposure. This model, developed by KRONICK (1976), WEBSTER et al. (1980), and SULIK et al. (1981), involves acute oral or, more typically, intraperitoneal administration of relatively high dosages of ethanol. Experimental models of alcohol teratogenesis and the advantages and disadvantages of various dosing regimens in animals have been discussed by BLAKLEY (1988). We will summarize the developmental toxicity of both chronic and acute ethanol exposures in this section, and discuss pathogenesis and underlying mechanisms (including in vitro studies) in the next section.

I. Chronic Exposures

As discussed above, mothers of children with FAS are chronic alcoholics. Alcoholism involves not only frequent high levels of alcohol intake, but often inadequate nutrition and other compromising maternal health conditions. In order to model an alcoholic pregnancy, many investigators administer ethanol to their test animals over multiple days in the drinking water or in a liquid diet. Since ethanol has caloric value, using a liquid diet allows the substitution of ethanol for other carbohydrate as a percentage of total dietary calories.

Table 2 summarizes studies in nonprimate animal models in which ethanol was administered chronically (at least several days) in the diet, in the drinking water, or by oral gavage. In one study, ethanol was administered via inhalation (NELSON 1985), a potential route of exposure in occupational settings. Ethanol has been shown to produce birth defects or other manifestations of developmental toxicity in many species. In some studies ethanol exposure started prior to mating, while in others shorter periods of exposure were used. In all cases, large dosages of ethanol were required to elicit developmental toxicity. Effective dosages generally ranged from 4 to 20 g/kg per day or 20% or more of calories derived from ethanol, and peak blood ethanol concentrations of 150 to more than 500 mg/dl have been reported after developmentally toxic exposures in various species. Induction of malformations usually requires peak blood alcohol levels of 200 mg/dl or more. While teratogenesis was observed in ferrets at a dosage of 1.5 g/kg per day, the reported blood alcohol level (207 mg/dl) was in the range of those reported in other species given higher dosages. Effects on fetal weight are usually seen at the low end of this blood alcohol range and, in a retrospective analysis of a large database, HANNIGAN et al. (1993) found that reduced birth weight was a consistent and robust effect of maternal consumption of 35% ethanol-derived calories in rats. Malformations of most organ systems can result from ethanol exposure during gestation, with the particular spectrum of effects depending on the timing of the exposure. CHERNOFF (1977) was among the first to report that ingestion of ethanol by pregnant mice could decrease litter size and fetal weight and cause malformations. Mice (CBA and C3H strains) were maintained on a liquid diet with 15%–35% ethanol-derived calories beginning 30 days prior to mating and continuing throughout gestation. These dietary levels of ethanol resulted in blood levels of 73–398 mg/dl in nonpregnant females after 10 days on the diets. Fetal weight near term was lower than controls at all dosage levels in both strains. Skeletal defects also occurred at all exposure levels, including incomplete or missing supraoccipital bone at the lowest dose level and sternebral and rib defects at higher levels. Soft tissue malformations observed included dilated cerebral ventricles, hemopericardium, and ventricular septal defects at low dosages and exencephaly and gastroschisis at the

Table 2. Effects of chronic prenatal ethanol exposure in nonprimate experimental animals

Species	Dose(s)	Route of exposure	Duration of exposure	Reported effects	Reference
CBA & C3H mouse	15%–35% EtOH-derived calories in LD (BAL, 73–398 mg/dl)	Dietary	30 days prior to and throughout gestation	↑ Resorptions, ↓ fetal weight ↑ Abnormalities	CHERNOFF 1977
C57BL/6J mouse	25% EtOH-derived calories	Dietary	GD 5–10	↑ Resorptions and malformations	RANDALL et al. 1977
Mouse	15% in H_2O	Drinking water	GD 6–15	↓ Fetal weight	SCHWETZ et al. 1978
Mouse	17%, 25% or 30% in LD	Dietary	GD 5–10	↑ Resorptions and malformations at two higher dose levels	RANDALL and TAYLOR 1979
C57BL/6J mouse	5.4% in LD (approx., 19–23 g/kg per day)	Dietary	GD 5–11	↓ Birth weight, litter size ↑ Incidence of hydro-nephrosis	BOGGAN et al. 1979
CBA/J mouse C3H/lg mouse C57BL/6J mouse	20% EtOH-derived calories	Dietary	Prior to and through-out gestation	↑ Resorptions in CBA ↓ Fetal weight in CBA and C3H. Abnormalities in all strains	CHERNOFF 1980
C3H mouse	10% or 20% in H_2O	Drinking water	Throughout gestation	↑ Resorptions at high dose ↓ Fetal weight; malforma-tions of brain, face, skeleton, eyes, lungs	RASMUSSEN and CHRISTENSEN 1980
CBA/J mouse	20% EtOH--derived calories	Dietary	4 weeks preconcep-tion and throughout gestation	↓Fetal weight ↑ Resorptions	EL BANNA et al. 1983
C3H mouse	4.1% in LD (approx. 15–25 g/kg per day)	Dietary	GD 0–17	↓ Fetal weight (↓ Fertility and litter size, but attribu-table to food intake)	GOAD et al. 1984
Mouse	30% in H_2O	Drinking water	Throughout gestation	↓ Offspring body weight and muscle weight, but not in muscle/body weight ratio	IHEMELANDU 1984

Swiss mouse	Approx 12.6 g/kg per day (20% EtOH-derived calories)	Dietary	GD 8–20	No effects on development relative to pair-fed controls; altered cardiac muscle ultrastructure at birth	Uphoff et al. 1984
Swiss mouse	20% in H_2O	Drinking water	Throughout gestation	↓ Fetal weight	Lee 1985
C57BL/6J mouse	25% EtOH--derived calories	Dietary	GD 6–21 (F_1 females on GD 5–11)	↓ Offspring weight (F_1) ↓ F_2 fetal weight, possibly enhanced by prenatal EtOH exposure of F_1, also possibly due to decreased F_1 weight	Becker and Randall 1987
CD-1 mouse	2.2, 3.6, 5.0, 6.4, or 7.8 g/kg per day	p.o.	GD 8–14	↑ Resorptions at 5.0 g/kg per day and higher	Wier et al. 1987
Rat	15% in H_2O	Drinking water	GD 6–15	↓ Fetal weight	Schwetz et al. 1978
Rat	6% in LD 10g/kg per day	Dietary p.o. (2×5 g/kg per day)	1 month prior to and throughout gestation GD 11–13 or 14–16	↑ Resorption in all groups ↑ fetal death with dietary exposure; ↓ fetal weight	Henderson et al. 1979
SD rat	5 g/kg per day (BAL, approx. 250 mg/dl) 6g/kg per day (BAL, approx. 310 mg/dl)	p.o.	GD 1–15 or 1–20	Polydactyly and polysyndactyly, hindlimbs more than forelimbs; growth retardation, smaller litters at 6 g/kg per day	West et al 1981
LE rat	0.4 or 4 ml/kg per day	p.o.	GD 6–15	↑ Abnormalities (but no effect on fetal weight or resorptions at low dose)	Mankes et al. 1982
Wistar rat	Approx. 4–7 g/kg per day	Drinking water	8 weeks prior to and throughout gestation	↓ Fetal weight, litter size ↑ Placenta weight, fetal Zn, Mg	Suh and Firek 1982
Rat (LE?)	Approx. 16 g/kg per day	Dietary	7 weeks prior to and throughout gestation	↓ Fetal liver gluconeogenic enzymes	Canny and Roe 1983
SD rat	8 g/kg per day	p.o. 2×4 g/kg per day)	GD 7–9, 10–12, or 13–15	↑ Resorptions ↓ Fetal weight	Fernandez et al. 1983

Table 2. (*Contd.*)

Species	Dose(s)	Route of exposure	Duration of exposure	Reported effects	Reference
SD rat	Approx. 17.5 ml/kg per day	Dietary	GD 2–20	↓ Fetal weight ↑ Placenta weight ↓ Fetal Zn, accumulation of Zn from maternal plasma	GHISHAN and GREENE 1983
SD rat	5% in LD	Dietary	GD 3–12	↑ Resorptions; no effect on fetal weight	PERSAUD 1983
LE rat (parity 1–4)	6 g/kg per day	p.o.	GD 7–22	No interactions of parity with EtOH effects	ABEL and DINTCHEFF 1984
SD rat	30% in H_2O	Drinking water	Prior to and throughout gestation	↓ Fetal weight not related to maternal water intake	LEICHTER and LEE 1984
SD rat	36% EtOH--derived calories	Dietary	Throughout gestation	↓ Fetal liver weight (no data on fetal weight) no effect on insulin binding to fetal liver cell membranes	SNYDER and SINGH 1983
SD rat	Approx. 11.3 g/kg per day	Dietary	10 days prior to and throughout gestation	↓ Fetal weight ↑ placenta weight ↓ Placenta folate receptor activity	FISHER et al. 1985
LE rat	Approx. 9–10 g/kg per day	Dietary	2 weeks prior to conception and throughout gestation	↑ Placenta weight; no other morphological effects ↓ ^{14}C-aa accumulation in fetuses (but was not normalized to ^{14}C-aa concentration in maternal serum)	B.H.J. GORDON et al. 1985
SD rat	Approx 11.6 g/kg per day (BAL, 140 mg/dl)	Dietary	Prior to and throughout gestation	Fetal lung hypoplasia	INSELMAN et al. 1985
SD rat	1.0%, 1.6%, or 2.0% in air.(BAL, 148–193 mg/dl at high dose)	Inhalation	7 h/day on GD 1–19	No significant effects	NELSON et al. 1985

SD rat	Approx. 12–14 g/kg per day (BAL, 144–192 mg/dl)	Dietary	GD 1–21	↑ Placenta weight ↓ Fetal weight ↓ Fetal brain weight (but not brain/body weight ratio)	WEINBERG 1985
LE rat	Approx. 12.9 g/kg per day	Dietary	GD 6–20	↓ Umbilical cord length	BARRON et al. 1986
Wistar Imanichi rat	30% in H_2O	Drinking water	4 Weeks prior to and throughout gestation	↓ Fetal weight; no effects on maternal pituitary or thyroid hormones	LEE and WAKA-BAYASKI, 1986
Wistar rat	Approx. 13.2 g/kg per day	Dietary	4–5 weeks prior to and throughout gestation	↓ Fetal and fetal liver weights ↑ Placenta weight ↓ Fetal aldehyde dehydrogenase activity	SANCHIS and GUERRI 1986
SD rat	5% in LD	Dietary	Throughout gestation	↓ Fetal blood glucose, liver glycogen, liver glycogen synthase and phosphorylase	SINGH et al. 1986
SD rat	Approx. 7.1 g/kg per day	Dietary	Throughout gestation	Transient perinatal effects on blood glucose levels	SINGH et al. 1986
Wistar rat	25% in H_2O (BAL 147 mg/dl)	Drinking water	Prior to and throughout gestation	↓ Fetal weight and length	TESTAR et al. 1986
Holtzman rat	Approx. 15–18.5 g/kg per day	Dietary	3 weeks prior to and throughout gestation	↓ Litter size, birth weight, neonatal liver glycogen	WITEK-JANUSEK 1986
Wistar rat	36% EtOH--derived calories	Dietary	4–5 weeks prior to and throughout gestation	Altered gap junctions in neonatal hepatocyte membranes	RENAU-PIQUERAS et al. 1987
SD rat	Approx. 12 g/kg per day	Dietary	3 weeks prior to and throughout gestation	↑ Resorptions, fetal weight; normal accumulation of Zn into fetus and placenta	ZIDENBERG-CHERR et al. 1988
Rat	6% in LD	Dietary	Males and females prior to mating, females throughout gestation	↓ Heart protein and RNA (but not DNA) ↓ Incorporation of leucine into fetal and neonatal heart protein	RAWAT 1979

Table 2. (*Contd.*)

Species	Dose(s)	Route of exposure	Duration of exposure	Reported effects	Reference
LE rat	35% EtOH--derived calories	Dietary	GD 6–20	↓ Parasagittal and coronal skull and jaw dimensions at adulthood	EDWARDS and DOW-EDWARDS 1991
Rat	35% EtOH--derived calories	Dietary	???	↓ Birth weight seen retrospectively across large database (not due to effect on food intake)	HANNIGAN et al. 1993
LE rat	35% EtOH--derived calories	Dietary	GD 10–20	↓ Birth weight and prepubertal growth	GAVIN et al. 1994
Wistar rat	36% EtOH--derived calories	Dietary	6 weeks prior to mating, throughout gestation to postnatal day 7	↓ Optic nerve diameter, axon diameter and axon number, glial cell nuclear area	STRÖMLAND and PINAZO-DURÁN 1994
Rabbit	15% in H_2O	Drinking water	GD 6–18	↓ Fetal weight ↑ Resorptions	SCHWETZ et al. 1978
Guinea pig	6 g/kg per day	p.o.	GD 35–70	Ultrastructural changes in skeletal muscle of neonates	NYQUIST-BATTLE et al. 1987
Guinea pig	3, 4, 5, 6 g/kg per day	p.o.	Throughout gestation	Abortion and perinatal death at 5 or 6 g/kg per day	CATLIN et al. 1993
Ferret	1.5 g/kg per day (BAL, 207 mg/dl)	p.o.	GD 15–35	↑ Malformations (especially cleft palate, facial abnormalities)	MCLAIN and ROE 1984
Sinclair miniature swine	> 3 g/kg per day	Drinking water	Entire life	↓ Litter size, piglet weight ↑ Perinatal death with increasing parity	DEXTER et al. 1983

↑, Increase; ↓ Decrease; LD, liquid diet; BAL, blood alcohol level; GD, gestational day; EtOH, ethanol; LE, Long-Evans; SD, Sprague-Dawley.

higher dosages. This well-designed study exemplifies the types of results obtained in many of the studies listed in Table 2.

A number of experiments have been carried out to examine the effects of prenatal ethanol exposure in primates (Table 3). These studies have most often used the oral route of exposure, although ethanol was administered intravenously in one study (MUKHERJEE and HODGEN 1982). Dosages in the range of 0.3–5.0 g/kg per day have been used, given throughout or during late gestation, often episodically (i.e., once per week). With a weekly dosage of 2.5 g/kg, reported blood alcohol levels were 200–300 mg/dl (CLARREN and BOWDEN 1982). In studies using multiple dosage levels, it was demonstrated that a weekly dosage of 1.8 g/kg or higher resulted in an increased frequency of abortion and anomalies, including facial dysmorphology (CLARREN et al. 1988; SHELLER et al. 1988). This dosage produced a peak blood alcohol level of 205 mg/dl or more.

Chronic administration of ethanol in the animal models discussed above results in decreased maternal food and water intake. It is imperative to use pair-fed controls in such studies to account for such effects, but it is difficult to match the pattern as well as the amount of food intake by the ethanol-exposed animals. Further, high levels of alcohol intake may interfere with absorption and/or utilization of nutrients, and pair-fed animals usually gain more weight than their ethanol-exposed cohorts (ABEL and HANNIGAN 1995). Although it is probably impossible in such studies to distinguish the indirect maternal effects of ethanol from direct effects on the conceptus, it is important to remember that both types of effects are important and both surely contribute to induction of FAS in humans.

II. Acute Exposures

KRONICK (1976) administered an ethanol dosage of 5.8 g/kg (0.03 ml of 25% ethanol per gram body weight) to pregnant mice by i.p. injection. When mice were treated on gestational day (GD) 8 and 9 or GD-10 and 11, an increase in fetal death was noted. Single treatments on one of GD-7, 8, 9, 10, 11, or 12 resulted in increased fetal death (treatment on GD-8) and increased fetal anomalies (treatment on GD-8, 9, or 10). The malformations most frequently observed were coloboma and limb defects. WEBSTER et al. (1980) administered the same dosage of ethanol to C57BL/6J mice by i.p. injection on one of GD-7, 8, 9, 10, or 11 and observed a high frequency of resorption after treatment on GD-9, 10, or 11. Treatment on GD-7 or 8 was less embryolethal and produced various brain and facial malformations, including exencephaly, maxillary hypoplasia, cleft lip/palate, and anophthalmia or microphthalmia.

SULIK et al. (1981), utilizing two doses (2.9 g/kg each) administered intraperitoneally 4 h apart on GD-7, demonstrated that this treatment regimen resulted in striking craniofacial malformations closely resembling those seen in human FAS infants, including short palpebral fissures, midfacial deficiencies, small nose, and long upper lip with deficient philtrum (Fig. 3). Eye mal-

Table. 3. Effects of prenatal ethanol exposure on primate development

Species	Dose(s)	Route of exposure	Duration of exposure	Reported effects	Reference
Macaca mulatta	To achieve a BAL of 150 or 75–100 mg/dl	p.o.	Throughout gestation	Two high-BAL animals aborted No low-BAL or control abortions	ALTSCHULER and SHIPPENBERG 1981
Macaca mulatta, Macaca fascicularis	3 g/kg	i.v.	Late gestation	Collapse of umbilical vasculature leading to fetal hypoxia	MUKHERJEE and HODGEN 1982
Macaca fascicularis	4 or 5 g/kg per day	p.o.	GD 20–150	At 5g/kg per day, ↑ abortions and stillbirths, ↓ birth weight; no effects at 4 g/kg per day	SCOTT and FRADKIN 1984
Macaca nemestrina	2.5 g/kg (BAL, 200–300 mg/dl)	p.o.	Once a week from GD 40 to term	One (of three) monkeys aborted. One (of two) fetuses had FAS-like facial anomalies	CLARREN and BOWDEN 1982
Macaca nemestrina	0.3, 0.6, 1.2, 1.8, 2.5, 3.3, 4.1 g/kg	p.o.	Once a week throughout gestation	↑ Spontaneous abortion at 1.8 g/kg and higher (BAL ≥ 205 mg/dl)	CLARREN et al. 1987b
Macaca nemestrina	0.3, 0.6, 1.2, 1.8, 2.5, 3.3, 4.1 g/kg	p.o.	Once a week throughout gestation	Minor dysmorphic and behavioral effects at 1.8 g/kg and higher, possibly at 1.2 g/kg (if BAL > 140 mg/dl)	CLARREN et al. 1988

Macaca nemestrina	0.3, 0.6, 1.2, 1.8, 2.5, 3.3, 4.1 g/kg	p.o.	Once a week throughout gestation	Definite craniofacial anomalies at 1.8 g/kg and above	SHELLER et al. 1988
Macaca nemestrina	0.3, 0.6, 1.2, 1.8, 2.5, 3.3, 4.1 g/kg 2.5, 3.3, or 4.1 g/kg (BAL, 24–549 mg/dl)	p.o. p.o.	Once a week throughout gestation Once a week from week 5 of gestation	Microphthalmia, retinal ganglion cell loss, ultrastructural alterations in caudate nucleus, altered striatal dopamine concentration	CLARREN et al. 1990
Macaca nemestrina	1.8 g/kg	p.o.	Once a week, weeks 2–19 of gestation	Case report of a fetus with holoprosencephaly	SIEBERT et al. 1991
Macaca nemestrina	1.8 g/kg	p.o.	Once a week for 3 or 6 weeks or throughout gestation (24 weeks)	Abnormal behaviour in 6 and 24 week groups	CLARREN et al. 1992
Macaca nemestrina	1.8 g/kg	p.o.	Once a week for 3 or 6 weeks or throughout gestation (24 weeks)	↑ Spontaneous abortion GD 30–160	CLARREN and ASTLEY 1992
Macaca nemestrina	1.8 g/kg	p.o.	Once a week for 3 or 6 week or throughout gestation (24 weeks)	↓ T cell response to tetanus toxin	GROSSMAN et al. 1993

↑, increase; ↓, decrease; BAL, blood alcohol level, GD, gestational day.

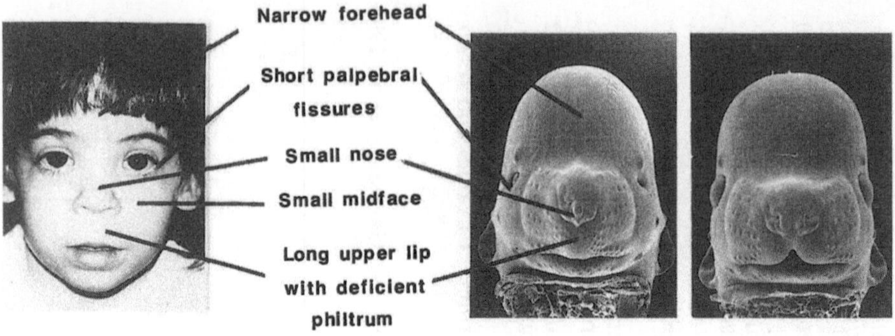

Fig. 3. Comparison of craniofacial features in human fetal alcohol syndrome (FAS) and the C57BL/6J mouse model. Mouse fetus is gestation day 14. (Modified from SULIK et al. 1981, with permission)

formations, including coloboma, microphthalmia, and apparent anophthalmia, were observed in 42% of the treated fetuses. Ethanol treatment resulted in alterations of the developing brain during the 24 h following administration, including deficiencies of the neural plate and its derivatives. Embryolethality using this dosing regimen was 18%. These authors stressed the importance of accurately timed exposures, as administration of the same dosage only a few hours later resulted in a greatly increased incidence of embryolethality. Since publication of this study by SULIK et al. (1981), the pregnant C57BL/6J mouse dosed intraperitoneally with 5.8 g/kg methanol (usually in two doses 4 h apart) has become a standard model of human FAS in studies of effects, pathogenesis, and mechanisms (see, e.g., WEBSTER 1989). Typically, 25% ethanol in saline is given at a dosage of 0.03 ml/g body weight (one dose of 5.8 g/kg) or 0.015 ml/g (two doses of 2.9 g/kg each). For convenience we have expressed all dosages in g/kg.

WEBSTER and coworkers (1983) administered a total dosage of 5.8 g ethanol/kg to pregnant C57BL/6J mice on a single day of gestation. This dosage was given as a single oral dose (GD-8 or 9), two i.p. injections 4 h apart (GD-7, 8, 9, or 10), or two i.p. injections 6 h apart (GD-7 or 10). Peak blood alcohol levels were highest after two i.p. injections 4 h apart (500–600 mg/dl BAL), while two i.p. injections 6 h apart or two oral doses 4 h apart resulted in peak blood alcohol levels of approximately 400 mg/dl. With i.p. administration, blood alcohol levels were back to baseline by 6 h after dosing. Elimination was somewhat slower following oral dosing, perhaps indicating continued absorption from the gastrointestinal tract during the elimination phase of the blood alcohol curve. As might be expected from these blood alcohol concentration data, when the same dosage and timing of exposure were used, i.p. administration of ethanol resulted in a higher incidence of malformations than did oral administration. However, both routes of exposure produced the same spectrum of effects in fetuses at term. Ethanol administration on GD-7 produced head defects exclusively. These included exencephaly, maxillary and/or mandibular hypoplasia, cleft lip, and cleft palate. Examination of the fetal brains revealed small or missing olfactory bulbs, reduced cerebral hemispheres with enlarged third ventricles, and dysgenesis of the hypophysis. Some fetuses were observed to have a narrow face, single nares, and an- or microphthalmia, a syndrome pathognomonic of holoprosencephaly. Exposure on GD-8 also resulted primarily in head defects, but at a lower incidence than was observed following dosing on GD-7. Dosing on GD-9 or 10 produced limb malformations, most commonly forelimb reduction defects. Absence of the fifth or fourth and fifth digits was observed, with the right limb being affected more often than the left. Forelimb syndactyly and polydactyly and a low incidence of hindlimb defects were also observed. When ethanol was administered as two i.p. doses 4 h apart, the incidence of limb defects was 8% with dosing on GD-9 and 56% with dosing on GD-10.

PADMANABHAN et al. (1984) administered 1.9, 3.8, or 5.8 g ethanol/kg to MF-1 mice on single days during GD-8–12 and found fetal mortality, growth reduction, and cranial and digit malformations. Postaxial polydactyly of the forepaw was the only malformation noted at the low dosage. In a study of skeletal dysmorphogenesis following exposure of pregnant mice to 3.8 or 5.8 g ethanol/kg on GD-8, PADMANABHAN and MUAWAD (1985) reported cranioschisis, basicranial malformations, maxillary hypoplasia, digit defects, and abnormalities of the vertebral centra and arches, ribs, and sternum.

WEBSTER et al. (1984) reported on the cardiac anomalies in fetuses from their two previous studies (WEBSTER et al. 1980, 1983), in which ethanol was administered to C57BL/6J mice on GD-7, 8, 9, or 10 (total dosage 5.8 g/kg in a single i.p. dose or two doses 4 h apart). Alcohol exposure on GD-8, 9, or 10 caused 60% , 75% and 15% incidences of ventricular septal defects, respectively. Defects of both the membranous and muscular parts of the septum were seen. In addition, involvement of the great vessels was noted in 43% of the ventricular septal defect cases. In a similar study, DAFT et al. (1986) dosed pregnant C57BL/6J mice with 5.8 g ethanol/kg in two doses on GD-8 at 12 h and 16 h and found heart and great vessel malformations. Size deficiencies and abnormalities of the developing cardiac tube were noted within 12 h of treatment, and by GD-12 deficiencies of the conal and atrioventricular endocardial cushions and abnormal position of the atrioventricular canal were evident. Deficiencies in ventricular septation were noted on GD-13, and these persisted to term. Other defects reported include double-outlet right ventricle, interrupted aortic arch, right aortic arch, and a vascular ring.

COOK et al. (1987) examined in depth the eye malformations produced by ethanol exposure on GD-7 in the C57BL/6J mouse. Development of the optic vesicle was deficient in embryos of exposed dams, resulting in induction of a small lens (microphakia) or no lens induction. Progressive corneal opacification and vascularization were noted, as was abnormal formation of the anterior chamber. These defects likely result from effects on the very early optic primordia, as treatment 1 day later (GD-8) did not result in eye defects.

These results suggest that the eye defects associated with FAS may have a narrow and specific time window of susceptibility during gastrulation. ASHWELL and ZHANG (1994) found that exposure to ethanol (5.8 g/kg i.p. in two doses 4 h apart) on GD-8 resulted in a decreased number of axons and deficient myelination in the optic nerve of offspring at 15 days of age. There was no effect on cellularity of the dorsal lateral geniculate nucleus or superior colliculus, suggesting that the optic nerve axonal deficit was due to direct retinal damage rather than postnatal axon attrition due to higher-level target cell deficits.

III. Pathogenesis of Ethanol-Induced Birth Defects

In this section we will describe abnormalities of embryogenesis and organogenesis observed in animal models of ARBD and relate these aberrant developmental processes to human phenotypes associated with maternal alcohol abuse. Putative biochemical mechanisms underlying these dysmorphogenetic processes will be reviewed in the next section. Although acute maternal ethanol exposure during any stage of pregnancy has the potential to be deleterious to the conceptus, the discussion in this section will be limited to pathogenesis of craniofacial and limb defects. Most of this work has been carried out using an acute i.p. dosing regimen in pregnant C57BL/6J, as described in the previous section, although a limited number of studies in the chick embryo have been informative as well.

1. Craniofacial Malformations

The processes of gastrulation and neurulation are sensitive to the toxic effects of ethanol. These developmental processes are discussed in Chap. 4 (Volume I). Ethanol administration to pregnant CD-1 mice at GD-6.5–7 slows gastrulation, causing retardation of mesodermal spreading as well as shrinking and disorganization of the overlying epiblast (NAKATSUJI and JOHNSON 1984).

Exposure of pregnant C57BL/6J mice to ethanol (two doses on GD-7 at 0 h and 4 h) during gastrulation affects forebrain derivatives (SULIK et al. 1981, 1984; SULIK and JOHNSTON 1983; WEBSTER et al. 1983), and pathologic alterations can be observed in the presomite embryo within hours of maternal ethanol administration. Narrowing of the anterior aspect of the embryonic disk is evident in treated embryos within 8–12 h, primarily at the expense of the midline (compare Fig. 4A,C) (SULIK and JOHNSTON 1983). Subsequently, the paired ectodermal primordia of the olfactory, lens, and otic placodes, normally positioned at the periphery of the embryonic disk, are brought closer to the midline (compare Fig. 4B,D). The olfactory placodes become bounded by elevations termed the medial nasal prominences (MNP) and lateral nasal prominences, the former giving rise to the tip of the nose and the philtrum. In moderately affected embryos, the MNP are abnormally narrow and close together, insufficient to form a normal nose or contribute to the philtrum. In

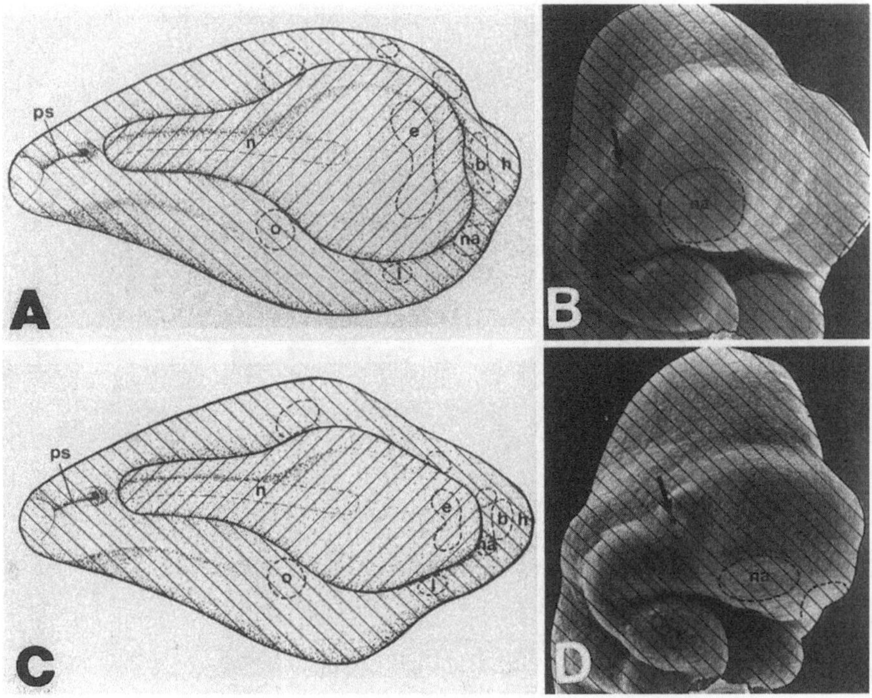

Fig. 4.A. Normal gastrulation-stage embryo. Mesodermal cells migrate cranially (to the *right* in this illustration) from the primitive streak (*ps*) to occupy the region subjacent to the ectoderm; prospective neural plate (▨) and surface ectoderm (▧). n, Notochord; *e*, eyefield; *p*, prochordal plate; *h*, heart; *o*, otic placode; *i*, lens placode; *na*, nasal (olfactory) placode; *pm*, prechordal mesoderm (between n and p). **B** Normal embryo (follwing neural tube closure). Note position of nasal placodes (*na*). *Arrow*, eye. **C** Abnormal gastrulation-stage embryo, illustrating the narrow cranial aspect with relatively close approximation of the nasal placodes and the reduction in the size of the anterior plate including the eye field. **D** Abnormal embryo (following neural tube closure), illustrating the relatively close positioning of the nasal placodes. (Modified from SUILK 1984, with permission)

this case, the maxillary prominences of the first visceral arch converge to the midline, resulting in the characteristic small nose, long upper lip with deficient philtrum, and thin vermilion border (in humans; Fig. 5B,D,F) or lack of the central notch (in mice; Fig. 6C,D). Derivatives of the MNP also form the alveolar ridge containing the upper incisors and the anterior hard palate, and development of these structures can be deficient in ethanol-treated embryos. In severely affected embryos, the olfactory placodes may be so close together that they converge and do not form any MNP, resulting in a cebocephalic facies, a phenotype in the holoprocencephaly series (Fig. 5A,B). Narrowing of the prospective forebrain region of the neural plate also results in varying degrees of ventromedial forebrain deficiencies consonant with holoprosencephaly which are apparent in GD-11 to -14 mouse embryos exposed to ethanol during

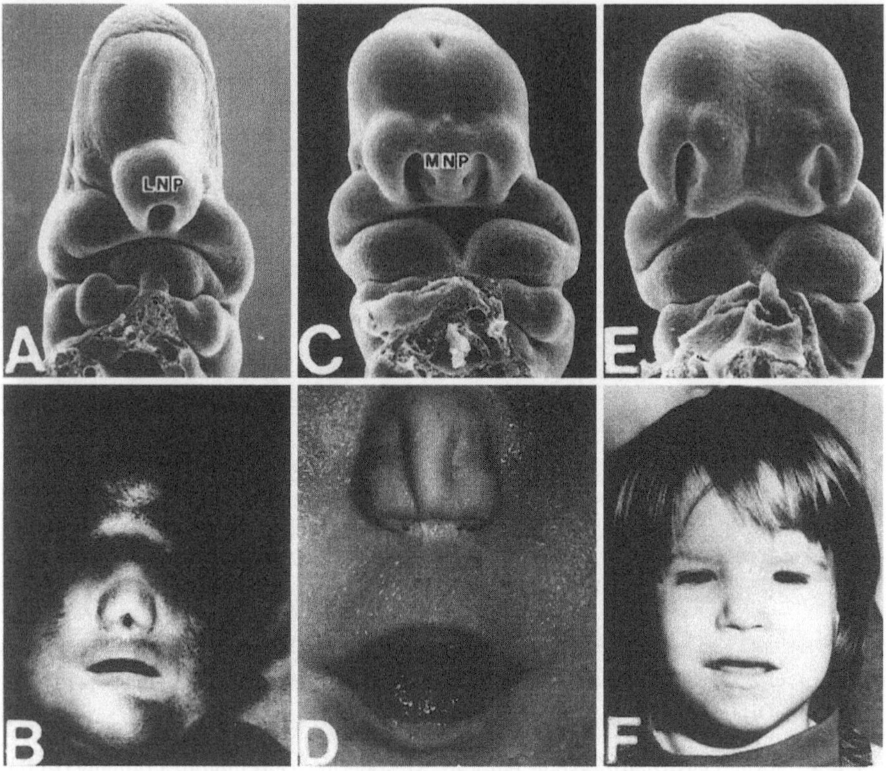

Fig. 5.A,C,E Abnormal facial prominence formation in mouse embryos and **B,D,F,** human facial malformations in the holoprosencephaly series. **A** One central nasal pit is surrounded by lateral nasal prominences (*LNP*) in this affected 36-somite C57BL/6J mouse embryo. Note the narrowness of the head. **B** Human with cebocephaly. **C** One central medial nasal prominence (*MNP*) separates the closely set olfactory pits in this affected 36-somite mouse embryo. **D** This human illustrates the phenotype which probably results from the embyonic morphology comparable to that illustrated in fig 5C. Note the midline ridge in the nose and the flattened philtral region (photograph originally from SIEBERT et al. **E** A mildly affected 36-somite mouse embryo has less distinct separation of the MNP than is seen in comparably staged controls. **F** This child has fetal alcohol syndrome (FAS). Note the presence of a flattened philtral region, small nose, flat nasal bridge, and short palpebral fissures. This morphology probably develops from an embryonic stage similar to that seen in Fig. 5E. (Plate modified SULIK and JOHNSTON 1982, with permission)

gastrulation. The severity of these brain malformations parallels the severity of the facial dysmorphology (SULIK and JOHNSTON 1982; SULIK 1984). Missing or defective optic sulci in the rostral part of the neural plate can result in anophthalmia or microphthalmia, as discussed earlier (COOK et al. 1987).

After establishment of the neural plate, further development of the central nervous system proceeds by elevations of the neural folds at the margins of the neural plate. About the same time, neural crest cells originating from the edges

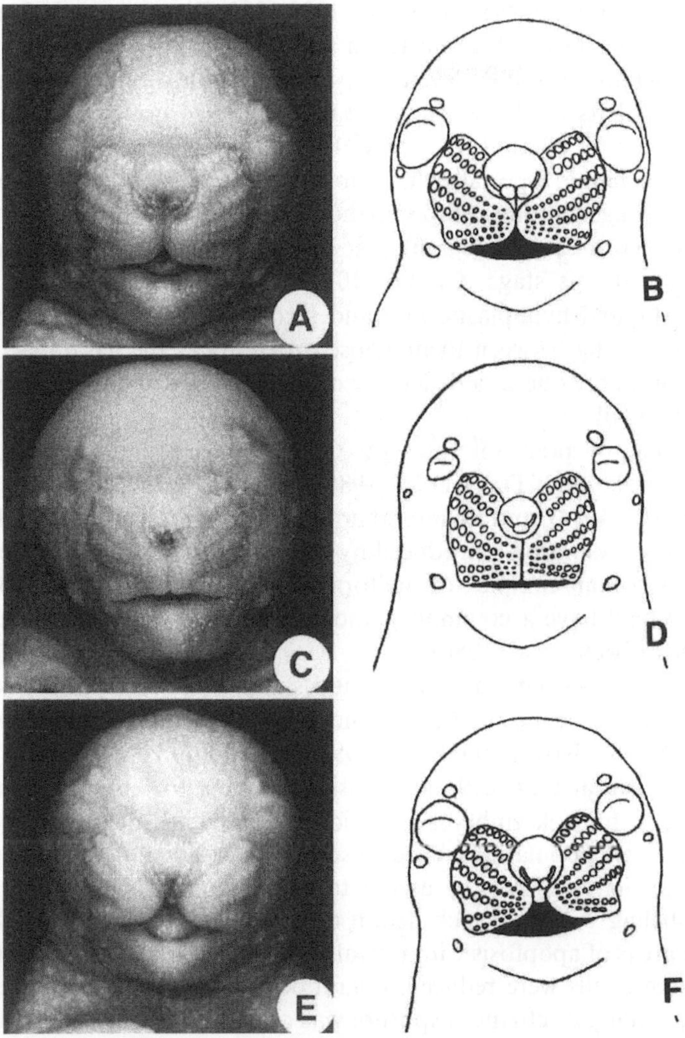

Fig. 6. A-F. Gestation day (GD)-18 mouse fetuses from **A** control dam or dams given a single intraperitoneal injection of 5.8 g/kg ethanol on **C** GD 7 or **E** GD 8. **B, D, F** Line drawings of these fetuses help to demonstrate the effects of ethanol on craniofacial development. Maternal exposure on GD 7 results in a narrow face, long smooth philtrum, and small eyes. Follwing maternal exposure on GD 8, the maxillary processes are underdeveloped and have not fused in the midline. (Modified from WEBSTER 1989, with permission)

of the neuroepithelium begin to migrate, in the cranial region forming the cranial ganglia and the bone, cartilage, and connective tissue of the face (NODEN 1991). Treatment of pregnant mice with ethanol at this stage (GD-8.5 and seven to ten somites, equivalent to 4 weeks of human gestation) results in a different type of facial abnormality, reflecting primarily deficiencies of the lateral facial primordia (WEBSTER et al. 1980,1983; WEBSTER 1989; SULIK et al. 1986). The upper face is foreshortened and midfacial derivatives of the maxillary prominence are underdeveloped, resulting in a short or cleft upper lip (Fig. 6E,F). Clefts of the secondary palate can also occur. Examination of embryos 12 h after ethanol treatment (13–16 somites) revealed deficiencies of cranial mesenchyme, especially that known to be of neural crest origin (SULIK et al. 1986). Degenerating cell populations were observed in the first visceral arch (neural crest-derived region), otic and epibranchial placodes, and neuroepithelium at this stage. On GD-10–11 (29–39 somites), ethanol-treated embryos exhibited hypoplastic first and second visceral arches, with abnormally wide spacing between them. Close proximity of the nasal placodes was observed, probably due to a deficiency of mesenchyme underlying the MNPs, which were small.

Treatment of pregnant mice on GD-8.5 also results in heart defects (WEBSTER et al. 1984; DAFT et al. 1986), abnormalities of the great vessels (WEBSTER et al. 1984), and thymus abnormalities (SULIK et al. 1986). Together with the facial dysmorphia produced by ethanol at this stage, these defects are similar to a human complex of malformations known as the DiGeorge syndrome and may have a common pathogenesis based on cranial neural crest insufficiency (SULIK et al. 1986).

SULIK and coworkers have used vital staining of whole embryos for cell death to provide evidence that cranial neural crest cells are sensitive to ethanol (SULIK et al. 1988; KOTCH and SULIK 1992a,b). Additional support for the idea that cranial neural crest cells are sensitive targets for ethanol comes from recent studies in chick embryos in which increased cell death and decreased neural crest cell populations were observed after ethanol treatment. CARTWRIGHT and SMITH (1995a,b) used vital staining with acridine orange and immunolabeling with the chick neural crest cell-specific antibody HNK-1 to colocalize areas of apoptosis with neural crest cell populations. While numbers of neural crest cells were reduced, migration patterns were unaffected. As in the mouse, timing of ethanol exposure was critical. Treatment at gastrulation resulted in craniofacial defects and cell death in more anterior regions compared to treatment at early somite stages.

2. Limb Malformations

Much less information is available on pathogenesis of the ethanol-induced limb malformations observed in mice. Limb defects produced by administration of ethanol to pregnant mice on GD-9, 10, 11, or 12 (WEBSTER et al. 1980, 1983) may also have their basis in increased cell death. KOTCH et al. (1992) treated mice with ethanol on GD-9, 6 h, producing forelimb malformations including postaxial ectrodactyly, preaxial syndactyly, and reduction defects of the intermediate digits. Excessive cell death localized in the apical ectodermal ridge and in the proximal mesenchyme was observed within

5–9 h of treatment, and patterns of cell death were predictive of the defects observed later.

PADMANABHAN and PALLOT (1995) observed digital malformations associated with vascular disruption in mice treated with ethanol on GD-11. Vascular disruption in distal structures can result from a number of chemical or physical teratogens, and such disruption may lead to loss of these structures due to poor nutrient delivery and gas exchange or due to hematomas that can physically impede normal development.

Overall, observations of pathogenesis suggest that embryonal cell proliferation and migration may be inhibited by ethanol and that cell death can occur in sensitive cell populations. These effects have been confirmed in studies using whole embryo culture to test the direct effects of ethanol during organogenesis, and potential biochemical lesions underlying this pathogenesis are discussed in the next section.

IV. Mechanisms Underlying Ethanol-Related Birth Defects

Many factors can contribute to the expression of FAS. The potential interrelationships among a number of these factors have been illustrated by ABEL and HANNIGAN (1995) (Fig. 7). These authors propose that sociobehavioral attributes can act as "permissive" factors to produce biological conditions which increase vulnerability to alcohol-induced damage. Among such permissive factors are alcohol intake pattern, race and SES, culture/ethnicity, and smoking. The biological/toxicological conditions associated with these permissive factors include high blood alcohol levels, under- or malnutrition, and intake of tobacco smoke. These conditions in turn produce physiological and biochemical sequelae which may ultimately cause damage and/or deficiencies at the cellular level leading to FAS, ARBD, and intrauterine growth retardation. These physiological and cellular mechanisms of ethanol developmental toxicity will be the subject of this section. The extent to which these different mechanisms contribute to FAS or ARBD is unknown and probably variable, particularly when comparing acute and chronic exposures. For other recent reviews, the reader is referred to WEST et al. (1994) and ABEL and HANNIGAN (1995).

1. Determination of the Proximate Teratogen – Ethanol or Acetaldehyde?

Determination of the relative roles in FAS of ethanol and its principal metabolite, acetaldehyde, requires an understanding of the magnitude of the exposure of the embryo to these chemicals after maternal ethanol consumption, as well as an understanding of the direct embryotoxic potential of the parent compound and metabolite. Both whole animal and in vitro approaches have been used to address these questions.

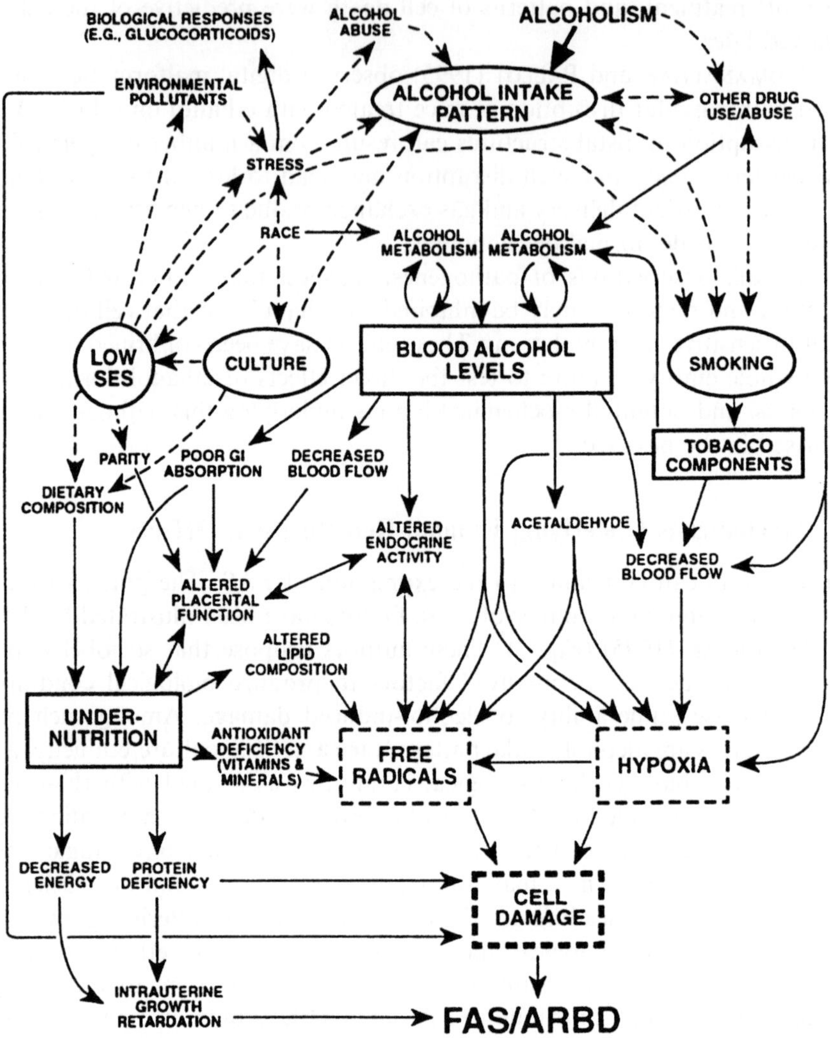

Fig. 7. Some of the relationships among permissive and provocative maternal risk factors and selected putative mechanisms underlying alcohol-related birth defects (*ARBD*) including fetal alcohol syndrome (*FAS*). Sociobehavioral permissive factors (*circled*) are highly correlated and increase risk of FAS/ARBD by establishing an environment for, and/or predisposing fetuses to the direct cellular toxicity of ethanol by exacerbating the provocative risk factors. The key biological prococative factors (inside *solid squares*) are blood alcohol levels (BAL), undernutrition, and tobacco smoke constituents. Several pathways by which the permissive and provocative risk factors act on the maternal/placental/fetal unit are shown. *Dotted lines* show recognised, sometimes bidirectional associations among various environmental, demographic, and behavioral variables. *Solid lines* indicate biological relationships and physiological pathways. In this model, the key teratogenic mechanisms of FAS operate via hypoxia and free radical damage (*dashed squares*) to produce the cell damage leading to FAS. Poor nutrition also directly contributes to intrauterine growth retardation, a cardinal feature of FAS/ARBD. *GI*, gastrointestinal; *SES*, socioeconomic status. (From ABEL and HANNIGAN 1995, with permission)

a) Whole Animal Studies with Acetaldehyde

Results of studies of the distribution of acetaldehyde following maternal exposure to ethanol or acetaldehyde suggest that acetaldehyde reaches the conceptus only after very high maternal exposures. KESÄNIEMI and SIPPEL (1975) examined ethanol and acetaldehyde concentrations in the rat fetus, placenta, and maternal blood after administration of a single i.p. dose of 2 g ethanol/kg. While ethanol concentrations in maternal blood, placenta, and fetus were similar 25 min after dosing, acetaldehyde concentrations were about four fold higher in maternal blood (100 nmol/ml) than in placenta, and no measurable acetaldehyde was detected in fetuses. SIPPEL and KESÄNIEMI (1975) demonstrated that, at a maternal blood concentration of 100 nmol/ml, the rat placenta oxidizes acetaldehyde at a rate sufficient to prevent diffusion to the fetus.

BLAKLEY and SCOTT (1984b) administered teratogenic dosages of 4, 6, or 7 g/kg ethanol to pregnant CD-1 mice on GD-10 and found detectable, but very low concentrations of acetaldehyde in the conceptus only at the highest dosage. Maternal i.p. treatment with 200 mg acetaldehyde/kg resulted in higher levels of acetaldehyde in the embryo, yet was not developmentally toxic. These results indicate that acetaldehyde probably does not play a primary role in ethanol teratogenesis in mice.

Embryofetal outcome was variable in experiments in which acetaldehyde was administered to pregnant animals. BANNIGAN and BURKE (1982) and BLAKLEY and SCOTT (1984a) did not observe increased embryolethality or teratogenicity following i.p. administration of acetaldehyde to pregnant mice on GD-9 or 10, respectively. In contrast, O'SHEA and KAUFMAN (1979, 1981) reported increased resorptions and neural tube defects and decreased growth following i.v. administration of acetaldehyde to pregnant CFLP mice. WEBSTER et al. (1983) administered acetaldehyde to pregnant C57BL/6J mice by i.p. injection on GD-10 and produced craniofacial and limb defects. However, maternal disulfiram (an inhibitor of aldehyde dehydrogenase which should cause acetaldehyde buildup) treatment either 24 h or 2 h prior to ethanol treatment did not exacerbate teratogenicity (WEBSTER et al. 1983). In rats, various dosing regimens have produced increased resorptions, craniofacial and limb defects, and reduced ossification (SREENATHAN et al. 1982, 1984) or increased resorptions and decreased fetal weight (DREOSTI et al. 1981). GARRO and colleagues (1991) demonstrated that inhibition of DNA methylase by acetaldehyde following acute maternal ethanol administration resulted in hypomethylation of fetal DNA.

b) In Vitro Studies with Ethanol and Acetaldehyde

Whole embryo culture has been used to address, at least in part, two questions pertaining to the teratogenicity of ethanol. First, can ethanol, at relevant concentrations, directly affect the conceptus in the absence of the concomitant maternal effects seen in whole animal studies? Second, can acetaldehyde affect development in vitro, and, if so, how do these effects compare to those produced by ethanol? A number of studies have been carried out to address these issues.

Results of studies in which rat embryos have been exposed to ethanol in whole embryo culture have been reasonably consistent. BROWN et al. (1979) exposed GD-10 rat embryos to 150 or 300 mg ethanol/dl culture medium for the full duration of a 48-h culture period. No gross structural malformations were observed, but embryo growth and development were retarded in a dose-related fashion. Embryo crown–rump length, head length, total protein and DNA content, and somite number were significantly reduced at the higher concentration compared to controls. These authors also reported small forebrains and estimated that embryos exposed to 300 mg ethanol/dl were delayed in development by 5–7 h and contained approximately 8.9×10^5 fewer cells compared to controls. Similar results were obtained by SANDOR et al. (1980), who also reported deficits in growth and development (decreased somite number and dorsiflexion) at ethanol concentrations above about 320 mg/dl. PRISCOTT (1982) observed decreased crown–rump length and somite number in embryos exposed to 390 mg ethanol/dl and in addition reported microencephaly and hydropericardium at 590 mg/dl. WYNTER et al. (1983) reported similar results following a 48-h incubation of rat embryos with ethanol, including reduced somite number, protein content, head length, and yolk sac circulation, as well as a decrease in the number of embryos undergoing normal rotation. These authors reported that only 6 h of exposure to 600–800 mg ethanol/dl followed by 42 h in normal medium resulted in persistent open anterior neuropores. Interestingly, 48-h incubation in the same ethanol concentrations did not produce this effect. CAMPBELL and FANTEL (1983) reported decreased DNA content in rat embryos cultured in 150 or 300 mg ethanol/dl for 24 h, while BECK et al. (1984) reported retarded growth and development of rat embryos cultured in human serum containing 120 mg ethanol/dl. CLODE et al. (1987) explored the hypothesis that variations in embryonal response to ethanol in whole embryo culture were due to small differences in developmental stage. These authors collected embryos and divided them into three groups according to developmental stage: early neural plate (presomite); early somite; or two to six somites. Exposing these embryos to 300 mg ethanol/dl for only 4 h, with a 48-h culture period in normal medium, resulted in stage-specific toxicity. Presomite embryos were the most affected, exhibiting retarded growth and severe abnormalities, including neural tube defects. Early somite-stage embryos showed less growth retardation, fewer abnormalities, and no open neural tubes. Two- to six- somite embryos showed only a slight growth effect evidenced by lower total protein content. GIAVINI et al. (1992) reported that 300 mg ethanol/dl was embryotoxic and teratogenic to rat embryos cultured from GD-9.5, but embryos cultured from GD-10 (four to six somites) were less sensitive, showing no effect at 300 mg/dl and retarded growth and development at 600 mg/dl.

The effect of ethanol on mouse embryo development has also been tested in whole embryo culture. THOMPSON and FOLB (1982) cultured GD-8 or -9 mouse embryos for 28 h in 150, 300, or 600 mg ethanol/dl. DNA synthesis was reduced at 600 mg/dl in GD-8 embryos and at 300 and 600 mg/dl in GD-9 embryos. GD-9 embryos were slightly more sensitive to ethanol-induced teratogenesis, exhibiting abnormal central nervous system development at all concentrations. HUNTER et al. (1994) exposed three- to six- somite mouse embryos to 300–800 mg ethanol/dl for varying lengths of time from 4 to 24 h and reported that effects were dependent on both duration and concentration. A 24-h exposure produced a concentration-dependent increase in neural tube defects and growth retardation. At concentrations of 600 or 800 mg/dl, exposures of 8 h or longer produced neural tube defects, but shorter exposures did not, while 450 mg ethanol/dl produced neural tube defects with exposures of 20-h duration. KOTCH et al. (1995) reported a high incidence of malformations, including open anterior neuropore, in GD-8 C57BL/6J mouse embryos cultured for 6 h in the presence of 500 mg ethanol/dl followed by further culture in normal medium.

The toxicity of acetaldehyde to embryos growing in whole embryo culture has been tested in rats (CAMPBELL and FANTEL 1983; PRISCOTT 1985; GIAVINI et al. 1992) and mice (THOMPSON and FOLB 1982). CAMPBELL and FANTEL (1983) reported growth retardation in rat embryos, including decreased total protein at 25 μM and above, decreased head length at 50 μM and above, and decreased DNA content at 75 μM. Uniform embryolethality was observed at 100 μM acetaldehyde. GIAVINI et al. (1992) found that 227 μM acetaldehyde resulted in retarded growth and development and malformations in rat embryos cultured from GD-9.5, but did not test lower concentrations. Acetaldehyde was uniformly embryolethal at 454 μM. In contrast, PRISCOTT (1985) saw no toxicity at concentrations up to 260 μM, but embryolethality at 815 μM. THOMPSON and FOLB (1982) found that mouse embryos showed dose-related decreases in growth and development when cultured in the presence of 170 μM acetaldehyde and above.

Although the direct embryotoxic and dysmorphogenic effect of ethanol is clearly established by the studies discussed above, results of studies with acetaldehyde are inconclusive with regard to its potential role in the developmental toxicity of ethanol. While studies in rodents suggest that little acetaldehyde reaches the conceptus at teratogenic dosages of ethanol, KARL et al. (1988) demonstrated that the perfused human placenta is capable of oxidizing ethanol to acetaldehyde as well as transferring acetaldehyde from the maternal to the fetal circulation such that fetal levels are about 50% of maternal perfusate levels. Blood acetaldehyde concentrations in alcoholics vary from 1 to 200 μM (SCHENKER et al. 1995), and results of exposure of rodent embryos to acetaldehyde in this concentration range are equivocal. The recent finding that methanol, which is not metabolized to acetaldehyde, produces the same effects as ethanol when pregnant C57BL/6J mice are treated during gastrulation (ROGERS 1995; see below) suggests that the alcohols, and not their metabolites, may be the proximate teratogens. Nevertheless, whether acetaldehyde plays a role in human FAS is still unclear.

2. Maternal Nutrition and Transfer of Nutrients to the Conceptus

Ethanol consumption can affect nutrition by decreasing food intake, displacing other nutrients in the diet, interfering with digestion or absorption of nutrients, or by gastrointestinal complications of alcoholism (LIEBER 1994). It is difficult, if not impossible, to experimentally separate the direct embryotoxic effects of ethanol from concomitant maternal and/or developmental nutritional deficiencies (KENNEDY 1984; ABEL and HANNIGAN 1995). GOAD et al. (1984) attempted to isolate these effects using a 2×2 study design. Groups of pregnant C3H mice were given a normal liquid diet ad libitum with or without 4.1% w/v ethanol or the normal diet without ethanol pair-fed to the 4.1% ethanol animals. An additional group received a 4.1% ethanol liquid diet fortified to provide nutrient intake similar to that of the ad libitum controls. This design allowed for reasonably good (but not complete) separation of the effects of ethanol consumption and nutrient deficiency. Their results indicate that effects on maternal gestational weight gain and litter size were largely due to decreased dietary intake, while both decreased dietary intake and ethanol consumption inhibited embryofetal growth and development. A lack of statistical interaction between these two factors indicated that ethanol could exert a fetotoxic effect regardless of maternal dietary intake.

Specific nutrients for which decreased nutritional status has been associated with excessive alcohol consumption include zinc, folate, vitamins A and D, pyridoxine, and magnesium, while tissue concentrations of iron and manganese have been reported to be increased with chronic alcoholism (DREOSTI 1993). Chronic alcoholism in rats also results in decreased fetal and/or placental amino acid (FISHER et al. 1981, 1982, 1986; LIN 1981; HENDERSON et al. 1982; MARQUIS et al. 1984; B.H.J. GORDON et al. 1985) and glucose (SNYDER et al. 1986) uptake late in gestation, and fetal weight in ethanol-treated litters correlated with placental glucose transfer (SNYDER et al. 1986). Pinocytosis and proteolysis by the rat visceral yolk sac are strongly inhibited by ethanol in vitro (STEVENTON and WILLIAMS 1987). Of these reported nutritional alterations, the best evidence for an important role in FAS exists for zinc, folate, and vitamin A. The biochemical roles of folate and the developmental effects of folate deficiency are reviewed by HANSEN (see Chap. 10, Volume I), including interactions with ethanol exposure during pregnancy. The role of zinc nutrition and vitamin A metabolism will be highlighted here, with a brief discussion of folate.

a) Zinc

Zinc is essential for normal development (KEEN and HURLEY 1989) and serves as a cofactor, coenzyme, or protein structural component for a wide range of biological functions, including DNA synthesis and sequence recognition, transcriptional regulation, cytoskeletal and membrane integrity, apoptosis, and cellular energy metabolism (VALLEE and FALCHUK 1993; FALCHUK 1993). Zinc also plays a role in several free radical defense mechanisms (DREOSTI

1991; see below) and is essential for the activity of alcohol dehydrogenases (VALLEE and GOLDES, 1984). An inverse relationship between maternal plasma zinc concentration and the expression of FAS in humans has been reported (FLYNN et al. 1981), and ASSADI and ZIAI (1986) reported low plasma zinc levels and hyperzincuria in FAS infants. As little as 4 days of maternal dietary zinc deficiency during organogenesis can result in apoptosis in the rat embryo (ROGERS et al. 1995).

The relationship between ethanol developmental toxicity and zinc nutriture has been investigated using a variety of approaches. Significantly lower levels of zinc were found in fetuses of rat dams given 6% ethanol in their drinking water during pregnancy (MENDELSON and HUBER 1980), and placental transport of zinc was reduced by 40% in ethanol-fed rats, resulting in fetal zinc concentrations 30% below control levels (GHISHAN et al. 1982). A single i.p. injection of 2.4 g ethanol/kg on GD-10 resulted in an increased incidence of terata and fetotoxicity in zinc-deficient pregnant rats compared to rats fed an adequate amount of zinc (RUTH and GOLDSMITH 1981). DREOSTI (1984) reported a markedly higher incidence of central nervous system malformations when zinc deficiency was combined with alcoholism in pregnant rats. YEH and CERKLEWSKI (1984) fed pregnant rats throughout pregnancy a liquid diet containing either 2 or 10 μg zinc/ml diet with or without 30% ethanol-derived calories. No adverse effects on litter size or early postnatal growth were observed, but ethanol depressed maternal and offspring plasma zinc levels in both dietary groups. KEPPEN et al. (1985) fed CBA mice 0%, 15%, or 20% ethanol-derived calories with 0.3 or 8.5 μg zinc/ml diet. Resorptions and external malformations were highest in the low-zinc ethanol groups, indicating that zinc deficiency potentiated the developmental toxicity of ethanol. KEPPEN et al. (1990) found that inadequate dietary zinc intake increased prenatal mortality in litters of dams fed 15% ethanol-derived calories. However, supplementing dietary zinc above the recommended daily allowance (RDA) did not reduce the severity of the ethanol effect compared to dams consuming diets with adequate zinc. ZIDENBERG-CHERR et al.(1988) found that maternal–fetal transfer of ^{65}Zn in rats on GD-14 was not affected by maternal ethanol consumption, despite adverse effects on litter outcome. Similarly, GREELY et al. (1990) found that maternal, placental, and fetal tissue zinc concentrations on GD-18 and 21 were similar in ethanol-fed and control rats.

Both ethanol and methanol can induce hepatic metallothionein (MT) synthesis in mice, although the mechanism may be indirect (BRACKEN and KLAASSEN 1987). Induction of maternal hepatic MT during pregnancy by acute administration of diverse toxicants, including ethanol, can reduce zinc availability to the embryo, possibly contributing to developmental toxicity (TAUBENECK et al. 1994). However, J.E. HARRIS (1990) reported that chronic ethanol feeding of pregnant rats did not affect maternal hepatic MT or glutathione concentrations.

Overall, the above findings suggest that zinc deficiency can exacerbate the teratogenic effects of ethanol. A direct link between zinc metabolism and FAS and the biochemical mechanisms underlying such an interaction remains to be demonstrated.

b) Vitamin A

Vitamin A (all-*trans*-retinol) is an essential nutrient, and retinol or retinoic acid (vitamin A acid) is required for normal growth, vision, spermatogenesis, and epithelial maintenence. Vitamin A deficiency was one of the first maternal conditions shown to be teratogenic in controlled experiments in mammals (HALE 1933; WILSON et al. 1953). Retinoic acid is a ligand for the retinoic acid receptors (RARs), transcriptional factors of the steroid receptor superfamily (MANGELSDORF et al. 1994), and probably plays a key role in embryonal pattern formation (CONLON 1995). Recently, it has been demonstrated that genetic ablation of multiple RARs in mice can produce all of the malformations observed after maternal vitamin A deficiency (LOHNES et al. 1994; MENDELSOHN et al. 1994). Chronic alcoholism depletes hepatic vitamin A in humans (LEO and LIEBER 1982), baboons (SATO and LIEBER 1981), and rats (SATO and LEIBER 1981; RYLE et al. 1986).

Hypervitaminosis A (GEELAN 1979) and a number of retinoids are also teratogenic in experimental animals (NAU et al. 1994), including low dosages of all-*trans* retinoic acid in the mouse (SULIK et al. 1995). Synthetic retinoids (e.g., isotretinoin) are teratogenic in humans (ROSA 1983; LAMMER et al. 1985), and maternal consumption of high levels of vitamin A in the diet and/or supplements has recently been associated with an increased prevalence of birth defects in humans (ROTHMAN et al. 1995).

The idea that ethanol may exert its teratogenicity at least in part by inhibiting retinol metabolism (DUESTER 1991; PULLARKAT 1991; KEIR 1991) is based on the knowledge that both ethanol and retinol are substrates for oxidation by alcohol dehydrogenases and that oxidation of retinol to retinoic acid can be competitively inhibited by ethanol (for a recent review, see DUESTER 1994b). Retinol dehydrogenase is considered identical to class I ADH in rat, mouse, and man (KEIR 1991). Human class I ADH isozymes have a K_m for ethanol of approximately 0.2–2.0 mM (LIEBER 1994) and a K_m for retinol of 0.028 (MEZEY and HOLT 1971). The K_i for competitive inhibition of retinol oxidation by ethanol was reported to be 0.36 mM (MEZEY and HOLT 1971). During intoxication, ethanol concentrations in tissues are such that ADH will not be available for oxidation of retinol to retinoic acid. In support of this hypothesis, GRUMMER et al. (1993) reported that, in rats, maternal ethanol consumption resulted in two- to three- fold higher levels of retinol in embryonal and fetal tissues, while nuclear retinoic acid receptor levels were lower in GD-10 embryos. DEJONGE and ZACHMAN (1995) reported that retinol and retinyl palmitate levels were higher while retinoic acid was lower in GD-20 fetal rat hearts from ethanol-exposed pregnancies.

Given its role in development, a deficiency of retinoic acid in embryonal tissues could result in teratogenesis. Using a *lacZ* reporter gene fused to a human ADH3 promoter, ŽGOMBIĆ-KNIGHT et al. (1994) demonstrated activation of this transgene in the neural tube of mouse embryos as early as GD-9.5. The promoter for human ADH3 contains a retinoic acid response element and is inducible by retinoic acid (DUESTER et al. 1991). *Adh-1*, the gene for mouse class I ADH, is active as early as GD-10.5, and expression is highly stage and tissue specific (VONESCH et al. 1994); this gene is induced by retinoic acid treatment in GD-10.5 mouse embryos or mouse F9 embryonal carcinoma cells (SHEAN and DUESTER 1993). Together, these findings suggest that ADH may act in developing tissues to catalyze the synthesis of retinoic acid.

In contrast, studies in the class I ADH-negative deer mouse demonstrate that embryos develop normally, indicating that other ADH isozymes or other dehydrogenases provide enough retinoic acid to support normal development. Further, ethanol does not inhibit retinol oxidation in kidney (POSCH et al. 1989) or testes (POSCH and NAPOLI 1992) of the class I ADH-negative deer mouse. The recent finding that methanol, which is not metabolized by class I ADH in the mouse, can produce the same effects as ethanol in the C57BL/6J mouse model (ROGERS 1995) also argues against a primary role for inhibition of class I ADH retinol metabolism in the expression of FAS. However, inhibition of alternate pathways of retinol oxidation by ethanol or methanol has not been explored.

c) Folate

Folic acid is an essential vitamin used in one-carbon metabolism, including nucleotide synthesis and DNA methylation. Folate deficiency is the most common vitamin deficiency among alcoholics (HALSTED 1992; SCHENKER et al. 1995). Data from the second National Health and Nutrition Survey (NHANES II) show that, in the United States, the highest incidence of low folate levels (serum level less than 3.0 ng/ml and erythrocyte levels less than 140 ng/ml), and hence the greatest risk of folate deficiency, occurs among women aged 20–44 years (LIFE SCIENCES RESEARCH OFFICE 1984). Folate supplements have been shown to be effective in reducing the recurrence of neural tube defects in humans (MRC VITAMIN STUDY RESEARCH GROUP 1991; WALD 1993), and TOLAROVA and HARRIS (1995) found that high doses of folic acid and multivitamin supplementation were linked to a reduction in the recurrence of orofacial clefts.

Acute ethanol ingestion has been shown to deplete serum folate in humans (EICHNER and HILLMAN 1973) and rats (MCMARTIN 1984; EISENGA and MCMARTIN 1987). There are probably several factors contributing to this depletion. Malabsorption of folate occurs in chronic alcoholics (HALSTED et al. 1967), and chronically alcohol-fed monkeys show increased urinary folate excretion (TAMURA and HALSTED 1983). Further, acute ethanol exposure inhibits carrier-mediated folate transport in everted sacs from rat intestine (SAID and STRUM 1986), and chronic ethanol exposure has been shown to decrease rat placental folate receptor activity (FISHER et al. 1985). These findings suggest that an interaction between folate and ethanol could exacerbate developmental toxicity either through the induction of a transient folate deficiency by ethanol or through altered ethanol metabolism in folate deficiency.

In a related finding, the effects of ethanol on rat embryos in culture were lessened by the addition of *S*-adenosyl methionine (SAM), a universal methyl donor (SEYOUM and PERSAUD 1994). SAM is not only the methyl donor for most transmethylation reactions, but also plays a role in polyamine synthesis and provides cysteine for glutathione synthesis.

3. Prostaglandins

Prostaglandins (PGs) are eicosanoids derived from arachidonic acid metabolism and are involved in stages of pregnancy from implantation to delivery (KEIRSE 1978). The decidua is a rich source of PGs and cytokines, and some vasoactive PGs play a role in the regulation of placental blood flow (KELLY 1994). The putative roles of altered PGs metabolism in developmental toxicity are discussed in Chap. 18 (this volume).

The PGs have been proposed as mediators of FAS, because PGE_2 concentrations are elevated in maternal and fetal tissues following maternal ethanol administration (ANTON et al. 1990; SINERVO et al. 1992; SMITH et al. 1990, 1991), ethanol infusion increases uteroplacental production of PGE_2 in sheep (BOCKING et al. 1993), and high dosages of PG are teratogenic (PERSAUD 1974, 1975; HILBELINK and PERSAUD 1981; LEIBGOTT and WILEY 1985). HORROBIN (1980) suggested that one of the biochemical lesions underlying FAS was interference of ethanol with essential fatty acid metabolism and PG synthesis in the conceptus. In support of this hypothesis, VARMA and PERSAUD (1982) found that pregnant rats treated with evening primrose oil (9% γ-linoleic acid, 72% *cis*-linoleic acid) were protected from the developmental toxicity or ethanol.

The approach taken to assess the role of PGs in ethanol teratogenesis has involved the use of the fatty acid cyclooxygenase (the first enzyme of the PG synthetase complex) inhibitors acetylsalicylic acid (ASA) and indomethacin. Results of these experiments have been equivocal. Randall and coworkers have shown convincingly that treatment with ASA (RANDALL and ANTON 1984; RANDALL et al. 1991a), indomethacin (RANDALL et al. 1987), or ibuprofen (RANDALL et al. 1991b) can decrease the incidence of malformations and improve the growth retardation induced by ethanol in C57BL/6J mice. Nonteratogenic dosages of these compounds were administered shortly prior to ethanol treatment. In a dose–response study to assess the degree of ASA amelioration of ethanol developmental toxicity, RANDALL et al. (1991a) found that 150 mg ASA/kg was as effective as 300 mg/kg, and the maximal reduction in malformations was approximately 50%. These authors found a good correlation between reduction in the incidence of malformations and PGE levels in uterine/embryo tissue. PENNINGTON et al. (1985) demonstrated that application of indomethacin decreased the hypoplastic effects of ethanol on the developing chick brain, and SMITH et al. (1991) demonstrated that ethanol-induced suppression of fetal breathing movements in sheep fetuses is reduced by indomethacin.

In contrast to the above findings, others have demonstrated that ASA can exacerbate ethanol-induced developmental toxicity (GUY and SUCHESTON 1986; BONTHIUS and WEST 1989; PADMANABHAN et al. 1994; PADMANABHAN and PALLOT 1995). Whether PGs are deleterious or protective in ethanol teratogenesis remains unclear. HORROBIN (1980) suggested that the effect of ethanol on PG levels was dependent on the length of exposure, stimulating PG synthesis with acute exposure and depleting PGs with chronic exposure. Although the relationship is complex, research to date suggests that PGs may be a significant component in the expression of FAS.

Changes in PG metabolism might also be secondary to alterations in cell membranes. Removal of hydrogen atoms from membrane fatty acids can induce membrane-bound phospholipase A_2 to produce arachidonic acid, which is converted to PG by cyclooxygenases. Lipid peroxidation can also induce thromboxane synthesis, and thromboxanes are potent vasoconstrictors which may contribute to free radical production via tissue ischemia/reperfusion (see below).

4. Hypoxia

Hypoxia has long been known to have deleterious effect on embryonic development (GRABOWSKI 1970), and BRONSKY et al. (1986) demonstrated that maternal hypoxia results in increased fetal mortality, malformations, and growth retardation in CL/Fr mice. Fetal hypoxemia has been reported to impair fetal growth (JONES and BATTAGLIA 1977) and to suppress fetal breathing movements (BODDY et al. 1974).

Delivery of oxygen to the fetus is linearly related to umbilical blood flow (ITSKOVITZ et al. 1983), and there is evidence that ethanol can cause uteroplacental ischemia. However, it has not been firmly established that this occurs at relevant ethanol concentrations. A bolus intravenous infusion of 3g ethanol/kg given over 2 min to pregnant Rhesus and cynomolgus monkeys resulted in collapse of umbilical vasculature within about 15 min (MUKHERJEE and HODGEN 1982). Peak maternal blood ethanol concentrations were approximately 250 mg/dl. The severe hypoxemia and acidosis which resulted in fetuses required about 1 h to resolve. When human umbilical vessels from spontaneously delivered full-term fetuses were incubated in vitro in the presence of 10–85 mg ethanol/dl, concentration-dependent contraction of the vessels was observed (ALTURA et al. 1983). SAVOY-MOORE et al. (1989) also reported that low concentration of ethanol contracts the human umbilical artery in vitro. Using microspheres, P.D.H. JONES et al. (1981) found that chronic exposure to ethanol reduced placental, but not renal blood flow or cardiac output.

Results of other studies argue against hypoxia as an important etiologic factor in FAS. PATRICK et al. (1985, 1988) have shown that fetal blood gas and acid–base disturbances do not occur with acute low-dose or multiple-dose ethanol infusion in near-term pregnant ewes. Maternal infusion of six doses of 0.5 g ethanol/kg administered over 8 h had no effect on fetal or maternal blood gases, even though peak blood ethanol levels were higher than those reported by MUKHERJEE and HODGEN (1982) to cause umbilical vascular collapse in monkeys. Although rapid bolus injection of 2 g ethanol/kg over 5 min did produce transient fetal hypoxemia, this was secondary to maternal respiratory depression (SMITH et al. 1991). ERSKINE and RITCHIE (1986) found no effect of a single dose of 0.25 g ethanol/kg on in vivo human umbilical artery blood flow measured by pulsed Doppler ultrasound.

The observation of increased PGE_2 levels following ethanol administration, discussed above, may be related to hypoxia induced by uteroplacental ischemia. Levels of PGE_2, a potent vasodilator, are known to be increased during ischemia (CHEMTOB et al. 1993; EGAN et al. 1981; SARDESAI 1992), and PGE_2 attenuates the fetal acidemia that occurs during hypoxia (HOOPER et al. 1992).

Although no attempt was made to assess uterine blood flow or embryonal oxygenation, the finding by PADMANABHAN and PALLOT (1995) of digital malformations associated with vascular disruption in mice treated with ethanol on GD-11 is suggestive of ischemia/reperfusion. Similar deformities have been observed following clamping of the uterine vasculature in pregnant rats (LEIST and GRAUWILER 1974) or the umbilical cord in rabbit embryos (MILLICOVSKY and DESESSO, 1980; see also Chap. 30, this volume).

Given the sensitivity of the developing organism to hypoxia, it is reasonable to hypothesize that embryofetal hypoxia may represent a primary developmental insult underlying FAS. However, this link remains to be demonstrated in an appropriate animal model. In addition to oxygen deficiency, ischemia/reperfusion can lead to generation of free radicals and tissue damage, discussed in the next section.

5. Free Radicals

Free radicals are short-lived, highly reactive molecules with one or more unpaired electrons. In excess, reactive oxygen species including superoxide (O_2^-), singlet oxygen ($^1O_2$), hydroxyl radical (OH^-), and hydrogen peroxide (H_2O_2), can be highly damaging to cells because of their ability to initiate chain reactions of peroxidative damage to lipids, proteins, DNA, or other biological molecules (YAGI, 1993). Free oxygen radicals are constantly produced in the course of normal aerobic metabolism and are scavanged by endogenous antioxidants such as superoxide dismutase (SOD), glutathione peroxidase, catalase, vitamin E, vitamin C, and reduced glutathione (DARGEL 1992).

Potential sources of increased free radical concentrations in alcoholic pregnancy include generation during ethanol (SLATER 1988; DAI et al. 1993; CASTILLO et al. 1992) and acetaldehyde metabolism (MÜLLER and SIES 1982) and uteroplacental ischemia/reperfusion (FANTEL et al. 1992; ZIMMERMAN et al. 1994), as well as depletion of antioxidant defenses. Alterations of ascorbic acid (BONJOUR 1979), reduced glutathione (SPEISKY et al. 1985; HIRANO et al. 1992), and vitamin E and selenium (TANNER et al. 1986) levels have been reported in alcoholics. Hepatic lipid peroxidation is greater after long-term ethanol feeding in rats receiving a low-vitamin E diet (KAWASE et al. 1989).

Superoxide has been demonstrated to be mutagenic and toxic to most organisms (HASSAN and FRIDOVICH 1980) and is generated during the oxidation of acetaldehyde. SOD is the principal defense against superoxide, catalyzing the conversion of O_2^- to H_2O_2 and O_2. Acute ethanol treatment inhibits SOD in cultured nerve cells (LEDIG et al. 1980), promoting the release of O_2^-. In vitro studies have shown that chicken neural crest cells, devoid of SOD, are particularly sensitive to ethanol toxicity and that addition of SOD to the culture medium was effective in reducing ethanol-induced cell death (W.L. DAVIS et al. 1990). KOTCH et al. (1995) demonstrated increased O_2^- levels in organogenesis-stage mouse embryos cultured in the presence of 500 mg ethanol/dl culture medium for 6 h. Addition of SOD to the culture medium resulted in a decrease in O_2^- concentration along with decreases in lipid peroxidation (malondialdehyde content), cell death, and dysmorphogenesis.

Given the generation of free radicals during ethanol metabolism, the potential for uteroplacental ischemia/reperfusion, and the deficiencies of antioxidant vitamins and enzymes associated with alcoholism, it is likely that free radical excess may play a role in ethanol teratogenesis. Further, catalase is a major free radical scavanger and also participates in the oxidation of both ethanol and methanol. Thus, free radicals may persist longer in the presence of these alcohols, which are competing substrates for catalase. The degree to which free radicals result in embryonal damage and the key targets for such damage remain to be elucidated. Membrane lipids may be one such critical target for peroxidation, as well as for other effects of ethanol on cell membranes.

6. Effects on Cell Membranes

Cell adhesions, cell–cell recognition and communication, cell migration, and signal transduction are all essential for normal development and depend on proper structure and function of the cell membrane. There is a large body of literature on the acute and chronic effects of ethanol on biological membranes, including effects on the phospholipid bilayer, on receptor and channel function, and on signal transduction. Results germane to mechanisms of FAS will be briefly reviewed in this section. The reader is referred to LITTLETON (1989), HOEK et al. (1992), and WEST et al. (1994) for additional coverage of this topic.

Ethanol can enter cell membranes. Because of its small size and polar nature, it most likely associates with the polar regions of phospholipid head groups or, to a lesser extent, with the hydrophobic membrane core. Estimates of the molar ratio of ethanol to lipids in the cell membrane under anesthetic conditions are on the order of 1:200 (LITTLETON 1989). Measurements of lipids at the center of the membrane bilayer indicate that the presence of ethanol increases disorder of these lipids, leading to increased membrane fluidity (CHIN and GOLDSTEIN 1977a; HITZEMANN et al. 1986). Ethanol-induced alteration of cell membrane fluidity has been recognized as a putative mechanism underlying FAS (LEONARD 1987; SCHENKER et al. 1990). Through poorly understood mechanisms, chronic alcohol exposure leads to adaptation by the cell membrane, decreasing the acute effects of ethanol on membrane fluidity (CHIN and GOLDSTEIN 1977b).

Ethanol administration acutely or chronically has been reported to affect a number of membrane neurotransmitter receptors. Acute ethanol exposure inhibits *N*-methyl-D-aspartate (NMDA)-stimulated calcium uptake in neuronal cells (HOFFMAN et al. 1989; DILDY and LESLIE 1989). SAVAGE et al. (1991) found reduced NMDA receptor binding after ethanol exposure during the last trimester of gestation in rats. On the other hand, IORIO et al. (1992) found that preexposing primary cerebellar granule cell cultures to ethanol greatly enhanced NMDA-stimulated calcium uptake and NMDA receptor function.

Membrane signal transduction is critical for normal development, and recent work has elucidated signal transduction pathways involved in cell proliferation, differentiation, and programmed cell death. The effects of ethanol on signal transduction have also received increased attention recently (HOEK et al. 1992). Acute ethanol exposure results in increased levels of adenosine receptor-stimulated cyclic adenosine monophosphate (cAMP) in a neuroblastoma-glioma hybrid cell line, while chronic ethanol exposure has the opposite effect in these cells (A.S. GORDON et al. 1986) or in PC12 cells (RABIN 1990). Ethanol may also alter levels of G proteins (RABIN 1993) and increases phosphorylation of protein kinase C (MESSING et al. 1991; ROIVAINEN et al. 1993).

Ethanol can perturb calcium channels and cytosolic calcium stores and concentrations. Acute ethanol exposure inhibits depolarization-induced calcium uptake through voltage-sensitive calcium channels, while chronic exposure upregulates membrane density of these channels (MESSING et al. 1986; LITTLETON et al. 1991). Acute high-dose ethanol administration causes an increase in calcium in synaptosomes (DANIELL et al. 1987) and in cytosol of hepatocytes (DANIELL et al. 1987; HOEK et al. 1987) and PC12 cells (RABE and WEIGHT 1988).

Ethanol can inhibit cell–cell or cell–substrate adhesion. Adhesion of embryonal carcinoma cells to laminin or type I collagen, but not fibronectin, is inhibited by ethanol (GATALICA and DAMJANOV 1990). Neuroblastoma-glioma hybrid NG108-15 cells can be induced to cluster by recombinant human osteogenic protein-1 (hOP-1), which increases cell–cell adhesion by strongly inducing nerve cell adhesion molecule (NCAM) and L1. CHARNESS et al. (1994) demonstrated that ethanol can inhibit this morphogenic process with an IC_{50} of 5–10 m*M* (a concentration range similar to that achieved in blood after only one or two drinks). This effect occurred in the absence of any effect on cell proliferation, the induction and cell surface expression of NCAM and L1, or the alternative splicing and sialylation of NCAM, indicating that cell adhesion may represent a very sensitive target for ethanol developmental toxicity.

Gangliosides are membrane-bound glycosphingolipids involved in cell migration, cell–cell interaction, neurite outgrowth, and other membrane-mediated events. There is evidence that gangliosides in the brain are involved in alcohol intoxication (HUNGUND and MAHADIK 1993), and gangliosides can reduce alcohol intoxication and membrane effects (HUNGUND et al. 1989, 1990). Pretreatment or cotreatment of primary cultures of neural crest cells derived from mouse embryos with ganglioside GM1 also reduces the toxicity of ethanol (CHEN et al. 1995). HUNGUND et al. (1994) found that ethanol administration to pregnant rats on GD-7 and 8 and/or GD-13 and 14 caused an increase in ganglioside GM1 in at least 50% of pup brains assayed on GD-20. Further, GM1 treatment of pregnant dams before ethanol treatment prevented the GD-20 increase, apparently by blocking the cellular membrane changes associated with acute embryo/fetal ethanol exposure.

At the high tissue concentrations of ethanol required to cause FAS, it seems likely that alterations of cell membrane structure and function in the embryo may occur. Membranes are capable of adaptive changes with chronic ethanol exposure, and acute and chronic exposures can have opposite effects on some parameters. Given the constant and critical requirement for normal membrane function during development, the role of ethanol-induced membrane alterations in FAS and ARBD certainly requires further study.

7. Interrelationship of Putative Mechanisms

All of the mechanisms discussed in this section probably play some role in the genesis of FAS and ARBD under varying circumstances. Together, these mechanisms, directly or indirectly, can result in impairment of cellular processes at critical periods of development. Indeed, one difficulty in sorting out these findings is that most of the mechanisms are biologically, and probably toxicologically, related. Changes in membrane arachidonic acid can affect PG levels, PG levels affect blood flow and thus oxygenation, ischemia/reperfusion can produce free radicals, and free radicals can damage membranes. The sequence or relationships of these putative mechanisms remain to be further elucidated. DREOSTI (1993) has summarized relationships among ethanol metabolism, free radical generation, mineral and vitamin antioxidants, and peroxidative membrane damage.

Ethanol is actually a weak teratogen, acting at high dosages with broad target specificity, and growth appears to be the most sensitive target. KENNEDY (1984) presented an integrated hypothesis tying together many of the findings concerning FAS and ARBD at that time, and it remains as a parsimonious model of ethanol developmental toxicity. In this model, cell membrane narcosis and depression of cellular metabolism, including protein, RNA, and DNA synthesis, result in hypoplasia and hypotrophy of cell populations most actively undergoing proliferation or differentiation at the time of exposure. Essentially all treatment paradigms used in animal models result in effects on growth in addition to dysmorphogenesis or other developmental toxicity. PENNINGTON et al. (1983), working with chick embryos, found that ethanol suppresses cell division resulting in fewer cells per embryo, and the results of BROWN et al. (1979) indicate the same is true in rats. Even very short periods of suppression of cell proliferation and growth during gastrulation can result in devastating craniofacial malformations, while effects on these processes later in development may have increasingly more subtle outcomes, including growth retardation and central nervous system functional deficiencies.

E. Developmental Toxicity of Methanol in Experimental Animals

I. Effects of Exposure During Pregnancy

1. Rats

The effects of inhaled methanol during pregnancy were first studied by NELSON and coworkers (1985). Sprague-Dawley rats were exposed to 5000, 10000, or 20000 ppm methanol for 7h/day. The two lower concentrations were administered daily from GD-1 to GD-19, while the 20000-ppm exposures were administered daily from GD-6 to GD-15. Blood methanol levels were assessed in nonpregnant females under the same exposure conditions. Signs of maternal toxicity were slight and were observed only at the highest concentration. Peak blood methanol levels were approximately 1.3, 2.0, and 8.7 mg/ml after exposure to 5000, 10000, or 20000 ppm, respectively. Maternal exposure to 20000 ppm methanol resulted in decreased fetal weight and significant increases in external, visceral, and skeletal malformations. Skeletal malformations were the most prevalent and included abnormalities of the basicranium and the vertebra, including an increase in the incidence of fetuses with a rib on the seventh cervical vertebra. There was also a low incidence of exencephaly, encephalocele, hydrocephalus, and various anomalies of the cardiovascular and urinary systems at the highest exposure level. At 10000 ppm there were slight, statistically insignificant increases in these same anomalies, as well as a significant decrease in fetal weight. No maternal or developmental parameters were affected at 5000 ppm methanol.

The effects of methanol exposure during early pregnancy in the rat were studied by CUMMINGS (1993). Rats were dosed orally with up to 3.2 g methanol/kg per day on GD-1–8, and groups of animals were killed on GD-9, 11, or 20. The decidual cell response (DCR) technique, in which the lining of the uterus is surgically stimulated to induce proliferation, was also applied. Reductions in pregnant uterus weight and implantation site weights were observed on GD-9, and effects on the DCR suggested that methanol impeded uterine decidualization. No other effects on viability or development of conceptuses were noted.

The postnatal effects of methanol exposure during late gestation in Long-Evans rats were studied by INFURNA and WEISS (1986). Methanol was administered at 2% in the drinking water for 3 days from GD-15–17 or 17–19. Daily methanol intake averaged about 2.5 g/kg body weight. On postnatal day (PND) 1, suckling behavior was tested by placing the pups with an anesthetized dam and recording the latency to finding the nipple and beginning to suckle. On PND 10, the ability of the pups to locate nesting material from their home cage was tested. Methanol exposure did not affect maternal weight gain or fluid intake, nor did it affect litter size, birth weight, postnatal survival or growth, or day of eye opening. Offspring of methanol-exposed dams required longer than controls to begin suckling on PND 1 (latency of approximately 90 s vs. 60 s in controls). Ability to locate material from the home nest in a plexiglass grid was also affected in methanol-exposed pups on PND 10. The methanol-exposed pups required about twice as long as controls to find their nest material, and their initial direction was more often incorrect than that of controls.

STANTON et al. (1995) exposed pregnant rats to 15000 ppm methanol via inhalation for 7 h/day on GD-7-19. Daily peak maternal blood methanol concentrations declined from 3.8 mg/ml after the first exposure to 3.1 mg/ml after the final exposure. Offspring body weight was reduced by about 5% on PND 1, but the behavioral tests used, including motor activity (PND 13-21, 30, 60), olfactory learning (PND 18), behavioral thermoregulation (PND 20–21), T-maze learning (PND 23–24), acoustic startle response (PND 24, 60), reflex modification audiometry (PND 60), passive avoidance (PND 72), and visual-evoked potentials (PND 160) did not detect any difference between methanol-exposed and control groups.

2. Mice

ROGERS et al. (1993a) studied the effects of methanol exposure (1000, 2000, 5000, 7500, 10000, 15000 ppm) during pregnancy in CD-1 mice under conditions similar to those used for rats by NELSON et al. (1985). No maternal toxicity was attributed to methanol, although the exposure procedure per se reduced maternal weight gain in all groups, including the filtered air-exposed mice, compared to unhandled controls. An additional unexposed control group was deprived of food for 7 h/day similar to the food deprivation experienced by the exposed groups. Food deprivation accounted for some, but not all, of the effect of exposure on maternal weight. The lack of overt methanol-induced maternal toxicity is somewhat surprising, given that these mice had higher peak blood levels of methanol than did rats similarly exposed (See ROGERS et al. 1993a and NELSON et al. 1985; see also Fig. 2).

CD-1 mice were found to be more sensitive to the developmental toxicity of methanol than were Sprague-Dawley rats. The incidence of fetuses with ribs on the seventh cervical vertebra was increased in a dose-related fashion at 2000 ppm and above. Cleft palate, exencephaly, and skeletal defects were observed at 5000 ppm and above. The skeletal anomalies observed, including cervical ribs, were similar to those observed in rats by NELSON et al. (1985). There was a significant increase in resorptions at 7500 ppm and above and an increase in complete litter loss at 10000 and 15000 ppm. Despite malformations at lower exposure levels, fetal weight was reduced only at 10000 and 15000 ppm. The induction of cleft palate and exencephaly by methanol were found to be non-independent events in this study. The incidence of cleft palate among fetuses without exencephaly was 21.9% (119 out of 543), while the incidence of cleft palate among fetuses with exencephaly was only 11.1% (six out of 54). YASUDA et al. (1991), studying 2, 3, 7, 8-tetrachlorodibenzo-p-dioxin (TCDD), reported a similar decrease in susceptibility to cleft palate among fetuses with exencephaly.

The effects of methanol administered by oral gavage during pregnancy in CD-1 mice were also studied by ROGERS et al. (1993a). Mice were given twice daily doses of 2 g methanol/kg, 7 h apart, on GD-6–15. Effects observed were similar to those observed following inhalation exposure (cleft palate, exencephaly, skeletal defects, and resorptions), and quantitatively the peak blood level and incidence of effects were similar to those observed after an inhalation exposure to 10 000 ppm methanol for 7 h/day.

The developmental phase specificity for the adverse effects of exposure to inhaled methanol in pregnant CD-1 mice has been examined. BOLON et al. (1993) exposed pregnant mice to 10 000 ppm methanol (7 h/day) for 10 days on GD-6-15, to 5000, 10 000, or 15 000 ppm for 3 days on 7–9, or to 10 000 or 15 000 ppm for 3 days on GD-9–11. GD-17 fetuses were examined for external and visceral anomalies, but skeletal examinations were not done. Exposure to 10 000 or 15 000 ppm on GD-6–15 resulted in cleft palate, exencephaly, hydronephrosis, tail and digit defects, and increased resorptions. Exposure to 10 000 or 15 000 ppm on GD-7–9 resulted in exencephaly, cleft palate, hydronephrosis, and ocular and tail defects. Exposure to these same concentrations on GD-9–11 resulted in cleft palate and digit and tail defects. Exposure to 15 000 ppm methanol on GD-7–8 also caused exencephaly. Single or 2-day inhalation exposures of pregnant CD-1 mice to 10 000 ppm methanol during the period of GD-5–13 were carried out by ROGERS and coworkers (ROGERS et al. 1993b; ROGERS and MOLE 1996). Two-day exposures were carried out beginning on each of GD-6–12 (e.g., the latest exposure was on GD-12–13), and single-day exposures were on each of GD-5–9. Peak blood methanol concentration after a single exposure was about 4 mg/ml, and blood levels returned to baseline within about 24 h. Two-day exposures on GD-6–7 or 7–8, and single-day exposure on GD-7 resulted in some fully resorbed litters. With single-day exposure, the number of resorptions per litter was highest on GD-7. The stage-specific patterns of developmental sensitivity to cleft palate and exencephaly are illustrated in Fig. 8. The period of susceptibility to methanol-induced cleft palate was broad, with single exposures on any of GD-5–9 or 2-day exposures on GD-6–7 to 11–12 eliciting this effect. Peak sensitivity to cleft palate occurred with 2-day exposure on GD-7–8 or

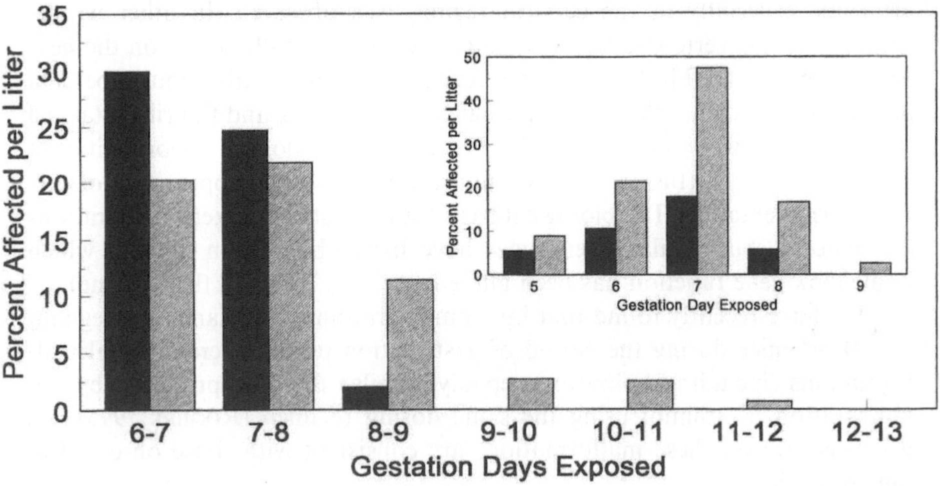

Fig. 8. Percentage of fetuses per litter affected by cleft palate (*shaded bars*) or exencephaly (*black bars*) following exposure on 1 (*inset*) or 2 days of gestation (GD) to 10 000 oom methanol (7 h/day) during GD 5–13

one-day exposure on GD-7. The basis for this rather early critical period (induction as early as GD-5) for methanol-induced cleft palate is unknown at present, but is unusual for an agent with a short biological half-life. Exencephaly occurred with 2-day exposure on GD-6–7 to GD-8–9 (peak GD-6–7) and one-day exposure on GD-5 to GD-8 (peak GD-7). The peak critical periods for skeletal defects were progressively later going from more anterior to more posterior structures. Thus the critical period for exoccipital defects was GD-5, for atlas defects GD-5 or 6, and for axis defects, lower cervical defects, and supernumerary (lumbar) ribs GD-7. Single-day exposure on GD-5 resulted in a significant increase in the incidence of fetuses with 25 presacral vertebrae (26 is normal), while single-day exposure on GD-7 resulted in an increased incidence of fetuses with 27 presacral vertebrae.

The skeletal abnormalities observed by Rogers and Mole (1996), including splits and duplications of the atlas and axis, ribs on cervical vertebra 7, and abnormal number of presacral vertebrae, are suggestive of disruption of embryo segmentation and/or segment identity. These skeletal malformations were examined in greater detail by Connelly and Rogers (1996). Methanol (5 g/kg) was administered orally to CD-1 mice on GD-7, and fetuses were collected on GD-18. Anatomical landmarks identifying specific cervical vertebrae were examined, including the tubercula anterior normally found on cervical vertebra 6 (C6) and various foramina and other features evident in disarticulated vertebrae. The number of free (e.g., ventrally unattached) ribs and ribs attached to the sternum were counted, and ribs found on C7 were categorized as partial or full (i.e., attached to the sternum). Methanol caused homeotic shifts of segment identity. Specifically, a posteriorization of vertebral elements, especially in the cervical region, was observed. In other words, certain of these vertebrae had structural features normally found on the next vertebra posteriad. Methanol-treated fetuses often had cartilaginous tubercula anterior on C5 rather than their normal position on C6, and full ribs attached to the sternum were observed on C7. Further, morphological abnormalities of the atlas and axis (fusions, splits, duplications) gave the appearance of disrupted segmentation. The biological basis for these striking skeletal alterations is unknown, but similar phenotypes have been observed in mice in which homeobox gene function has been altered (see, e.g., Chap.2, this volume).

We have recently found that i.p. administration of methanol to pregnant C57BL/6J mice during the period of gastrulation produces craniofacial malformations (including holoprosencephaly) similar to those produced by administration of ethanol using the same dosing regimen (ROGERS 1995). As discussed above, these malformations are consistent with those observed in human FAS.

II. Pathogenesis of Methanol-Induced Birth Defects

1. Whole Animal Studies

The pathogenesis of methanol-induced neural tube defects in CD-1 mouse embryos and fetuses was examined by BOLON et al. (1994). Following maternal exposure to 15 000 ppm methanol (6 h/day) on GD-7–9, 15% of fetuses exhibited cephalic dysraphism on GD-17. The severity of the defects ranged from encephalocele to exencephaly with or without facial clefting, anencephaly, and holoprosencephaly. Further, microcephaly was observed among methanol-exposed fetuses without neural tube defects. Embryos examined on GD-8.5 following maternal methanol exposure (15 000 ppm) on GD-7–8 had swollen and poorly elevated cephalic neural folds compared to controls. Reductions in the mitotic indices of the neuroepithelium (55% reduction) and the underlying mesoderm (47% reduction) were reported. Additional dams were exposed to 15 000 ppm methanol on GD-7–9, and embryos were examined on GD-9.5 and 10.5. The methanol-exposed embryos exhibited delayed rotation, microcephaly, and edema as well as anterior neural tube patency. Nile blue sulfate staining (visualizes apoptosis) of embryos on GD-8.5, 9.5, or 10.5 revealed no difference in the staining pattern between methanol-treated and control embryos.

We have recently examined the pathogenesis of craniofacial malformations induced by i.p. administration of methanol (4.9 /kg) during gastrulation (GD-7) in the C57BL/6J mouse (ROGERS 1995). Examination of embryos on GD-8 by confocal microscopy revealed edematous and poorly elevated neural folds (Fig. 9). In contrast to the findings of BOLON et al. (1994), using neutral red staining we observed some indications of increased cell death at the edges of the neural plate on GD-8, and clear evidence of cell death in the cranial neural folds on GD-9 (Fig. 9). Similar to the findings of BOLON et al. (1994),

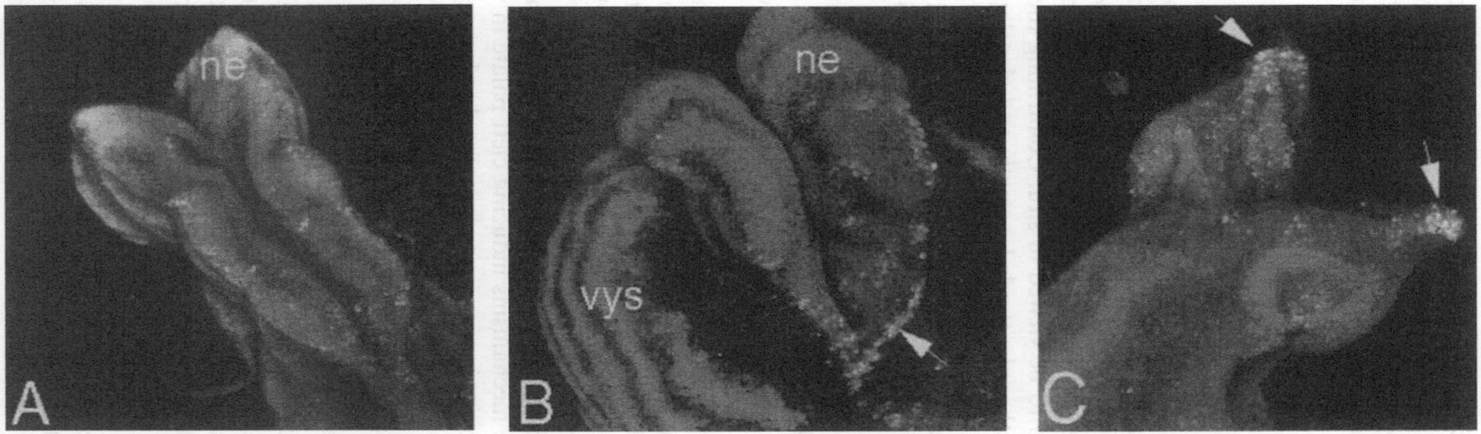

Fig. 9A-C. Three-dimensional computer reconstructions of confocal optical sections of C57BL6J mouse embryos following maternal intraperitoneal administration of two doses of 2.45 g/kg methanol (4 h apart) on gestational day (GD) 7. **A** Control embryo collected on GD 8. **B** Methanol-treated embryo collected on GD 8. **C** Methanol-treated embryo collected on GD 9. Embryos were stained with neutral red (stains areas of cell death, *bright stippling at arrows* and imaged by confocal laser scanning microscopy. Embryos of methanol-treated dams exhibited edematous and poorly elevated neural folds on GD 8, with an apparent increase in cell death along the edges of the folds (*arrow* in B), and increased cell death in the cranial neural elevations on GD 9 (*arrows* in C). Ne, neuroepithelium; *vys*, visceral yolk sac

optical sectioning in the transverse plane revealed sparse mesenchyme underlying the neural folds (not shown). The difference between our findings on cell death and those of BOLON et al. (1994) may be due to the different mouse strains used. In previous experiments we have not observed methanol-induced cell death using Nile blue sulfate in CD-1 mouse embryos (ROGERS, unpublished observations).

2. In Vitro Studies

ABBOTT et al. (1995) examined patterns of cell death in rat and mouse embryos exposed to methanol in culture. Early somite-stage embryos were exposed to dysmorphogenic concentrations of methanol and observed for cell death using a modified Feulgen whole-mount staining procedure which allows nuclei throughout the embryo to be viewed in situ. Confirming the results of ANDREWS et al. (1993a), methanol retarded growth and development (including delayed neural tube closure) of both rat and mouse embryos, and mouse embryos were affected at lower concentrations. Increased cell death was observed in specific regions of the forebrain, visceral arches, and otic placodes in both species.

The effects of methanol on palatogenesis in vitro were examined and compared to effects of ethanol by ABBOTT et al. (1994). Midcraniofacial tissues from GD-12 CD-1 mouse embryos were cultured in serum-free medium for 4 days. Exposure to 0–20 mg methanol /ml for 6 h, 12 h, 1 day, or 4 days resulted in a concentration and time-dependent decrease in the incidence and completeness of palatal fusion. A specific effect on cell proliferation was evidenced by reductions in total DNA content at concentrations of methanol that did not reduce protein content and by decreased incorporation of [^{3}H]-thymidine in palatal mesenchyme. Ethanol appeared to be a more potent inhibitor of palatogenesis in vitro than was methanol. This system probably does not, however, model the mechanisms inducing cleft palate in vivo, since the critical period of susceptibility in vivo is at an earlier developmental stage.

III. Mechanisms Underlying Methanol-Induced Birth Defects

1. Determination of Proximate Teratogen – Methanol or Formate?

ANDREWS and coworkers (1993a) examined the direct toxicity of methanol to CD-1 mouse and Sprague-Dawley rat embryos developing in culture. In rat embryos, methanol concentrations of 8 mg/ml culture medium and above resulted in decreased growth and development, and 12 mg/ml resulted in dysmorphogenesis in 66% of live embryos as well as 53% embryo mortality. In comparison, methanol concentrations of 4 mg/ml culture medium affected growth and development of mouse embryos, and dysmorphogenesis was observed in 58% of embryos cultured in the presence of 6 mg/ml. Importantly, in both the rat and the mouse, concentrations of methanol required to cause developmental toxicity in vitro were similar to peak maternal blood levels following developmentally toxic exposures in vivo (rats, NELSON et al. 1985; mice, ROGERS et al. 1993; see above). BROWN-WOODMAN et al. (1995) reported very similar findings for the effects of methanol on Sprague-Dawley rat embryos in culture, with concentrations of 286.5 μmol/ml (9.17 mg/ml) being developmentally toxic. These results demonstrated that methanol does not require maternal metabolism to elicit developmental toxicity and suggest that the greater sensitivity of the mouse to the developmental toxicity of methanol is, at least in part, due to an intrinsic difference in embryonal sensitivity. However, they do not address the question of whether the embryo itself might metabolize sufficient amounts of methanol to toxic metabolites.

The adult human toxicity of methanol has been ascribed to accumulation of the metabolite formate. The study of formate developmental toxicity in intact rodents is precluded because of rapid metabolic elimination, but the developmental toxicity or formate has been tested using rodent embryos in culture. ANDREWS et al. (1993b) cultured presomite rat embryos for 48 h in serum containing concentrations of Na^+ formate ranging from 0 to 1.6 mg/ml. The starting pH of the culture medium was 8.13 and was not affected by the addition of Na^+ formate. Since metabolism of methanol to formate in vivo can produce a metabolic acidosis, the effects of lowering pH in addition to formate exposure were tested. At pH 8.13, embryo development and growth was affected by Na^+ formate only at the highest concentration, 1.6 mg/ml (23.5 m*M* formate). However, decreasing media pH by the addition of HCl exacerbated the effect of formate, such that at pH 6.5 most developmental parameters were affected at 0.8 mg Na^+ formate/ml culture medium (11.8 m*M* formate). Confirming the results of ANDREWS et al. (1993b), BROWN-WOODMAN et al. (1995) reported that concentrations at or above 18.66 m*M* formic acid or Na^+ formate were developmentally toxic to rat embryos in vitro, but that Na^+ formate was less toxic due to the exacerbating effect of low pH in formic acid-treated cultures. The developmental toxicities of Na^+ formate and formic acid were subsequently compared in mouse and rat embryos (ANDREWS et al. 1995). Rat and mouse embryos exposed to either agent for 24 h exhibited concentration-dependent reductions in growth and development and increased incidences of malformed embryos. In contrast to results of previous studies with methanol, no difference in species sensitivity to Na^+ formate or formic acid was observed. Various developmental parameters were adversely affected at concentrations of 11.8 m*M* and above for both forms of this metabolite. The concentrations of formate found to be toxic to rat or mouse embryos growing in vitro are relevant to human risk assessment, because blood formate concentrations in excess of 7–10 m*M* have been reported to occur in humans following ingestion of methanol (MCMARTIN et al. 1980; SEJERSTED et al. 1983).

DORMAN and coworkers (1995) exposed pregnant CD-1 mice to methanol by inhalation at teratogenic concentrations (10 000 or 15 000 ppm for 6 h on GD-8) and reported no significant increase in blood or decidual formate, indicating that formate is probably not involved in methanol teratogenesis in the mouse. Dosing of pregnant mice by gavage with Na^+ formate (750 mg/

kg) was neither teratogenic nor raised maternal blood formate level. However, confirming the results of ANDREWS et al. (1993a, 1995), exposure of mouse embryos to either methanol (250 m*M*) or Na^+ formate (40 m*M*) in culture was developmentally toxic.

From the studies of formate in rodents in vivo and in vitro discussed above (ANDREWS et al. 1993b, 1995; BROWN-WOODMAN et al. 1995; DORMAN et al. 1995), it appears that formate is toxic to the developing embryo at concentrations far in excess of those achieved after teratogenic methanol exposure in the mouse, but within the range of levels achieved in humans after acute high-dose methanol poisoning. Thus, although methanol appears to be the proximate murine teratogen, formate is still of concern in terms of potential developmental toxicity in humans.

2. Role for Formaldehyde?

The potential for formaldehyde to contribute to methanol toxicity has been discounted because it is a reactive and short-lived metabolite. However, C. HARRIS et al. (1995) have recently reported that formaldehyde is toxic to mouse embryos developing in culture at concentrations below 10 μg/ml (0.3 m*M*). Further, formate or formaldehyde injected directly into the amniotic cavity (thus bypassing the yolk sac) is developmentally toxic at low doses (CONTRERAS and HARRIS 1995). These results indicate that it will be important to determine the intrinsic embryonal metabolism of methanol and the target tissue concentrations of metabolites.

3. Folate Deficiency – A Susceptibility Factor in Methanol Developmental Toxicity?

Methanol oxidation to CO_2 is dependent on tetrahydrofolate, and pregnant women are often of marginal or deficient folate status (see above, and Chap. 10, Volume I). SAKANASHI et al. (1996) examined the effect of maternal dietary folate level on the expression of developmental toxicity in CD-1 mice exposed to methanol by oral gavage. Female mice were fed one of three diets containing 400 (low), 600 (marginal), or 1200 (adequate) nmol folic acid/kg diet for 5 weeks prior to mating and throughout gestation. Twice-daily dosages of 2.0 or 2.5 g/kg per day were administered from GD-6–15. Maternal liver folate concentrations were approximately 40%–50% lower in the low-folate group than in the marginal or adequate-folate groups; methanol did not affect maternal liver folate concentrations at term. Methanol treatment produced 43%–58% cleft palate in the low-folate group, compared to 11%–15% in the marginal and adequate-folate groups. Exencephaly was also observed in methanol-treated litters, and the incidence was highest in the low dietary folate group. These results suggest that dietary folate intake may be an important consideration in assessing the risk of methanol-induced developmental toxicity.

F. Conclusions

Ethanol is clearly one of the most devastating causes of congenital morbidity in our society. Risk factors underlying the expression of FAS and ARBD are beginning to be understood, although their mutifactorial nature is difficult to unravel. Clearly, the many physical, psychological, and social conditions associated with poverty, including poor nutrition, poor antenatal care, smoking, and other drug use exacerbate the toxicity of ethanol and increase the expression of FAS. The cellular and biochemical mechanisms underlying FAS are also manifold and interrelated, yet much progress has been made in putting together plausible models of these mechanisms. Work with various nutrients associated with the expression of FAS as well as further elucidation of the possible protective effects of prostaglandin inhibitors may offer avenues for intervention and prevention of some of the prenatal effects of ethanol. However, none of these approaches will be as fruitful as continued efforts to curtail drinking during pregnancy and to increase access to antenatal care.

The magnitude of concern over the recent finding of developmental toxicity in animals treated with methanol will depend in part on whether use of this alcohol as an alternative automobile fuel is increased. Widespread use of methanol would obviously increase the incidence of human exposures by orders of magnitude, but the levels and durations of these potential exposures have not been well characterized. Methanol is currently used in industry, however, and accidental exposures should be avoided. Preliminary studies indicate that methanol and ethanol may have some common developmental effects and mechanisms in experimental animals, and further studies comparing the effects of these alcohols during development may help to determine the relative importance of putative mechanisms of action.

Acknowledgments. The authors would like to thank Dr. Kathy Sulik for providing photographs and critical reading of the manuscript, Dr. Gary Held for assistance with graphics reproduction, and Gertrude Lavinder and A.J. Rogers for help preparing the references.

References

Aase JM, Jones KL, Clarren SK (1995) Do we need the term "FAE"? Pediatrics 95: 428–430

Abbott BD, Logsdon TR, Wilke TS (1994) Effects of methanol on embryonic mouse palate in serum-free organ culture. Teratology 49: 122–134

Abbott BD, Ebron-McCoy M, Andrews JE (1995) Cell death in rat and mouse embryos exposed to methanol in whole embryo culture. Toxicology 97: 159–171

Abel EL (1982) Consumption of alcohol during pregnancy: a review of effects on growth and development of offspring. Hum Biol 54: 421–453

Abel EL (1990) Fetal alcohol syndrome. Medical Economics, Oradell

Abel EL (1995) An update on incidence of FAS: FAS is not an equal opportunity birth defect. Neurotoxicol Teratol 17: 445–462

Abel EL, Dintcheff BA (1984) Factors affecting the outcome of maternal alcohol exposure. Parity. Neurobehav Toxicol Teratol 6: 373–377

Abel EL, Hannigan JH (1995) Maternal risk factors in fetal alcohol syndrome: provocative and permissive influences. Neurotoxicol Teratol 17: 445–462

Allis JW, Simmons JE, Robinson BL, McDonald A, House DE (1992) Induction of rat hepatic cytochrome p450IIE1 by methanol: its role in the enhancement of carbon tetrachloride hepatotoxicity. Toxicologist 12: 85 (abstract)

Altschuler HL, Shippenberg TS (1981) A subhuman primate model for fetal alcohol syndrome research. Neurobehav Toxicol Teratol 3: 121–126

Altura BM, Altura BT, Carella A, Chaterjee M, Halevy S, Tejani N (1983) Alcohol produces spasms of human umbilical blood vessels: relationship to fetal alcohol syndrome (FAS). Eur J Pharmacol 86: 311–312

Andrews JE, Ebron-McCoy M, Logsdon TR, Mole LM, Kavlock RJ, Rogers JM (1993a) Developmental toxicity of methanol in whole embryo culture: a comparative study with mouse and rat embryos. Toxicology 81: 205–215

Andrews JE, Ebron-McCoy M, Kavlock RJ, Rogers JM (1993b) Lowering pH increases embryonic sensitivity to formate in whole embryo culture. Toxicol In Vitro 7: 757–762

Andrews JE, Ebron-McCoy M, Kavlock RJ, Rogers JM (1995) Developmental toxicity of formate and formic acid in whole embryo culture: a comparative study with mouse and rat embryos. Teratology 51: 243–251

Anton RF, Becker HC, Randall CL (1990) Ethanol increases PGE and thromboxane production in mouse pregnant uterine tissue. Life Sci 46: 1145–1153

Ashwell KW, Zhang LL (1994) Optic nerve hypoplasia in an acute exposure model of the fetal alcohol syndrome. Neurotoxicol Teratol 16: 161–167

Assadi FK, Ziai M (1986) Zinc status of infants with fetal alcohol syndrome. Pediatr Res 20: 551–554

Ayromlooi J, Tobias M, Berg PD, Desiderio D (1979) Effects of ethanol on the circulation and acid-base balance of pregnant sheep. Obstet Gynecol 54: 624–630

Bannigan J, Burke P (1982) Ethanol teratogenicity in mice: a light microscopic study. Teratology 26: 247–254

Barron S, Riley EP, Smotherman WP (1986) The effect of prenatal alcohol exposure on umbilical cord length in fetal rats. Alcohol Clin Exp Res 10: 493–495

Batt RD (1989) Absorption, distribution and elimination of alcohol. In: Crow KE, Batt RD (eds) Human metabolism of alcohol. 1. Pharmacokinetics, medicolegal aspects, and general interests. CRC Press, Boca Raton, pp 3–8

Beck F, Huxham IM, Gulamhusein AP (1984) Growth of rat embryos in the serum of alcohol drinkers. In: Porter R, O'Conner M (eds) Mechanisms of alcohol damage in utero. CIBA Found Symp 105: 218–233

Becker HC, Randall CL (1987) Two generations of maternal alcohol consumption in mice: effect on pregnancy outcome. Alcohol Clin Exp Res 11: 240–242

Bennett IL Jr, Cary FH, Mitchell GL Jr, Cooper MN (1953) Acute methanol poisoning: a review based on experiences in an outbreak of 323 cases. Medicine 32: 431–463

Blakley PM (1988) Experimental Teratology of ethanol.In: Kalter H (ed) Issues and reviews in teratology, vol 4. Plenum, New York, pp 237–282

Blakley PM, Scott WJ Jr (1984a) Determination of the proximate teratogen of the mouse fetal alcohol syndrome. II. Pharmacokinetics of the placental transfer of ethanol and acetaldehyde. Toxicol Appl Pharmacol 72: 364–371

Blakley PM, Scott WJ Jr (1984b) Determination of the proximate teratogen of the mouse fetal alcohol syndrome. I. Teratogenicity of ethanol and acetaldehyde. Toxicol Appl Pharmacol 72: 355–363

Bocking AD, Sinervo KR, Smith GN, Carmichael L, Challis JR, Olson DM, Brien JF (1993) Increased uteroplacental production of prostaglandin E2 during ethanol infusion. Am J Physiol 265: R640–645

Boddy K, Dawes GS, Fisher R, Pinter S, Robinson JS (1974) Foetal respiratory movements, electrocortical and cardiovascular responses to hypoxemia and hypercapnia in sheep. J Physiol (Lond) 243: 599–618

Boggan WO, Randall CM, DeBeukelaer M, Smith R (1979) Renal anomalies in mice prenatally exposed to ethanol. Res Commun Chem Pathol Pharmacol 23: 127–142

Bolon B, Dorman DC, Janszen D, Morgan KT, Welsch F (1993) Phase-specific developmental toxicity in mice following maternal methanol inhalation. Fundam Appl Toxicol 21: 508–516

Bolon B, Welsch F, Morgan KT (1994) Methanol-induced neural tube defects in mice: pathogenesis during neurulation. Teratology 49: 497–517

Bonjour JP (1979) Vitamins and alcoholism. In J Vit Nutr Res 49: 434–441

Bonthius DL, West JR (1989) Aspirin augments alcohol in restricting brain growth in the neonatal rat. Neurotoxicol Teratol 11: 135–143

Bosron WF, Ehrig T, Li T-K (1993) Genetic factors in alcohol metabolism and alcoholism. Semin Liver Dis 13: 126–135

Bracken WM, Klaassen CD (1987) Induction of hepatic metallothionein by alcohols: evidence for an indirect mechanism. Toxicol Appl Pharmacol 87: 257–263

Bradford BU, Forman DT, Thurman RG (1993a) 4-Methylpyrazole inhibits fatty acyl coenzyme synthetase and diminishes catalase-dependent alcohol metabolism: has the contribution of alcohol dehydrogenase to alcohol metabolism been previously overestimated? Mol Pharmacol 43: 115–119

Bradford BU, Seed CB, Handler JA, Forman DT, Thruman RG (1993b) Evidence that catalase is a major pathway of ethanol oxidation in vivo: dose-response studies in deer mice using methanol as a selecive substrate. Arch Biochem Biophys 303: 172–176

Brien JF, Loomis CW, Tranmer J, McGrath M (1983) Disposition of ethanol in human maternal venous blood and amniotic fluid. Am J Obstet Gynecol 146: 181–186

Brien JF, Clarke DW, Richardson B, Patrick J (1985) Disposition of ethanol in maternal blood, fetal blood and amniotic fluid of third-trimester pregnant ewes. Am J Obstet Gynecol 152: 583–590

Bronsky PT, Johnston MC, Sulik KK (1986) Morphogenesis of hypoxia-induced cleft lip in CL/Fr mice. J Craniofac Genet Dev Biol [Suppl] 2. 113–128

Brown NA, Goulding EH, Fabro S (1979) Ethanol embryotoxicity: direct effects on mammalian embryos in vitro. Science 206: 573–575

Brown Woodman PDC, Huq F, Hayes L, Herlihy C, Picker K, Webster WS (1995) In vitro assessment of the effect of methanol and the metabolite, formic acid, on embryonic development of the rat. Teratology 52: 233–243

Buller F, Wood CA (1904) Poisoning by wood alcohol: cases of death and blindness from Columbian spirits and other methylated preparations. JAMA 43: 1117, 1132, 1289–1296

Campbell MA, Fantel AC (1983) Teratogenicity of acetaldehyde in vitro: relevance to the fetal alcohol syndrome. Life Sci 32: 2641–2647

Canny NE, Roe DA (1983) Effects of maternal ethanol consumption on the ontogeny of gluconeogenesis in the perinatal period. Drug Nutrient Interact 2: 193–202

Cartwright MM, Smith SM (1995a) Increased cell death and reduced neural crest cell numbers in ethanol-exposed embryos: partial basis for the fetal alcohol syndrome phenotype. Alcohol Clin Exp Res 19: 378–386

Cartwright MM, Smith SM (1995b) Stage-dependent effects of ethanol on cranial neural crest cell development: partial basis for the phenotypic variations observed in fetal alcohol syndrome. Alcohol Clin Exp Res 19: 1454–1462

Castillo T, Koop DR, Kamimura S, Triadafilopoulos G, Tsukamoto H (1992) Role of cytochrome P-450 2E1 in ethanol-, carbon tetrachloride-, and iron-dependent microsomal lipid peroxidation. Hepatology 16: 992–996

Catlin MC, Abdollah S, Brien JF(1993) Dose-dependent effects of prenatal ethanol exposure in the guinea pig. Alcohol 10: 109–115

Centers for Disease Control and Prevention (1995) Update: trends in fetal alcohol syndrome – United States, 1979–1993. MMWR Morb Mortal Wkly Rep 44: 249–251

Charness ME, Safran RM, Perides G (1994) Ethanol inhibits neural cell-cell adhesion. J Biol Chem 269: 9304–9309

Checiu M (1993) The effect of ethanol upon early development in mice and rats. XVII. In vivo effect of chronic biparental alcoholization upon preimplantation development in rats. Rom J Morphol Embryol 39: 21–25
Checiu M, Sandor S (1981) The effect of ethanol upon early development in mice and rats. III. In vivo effect of acute ethanol intoxication upon implantation and early postimplantation stages in mice. Morphol Embryol (Bucur) 27: 117–122
Checiu M, Sandor S (1982) The effect of ethanol upon early development in mice and rats. IV. The effect of acute ethanol intoxication of day 4 of pregnancy upon implantation and early postimplantation development in mice. Morphol Embryol (Bucur) 28: 127–133
Cheicu M, Sandor S (1983) The effect of ethanol upon early development in mice and rats. V. In vivo effect of acute preimplantation intoxication with or without previous chronic alcoholization. Morphol Embryol (Bucur) 29: 149–157
Checiu M, Sandor S (1986) The effect of ethanol upon early development in mice and rats. IX. Late effect of acute preimplantation intoxication in mice. Morphol Embryol (Bucur) 32: 5–11
Checiu M, Sandor S (1987) The effect of ethanol upon early development in mice and rats. XI. Pathogenetic aspects studied by reciprocal transfer of preimplantation embryos. Morphol Embryol (Bucur) 33: 13–18
Checiu M, Sandor S, Garban Z (1984) The effect of ethanol upon early development in mice and rats. VI. In vivo effect of acetaldehyde upon preimplantation stages in rats. Morphol Embryol (Bucur) 30: 175–184
Chemtob S, Roy M-S, Abran D, Fernandez H, Varma DR (1993) Prevention of postasphyxial increase in lipid peroxides and retinal function deterioration in the newborn pig by inhibition of cyclooxygenase activity and free radical generation. Pediatr Res 33: 336–340
Chen S-Y, Yang B, Jacobson K, Sulik KK (1995) Ethanol-induced neural crest cell death: changes in membrane fluidity and the protective role of ganglioide GM1. Teratology 51: 157 (abstr)
Chernoff GF (1977) The fetal alcohol syndrome in mice: an animal model. Teratology 15: 223–230
Chernoff GF (1980) The fetal alcohol syndrome in mice: maternal variables. Teratology 22: 71–75
Chew WB, Berger EH, Brines OA, Capron MJ (1946) Alkali treatment of methyl alcohol poisoning. JAMA 130: 61–64
Chin JH, Goldstein, DB (1977a) Effects of low concentrations of alcohol on the fluidity of spin-labelled erythrocyte and brain membranes. Mol Pharmacol 13: 435–441
Chin JH, Goldstein DB (1977b) Drug tolerance in biomembranes. A spin label study of the effects of ethanol. Science 196: 684–685
Clarke DW, Steenaart NAE, Breedon TH, Brien JF (1985) Differential pharmacokinetics for oral and intraperitoneal administration of ethanol to the pregnant guinea pig. Can J Physiol Pharmacol 63: 169–172
Clarke DW, Steenaart NAE, Slack CJ, Brien JF (1986) Pharmacokinetics of ethanol and its metabolite, acetaldehyde, and fetolethality in the third-trimester pregnant guinea pig for oral administration of acute, mutiple-dose ethanol. Can J Physiol Pharmacol 64: 1060–1067
Clarren SK, Astley SJ (1992) Pregnancy outcomes after weekly oral administration of ethanol during gestation in the pig-tailed macaque: comparing early gestational exposure to full gestational exposure. Teratology 45: 1–9
Clarren SK, Bowden DM (1982) Fetal alcohol syndrome: a new primate model for binge drinking and its relevance to human ethanol teratogenesis. J Pediatr 101: 819–824
Clarren SK, Sampson PD, Larsen J, Donnell DJ, Barr HM, Bookstein FL, Martin DC, Streissguth AP (1987a) Facial effects of fetal alcohol exposure: assessment by photographs and morphometric analysis. Am J Med Genet 26: 651–666

Clarren SK, Bowden DM, Astley SJ (1987b) Pregnancy outcomes after weekly oral administration of ethanol during gestation in the pig-tailed macaque (Macaca nemestrina). Teratology 35: 345–354

Clarren SK, Astley SJ, Bowden DM (1988) Physical anomalies and developmental delays in nonhuman primate infants exposed to weekly doses of ethanol during gestation. Teratology 37: 561–569

Clarren SK, Astley SJ, Bowden DM, Lai H, Milam AH, Rudeen PK, Shoemaker WJ (1990) Neuroanatomic and neurochemical abnormalities in nonhuman primate infants exposed to weekly doses of ethanol during gestation. Alcohol Clin Exp Res 14: 674–683

Clarren SK, Astley SJ, Gunderson VM Spellman D (1992) Cognitive and behavioral deficits in nonhuman primates associated with very early embryonic binge exposures to ethanol. J Pediatr 121: 789–796

Clode AM, Pratten MK, Beck F (1987) The effect of ethanol on the growth of rat embryos: the role of stage dependency and hyperosmolality. Arch Toxicol [Suppl] 11: 163–167

Conibear SA (1988) The absorption, metabolism, excretion and health effects of industrially useful alcohols. Dangerous Properties Indust Mater 8: 2–9

Conlon RA (1995) Retinoic acid and pattern formation in vertebrates. Trends Genet 11: 314–319

Connelly LE, Rogers JM (1996) Methanol causes posteriorization of cervical vertebrae in mice. Teratology (submitted)

Contreras KM, Harris C (1995) Embryotoxicity of methanol, fomaldehye, and sodium formate in the rat conceptus in vitro: exposure by direct intraamniotic microinjection. Toxicologist 15: 163 (abstr)

Cook PS, Abrams RM, Notelovitz M, Frisinger JE (1981) Effect of ethyl alcohol on maternal and fetal acid-base balance and cardiovascular status in chronic sheep preparations. Br J Obstet Gynaecol 88: 188–194

Cook CS, Nowotny AZ, Sulik KK (1987) Fetal alcohol syndrome. Eye malformations in a mouse model. Arch Ophthalmol 105: 1576–1581

Crow KE, Greenaway RM (1989) Metabolic effects of alcohol on the liver. In: Crow KE, Batt RD (eds) Human metabolism of alcohol. II. Metabolic and physiological effects of alcohol. CRC Press, Boca Raton, pp 3–18

Crow KE, Hardman MJ (1989) Regulation of rates of ethanol metabolism. In: Crow KE, Batt RD (eds) Human metabolism of alcohol. II. Regulation, enzymology, and metabolites of ethanol. CRC Press, Boca Raton, pp 3–16

Cumming ME, Ong BY , Wade JG, Sitar DS (1984) Maternal and fetal ethanol pharmacokinetics and cardiovascular responses in near-term pregnant sheep. Can J Physiol Pharmacol 62: 1435–1439

Cummings AM (1993) Evaluation of the effects of methanol during early pregnancy in the rat. Toxicology 79: 205–214

Daft PA, Johnston MC, Sulik KK (1986) Abnormal heart and great vessel development following acute ethanol exposure in mice. Teratology 33: 93–104

Dai Y, Rashba-Step J, Cederbaum AI (1993) Stable expression of human cytochrome P4502E1 in HepG2 cells: characterization of catalytic activities and production of reactive oxygen intermediates. Biochemistry 32: 6928–6937

Daniell LC, Brass EP, Harris RA (1987) Effect of ethanol on intracellular ionized calcium concentrations in synaptosomes and hepatocytes. Mol Pharmacol 32: 831–837

Dargel R (1992) Lipid peroxidation – a common pathogenetic mechanism? Exp Toxicol Pathol 44: 169–181

Davis WL, Crawford LA, Cooper OJ, Farmer GR, Thomas D, Freeman BL (1990) Ethanol unduces the generation of reactive free radicals by neural crest cells in vitro. J Craniofac Genet Dev Biol 10: 277–293

DeJonge MH, Zachman RD (1995) The effect of maternal ethanol ingestion on fetal rat heart vitamin A: a model for fetal alcohol syndrome. Pediatr Res 37: 418–423

Dexter JD, Tumbleson ME, Decker JD, Middleton CC (1983) Comparison of the offspring of three serial pregnancies during voluntary alcohol consumption in Sinclair (S-1) miniature swine. Neurobehav Toxicol Teratol 5: 229–231
Dildy JE, Leslie SW (1989) Ethanol inhibits NMDA-induced increases in free intracellular Ca^{2+} in dissociated brain cells. Brain Res 499: 383–387
Dilts PV (1970) Placental transfer of ethanol. Am J Obstet Gynecol 107: 1018–1021
Dorman DC, Bolon B, Struve MF, LaPerle KMD., Wong BA, Elswick B, Welsch F (1995) Role of formate in methanol-induced exencephaly in CD-1 mice. Teratology 52: 30–40
Dreosti IE (1984) Interactions between trace elements and alcohol in rats. In: Mechanisms of alcohol damage in utero. Ciba Found Symp 105: 103–123
Dreosti IE (1991) Free radical damage and the genome. In: Dreosti IE (ed) Trace elements, micronutrients and free radicals. Humana, New Jersey, pp 149–169
Dreosti IE (1993) Nutritional factors underlying the expression of the fetal alcohol syndrome. In: Keen CL, Bendich A, Whillhite CC (eds) Maternal nutrition and pregnancy outcome. Ann NY Acad Sci 678: 193–214
Dreosti IE, Ballard FJ, Belling GB, Record IR, Manuel SJ, Hetzel BS (1981) The effect of ethanol and acetaldehyde on DNA synthesis in growing cells and on fetal development in the rat. Alcohol Clin Exp Res 5: 357–362
Driscoll CD, Streissguth AP, Riley EP (1990) Prenatal alcohol exposure: comparability of effects in humans and animal models. Neurotoxicol Teratol 12: 231-237
Duester G (1991) A hypothetical mechanism for fetal alchol syndrome involving ethanol inhibition of retinoic acid synthesis at the alcohol dehydrogenase step. Alcohol Clin Exp Res 15: 568–572
Duester G (1994a) Retinoids and the alcohol dehydrogenase gene family EXS 71: 279–290
Duester G (1994b) Are ethanol-induced birth defects caused by functional retinoic acid deficiency? In: Blomhoff R (ed) Vitamin A in health and disease. Dekker, New York, pp 343–363
Duester G, Shean ML, McBride MS, Stewart MJ (1991) retitnoic acid response element in the human alcohol dehydrogenase gene ADH3: implications for regulation of retinoic acid synthesis. Mol Cell Biol 11: 1638–1646
Edwards HG, Dow-Edwards DL (1991) Craniofacial alterations in adult rats prenatally exposed to ethanol. Teratology 44: 373–378
Eells JT, Makar AB, Noker PE, Tephly TR (1981) Methanol poisoning and formate oxidation in nitrous oxide -treated rats. J Pharmacol Exp Ther 217: 57–61
Eells JT, Black KA, Makar AB, Tedford CE, Tephly TR (1982) The regulation of one-carbon oxidation in the rat by nitrous oxide and methionine. Arch Biochem Biophys 219: 316–326
Egan RW, Gale PH, Baptista EM, Kennicott KL, Van den Heuvel WJA, Walker RW, Fagerness PE, Kuehl, FA (1981) Oxidation reactions by prostaglandin cyclooxygenase-hydroperoxidase. J Biol Chem 256: 7352
Eichner ER, Hillman RS (1973) Effect of alcohol on serum folate level. J Clin Invest 52: 584–591
Eisenga BH, McMartin KE (1987) The effect of acute ethanol administration on the clearance of ^{3}H-PteGlu by the rat kidney. Nutr Res 7: 1051–1060
El Banna N, Picciano MF, Simon J (1983) Effects of chronic alcohol consumption and iron deficiency on maternal folate status and reproductive outcome in mice. J Nutr 113: 2059–2070
Ernhart CB, Sokol RJ, Martier S, Moron P, Nadler D, Ager JW, Wolf A (1987) Alcohol teratogenicity in the human: a detailed assessment of specificity, critical period and threshold. Am J Obstet Gynecol 159: 33–39
Erskine RLA, Ritchie JWK (1986) The effect of maternal consumption of alcohol on human umbilical artery blood flow. Am J Obstet Gynecol 154: 318–321
Ewald SJ (1989) T lymphocyte populations in fetal alcohol syndrome. Alcohol Clin Exp Res 13: 485–489

Falchuk KH (1993) Zinc in developmental biology: the role of metal dependent transcription regulation. Progr Clin Biol Res 380: 91–111

Fantel AG, Barber CV, Carda MB, Tumbic RW, Mackler B (1992) Studies of the role of ischemia/reperfusion and superoxide anion radical production in the teratogenicity of cocaine. Teratology 46: 293–300

Fazakas-Todea I (1993a) The effect of ethanol upon early development in mice and rats. XVI. In vivo effect of acute preimplantation intoxication with beer and cognac on the background of chronic consumption. Rom J Morphol Embryol 39: 13–19

Fazakas-Todea I (1993b) The effect of ethanol upon early development in mice and rats. XVIII. In vivo effect of acute preimplantation intoxication with beer and cognac on the background of chronic biparental intake (in mice). Rom J Morphol Embryol 39: 27–32

Fazakas-Todea I, Checiu M, Sandor S (1985) The effect of ethanol upon early development in mice and rats VIII. The effect of chronic consumption of some beverages upon preimplantation development in rats. Morphol Embryol (Bucur) 31: 249–256

Felder MR, Burnett KG, Balak KJ (1983) Genetics, biochemistry, and developmental regulation of alcohol dehydrogenase in peromyscus and laboratory mice. Curr Top Biol Med Res 9: 143–161

Feré CH (1895) Etudes experimentales sur l'influence teratogene ou degénérative des alcools et des essences. J Anat Physiol 31: 161–186

Fernandez K, Caul WF, Boyd JE, Henderson GI, Michaelis RC (1983) Malformations and growth of rat fetuses exposed to brief periods of alcohol in utero. Teratogen Carcinogen Mutag 3: 457–460

Fisher SE, Atkinson M, Holzman I, David R, Van Thiel DH (1981) Effect of ethanol upon placental uptake of amino acids. Prog Biochem Pharmacol 18: 216–223

Fisher SE , Atkinson MS, Burnao JK, Jacobson S, Sehgal P, Scott W, Van Thiel DH (1982) Ethanol-associated selective fetal malnutrition: a contributing factor in the fetal alcohol syndrome. Alcohol Clin Exp Res 6: 197–201

Fisher SE, Inselman LS, Duffy L, Atkinson M, Spencer H, Chang B (1985) Ethanol and fetal nutrition: effect of chronic ethanol exposure on rat placental growth and membrane-associated folic acid receptor binding activity. J Ped Gastroenterol Nutr 4: 645–649

Fisher SE, Duffy L, Atkinson M (1986) Selective fetal malnutrition: effect of acute and chronic ethanol exposure upon rat placental Na, K-ATPase activity. Alcohol Clin Exp Res 10: 150–153

Flynn A, Martier SS, Sokol RI, Miller SL, Golden NL Villano BC (1981) Zinc status of pregnant alcoholic women: a determinant of fetal outcome. Lancet 1: 572–574

Froster UG, Baird PA (1992) Congenital defects of the limbs and alcohol exposure in pregnancy: data from a population based study. Am J Med Genet 44: 782–785

Gavin CE, Kates B, Gerken LA, Rodier PM (1994) Patterns of growth deficiency in rats exposed in utero to undernutrition, ethanol, or the neuroteratogen methylazoxymethanol (MAM). Teratology 49: 113–121

Garro AJ, McBeth DL, Lima V, Lieber CS (1991). Ethanol consumption inhibits fetal DNA methylation in mice: implications for the fetal alcohol syndrome. Alcohol Clin Exp Res 15: 395–398

Gatalica Z, Damjanov I (1990) Effects of alcohols on mouse embryonal carcinoma-substrate adhesion. Histochemistry 95: 189–193

Geelan JA (1979) Hypervitaminosis A induced teratogenesis. Crit Rev Toxicol 6: 351–375

Ghishan FK, Greene HL (1983) Fetal alcohol syndrome: failure of zinc supplementation to reverse the effect of ethanol on placental transport of zinc. Pediatr Res 17: 529–531

Ghishan FK, Patwardhan R, Greene HL (1982) Fetal alcohol syndrome: inhibition of placental zinc transport as a potential mechanism for fetal growth retardation in the rat. J Lab Clin Med 100: 45–52
Giavini E, Broccia ML, Prati M, Bellomo D, Menegola E (1992) Effects of ethanol and acetaldehyde on rat embryos developing in vitro. In Vitro Dev Biol 28A: 205–210
Goad PT, Hill DE, Slikker W, Kimmel CA, Gaylor DW (1984) The role of maternal diet in the developmental toxicology of ethanol. Toxicol Appl Pharmacol 73: 256–267
Goodman JI, Tephly TR (1970) Peroxidative oxidation of methanol in human liver: the role of hepatic microbody and soluble oxidases. Res Commun Chem Pathol Pharmacol 1: 441–450
Gordon AS, Collier K, Diamond I (1986) Ethanol regulation of adenosine receptor-stimulated cAMP levels in a clonal neural cell line: an in vitro model of cellular tolerance to ethanol. Proc Natl Acad Sci USA 83: 2105–2108
Gordon BHJ, Streeter ML, Rosso P, Winick M (1985) Prenatal alcohol exposure: abnormalities in placental growth and fetal amino acid uptake in the rat. Biol Neonate 47: 113–119
Grabowski CT (1970) Embryonic oxygen deficiency: a physiological approach to analysis of teratological mechanisms. Adv Teratol 4: 137–167
Greeley S, Johnson WT, Schafer D, Johnson PE (1990) Gestational alcoholism and fetal zinc accretion in Long-Evans rats. J Am Coll Nutr 9: 265–271
Grossmann A, Astley SJ, Liggitt HD, Clarren SK, Shiota F, Kennedy B, Thouless ME, Maggio-Price L (1993) Immune function in offspring of nonhuman primates (Macaca nemestrina) exposed weekly to 1.8 g/kg ethanol during pregnancy: preliminary observations. Alcohol Clin Exp Res 17: 822–827
Grummer MA, Langough RE, Zachman RD (1993) Maternal ethanol ingestion effects on fetal rat brain vitamin A as a model for fetal alcohol syndrome. Alcohol Clin Exp Res 17: 592–597
Gummer CL, Maibach HI (1986) The penetration of [14C]ethanol and [14C]methanol through excised guinea pig skin in vitro. Food Chem Toxicol 24: 305–309
Guy JF, Sucheston ME (1986) Teratogenic effects on CD-1 mouse embryos exposed to concurrent doses of ethanol and aspirin. Teratology 34: 249–261
Hale F (1933) Pigs born without eyeballs. J Hered 24: 105–106
Halsted CH (1992) Folate metabolism in alcoholism. In: Watson RR, Watzl B (eds) Nutrition and alcohol. CRC Press, Boca Raton, pp 255–267
Halsted CH, Griggs RC, Harris JW (1967) The effect of alcoholism on the absorption of folic acid (H3-PGA) evaluated by plasma levels and urine excretion. J Lab Clin Med 69: 116–131
Hannigan JH, Abel EL, Kruger ML (1993) Population characteristics of birthweight in an animal model of alcohol-related developmental effects. Neurotoxicol Teratol 15: 97–105
Harris C, Kenneway M, Lee E, Larsen SJ, Thorsrud BA (1995) Formaldehyde embryotoxicity in mouse conceptuses grown in whole embryo culture. Toxicologist 15: 162 (abstract)
Harris JE (1990) Hepatic glutathione, metallothionein and zinc in the rat on gestational day 19 during chronic ethanol administration. J Nutr 120: 1080–1086
Harrop GA, Benedict EM (1920) Acute methyl alcohol poisoning associated with acidosis. JAMA 74: 25–27
Hassan HM, Fridovich I (1980) Superoxide dismutase: detoxication of a free radical. In: Jacoby WB (ed) Enzymatic basis of detoxication. vol 1, pp 311–332
Henderson GI, Hoyumpa AM, McClain C, Schenker S (1979) Effects of chronic and acute alcohol administration on fetal development in the rat. Alcohol Clin Exp Res 3: 99–1006

Henderson GI, Patwardhan RV, McLeroy S, Schenker S (1982) Inhibition of placental amino acid uptake in rats following acute and chronic ethanol exposure. Alcohol Clin Exp Res 6: 495–505
Hilbelink DR, Persaud TVN (1981) Teratogenic effects of prostaglandin E2 in hamsters. Prog Lipid Res 20: 241–242
Hill DE, Slikker W Jr, Goad PT, Bailey JR, Sziszak TJ, Hendrickx AG (1983) Maternal, fetal, and neonatal elimination of ethanol in nonhuman primates. Dev Pharmacol Ther 6: 259–268
Hingson R, Alpert JJ, Day N, Dooling E, Kayne H, Morelock S, Oppenheimer E, Zuckerman B (1982) Effects of maternal drinking and marijuana use on fetal growth and development. Pediatrics 70: 539–546
Hirano T, Kaplowitz N, Tsukamoto H, Kamimura S, Fernandez-Checa JC (1992) Hepatic mitochondrial glutathione depletion and progession of experimental alcoholic liver disease in rats. Hepatology 6: 1423–1427
Hitzemann RJ, Schueler HE, Graham-Brittain C, Kreishman GP (1986) Ethanol-induced changes in neuronal membrane order. An NMR study. Biochim Biophys Acta 859: 189–197
Hoek JB, Thomas AP, Rubin R, Rubin E (1987) Ethanol-induced mobilization of calcium by activation of phosphoinositide-specific phospholipase C in intact hepatocytes. J Biol Chem 262: 682–691
Hoek JB, Thomas AP, Rooney TA, Higashi K, Rubin E (1992) Ethanol and signal transduction in the liver. FASEB J 6: 2386–2396
Hoffman PL, Rabe CS, Moses F, Tabakoff B (1989) NMDA receptors and ethanol: inhibition of calcium flux and cyclic-GMP production. J Neurochem 52: 1937–1940
Hooper SH, Harding R, Deayton J, Thorburn GD (1992) Role of prostaglandins in the metabolic responses of the fetus to hypoxia. Am J Obstet Gynecol 166: 1568–1575
Horrobin DF (1980) A biochemical basis for alcoholism and alcohol-induced damage including the fetal alcohol syndrome and cirrhosis: interference with essential fatty acid and prostaglandin metabolism. Med Hypoth 6: 929–942
Horton VL, Higuchi MA, Rickert DE (1992) Physiologically based pharmacokinetic model for methanol in rats, monkeys and humans. Toxicol Appl Pharmacol 117: 26–36
Hungund BL, Mahadik SP (1993) Role of gangliosides in behavioral and biochemical actions of alcohol: cell membrane structure and function. Alcohol Clin Exp Res 17: 329–339
Hungund BL, Reddy MV, Mahadik SP (1989) Gangliosides protect against acute alcohol action, mortality and plasma membrane changes. Trans Am Soc Neurochem 20: 108
Hungund BL, Reddy MV, Bharucha VA, Mahadik SP (1990) Monosialogangliosides (GM1 and AGF2) reduce acute alcohol intoxication: sleep time, mortality and cerebral cortical Na^+, K^+-ATPase. Drug Dev Res 19: 443–451
Hungund BL, Ross DC, Gokhale VS (1994) Ganglioside GM1 reduces fetal alcohol effects in rat pups exposed to ethanol in utero. Alcohol Clin Exp Res 18: 1248–1251
Hunter ES III, Tugman JA, Sulik KK, Sadler TW (1994) Effects of short-term exposure to ethanol on mouse embryos in vitro. Toxicol in Vitro 8: 413–421
Idanpaan-Heikkila J, Jouppila P, Akerblom HK, Isoaho R, Kauppila E, Koivisto M (1972) Elimination and metabolic effects of ethanol in mother, fetus, and newborn infant. Am J Obstet Gynecol 112: 387–392
Ihemelandu EC (1984) Effect of maternal alcohol consumption on pre- and postnatal muscle development of mice. Growth 48: 35–43
Infurna R, Weiss B (1986) Neonatal behavioral toxicity in rats following prenatal exposure to methanol. Teratology 33: 259–265
Inselman LS, Fisher SE, Spencer H, Atkinson M (1985) Effect of intrauterine ethanol exposure on fetal lung growth. Pediatrics 19: 12–14

Iorio KR, Reinlib L, Tabakoff B, Hoffman PL (1992) Chronic exposure of cerebellar granule cells to ethanol results in increased N-methyl-D-asparatate receptor function. Mol Pharmacol 41: 1142–1148
Itskovitz J, LaGamma EF, Rudolph AM (1983) The effect of reducing umbilical blood flow on fetal oxygenation. Am J Obstet Gynecol 145: 813–818
Jones KL, Smith DW (1973) Recognition of the fetal alcohol syndrome in early infancy. Lancet 2: 999–1001
Jones KL, Smith DW, Ulleland CN, Streissguth AP (1973) Pattern of malformation in offspring in chronic alcoholic women. Lancet 1: 1267–1271
Jones MD, Battaglia FC (1977) Intrauterine growth retardation. Am J Obstet Gynecol 5: 540–549
Jones PDH, Leichter J, Lee M (1981) Placental blood flow in rats fed alcohol before or during gestation. Life Sci 29: 1153–1159
Karl PI, Gordon BHJ, Lieber CS, Fisher SE (1988) Acetaldehyde production and transfer by the perfused human placental cotyledon. Science 242: 273–275
Kavet R, Nauss K (1990) The toxicity of inhaled methanol vapors. Crit Rev Toxicol 21: 21–50
Kawase T, Kato S, Lieber CS (1989) Lipid peroxidation and antioxiant defense systems in rat liver after chronic ethanol feeding. Hepatology 10: 815–821
Keen CL, Hurley LS (1989) Zinc and reproduction: effects of deficiency on foetal and postnatal development. In: Mills CF (ed) Zinc in human biology. Springer; Berlin Heidelberg New York, pp 183–220
Keir WJ (1991) Inhibition of retinoic acid synthesis and its implications in fetal alcohol syndrome. Alcohol Clin Exp Res 15: 560–564
Keirse MJ (1978) Biosynthesis and metabolism of prostaglandins in the pregnant human uterus. Adv Prostagland Thrombox Leukotriene Res 4: 87–102
Kelly RW (1994) Pregnancy maintenance and parturition: the role of prostaglandin in manipulating the immune and inflammatory response. Endocr Rev 15: 684–706
Kennedy LA (1984) The pathogenesis of brain abnormalities in the fetal alcohol syndrome: an integrating hypothesis. Teratology 29: 363–368
Keppen LD, Pysher T, Rennert OM (1985) Zinc deficiency acts as a co-teratogen with alcohol in fetal alcohol syndrome. Pediatr Res 19: 944–947
Keppen LD, Moore DJ, Cannon DJ (1990) Zinc nutrition in fetal alcohol syndrome. Neurotoxicology 11: 375–380
Kesäniemi Y, Sippel H (1975) Placental and foetal metabolism of acetaldehyde in rat. I. Contents of ethanol and acetaldehyde in placenta and foetus of the pregnant rat during ethanol oxidation. Acta Pharmacol Toxicol 37: 43–48
Koop DR (1989) Minor pathways of ethanol metabolism. In: Crow KE, Batt RD (eds) Human metabolism of alcohol. II. Regulation, enzymology, and metabolites of ethanol. CRC Press, Boca Raton, pp 133–145
Koop SE, Tierney DJ (1990) Multiple mechanisms in the regulation of ethanol-inducible cytochrome. Bioessays 12: 429–435
Kotch LE, Sulik KK (1992a) Patterns of ethanol-induced cell death in the developing nervous system of mice: neural fold stages through the time of anterior neural tube closure. Int J Dev Neurosci 10: 273–279
Kotch LE, Sulik KK (1992b) Experimental fetal alcohol syndrome: proposed pathogenic basis for variety of associated facial and brain anomalies. Am J Med Genet 44: 168–176
Kotch LE, Dehart DB, Alles AJ, Chernoff N, Sulik KK (1992) Pathogenesis of ethanol-induced limb reduction defects in mice. Teratology 46: 323–332
Kotch LE, Chen S-Y, Sulik KK (1995) Ethanol-induced teratogenesis: free radical damage as a possible mechanism. Teratology 52: 128–136
Krebs HA, Perkins JR (1970) The physiological role of liver alcohol dehydrogenase, Biochem J 118: 635–644
Kronick JB (1976) Teratogenic effects of ethyl alcohol administered to pregnant mice. Am J Obstet Gynecol 124: 676–80

Lammer EJ, Chen DT, Hoar RM, Agnish ND, Benke PJ, Braun JT, Curry CJ, Fernhoff PM, Grix AW, Lott IT, Richard JM, Sun SC (1985) Retinoic acid embryopathy. New Engl J Med 313: 837–841

Ledig ML, M'Paria JR, Louis J-C, Fried R, Mandel P (1980) Effect of alcohol on superoxide dismutase activity in cultured neural cells. Neurochem Res 5: 1155–1162

Lee M (1985) Potentiation of chemically induced cleft palate by ethanol ingestion during gestation in the mouse. Teratog Carcinog Mutag 5: 433–440

Lee M, Wakabayashi K (1986) Pituitary and thryoid hormones in pregnant alcohol-fed rats and their fetuses. Alcohol Clin Exp Res 10: 428–431

Leibgott B, Wiley MJ (1985) Prenatal hamster development following maternal administration of PGE2 at midterm. Prostaglandins Leukot Med 17: 309–328

Leichter J, Lee M (1984) Does dehydration contribute to retarded fetal growth in rats exposed to alcohol during gestation? Life Sci 35: 2105–2111

Leist KH, Grauwiler J (1974) Fetal pathology in rats following uterine vessel clamping on day 14 of gestation. Teratology 10: 55–68

Lemoine P, Harousseau H, Borteyru JP (1968) Les enfants de parents alcooliques: anomalies observées a propos de 127 cas. Ouest Med 21: 476–482

Leo MA, Lieber CS (1982) Hepatic vitamin A depletion in alcoholic liver injury. New Engl J Med 307: 597–601

Leonard BE (1987) Ethanol as a neurotoxin. Biochem Pharmacol 36: 2055–2059

Li T-K (1977) Enzymology of human alcohol metabolism. Adv Enzymol 45: 427–483

Lieber CS (1994) Alcohol and the liver: 1994 update. Gastroenterology 106: 1085–1105

Lieber CS, DeCarli LM (1963) Fatty liver, hyperlipemia and hyperuricemia produced by prolonged alcohol consumption despite adequate dietary intake. Trans Assoc Am Physicians 76: 289–300

Lieber CS, DeCarli LM (1982) The feeding of alcohol in liquid diets: two decades of applications and 1982 update. Alcohol Clin Exp Res 6: 523–531

Life Sciences Research Office (LSRO) (1984) Assessment of the folate nutritional status of the U.S. population based on data colected in the Second National Health and Nutrition Examination Survey, 1976–1980. Federation of American Societies for Experimental Biology, Bethesda

Lin GW-J (1981) Fetal malnutrition: a possible cause of the fetal alcohol syndrome. Prog Biochem Pharmacol 18: 115–121

Litovitz TL (1988) Acute exposure to methanol and fuels: a prediction of ingestion incidence and toxicity. Proceedings of the Methanol Health Safety Workshop, South Coast Air Quality Managment District, October 1988

Little RE (1977) Moderate alcohol use during pregnancy and decreased infant birth weight. Am J Publ Health 67: 1154–1156

Littleton JM (1989) Effects of ethanol on membranes and their associated functions. In: Crow KE, Batt RD (eds) Human metabolism of alcohol. III. Metabolic and physiological effects of alcohol. CRC Press, Boca Raton, pp 161–173

Littleton JM, Brennan C, Bouchenafa O (1991) The role of calcium flux in the central nervous system actions of ethanol. Ann NY Acad Sci 625: 388–394

Lohnes D, Mark M, Mendelsohn C, Dollé P, Dierich A, Gorry P, Gansmuller A, Chambon P (1994) Function of the retinoic acid receptors (RARs) during development. 1. Craniofacial and skeletal abnormalities in RAR double mutants. Development 120: 2723–2748

Makar AB, Tephly TR (1976) Methanol poisoning in the folate-deficient rat. Nature 261: 715–716

Makar AB, Tephly TR, Mannering GJ (1968) Methanol metabolism in the monkey. Mol Pharmacol 4: 471–483

Manglesdorf DJ, Umesono K, Evans RM (1994) In: Sporn MB, Robers AB, Goodman DS (eds) The retinoids: biology, chemistry and medicine, : 2nd edn. Raven, New York, pp 319–349

Mankes RF, LeFevre R, Benitz KF, Rosenblum I, Bates H, Walker AIT, Abraham R (1982) Paternal effects of ethanol in the Long-Evans rat. J Toxicol Environ Health 10: 871–878
Marquis S, Leichter J, Lee M (1984) Plasma amino acids and glucose levels in the rat fetus and dam after chronic maternal ethanol consumption. Biol Neonate 46: 36–43
McLain DE, Roe DA (1984) Fetal alcohol syndrome in the ferret (Mustela putorius). Teratology 30: 203–210
McMartin KE (1984) Increased urinary folate excretion and decreased plasma folate levels in the rat after acute ethanol treatment. Alcohol Clin Exp Res 8: 172–178
McMartin KE, Martin-Amat G, Makar AB, Tephly TR (1977) Methanol poisoning. V. Role of formate metabolism in the monkey. J Pharmacol Exp Ther 201: 564–572
McMartin KE, Ambre JJ, Tephly TR (1980) Methanol poisoning in human subjects. Role of formic acid accumulation in the metabolic acidosis. Am J Med 68: 414–418
Mendelsohn C, Lohnes D, Décimo D, Lufkin T, LeMeur M, Chambon P (1994) Function of the retinoic acid receptors during development. II. Multiple abnormalities at various stages of organogenesis in RAR double mutants. Development 120: 2749–2771
Mendelson RA, Huber AM (1980) The effect of ethanol on trace elements in the fetal rat. In: Galanter M (ed) Currents in alcoholism: recent advances in research and treatment. Greene and Stratton, New York, pp 38–48
Messing RO, Carpenter CL, Diamond I, Greenberg DA (1986) Ethanol regulates calcium channels in clonal neural cells. Proc Natl Acad Sci USA 83: 6213–6215
Messing RO, Peterson PJ, Henrich CJ (1991) Chronic ethanol exposure increases levels of protein kinase C, D and E and protein kinase C-mediated phosphorylation in cultured neural cells. J Biol Chem 266: 23428–23432
Mezey E, Holt PR (1971) The inhibitory effect of ethanol on retinol oxidation by human liver and cattle retina. Exp Mol Pathol 15: 148–156
Millicovsky G, DeSesso JM (1980) Differential embryonic cardiovascular responses to acute maternal uterine ischemia: an in vivo microscopic study of rabbit embryos with either intact or clamped umbilical cords. Teratology 22: 335–343
Moreno A, Parés X (1991) Purification and characterization of a new alcohol dehydrogenase from human stomach. J Biochem 266: 1128–1133
MRC Vitamin Study Research Group (1991) Prevention of neural tube defects: results of the Medical Research Council Vitamin Study. Lancet 338: 131–137
Mukherjee AB, Hodgen GD (1982) Maternal ethanol exposure induces transient impairment of umbilical circulation and fetal hypoxia in monkeys. Science 218: 700–701
Müller A, Sies H (1982) Role of alcohol dehydrogenase activity and of acetaldehyde in ethanol-induced ethane and pentane production by isolated perfused rat liver Biochem J 206: 153–156
Nakatsuji N, Johnson KE (1984) Effects of ethanol on the primitive streak stage mouse embryo. Teratology 29: 369–375
Nau H, Chahoud I, Dencker L, Lammer EJ, Scott WJ (1994) Teratogenicity of vitamin A and retinoids. In: Blomhoff R (ed) Vitamin A in health and disease. Dekker, New York, pp 615–663
Nelson BK, Brightwell WS, MacKenzie OR, Khan A, Burg JR, Weigel WW, Goad PT (1985) Teratological assessment of methanol and ethanol at high inhalation levels in rats. Fundam Appl Toxicol 5: 727–736
Ng PK Cottle MK, Baker JM, Johnson B, Van Muyden P,Van Petten GR (1982) Ethanol kinetics during pregnancy. Study in ewes and their fetuses. Prog Neuropsychopharmacol Biol Psychiatry 6: 37–42
Noden DM (1991) Vertebrate craniofacial development: the relation between ontogenetic process and morphological outcome. Brain Behav Evol 38: 190–225
Noker PE, Eells JT, Tephly TR (1990) Methanol toxicity: treatment with folic acid and 5-formyl tetrahydrofolic acid. Alcohol Clin Exp Res 4: 378–383

Nyquist-Battle C, Uphoff C, Cole TB (1987) Maternal ethanol consumption: effect on skeletal muscle development in guinea pig offspring. Alcohol 4: 11–16

Opperman JA (1984) Aspartame metabolism in animals, In: Stegink LD, Filer LJ Jr (eds) Aspartame. Physiology and biochemistry. Dekker, New York, pp 141–159

O'Shea KS, Kaufman MH (1979) The teratogenic effect of acetaldehyde: implications for the study of the fetal alcohol syndrome. J Anat 128: 65–76

O'Shea KS, Kaufman MH (1981) Effect of acetaldehyde on the neuroepithelium of early mouse embryos. J Anat 132: 107–118

Padmanabhan R, Muawad WM (1985) Exencephaly and axial skeletal dysmorphogenesis induced by acute doses of ethanol in mouse fetuses. Drug Alcohol Depend 16: 215–227

Padmanabhan R, Pallot DJ (1995) Aspirin-alcohol interaction in the production of cleft palate and limb malformations in the TO mouse. Teratology 51: 404–417

Padmanabhan R, Hameed MS, Sugathan TN (1984) Effects of acute doses of ethanol on pre- and postnatal development in the mouse. Drug Alcohol Depend 14: 197–208

Padmanabhan R, Wasfi IA, Craigmyle MB (1994) Effect of pretreatment with aspirin on th neural tube defects induced by maternal exposure to alcohol in the TO mouse fetuses. Drug Alcohol Depend 36: 175–186

Palese M, Tephly TR (1975) Metabolism of formate in the rat. J Toxicol Environ Health 1: 13–24

Parés X, Moreno A, Cederlund E, Hoog J-O Jornvall J (1990) Class IV mammalian alocohol dehydrogenase: structural data of the rat stomach enzyme reveal a new class well separated from those already characterized. FEBS Lett 277: 115–118

Patrick J, Richardson B, Hanson G, Clarke DW, Wlodek M, Bousquet J, Brien J (1985) Effects of maternal ethanol infusion on fetal cardiovascular and brain activity in lambs. Am J Obstet Gynecol 151: 859–867

Patrick J, Carmichael L, Richardson B, Smith G, Homan J, Brien J (1988) Effects of multiple-dose maternal ethanol infusion on fetal cardiovascualr and brain activity in lambs. Am J Obstet Gynecol 159: 1424–1429

Pearl R (1916) the effect of parental alcoholism (and certain other drug intoxications) upon the progeny in the domestic fowl. Proc Natl Acad Sci USA 2: 675–683

Pennington SN, Boyd JW, Kalmus GW, Wilson RW (1983) The molecular mechanism of fetal alcohol syndrome (FAS). I. Ethanol-induced growth suppression. Neurobehav Toxicol Teratol 5: 259–262

Pennington S, Allen Z, Runion J, Farmer P, Rowland L, Kalmus G (1985) Prostaglandin synthesis inhibitors block alcohol-induced fetal hypoplasia. Alcohol Clin Exp Res 9: 433–437

Perkins RA, Ward KW, Pollock GM (1995a) Comparative toxicokinetics of inhaled methanol in the female CD-1 mouse and Sprague-Dawley rat. Fundam Appl Toxicol 28: 245–254

Perkins RA, Ward KW, Pollock GM (1995b) A pharmacokinetic model of inhaled methanol in humans and comparison to methanol disposition in mice and rats. Environ Health Perspect 103: 726–733

Persaud TVN (1974) Fetal development in rabbits following intraamniotic administration of prostaglandin E2. Exp Pathol 9: S336–S341

Persaud TVN (1975) The effects of prostaglandin E2 on pregnancy and embryonic development in mice. Toxicology 5: 97-101

Persaud TVN (1983) Prenatal loss in the rat following moderate consumption of alcohol incorporated in a liquid diet. Anat Anz 153: 169–174

Posch KC, Napoli JL (1992) Multiple retinoid dehydrogenases in testes cytosol from alcohol dehydrogenase negative or positive deermice. Biochem Pharmacol 43: 2296–2298

Posch KC, Enright WJ, Napoli JL (1989) Retinoic acid synthesis by cytosol from the alcohol dehydrogenase negative deermouse. Arch Biochem Biophys 274: 171–178

Priscott PK (1982) The effects of ethanol on rat embryos developing in vitro. Biochem Pharmacol 31: 3641–3643
Priscott PK (1985) The effects of acetaldehyde and 2, 3-butanediol on rat embryos developing in vitro. Biochem Pharmacol 34: 529–532
Pullarkat RK (1991) Hypothesis: prenatal ethanol-induced birth defects and retinoic acid. Alcohol Clin Exp Res 15: 565–567
Rabe CS, Weight FF (1988) Effects of alcohol on neurotransmitter release and intracellular free calcium in PC12 cells. J Pharmacol Exp Ther 244: 417–422
Rabin RA (1990) Chronic ethanol exposure of PC12 cells alters adenlyate cyclase activity and intracellular cyclic AMP content. J Pharmacol Exp Ther 252: 1021–1027
Rabin RA (1993) Ethanol-induced desensitization of adenylate cyclase: role of the adenosine receptor and GTP-binding proteins. J Pharmacol Exp Ther 264: 977–983
Randall CL, Anton RF (1984) Aspirin reduces alcohol-induced prenatal mortality and malformations in mice. Alcohol Clin Exp Res 8: 513–515
Randall CL, Taylor WJ (1979) Prenatal ethanol exposure in mice: teratogenic effects. Teratology 19: 305–311
Randall CL, Taylor WJ, Walker DW (1977) Ethanol-induced malformations in mice. Alcohol Clin Exp Res 1: 219–224
Randall CL, Anton RF, Becker HC (1987) Effect of indomethacin on alcohol-induced morphological anomalies in mice Life Sci 41: 361–369
Randall CL, Anton RF, Becker HC, Hale RL, Ekblad U (1991a) Aspirin dose-dependently reduces alcohol-induced defects and prostaglandin E levels in mice. Teratology 44: 521–529
Randall CL, Becker HC, Anton RF (1991b) Effect of ibuprofen on alcohol-induced teratogenesis in mice. Alcohol Clin Exp Res 15: 673–677
Rasmussen BB, Christensen N (1980) Teratogenic effect of maternal alcohol consumption on the mouse fetus. A histopathological study. Acta Path Scand Sect A 88: 285–289
Rawat AK (1979) Derangement in cardiac protein metabolism in fetal alcohol syndrome. Res Commun Chem Pathol Pharmacol 25: 365–375
Reitbock N (1965) Formaldehydroxydation bei der Ratte. Naunyn Schmeidebergs Arch Exp Pathol Pharmakol 251: 189–201
Renau-Piqueras J, Miragall F, Guerrit C, Sancho-Tello M, Baguena-Cervellerat R (1987) Prenatal exposure to ethanol alters lateral plasma membranes and gap junctions of newborn rat hepatocytes as revealed by freeze-fracture. J Submicrosc Cytol 19: 397–404
Riley EP (1990) The long-term behavioral effects of prenatal alcohol exposure in rats. Alcohol Clin Exp Res 14: 670–673
Riley EP, Barron S (1989) The behavioral and neuroanatomical effects of prenatal alcohol exposure. In: Hutchings DE (ed) Prenatal abuse of licit and illicit drugs. Ann NY Acad Sci 562: 173–177
Rogers JM (1995) Methanol causes the holoprosencephaly spectrum of malformations at lower dosages than ethanol in C57BL/6J mice. Teratology 51: 195 (abstract)
Rogers JM, Mole ML (1996) Critical periods of sensitivity to the developmental toxicity of inhaled methanol in the CD-1 mouse. Fundam Appl Toxicol (submitted)
Rogers JM, Mole ML, Chernoff N, Barbee BD, Turner CI, Logsdon TR, Kavlock RJ (1993a) The developmental toxicity of inhaled methanol in the CD-1 mouse, with quantitative dose-response modeling for estimation of benchmark doses. Teratology 47: 175–188
Rogers JM Barbee BD, Rehnberg BF (1993b) Critical periods of sensitivity for the developmental toxicity of inhaled methanol. Teratology 47: 395 (abstract)
Rogers JM, Taubeneck MW, Daston GP, Sulik KK, Zucker RM, Elstein KH, Jankowski MA, Keen CL (1995) Zinc deficiency causes apoptosis but not cell cycle

alterations in organogenesis-stage rat embryos: effect of varying duration of deficiency. Teratology 52: 149–159
Roivainen R, McMahon T, Messing RO (1993) Protein kinase C isozymes that mediate enhancement of neurite outgrowth by ethanol and phorbol esters in PC12 cells. Brain Res 624: 85–93
Rosa FW (1983) Teratogenicity of isotretinoin. Lancet 2: 513
Rosett HL (1980) A clinical perspective of the fetal alcohol syndrome. Alcohol Clin Exp Res 4: 119–122
Rosett HL, Weiner L, Edelin KC (1983) Treatment experience with pregnancy problem drinkers. JAMA 249(15): 2029–2033
Rothman KJ, Moore LL,Singer MR, Nguyen U-SDT, Mannino S, Milunsky A (1995) Teratogenicity of high vitamin A intake. New Engl J Med 333: 1369–1373
Ruth RE, Goldsmith SK (1981) Interaction between zinc deprivation and acute ethanol intoxication during pregnancy in rats. J Nutr 111: 2034–2038
Ryle PR, Hodges A, Thomson AD (1986) Inhibition of ethanol induced vitamin A depletion by administration of N,N'-diphenyl-p-phenylene-diamine (DPPD). Life Sci 38: 695–702
Said HM, Strum WB (1986) Effect of ethanol and other aliphatic alcohols on the intestinal transport of folates. Digestion 35: 129–135
Sakanashi TM, Rogers JM, Fu SS, Connelly LE, Keen CL (1996) Influence of maternal folate status on the developmental toxicity of methanol in the CD-1 mouse. Teratology (In Press)
Sanchis R, Guerri C (1986) Alcohol-metabolizing enzymes in placenta and fetal liver: effect of chronic ethanol intake. Alcohol Clin Exp Res 10: 39–44
Sandor S (1979) The prenatal noxious effect of ethanol. Morphol Embryol (Bucur) 25: 211–223
Sandor S (1988) Experimental alcohol blastopathy. Acta Biol Hung 39: 419–439
Sandor S, Checiu M, Fazakas-Todea I, Garban Z (1980) The effect of ethanol upon early development in mice and rats. I. In vivo effect upon preimplantation and early postimplantation stages. Morphol Embryol (Bucur) 26: 265–274
Sardesai VM (1992) Biochemical and nutritional aspects of eicosanoids. J Nutr Biochem 3: 562 579
Sato M, Lieber CS (1981) Hepatic vitamin A depletion after chronic ethanol consumption in baboons and rats. J Nutr 111: 2015–2023
Savage DD, Queen SA, Sanchez CF, Paxton LL. Mahoney JC, Goodlett CR, West JR (1991) Prenatal ethanol exposure during the last third of gestation in rats reduces hippocampal NMDA agonist binding site density in 45-day-old offspring. Alcohol 9: 37–41
Savoy-Moore RT, Dombrowski MP, Cheng A, Abel EL, Sokol RJ (1989) Low dose alcohol contracts the human umbilical artery in vitro. Alcohol Clin Exp Res 13: 40–42
Schenker S, Becker HC, Randall CL, Phillips DK, Baskin GS, Henderson GI (1990) Fetal alcohol syndrome: current status of pathogenesis. Alcohol Clin Exp Res 14: 635–647
Schwetz BA, Smith FA, Staples RE (1978) Teratogenic potential of ethanol in mice, rats and rabbits. Teratology 18: 385–392
Scott VJ, Fradkin R(1984) The effects of prenatal ethanol in cynomolgus monkeys. Teratology 29: 49–56
Sejersted OM, Jacobsen D, Ovrebo S, Jansen H (1983) Formate concentrations in plasma from patients poisoned with methanol. Acta Med Scand 213: 105–110
Seyoum G, Persaud TV (1994) In vitro effect of S-adenosyl methionine on ethanol embryopathy in the rat. Exp Toxicol Pathol 46: 177–181
Shean ML, Duester G (1993) The role of alcohol dehydrogenase in retinoic acid homeostasis and fetal alcohol syndrome. Alcohol Alcohol Suppl 2: 51–56
Sheller B, Clarren SK, Astley SJ, Sampson PD(1988) Morphometric analysis of Macaca nemestrina exposed to ethanol during gestation. Teratology 38: 411–417

Siebert JR, Kokoch VG, Beckwith BJ,Cohen MM Jr, Lemire RJ (1981) The facial features of holoprosencephaly in anencephalic human specimens. II. Craniofacial anatomy. Teratology 23: 305–315
Siebert JR, Astley SJ, Clarren SK (1991) Holoprosencephaly in a fetal macaque (Macaca nemestrina) following weekly exposure to ethanol. Teratology 44: 29–36
Sinervo KR, Smith GN, Bocking AD, Patrick J, Brien JF (1992) Effect of ethanol on the release of prostaglandins from ovine fetal brain stem during gestation. Alcohol Clin Exp Res 16: 443–448
Singh SP, Snyder AK, Pullen GL (1986) Fetal alcohol syndrome: glucose and liver metabolism in term rats fetus and neonate. Alcohol Clin Exp Res 10: 54–58
Sippel HW, Kesäniemi YA (1975) Placental and foetal metabolism of acetaldehyde in rat. II. Studies on metabolism of acetaldehyde in the isolated placenta and foetus. Acta Pharmacol Toxicol 37: 49–55
Slater TF (1988) Free radical mechanisms in tissue injury with special reference to the cytotoxic effects of ethanol and related alcohols. In: Nordman R, Ribiere C, Rouach H (eds) Alcohol toxicity and free radical mechanisms. Advances in the biosciences, vol 71. Pergammon, Oxford, pp 1–9
Smith GN, Brien JF, Homan J, Carmichael J, Treissman D, Patick J (1990) Effect of ethanol on ovine fetal and maternal plasma prostaglandin E2 concentrations and fetal breathing movements. J Dev Physiol 14: 23–28
Smith GN, Patrick J, Sinervo KR (1991) Effect of ethanol exposure on the embryo-fetus: experimental considerations, mechanisms, and the role of prostaglandins. Can J Physiol Pharmacol 69: 550–569
Snyder AK, Singh SP (1983) Hepatic insulin binding following in utero exposure to ethanol. Horm Metabol Res 16: 175–178
Snyder AK, Singh SP, Gordon LP (1986) Ethanol-induced intrauterine growth retardation: correlation with placental glucose transfer. Alcohol Clin Exp Res 10: 167–170
Sokol RJ, Miller SI, Reed G (1980) Alcohol abuse during pregnancy: an epidemiologic study. Alcohol Clin Exp Res 4: 135–145
Sokol RJ, Ager J, Martier S (1986) Significant determinants of susceptibility to alcohol teratogenicity. Ann NY Acad Sci 477: 87–102
Speisky H, MacDonald A, Giles G, Orrego H, Israel Y (1985) Increased loss and decreased synthesis of hepatic glutathione after acute ethanol administration. Biochem J 206: 153–156
Sreenathan RN, Padmanabhan R, Singh S (1982) Teratogenic effects of acetaldehyde in the rat. Drug Alcohol Depend 9: 339–350
Sreenathan RN, Padmanabhan R, Singh S (1984) Structural changes of placenta following maternal administration of acetaldehyde in the rat. Z Mikrosk Anat Forsch 98: 597–604
Stanton ME, Crofton KM, Gray LE, Gordon CJ, Boyes WK, Mole ML, Peele DB, Bushnell PJ (1995) Assessment of offspring development and behavior following gestational exposure to inhaled methanol in the rat. Fundam Appl Toxicol 28: 100–110
Steventon GB, Williams KE (1987) Ethanol-induced inhibition of pinocytosis and proteolysis in rat yolk sac in vitro. Development 99: 247–253
Stockard CR (1910) The influence of alcohol and other anaesthetics on embryonic development. Am J Anat 10: 369–392
Streissguth AP, Clarren SK, Jones KL (1985) Natural history of the fetal alcohol syndrome: a ten-year follow-up of eleven patients. Lancet 2: 85–91
Streissguth AP, Aase JM, Clarren SK, Randels SP, LaDue RA, Smith DF (1991a) Fetal alcohol syndrome in adolescents and adults. JAMA 265: 1961–1967
Streissguth AP, Randels SP, Smith DF (1991b) A test-retest study of intelligence in patients with fetal alcohol syndrome: implications for care. J Am Acad Child Adolesc Psychiatry 30: 584–587

Streissguth AP, Barr HM, Sampson PD, Bookstein FL (1994a) Prenatal alcohol and offspring development: the first fourteen years. Drug Alcohol Depend 36: 89–99
Streissguth AP, Sampson PD, Olson HC, Bookstein FL, Barr HM, Scott M, Feldman J, Mirsky AF (1994b) Maternal drinking during pregnancy: attention and short-term memory in 14-year-old offspring – a longitudinal prospective study. Alcohol Clin Exp Res 18: 202–218
Streissguth AP, Barr HM, Olson HC, Sampson PD, Bookstein FL, Burgess DM (1994c) Drinking during pregnancy decreases word attack and arithmetic scores on standardized tests: adolescent data from a population-based prospective study. Alcohol Clin Exp Res 18: 248–254
Strömland K, Pinazo-Durán MD (1994) Optic nerve hypoplasia: comparative effects in children and rats exposed to alcohol during pregnancy. Teratology 50: 100–111
Suh SM, Firek AF (1982) Magnesium and zinc deficiency and growth retardation in offspring of alcoholic rats. J Am Coll Nutr 1: 193–198
Sulik KK (1984) Critical periods for alcohol teratogenesis in mice, with special reference to the gastrulation stage of embryogenesis. CIBA Found Symp 105: 124–141
Sulik KK, Johnston MC (1982) Embryonic origin of holoprosencephaly: interrelationship of the developing brain and face. Scan Electron Microsc 1: 309–322
Sulik KK, Johnston MC (1983) Sequence of developmental alterations following acute ethanol exposure in mice: craniofacal features of the fetal alcohol syndrome. Am J Anat 166: 257–269
Sulik KK,Johnston MC, Webb MA (1981) Fetal alcohol syndrome: embryogenesis in a mouse model. Science 214: 936–938
Sulik KK, Lauder JM, Dehart DB (1984) Brain malformations in prenatal mice following acute maternal ethanol administration. Int J Dev Neurosci 2: 203–214
Sulik KK, Johnston MC, Daft PA, Russell WE, Dehart DB (1986) Fetal alcohol syndrome and DiGeorge anomaly: critical ethanol exposure periods for craniofacial malformations illustrated in an animal model. Am J Med Genet Suppl 2: 97–112
Sulik KK, Cook CS, Webster WS (1988) Teratogens and craniofacial malformations: relationships to cell death. Development 103 Suppl: 213–231
Sulik KK, Dehart DB, Rogers JM, Chernoff N (1995) Teratogenicity of low doses of all-trans retinoic acid in presomite embryos. Teratology 51: 398–403
Tamura T, Halsted CH (1983) Folate turnover in chronically alcoholic monkeys. J Lab Clin Med 101: 623–628
Tanner AR, Bantock I, Hinks L, Lloyd B, Turner NR, Wright R (1986) Depressed selenium and vitamin E levels in an alcoholic population: possible relationship to hepatic injury through increased lipid peroxidation. Dig Dis Sci 31: 1307–1312
Taubeneck MW, Daston GP, Rogers JM, Keen CL (1994) Altered maternal zinc metabolism following exposure to diverse developmental toxicants. Reprod Toxicol 8: 25–40
Tephly TR, McMartin KE (1984) Methanol metabolism and toxicity. In: Stegink LD, Filer LJ Jr (eds) Aspartame. Physiology and biochemistry. Dekker, New York, pp 111–140
Tephly TR, Parks RE, Mannering GJ (1964) Methanol metabolism in the rat. J Pharmacol Exp Ther 143: 292–300
Testar X, Lopez D, Llobera M, Herrera E (1986) Ethanol administration in the drinking fluid to pregnant rats as a model for the fetal alcohol syndrome. Pharmacol Biochem Behav 24: 625–630
Thompson PA, Folb Pl (1982) An in vitro model of alcohol and acetaldehyde teratogenicity. J Appl Toxicol 2: 190–195
Thurman RG, Glassman EB, Handler JA, Forman D (1989) The swift increase in alcohol metabolism (SIAM): a commentary on the regulation of alcohol meta-

bolism in mammals. In: Crow KE, Batt RD (eds) Human metabolism of alcohol. II. Regulation, enzymology, and metabolites of ethanol. CRC Press, Boca Raton, pp 17–30

Tolarova M, Harris J (1995) Reduced recurrence of orofacial clefts after periconceptional supplementation with high-dose folic acid and multivitamins. Teratology 51: 71–78

United States Environmental Protection Agency (1994) 1992 Toxics release inventory – public data. EPA-745-R-94-001, Washington

Uphoff C, Nyquist-Battle C, Toth R(1984) Cardiac muscle development in mice exposed to ethanol in utero. Teratology 30: 119–129

Urfer FN, Fouron JC, Bard H, DeMuylder X (1984) Low levels of blood alchol and fetal myocardial function. Obstet Gynecol 63: 401–404

Vallee BL, Bazzone TJ (1983) Isozymes of human liver ADH. In: Rattazzi MC, Scandalios JG, Witt GS (eds) Isozymes: current topics in biological and medical research, vol 8. Liss, New York, pp 219-244

Vallee BL, Falchuk KH (1993) The biochemical basis of zinc physiology. Physiol Rev 73: 79–118

Vallee BL, Goldes A (1984) The metallobiochemistry of zinc enzymes. Adv Enzymol 56: 284–430

Varma PK, Persaud TVN (1982) Protection against ethanol-induced damage by administering gamma-linolenic and linoleic acid. Prostaglandins Leukot Med 8: 641–645

von Wartburg J-P (1989) Pharmacokinetics of alcohol. In: Crow KE, Batt RD (eds) Human metabolism of alcohol. I. Pharmacokinetics, medicolegal aspects, and general interests. CRC Press, Boca Raton, pp 9–22

Vonesch J-L, Nakshatri H, Phillipe M, Chambon P, Dollé P (1994) Stage and tissue-specific expression of the alcohol dehydrogenase 1 (Adh-1) gene during mouse development. Dev Dynam 199: 199–213

Wald N (1993) Folic acid and the prevention of neural tube defects. In: Keen CL, Bendich A, Willhite CC (eds) Maternal nutrition and pregnancy outcome. Ann NY Acad Sci 678: 112–129

Ward KW, Pollack GM (1996a) Comparative toxicokinetics of methanol in rodents: effect of pregnancy on methanol disposition and contribution of fetal metabolism to methanol elimination in rats and mice. Drug Metab Dispos (in press)

Ward KW, Pollack GM (1996b) Use of intrauterine microdialysis to investigate methanol-induced alterations in uteroplacental blood flow. Toxicol Appl Pharmacol (in press)

Ward KW, Perkins RA, Kawagoe JL, Pollack GM (1995) Comparative toxicokinetics of methanol in the female mouse and rat. Fundam Appl Toxicol 26: 258–264

Webster WS (1989) Alcohol as a teratogen: a teratological perspective of the fetal alcohol syndrome. In: Crow KE, Batt RD (eds) Human metabolism of alcohol. I. Pharmacokinetics, medicolegal aspects, and general interests. CRC Press, Boca Raton, pp 133–155

Webster WS, Walsh DA, Lipson AH, McEwen SE (1980) Teratogenesis after acute alcohol exposure in inbred and outbred mice. Neurobehav Toxicol 2: 227–234

Webster WS, Walsh DA, McEwen SE, Lipson AH (1983) Some teratogenic properties of ethanol and acetaldehyde in C57BL/6J mice: implications for the study of the fetal alcohol syndrome. Teratology 27: 231–243

Webster WS, Germain MA, Lipson A, Walsh D (1984) Alcohol and congenital heart defects: an experimental study in mice. Cardiovascular Res 18: 335–338

Wegman ME (1987) Annual summary of vital statistics – 1986. Pediatrics 80: 817–827

Weinberg J (1985) Effects of ethanol and maternal nutritional status on fetal development. Alcohol Clin Exp Res 9: 49–55

West JR, Black AC, Reimann PC, Alkana RL (1981) Polydactyly and polysyndactyly induced by prenatal exposure to ethanol. Teratology 24: 13–18

West JR, Chen WJ, Pantazis NJ (1994) Fetal alcohol syndrome: the vulnerability of the developing brain and possible mechanisms of damage. Metab Brain Dis 9: 291–322
Wier PJ, Lewis SC, Traul KA (1987) A comparison of developmental toxicity evident at term to postnatal growth and survival using ethylene glycol monoethyl ether, ethylene glycol monobutyl ether, and ethanol. Teratogen Carcinogen Mutag 7: 55–64
Wilson JG, Roth CB, Warkany J (1953) An analysis of the syndrome of malformations induced by maternal vitamin A deficiency. Effects of restoration of vitamin A at various times during gestation. Am J Anat 92: 189–217
Witek-Janusek L (1986) Maternal ethanol ingestion: effect on maternal and neonatal glucose balance. Am J Physiol 251: E178–E184
Wood CA, Buller F (1904) Poisoning by wood alcohol: cases of death and blindness from Columbian spirits and other methylated preparations. JAMA 43: 972–977
Wynter JM, Walsh DA, Webster WS, McEwen SE (1983) Teratogenesis after acute alcohol exposure in cultured rat embryos. Teratogen Carcinog Mutag 3: 421–428
Yagi K (1993) Active oxygens, lipid peroxides, and antioxidants. CRC Press, Boca Raton
Yasuda M, Sato TJ, Sumida H (1991) Excencephalic mouse fetuses are resistant to cleft palate induction with 2,3,7,8-tetrachlorodibenzo-p-dioxin (TCDD). Teratology 43: 445 (abstract)
Yasunami M, Chen C-S, Yoshida A (1991) A human alcohol dehydrogenase gene (ADH6) encoding an additional class of isozyme. Proc Natl cad Sci 88: 7610–7614
Yeh LCC, Cerklewski FL (1984) Interaction between ethanol and low dietary zinc during gestation and lactation in the rat. J Nutr 114: 2027–2033
Yin S-J, Wang M-F, Liao C-S, Chen C-M, Wu C-W (1990) Identification of a human stomach alcohol dehydrogenase with distinctive kinetic properties. Biochem Int 22: 829–835
Žgombić-Knight M, Satre MA, Duester G (1994) Differential activity of the promotor for the human alcohol dehydrogenase (retinol dehydrogenase) gene ADH3 in neural tube of transgenic mouse embryos. J Biol Chem 269: 6790–6795
Zidenberg-Cherr S, Rosenbaum J, Keen CL (1988) Influence of ethanol consumption on maternal-fetal transfer of zinc in pregnant rats on day 14 of pregnancy. J Nutr 118: 865–870
Zimmerman EF, Potturi RB, Resnick E, Fisher JE (1994) Role of oxygen free radicals in cocaine-induced vascular disruption in mice. Teratology 49: 192–201

CHAPTER 30

Developmental Toxicity of Dioxin: Searching for the Cellular and Molecular Basis of Morphological Responses*

B.D. ABBOTT

A. Introduction

I. Overview

2,3,7,8-Tetrachlorodibenzo-*p*-dioxin (TCDD) is a member of a class of chemicals, the polycyclic aromatic halogenated hydrocarbons, which includes dibenzo-*p*-dioxins, dibenzofurans, and coplanar polychlorinated biphenyls (PCB). TCDD is developmentally toxic, and in embryonic mice its teratogenicity is manifested as cleft palate and hydronephrosis. The relatively specific targeting of the developing secondary palate prompted intensive investigation spanning more than 20 years. Early studies focused on the genetics of strain susceptibility for the response and linked the responsiveness to expression of the Ah gene. Subsequent studies revealed details of the morphological, cellular, and molecular basis for the induction of cleft palate. This chapter will discuss the pathogenesis of TCDD-induced clefting and the cellular and molecular mechanisms which are the basis for the morphological responses. Prenatal exposure to TCDD adversely affects other developing systems, including the urogenital tract and the immune system, and a brief discussion of these topics will be presented. Reviews of reproductive and immunological responses are available in PETERSON et al. (1993) and HOLLADAY and LUSTER (1994).

II. Dioxin in the Environment

TCDD is a polychlorinated aromatic compound found throughout the environment. There are multiple sources of new environmental burdens of this contaminant, including medical waste incinerators, municipal incinerators, automobile exhaust, cigarette smoke, and industrial chlorine bleaching. This compound persists in the environment and bioaccumulates in biological systems, as it is not readily degraded or metabolized. The sources of release,

---

*This paper has been reviewed by the Health Effects Research Laboratory, U.S. Environmental Protection Agency, and approved for publication. Mention of trade names of commercial products does not constitute endorsement or recommendation for use.

bioaccumulation, body burdens, and biological effects of this compound are discussed in several recent reviews (Birnbaum 1991, 1994a,b; Whitlock 1990).

III. General Biological Effects

TCDD produces biological responses through binding to a cytoplasmic receptor, the aryl hydrocarbon receptor (AhR). Polycyclic aromatic halogenated hydrocarbons and coplanar PCB also bind to the AhR, although TCDD has the greatest affinity for the ligand-binding site (as reviewed in Poland and Knutson 1982; Exner 1987; Whitlock 1987; Birnbaum 1994b). As the most potent member of the chemical class, TCDD has been used extensively as a model compound to study the many biological effects which are mediated by the AhR. In animals, the adverse biological effects include carcinogenesis, immune function alteration, hyperkeratosis, hepatotoxicity, thymic involution, reproductive toxicity, and teratogenesis. In humans, the most commonly observed effect after a high-level exposure is chloracne, a skin disorder involving hyperplasisa and hyperkeratosis of the epithelial cells. The types of effects and exposure level required to produce effects are highly dependent on the species and target tissue. There are also differences in potency among members of the chemical class, and structure–activity relationships demonstrate that affinity for the AhR correlates with induction of developmental toxicity (Schwetz et al. 1973).

B. Pathogenesis and Mechanisms of Developmental Toxicity

I. Overview

In our laboratory, the search for cellular and molecular mechanisms to explain the pathogenesis of TCDD-induced cleft palate focused on the morphological etiology of the clefts and also revealed that TCDD altered the proliferation and differentiation of palatal cells. These effects are correlated with changes in growth factors. In other laboratories, studies of cultured hepatocytes and keratinocytes also found similar effects on growth and differentiation. These responses are mediated through binding of TCDD to the cytoplasmic AhR, which translocates to the nucleus, where the receptor complex regulates gene expression. Details of the receptor model were obtained predominantly from studies performed in cultured hepatoma cells, and an overview of the current understanding of the AhR model is presented.

II. The Dioxin Receptor

The biological activity of TCDD is mediated through binding to a cytoplasmic receptor. This receptor is a ligand-activated transcription factor that is a member of the Per/ARNT/Sim (PAS) superfamily. This gene family consists of transcription factors which dimerize at the helical regions of a helix–loop–helix (HLH) motif prior to binding to nuclear DNA response elements and regulating gene expression. Since a number of reviews of the genetics, structure, and function of the AhR are available (NEBERT et al. 1993; SWANSON and BRADFIELD 1993; WHITLOCK 1993; OKEY et al. 1994; BIRNBAUM 1994b), only a synopsis will be presented here.

The AhR is a soluble protein with a molecular weight of 95 kDa in the C57 mouse, but the size of the protein ranges in various species and strains from 95 to 145 kDa (DENISON et al. 1986; OKEY et al. 1989; POLAND and GLOVER 1990; POLAND et al. 1991; SWANSON and PERDEW 1991). The AhR has been found in mammalian and nonmammalian vertebrate species and is expressed in a variety of tissues (HENRY et al. 1989; SWANSON and BRADFIELD 1993; HAHN et al. 1994). The mouse and human AhR have been purified, cloned, and sequenced (BURBACH et al. 1992; DOLWICK et al. 1993). Analysis of these genes reveals homology and structural similarity with the *Drosophila* neurogenic protein Sim and circadian rhythm protein Per in a 250-amino acid region referred to as the PAS (*P*er, ARNT, *S*im) domain (BURBACH et al. 1992; DOLWICK et al. 1993; MASON et al. 1994; REYES et al. 1992; SCHMIDT et al. 1993; SWANSON and BRADFIELD 1993; WHITELAW et al. 1993a).

The AhR exists in the cytoplasm as a complex with two heat shock protein (HSP) 90 molecules and one or more other small peptides (CUTHILL et al. 1987; DENIS et al. 1988; PERDEW 1988, 1992; PERDEW and POLAND 1988; WHITELAW et al. 1993b; DE MORAIS et al. 1994). The endogenous ligand for this receptor remains to be identified; however, the receptor does bind indole carbinols and other flavinoids of dietary origin (JELLINCK et al. 1993; BJELDANES et al. 1991). In animals exposed to TCDD, the compound diffuses into cells, where it binds to the ligand-binding site of the AhR. Ligand binding triggers release of the HSP90 and activates or transforms the receptor to be receptive to dimerization with the AhR nuclear translocator (ARNT) protein (HOFFMAN et al. 1991; PONGRATZ et al. 1992; REYES et al. 1992). ARNT is structurally similar to AhR, and dimerization between these proteins involves interactions at the PAS domain (HOFFMAN et al. 1991). The ligand–AhR–ARNT complex is the activated, nuclear form of the AhR complex. This complex binds to *cis*-acting DNA target sequences, termed xenobiotic response elements (XRE), present in one or several copies in the 5′ regulatory region of AhR-controlled genes (DENISON et al. 1988; FUJISAWA-SEHARA et al. 1987; HANKINSON et al. 1991). AhR binds to the XRE, and ARNT is required for this interaction and subsequent transcriptional activation to occur (WHITELAW et al. 1993a; REISZ-PORSAZSZ et al. 1994). ARNT alone has no affinity for the XRE; however, PROBST et al. (1993) present evidence that ARNT interacts with the XRE sequence when heterodimerized with AhR.

AhR binding to the XRE regulates gene expression. A family of related, metabolic genes (cytochrome P450 enzymes), as well as several growth-regulatory genes, are known to be regulated by TCDD, and this group has been defined as an Ah gene battery (KUMAKI et al. 1977; NEBERT et al. 1993; SUTTER and Greenlee 1992). A classification scheme for the gene battery identifies 26 genes which may be regulated, based on evidence of gene transcriptional activation, accumulation of mRNA, or altered enzyme activity (Sutter and Greenlee 1992). Included in the list are phase I and II metabolic enzymes (cytochrome P450 1A1and 1A2, uridine diphosphate (UDP)-glucuronyl transferase, glutathione-*S*-transferase) ornithine decarboxylase, and several growth-regulatory proteins (transforming growth factor (TGF)-α, the proteinase inhibitor PAI-2, the cytokine interleukin (IL)-1β, and the glycoprotein hormone hCG). Regulation of secreted growth regulators provides a mechanism for amplification of the responses to TCDD through a cascade of secondary responses and may include complex, coordinated regulation of multiple growth factors. In human keratinocytes, TCDD alters mRNA and protein for TGF-α, TGF-β, PAI-2, and IL-1β with consequences for both growth and differentiation of the cultured cells (Gaido and Maness 1994; Sutter et al. 1991). The pattern of gene regulation in rat liver varies somewhat from that observed in cultured keratinocytes. TCDD regulates CYP1A and UGT1 mRNA levels and enzymatic activity in liver, but TGF-α and PAI-2 are not induced (Vanden Heuvel et al. 1994). Thus regulation of a particular gene varies between tissues, and dose-related responses are observed. In mouse embryos exposed to TCDD in vivo and in human palatal shelves exposed in vitro, changes in growth factor expression are observed in the developing palatal shelves. The type and degree of responses (increased versus decreased expression) depends on both stage of development and cell type (Abbott and Birnbaum 1990a, 1991). Thus the AhR mediates responses to TCDD through direct regulation of gene transcription as well as by secondary cascades of responses involving growth and differentiation regulators. These mechanisms are involved in the responses of the developing embryo to TCDD and specifically in the secondary palatal shelf, which is a target of TCDD-induced teratogenicity.

III. Cleft Palate

Palatogenesis is discussed in Chap. 5 of this volume in detail, and only a brief overview of the processes involved is provided here. During normal development, the palatal shelves appear as outgrowths of the maxillary arches, grow vertically beside the tongue, elevate above the tongue, and contact and fuse along the medial epithelium (Greene and Pratt 1976). Just prior to contact, the medial epithelium loses the peridermal cell layer (Fig. 1), the basal epithelial cells stop proliferating, opposing palatal shelves adhere along the medial edge, and the medial seam is gradually eliminated as fusion progresses (Pratt and Hassell 1975; Pratt and Martin 1975). Some medial cells undergo apoptotic cell death or migrate to the oral or nasal epithelial surface. However, most of the cells of the medial seam transform to mesenchymal cells and migrate away from the midline (Shuler et al. 1992; Fitchett and Hay 1989).

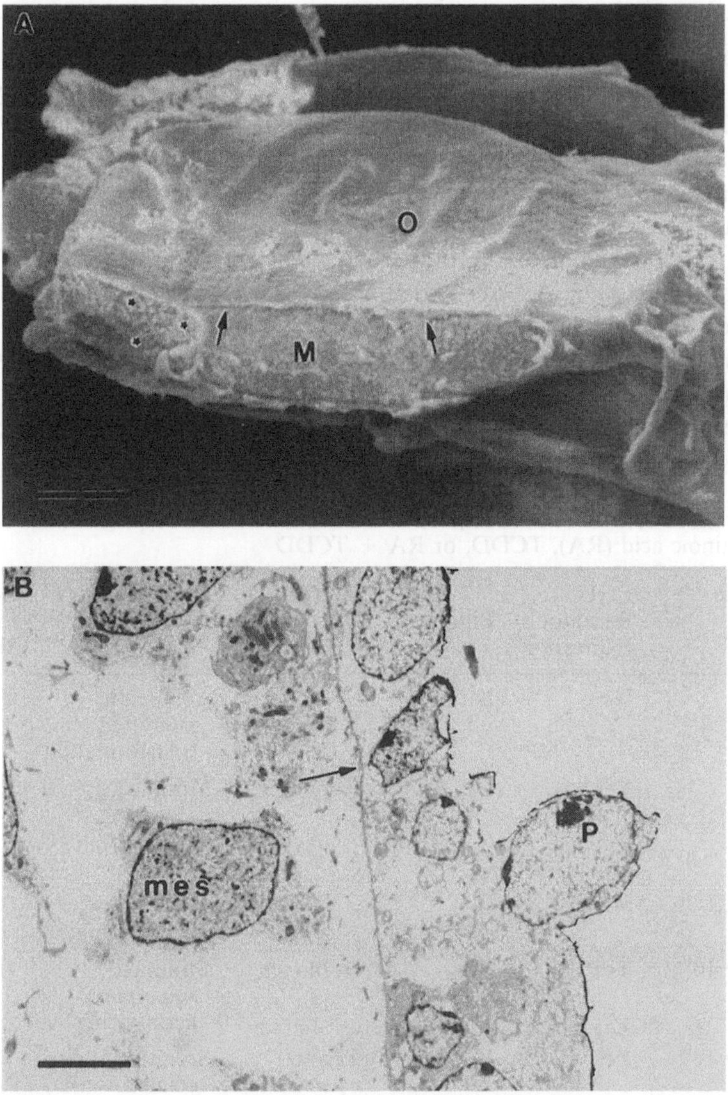

Fig. 1. A In a control palate, the opposing shelves are in contact by gestational day (GD) 14 along the medical edge (*M*). The basal cells are exposed and a clear border (*arrow*) can be seen between the oral (*O*) and medial regions. In the anterior of the shelf, the mesenchyme is exposed (asterisks). *Bar*, 200 μm. **B** The basal epithelial cells are present on the medial edge with occasional peridermal cells detaching (*P*). Mesenchyme lies in close apposition to the basement membrane (*mes*). *Bar*, 2 μm. (From ABBOTT and BIRNBAUM 1989a)

The early stages of palatogenesis appear to proceed normally in embryonic mice exposed to teratogenic doses of TCDD. Palatal shelves form, elevate, grow, and come into contact, but fail to fuse (PRATT et al. 1984; BIRNBAUM et al. 1986). The failure to fuse is attributed to TCDD-stimulated proliferation of the medial epithelial cells and formation of a stratified, squamous, oral-like epithelium along the medial edge (Fig. 2). As shown in Table 1, these responses occur after exposure on either gestational day (GD) 10 or 12. These responses are a result of direct action of TCDD in the palatal tissues and are not due to indirect effects induced through maternal or placental toxicities. This was demonstrated by induction of altered epithelial differentiation and proliferation in a palatal organ culture model in which mouse, rat, or human embryonic palatal shelves were exposed to TCDD (Abbott et al. 1989a, Abbott and Birnbaum 1990b, 1991).

In embryonic mice exposed to TCDD in vivo, the effects on epithelial cell differentiation were accompanied by altered growth factor expression (Abbott and Birnbaum 1989a,b; Abbott et al. 1992). The specific effects on growth factor expression depended on the stage of development exposed, the cell type,

Table 1. A comparison of the responses of medial cells exposed on gestation day 10 or 12 to retinoic acid (RA), TCDD, or RA + TCDD

Exposure Group	EGF Receptor Expression	^{125}I-EGF Binding	^{3}H-TdR Uptake (GD 14/16)	Differentiation Pathway	Contact/ Fusion of Shelves
Control	No	No	No/no	Cell death, migration, transformation	Yes/yes
TCDD GD 10[a]	Yes	Yes	No/no	Stratified, squamous, keratinizing	Yes/no
TCDD GD 12[a]	Yes	Yes	No/no	Stratified, squamous, keratinizing	Yes/no
RA GD 10[b]	Yes	Yes	No/no	Stratified, squamous, keratinizing	No/no
RA GD 12[b]	Yes	Yes	Yes/no	Secretory, ciliated, microvilli	Yes/no
TCDD + RA GD 10[c]	Yes	Yes	Yes/no	Stratified, squamous, keratinizing	No/no
TCDD + RA GD 12[d]	Yes	Yes	No/no	Stratified, squamous, keratinizing	Yes/no

Parameters were assessed on gestational day (GD) 14 except for ^{3}H-Tdr uptake, which is shown for GD 14 and GD 16. From Abbott and Birnbaum (1989b).
EGF, epidermal growth factor; ^{3}H-Tdr, [^{3}H] thymidine
[a]24 µg/kg. [b]100 mg/kg.
[c]6 µg TCDD/kg, 40 mg RA/kg. [d]6 µg TCDD/kg, 80 mg RA/kg.

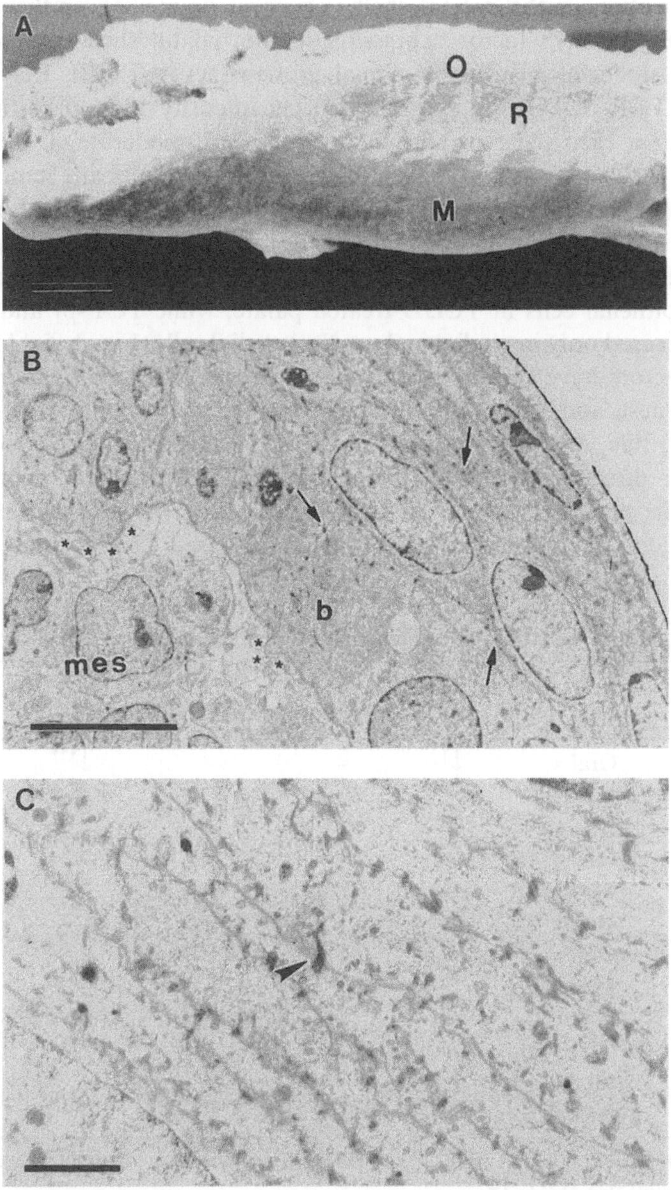

Fig. 2. **A** A palatal shelf from an embryo exposed to TCDD at 24 μg/kg on gestational day (GD) 12 has an intact medial epithelium on GD 14. *O*, oral surface; *R*, rugae; *M*, medial epithelium *Bar, 200* μm. **B** By GD 15 the medial cells have differentiated to form a stratified, squamous epithelium. *Arrows* indicate desmosomes, *asterisks* basal lamina. b, basal epithelial cells; *mes*, mesenchymal cells. *Bar*, 5 μm. **C** The medial epithelium has large numbers of desmosomes (*arrowhead*) linking the squamous cells, and this is not observed even in the normally differentiated oral epithelium of control fetuses. *Bar*, 1 μm. (From ABBOTT and BIRNBAUM 1989a)

and its position in the palatal shelf (e.g., oral or medial, epithelial or mesenchymal). Growth factor expression in the palatal shelf was determined immunohistochemically for epidermal growth factor (EGF), EGF receptor (EGFR), TGF- α, TGF-β_1, and TGF-β_2. The intensity of immunostaining was evaluated for each cell type and region (1, low; 2, moderate; 3, high) and in some studies was measured densitometrically. The results are summarized in Table 2. The medial epithelial cells continue to express EGFR and bind EGF, although unexposed medial palatal cells do not have EGFR or bind EGF at this developmental stage (Table 1). Expression of TGF-α is decreased in all palatal epithelial cells in TCDD-treated palate, while TGF-β_1 and TGF- β_2 levels increased only in medial and nasal epithelial cells (Fig. 3; Table 2). These growth factors have specific spatial and temporal expression patterns during embryogenesis and are important regulators of growth and differentiation (see Chap. 14, this volume). Although there is no experimental evidence in the palate for direct transcriptional regulation of growth factor expression by TCDD, the possibility cannot be completely disregarded. In cultured human keratinocytes, TCDD- exposure results in increased TGF-α mRNA and se-

Table 2. Shifts in growth factor expression relative to controls

Growth factor	Epithelial cell type	TCDD[a]	HC[b]	HC + TCDD[c]	RA[d]	RA + TCDD[e]
TGF-α	Oral	↓*	–	–	↓***	↓**
	Medial	↓**	–	–	↓*	↓**
	Nasal	↓*	–	–	–	↓*
EGF	Oral	–	↑*	↑	–	↓**
	Medial	–	↑*	↑	–	–
	Nasal	–	↑***	↑**	–	–
TGF-β1	Oral	–	↑***	↑*	–	↓**
	Medial	↑*	↑***	↑***	–	↓**
	Nasal	↑*	↑***	↑***	–	↓*
TGF-β2	Oral	–	–	–	–	–
	Medial	↑*	–	–	–	–
	Nasal	↑*	↑*	↑*	↑*	–

From ABBOTT et al. (1992); ABBOTT and BIRNBAUM (1990a,c). Parameters were assessed on gestational day (GD) 14.
– no change; ↑, trend toward increased expression; ↓, trend toward decreased expression; HC, hydrocortisone; RA, retinoic acid; TGF, transforming growth factor; EGF, epidermal growth factor.
[a] 24 μg/kg, GD 10.
[b] 100 mg/kg, GD 10–13.
[c] 25 mg HC/kg, 3 μg TCDD/kg, GD 10–13.
[d] 100 mg/kg, GD 10.
[e] 40 mg RA/kg, 6 μg TCDD/kg, GD 10.
*$p < 0.05$.
**$p < 0.01$.
***$p < 0.001$.

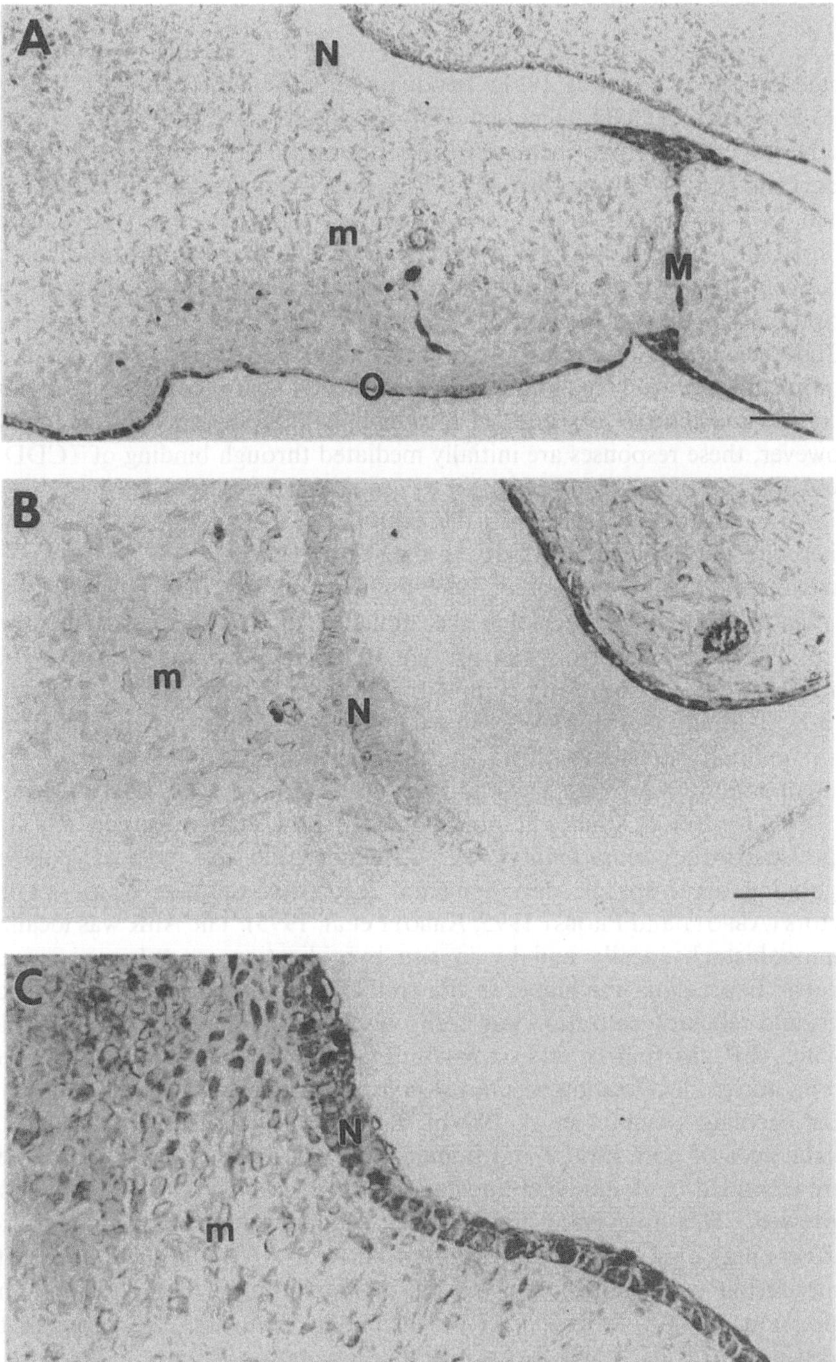

Fig. 3. A In the control gestational day (GD) 14 palatal shelf, transforming growth factor (TGF)-β2 is detected immunohistochemically predominantly in medial (*M*) and oral (*O*) epithelia. *M*, mesenchyme. *Bar*, 50 μm. **B** In the nasal epithelium (*N*) and adjacent mesenchyme (*m*), little or no TGF-β_2 can be detected in the control palate. *Bar*, 25 μm. **C** Exposure to hydrocortisone (HC) + TCDD (25 mg/kg + 3 μg/kg on GD 10–13) increases the expression of TGF-β_2 in the nasal epithelial cells and in the underlying mesenchyme. *Bar*, 25 μm

creted peptide (CHOI et al. 1991). In squamous carcinoma cells (SCC-12F cell line) exposed to TCDD, increased TGF-α mRNA was due to stabilization of the message rather than increased transcription (GAIDO et al. 1992). Also in these cells TCDD downregulated TGF-β_2 mRNA at the transcriptional level, while TGF-β_1 mRNA levels were unchanged. Similarly, other investigators report that repression of EGFR by TCDD in this transformed cell line does not occur at the transcriptional level and may be the result of post-transcriptional modifications (OSBORNE et al. 1988). The emerging picture of TCDD regulation of expression of growth factors suggests that there are multiple avenues for regulation and consequently that a complex mechanism of direct and indirect cascades of responses is likely to exist in the embryo. However, these responses are initially mediated through binding of TCDD to the AhR.

An extensive literature links the expression of high-affinity Ah receptor genes with developmental toxicity (LAMBERT and NEBERT 1977; POLAND and GLOVER 1980; SILKWORTH et al. 1989; as reviewed by COUTURE et al. 1990a; PETERSON et al. 1993), including the induction of cleft palate and hydronephrosis in mice (COURTNEY and MOORE 1971; NEUBERT and DILLMAN 1972; BIRNBAUM et al. 1986). Early competitive binding studies revealed expression of a high-affinity receptor (AhR) in palatal cells, and this was later confirmed with immunohistochemical and in situ hybridization localization of AhR protein and mRNA (DENCKER and PRATT 1981; PRATT et al. 1984; ABBOTT et al. 1994a,b). Recent studies in our laboratory show that AhR and ARNT are expressed in the mouse embryo with specific spatial and temporal patterns, suggesting organ-specific developmental regulation of these transcriptional factors (ABBOTT and PROBST 1995; ABBOTT et al. 1995). The AhR was localized immunohistochemically and by in situ hybridization in embryonic mouse palates. Expression was higher in the epithelial cells relative to the mesenchyme, and cellular localization was both cytoplasmic and nuclear. In the human palate, AhR distribution was similar to that observed in the mouse; however, strong nuclear localization was noted in mesenchymal cells and in developing nasal cartilage (ABBOTT et al. 1994b). ARNT was exclusively nuclear in all palatal cells of both mouse and human embryos. In the human palate, cells were identified by double immunostaining in which only AhR or ARNT was expressed. This interesting observation suggests that other dimerization partners may exist or that expression of AhR or ARNT in the absence of a dimerization partner may be a means of regulating gene expression. Thus expression of both AhR and ARNT in palatal epithelial cells provides the necessary molecular components for gene regulation following exposure to TCDD.

IV. Synergistic Interactions with Dioxin

Synergistic interactions for induction of cleft palate are known to occur between TCDD and glucocorticoids (GC) and between TCDD and retinoic acid (RA) (BIRNBAUM et al. 1986, 1989). Both glucocortiocids and retinoids occur endogenously and have significant roles in normal development (SUGIMOTO et al. 1976; SALOMON et al. 1978; SALOMON and PRATT 1979). Both glucocorticoids and RA act through cytoplasmic receptors, the glucocorticoid receptor (GR), and RA receptors RXR/RAR, respectively. Although glucocorticoids and RA occur endogenously in the embryo, exposure to pharmacological levels of either compound can induce birth defects, including cleft palate. Examination of the interaction between low, nonteratogenic levels of TCDD, glucocorticoids, and RA could provide information about possible common pathways for regulation of development through these receptor-mediated responses.

1. Hydrocortisone

A potent, synergistic interaction has been observed between TCDD and hydrocortisone (HC), a glucocorticoid (NEUBERT et al. 1973; BIRNBAUM et al. 1986). Doses which alone do not induce clefting produce cleft palate in 100% of the exposed embryos when administered together. This interaction was first noted in NMRI embryonic mice exposed to both TCDD and dexamethasone (a synthetic glucocorticoid) (NEUBERT et al. 1973). This phenomena was further examined in C57BL/6N mice by BIRNBAUM et al. (1986). TCDD at 0 or 3 µg/kg per day, p.o., HC at 0, 25, 50, or 100 mg/kg per day, s.c., and combinations of these dosages were given on GD 10-13. A further study was performed to extend the dosage range to include HC at 1 and 10 mg/kg per day, with or without 3 µg TCDD/kg/per day (ABBOTT et al. 1992). The induction of cleft palate by these exposures is summarized in Table 3. Exposure to TCDD at a dosage of 3 µg/kg per day from GD 10-13 did not affect litter size, fetal weight or viability, or maternal weight gain. Hydronephrosis was induced in all exposed fetuses, but no cleft palate occurred. HC produced dose-related decreases in fetal weight and maternal liver to body weight ratios, and there was a dose-dependent increase in cleft palate incidence (up to 30% at 100 mg/kg per day). Exposure to TCDD + HC resulted in a significant, synergistic increase in the incidence of cleft palate, and this occurred even at the lowest HC dosage of 1 mg/kg per day (53% cleft palate on a per litter basis). In both of these studies, the morphological basis of clefting was attributed to formation of small palatal shelves which could not contact and fuse.

TCDD and glucocorticoids, such as HC, induce clefting through distinctly different morphological etiologies (reviewed by PRATT 1985). As discussed earlier in this chapter, exposure to TCDD results in normally sized palatal shelves which come into contact but fail to fuse (PRATT et al. 1984). In contrast, after HC exposure the palatal shelves are small and do not make contact

Table 3. HC and TCDD induction of cleft palate

Group	HC (mg/kg)	TCDD (μg/kg)	Litters(*n*)	Cleft palate (% per litter basis)[a]
Control	–	–	10	0
HC	100	–	20	29.4 ± 6.5*
	25	–	14	4.5 ± 3.6
	10	–	8	0
	1	–	10	0
TCDD	–	24	13	100
	–	3	13	0
HC + TCDD	25	3	13	98.9 ± 1.1*
	10	3	8	88.1 ± 5.1*
	1	3	8	53.2 ± 8.9*

From Abbott et al. (1992). Dosing: Gestational day (GD) 10–13 except TCDD (24 μg/kg), which was dosed on GD 10 only. Cleft palate determined on GD 18.
HC, hydrocortisone.
[a]Mean + S.E.
*$p < 0.001$ vs. control.

at the time fusion normally occurs (Jelinek and Dostal 1975; Diewert and Pratt 1981). Cleft palate following exposure to HC + TCDD is morphologically similar to that seen after exposure to HC alone, as small shelves form which do not make contact (Birnbaum et al. 1986; Abbott et al. 1992). Further evidence that exposure to TCDD + HC produces HC-like clefts can be found in the effects of these compounds on growth factor expression. As shown in Table 2, exposure to HC or HC + TCDD produces similar shifts in expression of EGF and TGF-β_1. These patterns of growth factor expression differ from the effects occurring after TCDD alone. This suggests that a linkage exists between the effects on growth factors and the divergent morphological outcomes.

Both TCDD and HC act through distinct receptor-mediated mechanisms. AhR and GR exist as cytoplasmic protein complexes, and binding of their respective ligands results in activation and translocation to the nucleus, where binding to specific DNA sequences (XRE and GRE, respectively) alters transcription of specific genes (for AhR, see Poland et al. 1976; Denison et al. 1988; Perdew 1988; Hoffman et al. 1991; Reyes et al. 1992; Pollenz et al. 1994; Probst et al. 1993; for GR, see Gehring 1993; Muller and Renkawitz 1991). Palatal expression of AhR and its involvement in mediating cleft palate was previously discussed in this chapter. The glucocorticoid receptor is also linked to induction of cleft palate by glucocorticoids (Baxter and Fraser 1950; Walker 1967; Nanda et al. 1970; Shah and Kilistoff 1976; Goldman 1984). Strain-specific sensitivity to glucocorticoid-induced teratogenesis is dependent on expression of high-affinity GR (Pinsky and DiGeorge 1965; Saxen 1973; Biddle and Fraser 1976; Vekemans 1982; Pratt and Salomon 1981; Salomon and Pratt 1976). The glucocorticoid receptor is expressed in the palate with a specific expression pattern (Kim et al. 1984; Abbott et al. 1994b,c).

In order to understand the mechanisms through which HC and TCDD interact, it was important to examine AhR and GR expression in the treated palates. These studies revealed that AhR-mediated palatal responses include effects on GR expression and GR-mediated responses include effects on AhR expression (Abbott et al. 1994b). Table 4 summarizes the effects of TCDD, HC, or HC + TCDD exposure on AhR expression. TCDD downregulated mRNA and protein for AhR. HC increased AhR protein and mRNA in the palatal shelves. HC + TCDD elevated AhR protein levels, although mRNA decreased, as shown in both northern analysis and in situ hybridization. These analyses were performed on GD 14 palates, and it is possible that downregulation of mRNA in this group occurred prior to collection of the tissue and that the effects on protein were not yet detectable. Alternatively, the half-life of the AhR may have been affected. Also shown in Table 4 are the effects of TCDD, HC, or HC + TCDD on GR expression. In GD 14 palatal shelves, HC exposure downregulated GR protein and mRNA was decreased on the northern blots which present the overall effect on midfacial tissues. This is consistent with many other reports of decreased GR following ligand binding (McIntyre and Samuels 1985; Okret et al. 1986; Kalinyak et al. 1987; Rosewicz et al. 1988; Dong et al. 1988; Vedeckis et al. 1989; Burnstein et al.

Table 4. Summary of treatment effects

Treatment group	Effects on Ah receptor expression			Effects on glucocorticoid receptor expression		
	Immuno-Assay: protein	In situ: mRNA	Northern blot: mRNA	Immuno-Assay: protein	In situ: mRNA	Northern blot: mRNA
HC						
25 mg/kg[a]	–	↑	–	↓	–	↓*
100 mg/kg[a]	↑*	↑	↑	↓	↑	↓
TCDD						
3 μg/kg[a]	↓***	↓	↓	↑*	–	–
24 μg/kg[b]	↓***	↓	↓***	–	↑	↓***
HC + TCDD 25 mg/kg + 3 μg/kg[a]	↑	↓	↓*	↑*	↑	↓**

From Abbott et al. (1994b). All parameters measured on gestational day (GD) 14.
– no change; ↑, trend toward increased expression; ↓, trend toward decreased expression.
HC, hydrocortisone.
[a]GD 10–13.
[b]GD 10.
*$p < 0.05$.
**$p < 0.01$.
***$p < 0.001$.

1991). GR mRNA was elevated in some sections examined by in situ hybridization following the high dose of HC. Correlating the responses of mRNA and protein in these embryos is complicated by the fact that individual embryos cannot be examined for both end points and that in this strain cleft palate can only be produced in 30% of the HC-exposed fetuses. After TCDD or HC+TCDD exposure, GR protein levels increased and were correlated with increased mRNA localized by in situ hybridization. However, the average expression over the entire midcraniofacial region decreased (as shown by northern blot). Based on the observed data, an interaction cycle is proposed in which cross-regulation of the receptors contributes to the synergism between HC and TCDD for induction of cleft palate. This model, as diagrammed in Fig. 4, incorporates, for each exposure regimen, the distinctly different morphological outcomes, the patterns of effects on growth factors, and the effects of each compound on expression of AhR and GR. TCDD binds to the AhR, forming the AhR–ligand (Ah:L) complex. Responses to this complex include down-regulation of mRNA and protein for the receptor and a cascade of

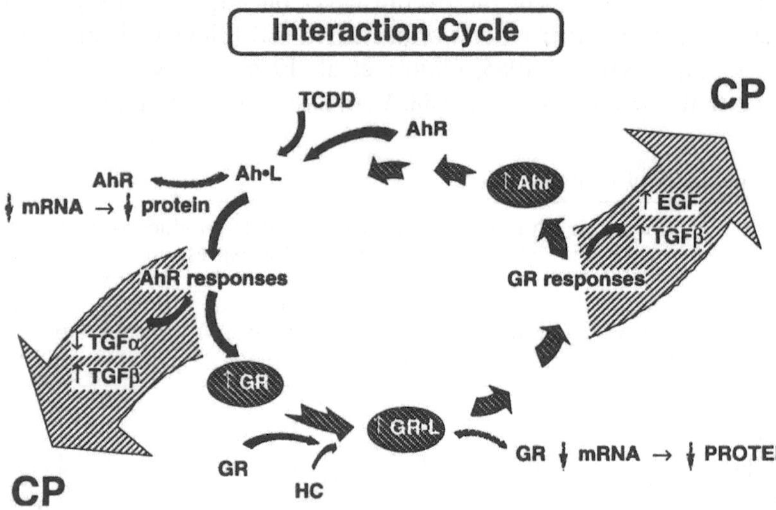

Fig. 4. The interaction between hydrocartisone (HC) and TCDD is mediated through their respective receptors. The induction of glucocorticoid receptor (*GR)* following TCDD exposure and the upregulation of the aryl hydrocarbon receptor (AhR) after HC exposure can lead to an ongoing cycle of receptor upregulation that promotes and amplifies responses to HC and TCDD. Exposure to doses of HC and TCDD which alone do not induce cleft palate (*cp*) would, through this process, amplify the levels of both AhR and GR, with a resulting cascade of GR-mediated responses. This model provides an explanation for the HC-like clefts observed after HC + TCDD, as HC-induced effects on palatal size would be the predominant cause of clefting and effects of TCDD on medial epithelial proliferation and differentiation would be secondary to the fact that small shelves cannot contact and fuse. *Ah.L*, ligand (TCDD) bound to Ah receptor; *GR.L* ligand (hydrocortisone) bound to GR; transforming growth factor (TGF)-α and TGF-β are growth factors expressed in the palate that are affected by HC, TCDD, and HC + TCDD exposure

responses that contribute to cleft palate. An important component of the interaction cycle is the increased expression of GR induced in the palatal cells by TCDD. The excess GR induced by TCDD exposure binds HC and a series of GR-mediated responses occur, including upregulation of AhR. Exposure to doses of HC and TCDD which alone do not induce cleft palate would, through this process, amplify the levels of both AhR and GR with the resulting cascade of GR-mediated responses. This model provides an explanation for the HC-like clefts observed after HC + TCDD, as HC-induced effects on palatal size would be the predominant cause of clefting and effects of TCDD on medial epithelial proliferation and differentiation would be secondary to the fact that small shelves cannot contact and fuse. TCDD persists in the tissues; continued exposure would be expected to increase the pool of ligand available for binding to AhR, and HC administered daily (GD 10-13) would maintain the cycle. With exposure to high levels of HC alone, the right side of the diagram would apply, and with exposure to TCDD alone the left side would be applicable.

In summary, AhR-mediated responses included effects on GR, and GR-mediated responses included effects on AhR. The observed cross-regulation of the receptors is believed to be important in the synergistic interaction between TCDD and HC for the induction of cleft palate.

2. Retinoic Acid

RA and TCDD also interact synergistically to induce cleft palate (Birnbaum et al. 1989). Exogenous RA induces cleft palate, and the mechanism through which this occurs is dependent on the stage of embryonic development at the time of exposure (Kochhar 1973; Geelen 1979; Kochhar et al. 1984; Abbott et al. 1989b; Abbott and Birnbaum 1990a,c). The morphological basis for clefting following exposure of embryonic mice on GD 10 involves reduced growth of the palatal shelves, thus preventing fusion through lack of contact, and following exposure on GD 12 the shelves are of normal size, yet fail to fuse due to abnormal differentiation of the medial epithelial cells. Both of these outcomes correlate with somewhat different effects of the exposure on expression of growth factors (Abbott and Birnbaum 1989b, 1990a,c). The effects of RA on differentiation, proliferation, and growth factor expression differ from the effects observed following TCDD exposure on either GD 10 or 12. The interaction between low doses of RA and TCDD (administered on either GD 10 or 12) produces distinct overall patterns of morphological and growth factor responses for each exposure regimen as well as overall patterns of response which are distinct from that occurring after either RA or TCDD alone (Tables 1, 2). Based on the morphological outcomes of the interactive exposures (RA + TCDD), it appeared that RA stimulated a dioxin-like response following GD 12 exposure, while GD 10 exposure produced an RA-like response. The complexity of this interaction may be due in part to the multiplicity of RA actions in the developing embryo. The receptor model for

RA-mediated responses becomes increasingly complex as additional isoforms for RAR and RXR are identified and as interactions between isoforms and other receptors are revealed. Although the interaction of TCDD and RA is intensely interesting with respect to development of the embryo, cell culture models may be a more practical model in which to study this interaction. In studies of the interaction between TCDD and RA in cultured human breast cancer cells, a link between signal transduction pathways is suggested (LU et al. 1994), and in keratinocytes RA is a potent antagonist of TCDD-induced differentiation (BERKERS et al. 1994).

C. Hydronephrosis, Reproductive Toxicity, and Immunosuppression

I. Hydronephrosis

Although this discussion has mainly centered on the mechanisms of induction of cleft palate by TCDD, prenatal exposure produces other adverse effects as well. The range of developmental responses includes embryo/fetal mortality, cleft palate, thymic hypoplasia, hydronephrosis, renal congestion, renal anomalies, subcutaneous edema, decreased fetal weight, gastrointestinal hemorrhage, reproductive toxicity including altered development of male and female reproductive function, and immunotoxicity. The manifestation of most of these responses is highly species and dose dependent, as discussed in the review of teratogenicity in various species by COUTURE et al. (1990a). For most of the teratogenic responses the etiologies are unknown and mechanistic studies are generally not available. However, the mechanism of induction of hydronephrosis has been examined in mice. Hydronephrosis can be induced at levels of exposure which are insufficient to cause cleft palate (MOORE 1973; BIRNBAUM et al. 1989). The sensitive period for induction extends throughout gestation and into the neonatal period (COUTURE-HAWS et al. 1991; COUTURE et al. 1990b). The prolonged susceptibility may be linked to the presence of responsive, undifferentiated ureteric epithelial cells. The epithelial cells lining the ureter respond to TCDD with increased proliferation, which results in a thickening of the ureteric epithelium, thus occluding flow of urine from the kidney to the bladder. If exposure occurs early in development of the urinary tract, before urine is produced by the kidney, the ureteric lumen can be completely filled with epithelial cells. The increased proliferation was detected using autoradiography to determine uptake of [^{3}H]-thymidine by the ureteric epithelia, and the increased number of cells in the ureteric epithelium was observed histologically and by transmission electron microscopy (ABBOTT et al. 1987). These responses correlated with altered expression of the EGFR (ABBOTT and BIRNBAUM 1990d). The prolonged susceptibility to TCDD-induced hydronephrosis in late gestation and following lactational exposure in the first few days of neonatal life may be due to undifferentiated stem cells in the basal ureteric epithelium which remain responsive to disruption of growth regulators. There are similarities in the mechanism through which TCDD induces cleft palate and hydronephrosis. In both cases, TCDD disrupts the expression of growth factors which regulate epithelial cell proliferation and/or differentiation. In the palate this produces an abnormal medial epithelium, which prevents fusion of opposing shelves, and in the urinary tract it induces hyperplasia of the luminal epithelium, which occludes urine flow.

II. Developmental Reproductive Toxicity

Recently there has been increasing interest in the reproductive effects of prenatal exposure to TCDD. Male rats and hamsters exposed prenatally and lactationally to TCDD have decreased epididymal and ejaculated sperm counts (Mabley et al. 1992c; Gray et al. 1995). In Holtzman rats, TCDD affected anogenital distance, testes descent, sex organ weights, and sexual behaviors, and partial demasculinization and feminization occurred (Mabley et al. 1992a–c; Bjerke and Peterson 1994; Bjerke et al. 1994a,b; Gray et al. 1995). Some of these responses were not produced in Long Evans rats and Syrian hamsters, and the variations in responses as well as reproducibility of observations are examined in Peterson and Gray (1995). In female rats and hamsters, the effects on the reproductive tract include cleft phallus and retention of a tissue strand across the vaginal orifice at the time of vaginal opening (Gray and Ostby 1995). Thus the developing reproductive tract exhibits sensitivity to prenatal and postnatal exposure to TCDD, which produces effects that are manifested at puberty and in the adult.

III. Immunotoxicity

TCDD-induced immunosuppression is of considerable interest, and a comprehensive discussion of developmental immunotoxicity and dioxin-induced immunosuppression was recently published by Holladay and Luster (1994). Prenatal exposure to TCDD can alter immune system development and induce changes in immune function which persist into adulthood (Vos and Moore 1974; Luster et al. 1980; Thomas and Hinsdale 1979).Indicators of TCDD-induced immunotoxicity include growth of transplanted tumor cells, mortality after endotoxin challenge, allograft rejection time,bone marrow cellularity, hypersensitivity responses, and spleen and thymus weights. Studies of fetal thymus, thymus cell markers, and fetal liver can provide information about the particular stages of immune development that are targets for TCDD action. The expression of thymocyte antigens and postnatal immune function are altered by prenatal exposure, and there is evidence that thymic atrophy is the result of reducing the number of prothymocytes in the fetal liver (Holladay et al. 1991; Blaylock et al. 1992; Fine et al. 1989).

D. Nonhuman Primate and Human Developmental Toxicity

In nonhuman primates, exposure in the diet to TCDD can result in perinatal mortality at levels which are not overtly maternally toxic. TCDD at 25 and 50 ppt before and during pregnancy resulted in increased spontaneous abortions, stillbirths, and decreased fertility (Allen et al. 1979; Barsotti et al. 1979; Schantz et al. 1979; Schantz and Bowman 1989). Neurobehavioral effects were reported in juvenile monkeys exposed prenatally and lactionally to TCDD (Schantz and Bowman 1989). At 5 or 25 ppt TCDD in maternal diet during pregnancy, offspring exhibit specific reproducible deficits in cognitive function as well as changes in social interactions.

The induction of cleft palate by levels of TCDD that are not overtly maternally toxic has only been observed in the mouse. Cleft palate can be induced in rat and hamster at maternally toxic doses (Olson et al. 1980; Schwetz et al. 1973). However, in humans and nonhuman primates, cleft palate was not seen after perinatal exposure to TCDD or a mixture of PCB and chlorinated furans. Human exposures have typically been to complex mixtures of PCB, furans, and dioxins. In the Yusho and Yu-Cheng incidents in Japan and Taiwan, children were exposed in utero and lactionally to PCB, furans, and quarterphenyls in rice oil consumed by the mothers. These children showed low birth weights and increased perinatal mortality and at birth had hyperpigmentation of the skin and abnormalities of the hair, nails, teeth, and gums. These children also exhibit persistent growth delay and deficits in cognitive and behavioral development (Hsu et al. 1985; Rogan et al. 1988).

Epidemiological assessment of human populations exposed to TCDD has not revealed a clear association between TCDD exposure and teratogenicity or other adverse reproductive outcomes. Following an industrial accident in 1976 in Seveso, Italy, pregnancies were monitored, but an increased risk of birth defects was not detected (Mastroiacovo et al. 1988; Reggiani 1989). The retrospective cohort study to examine reproductive risk associated with exposure to TCDD-contaminated soil in Missouri also did not detect a significant increase in birth defects or fetal/infant death (Hoffman and Stehr-Green 1989). Finally, no correlation has been found between exposure of populations to 2,4,5-trichlorophenoxyacetic acid (2,4,5-T) in Vietnam and increased birth defects (Kunstadter 1982). The ability to detect a statistically significant increase above the background incidence of human birth defects would require a much larger cohort than is generally available from such accidentally exposed populations.

The failure to detect a statistically significant increased risk for adverse human reproductive outcomes following the exposures at Seveso, Italy, and Times Beach, Missouri, may be due to the small numbers of affected pregnancies, but may also be related to the actual exposure level. The palatal organ culture model allows the responses of rat, mouse, and human palatal medial epithelial cells to be compared following TCDD exposure (Abbott et al. 1989a; Abbott and Birnbaum 1990b, 1991). The human embryonic secondary palate responds to TCDD in vitro, but the concentration required is considerably higher than that required to elicit a similar response in embryonic mouse palatal shelves. Rat and human palates respond at similar levels in vitro (ABBOTT and BIRNBAUM 1991). The effects on epithelial differentiation and proliferation are very similar between species and are in agreement with previously described effects following in vivo exposure of the mouse. The human palatal shelves express AhR and its dimerization partner ARNT in the epithelial cells, although a more heterogeneous pattern is found in the mesenchyme (ABBOT et al. 1994d). Thus in both rodent and human, the receptor and its partner are expressed in the target cells; however, a higher concentration of TCDD is required to elicit the response in rat and human cells than in the mouse. The reasons for difference in sensitivity between species continues to challenge investigators, but increasingly detailed knowledge of the molecular components of response should provide additional insight.

E. Future Directions

The developmental and reproductive toxicity of TCDD and other compounds which bind to the AhR continues to be of interest and concern. The relationship between these compounds and endogenous growth regulators, morphogens, and steroid hormones could be helpful in deciphering the mechanism of action of these compounds. Current and future studies focus on the various components of receptor response to determine the presence of each peptide and its functionality in the developing embryo. Of further interest will be the role, if any, of the AhR, ARNT, and related peptides in normal development. These proteins are expressed in specific spatial and temporal patterns throughout development and are found at relatively high levels in critical organs, including the neural tube, adrenal, bone, muscle, lung, and liver. The study of directly regulated genes will continue to be important in revealing mechanisms; however, the complexity of the cascade of responses seen for growth regulators may be equally significant in the final scheme to explain developmental dysregulation by these compounds. There are multiple research stratagies available to address these issues, and the progress made utilizing molecular approaches should be substantial in the near future.

References

Abbott BD, Birnbaum LS (1989a) TCDD alters medial epithelial cell differentiation during palatogenesis. Toxicol Appl Pharmacol 99: 276–286

Abbott BD, Birnbaum LS (1989b) Cellular alterations and enhanced induction of cleft palate after coadministration of retinoic acid and TCDD. Toxicol Appl Pharmacol 99: 287–301

Abbott BD, Birnbaum LS (1990a) TCDD-induced altered expression of growth factors may have a role in producing cleft palate and enhancing the incidence of clefts after coadministration of retinoic acid and TCDD. Toxicol Appl Pharmacol 106: 418–432

Abbott BD, Birnbaum LS (1990b) Rat embryonic palatal shelves respond to TCDD in organ culture. Toxicol Appl Pharmacol 103: 441–451
Abbott BD, Birnbaum LS (1990c) Retinoic acid induced alterations in expression of growth factors in embryonic mouse palatal shelves. Teratology 42: 597–610
Abbott BD, Birnbaum LS (1990d) Effects of TCDD on embryonic ureteric epithelial EGF receptor expression and cell proliferation. Teratology 41: 71–84
Abbott BD, Birnbaum LS (1991) TCDD exposure of human embryonic palatal shelves in organ culture alters the differentiation of medial epithelial cells. Teratology 43: 119–132
Abbott BD, Probst MR (1995) Developmental expression of two members of a new class of transcriptional factors. Expression of Ah receptor nuclear translocator (ARNT) in the C57BL/6N mouse embryo. Dev Dynam 204: 144–155
Abbott BD, Birnbaum LS, Pratt RM (1987)TCDD-induced hyperplasia of the ureteral epithelium produces hydronephrosis in murine fetuses. Teratology 35: 329–334
Abbott BD, Diliberto JJ, Birnbaum LS (1989a) TCDD alters embryonic palatal medial epithelial cell differentiation in vitro. Toxicol Appl Pharmacol 100: 119–131
Abbott BD, Harris MW, Birnbaum LS (1989b) Etiology of retinoic acid induced cleft palate varies with the embryonic stage exposed. Teratology 40: 533–554
Abbott BD, Harris MW, Birnbaum LS (1992) Comparisons of the effects of TCDD and hydrocortisone on growth factor expression provide insight into the synergistic interaction occurring in embryonic palates. Teratology 45: 35–53
Abbott BD, Perdew GH, Birnbaum LS (1994a) Ah receptor in embryonic mouse palate and effects of TCDD on receptor expression. Toxicol Appl Pharmacol 126: 16–25
Abbott BD, Perdew GH, Buckalew AR, Birnbaum LS (1994b) Interactive regulation of Ah and glucocorticoid receptors in the synergistic induction of cleft palate by TCDD and hydrocortisone. Toxicol Appl Pharmacol 128: 138–150
Abbott BD, McNabb FMA, Lau C (1994c) Glucocorticoid receptor expression during the development of the embryonic mouse secondary palate. J. Craniofac Genet Dev Biol 14: 87–96
Abbott BD, Probst MR, Perdew GH (1994d) Immunohistochemical double-staining for Ah receptor and ARNT in human embryonic palatal shelves. Teratology 50: 361–366
Abbott BD, Birnbaum LS, Perdew GH (1995) Developmental expression of two members of a new class of transcriptional factors. I. Expression of Ah receptor in the C57BL/6N mouse embryo. Dev Dynam 204: 133–143
Allen JR, Barsotti DA, Lambrecht LK, van Miller JP (1979) Reproductive effects of halogenated aromatic hydrocarbons on nonhuman promates. Ann N Y Acad Sci 320: 419–425
Barsotti DA, Abrahamson LJ, Allen JR (1979) Hormonal alterations in female rhesus monkeys fed a diet containing 2,3,7,8-tetrachlorodibenzo-p-dioxin. Bull Environ Contam Toxicol 21: 463–469
Baxter H, Fraser FC (1950) Production of congenital defects in the offspring of female mice treated with cortisone. McGill Med J 19: 245–249
Berkers JAM, Hassing GAM, Brouwer A, Blaauboer BJ (1994) Effects of retinoic acid on the 2,3,7,8-tetrachlorodibenzo-p-dioxin-induced differentiation of in vitro cultured human keratinocytes. Toxicol In Vitro 84: 605–608
Biddle FG, Fraser FC (1976) Genetics of cortisone-induced cleft palate in the mouse-embryonic and maternal effects. Genetics 84: 743–754
Birnbaum LS (1991) Developmental toxicity of TCDD and related compounds: species sensitivities and differences. In: Gallo MA, Scheuplein RJ, Van der Heijden KA (eds) Biological basis for risk assessment of dioxins and related compounds. Cold Spring Harbor Laboratory Press, (Cold Spring Harbor) pp 51–67 (Banbury report no 35).
Birnbaum LS (1994a) The mechanism of dioxin toxicity: relationship to risk assessment. Environ Health Perspect 102: 157–167

Birnbaum LS (1994b) Evidence for the role of the Ah receptor in responses to dioxin. Prog Clin Biol Res 387: 139–154

Birnbaum LS, Harris MW, Miller CP, Pratt RM, Lamb JC (1986) Synergistic interaction of 2,3,7,8-tetrachlorodibenzo-p-dioxin and hydrocortisone in the induction of cleft palate in mice. Teratology 33: 29–35

Birnbaum LS, Harris MW, Stocking L, Clark AM, Morrissey RE (1989) Retinoic acid and 2,3,7,8-tetrachlorodibenzo-p-dioxin (TCDD) selectively enhance teratogenesis in C57BL/6N mice. Toxicol Appl Pharmacol 98: 487–500

Bjeldanes LF, Kim J-Y, Grose KR, Bartholomew JC, Bradfield CA (1991) Aromatic hydrocarbon responsiveness-receptor agonists generated from indole-3-carbinol in vitro and in vivo: comparisons with 2,3,7,8-tetrachlorodibenzo-p-dioxin. Proc Natl Acad Sci USA 88: 9543–9547

Bjerke DL, Peterson RE (1994) Reproductive toxicity of 2,3,7,8-tetrachlorodibenzo-p-dioxin in male rats: different effects of in utero versus lactational exposure. Toxicol Appl Pharmacol 127: 241–249

Bjerke DL, Sommer RJ, Moore RW, Peterson RE (1994a) Effects of in utero and lactational 2,3,7,8-tetrachlorodibenzo-p-dioxin exposure on responsiveness of the male rat reproductive system to testosterone stimulation in adulthood. Toxicol Appl Pharmacol 127: 250–257

Bjerke DL, Brown TJ, MacLusky NJ, Hochberg RB, Peterson RE (1994b) Partial demasculinization and feminization of sex behavior in male rats by in utero and lactational exposure to 2,3,7,8-tetrachlorodibenzo-p-dioxin is not associated with alterations in estrogen receptor binding or volumes of sexually differentiated brain nuclei. Toxicol Appl Pharmacol 127: 258–267

Blaylock BL, Holladay SD, Comment CE, Heindel JJ, Luster MI (1992) Exposure to tetrachlorodibenzo-p-dioxin (TCDD) alters fetal thymocyte maturation. Toxicol Appl Pharmacol 112: 207–213

Burbach KM, Poland A, Bradfield CA (1992) Cloning of the Ah-receptor cDNA reveals a distinctive ligand-activated transcription factor. Proc Natl Acad Sci USA 89: 8185–8189

Burnstein KL, Bellingham DL, Jewell CM, Powell-Oliver FE, Cidlowski JA (1991) Autoregulation of glucocorticoid receptor gene expression. Steroids 56: 52–58

Choi EJ, Toscano DG, Ryan JA, Reidel N, Toscano WA (1991) Dioxin induces transforming growth factor-α in human keratinocytes. J Biol Chem 266: 9591–9597

Courtney KD, Moore JA (1971) Teratology studies with 2,4,5-trichlorophenoxy-acetic acid and 2,3,7,8-tetrachlorodibenzo-p-dioxin. Toxicol Appl Pharmacol 20: 396–403

Couture LA, Abbott BD, Birnbaum LS (1990a) A critical review of the developmental toxicity and teratogenicity of 2,3,7,8-tetrachlorodibenzo-p-dioxin: recent advances toward understanding the mechanism. Teratology 42: 619–627

Couture LA, Harris MW, Birnbaum LS (1990b) Characterization of the peak period of sensitivity for the induction of hydronephrosis in C57BL/6N mice following exposure to 2,3,7,8-tetrachlorodibenzo-p-dioxin. Fundam Appl Toxicol 15: 142–150

Couture-Haws LA, Harris MW, Lockhart AC, Birnbaum LS (1991) Evaluation of the persistence of hydronephrosis induced in mice following in utero and/or lactational exposure to 2,3,7,8-tetrachlorodibenzo-p-dioxin (TCDD). Toxicol Appl Pharmacol 107: 402–412

Cuthill S, Poellinger L, Gustafasson J-A (1987) The receptor for 2,3,7,8-tetrachlorodibenzo-p-dioxin in the mouse hepatoma cell line Hepa lclc7. A comparison with the glucocorticoid receptor and the mouse and rat hepatic dioxin receptors. J Biol Chem 262: 3477-3481

de Morais SMF, Giannone JV, Okey AB (1994) Photoaffinity labeling of the Ah receptor with 3-[^{3}H]methylcholanthrene and formation of a 165-kDa complex between the ligand-binding subunit and a novel cytosolic protein. J Biol Chem 269: 12129–12136

Dencker L, Pratt RM (1981) Association between the presence of the Ah receptor in embryonic murine tissues and sensitivity to TCDD-induced cleft palate. Teratog Carcinog Mutagen 1: 399–406

Denis M, Cuthill S, Wikstrom A-C, Poellinger L, Gustafsson J-Å (1988) Association of the dioxin receptor with the M_r 90,000 heat shock protein: a structural kinship with the glucocorticoid receptor. Biochem Biophys Res Commun 155: 801–807

Denison MS, Vella LM, Okey AB (1986) Structure and function of the Ah receptor for 2,3,7,8-tetrachlorodibenzo-p-dioxin: species difference in molecular properties of the receptors from mouse and rat hepatic cytosols. J Biol Chem 261: 3987–3995

Denison MS, Fisher JM, Whitlock JP (1988) The DNA recognition site for the dioxin-Ah receptor complex. Nucleotide sequence and functional analysis. J Biol Chem 263: 17221–17224

Diewert VM, Pratt RM (1981) Cortisone-induced cleft palate in the A/J mouse: failure of palatal shelf contact. Teratology 24: 149–162

Dolwick KM, Schmidt JV, Carver LA, Swanson HI, Bradfield CA (1993) Cloning and expression of a human Ah receptor cDNA. Mol Pharmacol 44: 911–917

Dong Y, Poellinger L, Gustafsson J, Okret S (1988) Regulation of glucocorticoid receptor expression: evidence for transcriptional and posttranslational mechanisms. Mol Endocrinology 2: 1256–1264

Exner J (1987) Perspective on hazardous waste problems related to dioxins: In: Exner J (ed) Solving hazardous waste problems. American Chemical Society, Washington, pp 1–19

Fine JS, Gasiewicz TA, Silverstone AE (1989) Lymphocyte stem cell alterations following perinatal exposure to 2,3,7,8-tetrachlorodibenzo-p-dioxin. Mol Pharmacol 35: 18–25

Fitchett JE, Hay ED (1989) Medial edge epithelium transforms to mesenchyme after embryonic palatal shelves fuse. Dev Biol 131: 455–474

Fujisawa-Sehara A, Sogawa M, Yamane M, Fujii-Kuriyama Y (1987) Characterization of xenobiotic responsive elements upstream from the drug-metabolizing cytochrome P-450c gene: a similarity to glucocorticoid regulatory elements. Nucleic Acid Res 15: 4179–4191

Gaido KW, Maness SC (1994) Regulation of gene expression and acceleration of differentiation in human keratinocytes by 2,3,7,8-tetrachlorodibenzo-p-dioxin. Toxicol Appl Pharmacol 127: 199–208

Gaido KW, Maness SC, Leonard LS, Greenlee WF (1992) 2,3,7,8-tetrachlorodibenzo-p-dioxin-dependent regulation of transforming growth factors-α and -β_2 expression in a human keratinocyte cell line involves both transcriptional and post-transcriptional control. J Biol Chem 267: 24591–24595

Geelen JAG (1979) Hypervitaminosis A induced teratogenesis. CRC Crit Rev Toxicol 6: 351–376

Gehring U (1993) The structure of glucocorticoid receptors. J Steroid Biochem Mol Biol 45: 183–190

Goldman AS (1984) Biochemical mechanism of glucocorticoid- and phenytoin-induced cleft palate. In: Zimmerman EF (ed) Current topics in developmental biology, (vol) 19. Academic, New York, pp 217–239

Gray LE, Ostby JS (1995) TCDD alters morphological but not behavioral sex differentiation in female rats. Toxicol Appl Pharmacol 133: 285–294

Gray LE Jr, Kelce WR, Monosson E, Ostby JS, Birnbaum LS (1995) Exposure to TCDD during development permanently alters reproductive function in male Long Evans rats and hamsters: reduced ejaculated and epididymal sperm numbers and sex accessory gland weights in offspring with normal androgenic status. Toxicol Appl Pharmacol 131: 108–118

Greene RM, Pratt RM (1976) Developmental aspects of secondary palate formation. J Embryol Exp Morph 36: 225–245

Hahn ME, Poland A, Glover E, Stegeman JJ (1994) Photoaffinity labeling of the Ah receptor – phylogenetic survey of diverse vertebrate and invertebrate species. Arch Biochem Biophys 310: 218–228

Hankinson O, Brooks BA, Weir-Brown KI, Hoffman EC, Johnson BS, Nanthur J, Reyes H, Watson AJ (1991) Genetic and molecular analysis of the Ah receptor and of CYP1A1 gene expression. Biochimie (Paris) 73: 61–66

Henry ED, Rucci G, Gasiewicz TA (1989) Characterization of multiple forms of the Ah receptor: comparison of species and tissues. Biochemistry 28: 6430–6440

Hoffman EC, Reyes H, Chu F, Sander F, Conley LH, Brooks BA, Hankinson O (1991) Cloning of a factor required for activity of the Ah (dioxin) receptor. Science 252: 954–958

Hoffman RE, Stehr-Green PA (1989) Localized contamination with 2,3,7,8-tetrachlorodibenzo-p-dioxin: the Missouri episode. In: Kinbrough K, Jensen A (eds.) Halogenated biphenysl, terphenyls, naphthalenes, dibenzodioxins and related products. Elsevier Science, New York, pp 471–484

Holladay SD, Luster MI (1994) Developmental immunotoxicology. In: Kimmel CA, Buelke-Sam J (eds) Developmental toxicology. Raven, New York, pp 93–118

Holladay SD, Lindstrom P, Blaylock BL, Comment CE, Germolec DR, Heindel JJ, Luster MI (1991) Perinatal thymocyte antigen expression and postnatal immune development altered by gestational exposure to tetrachlorodibenzo-p-dioxin (TCDD). Teratology 44: 385–393

Hsu ST, Ma CI, Hsu SKH, Wu SS, Hsu NHM, Yeh CC, Wu SB (1985) Discovery and epidemiology of PCB poisoning in Taiwan: a four-year followup. Environ Health Perspect 59: 5–10

Jelinek R, Dostal M (1975) Inhibitory effect of corticoids on the proliferative pattern in mouse palatal processes. Teratology 11: 193–198

Jellinck PH, Forkert PG, Riddick DS, Okey AB, Michnovicz JJ, Bradlow HL (1993) Ah receptor binding properties of indole carbinols and induction of hepatic estradiol hydroxylation. Biochem Pharmacol 45: 1129–1136

Kalinyak JE, Dorin RI, Hoffman AR, Perlman AJ (1987) Tissue-specific regulation of glucocorticoid receptor mRNA by dexamethasone. J Biol Chem 262: 10441–10444

Kim CS, Lauder JM, Joh TH, Pratt RM (1984) Immunocytochemical localization of glucocorticoid receptors in midgestation murine embryos and human embryonic cultured cells. J Histochem Cytochem 32: 1234–1237

Kochhar DM (1973) Limb development in mouse embryos. I. Analysis of teratogenic effects of retinoic acid. Teratology 7: 289–298

Kochhar DM, Penner JD, Tellone CI (1984) Comparative teratogenic activities of two retinoids: effects on palate and limb development. Teratogen Carcinogen Mutagen 4: 377–387

Kumaki K, Jensen NM, Shire JGM, Nebert DW (1977) Genetic differences in induction of cytosol reduced-NAD(P): menadione oxidoreductase and microsomal aryl hydrocarbon hydroxylase in the mouse. J Biol Chem 252: 157–165

Kunstadter P (1982) A study of herbicides and birth defects in the republic of Vietnam. An analysis of hospital records. National Academy, Washington D C

Lambert GH, Nebert DW (1977) Genetically mediated induction of drug metabolizing enzymes associated with congenital defects in the mouse. Teratology 16: 147–153

Lu Y, Wang X, Safe S (1994) Interaction of 2,3,7,8-tetrachlorodibenzo-p-dioxin and retinoic acid in MCF-7 human breast cancer cells. Toxicol Appl Pharmacol 127: 1-8

Luster MI, Boorman GA, Dean JH, Harris MW, Luebke RW, Padarathsingh ML, Moore JA (1980) Examination of bone marrow, immunologic parameters and host susceptibility following pre-and postnatal exposure to 2,3,7,8-tetrachlorodibenzo-p-dioxin (TCDD). Int J Immunopharamacol 2: 301–310

Mably TA, Moore RW, Peterson RE (1992a) In utero and lactational exposure of male rats to 2,3,7,8-tetrachlorodibenzo-p-dioxin. 1. Effects on androgenic status. Toxicol Appl Pharmacol 114: 97–107

Mably TA, Moore RW, Goy RW, Peterson RE (1992b) In utero and lactational exposure of male rats to 2,3,7,8-tetrachlorodibenzo-p-dioxin. 2. Effects on sexual behavior and the regulation of luteinizing hormone secretion in adulthood. Toxicol Appl Pharmacol 114: 108–117

Mably TA, Bjerke DL, Moore RW, Gendron-Fitzpatrick A, Peterson RE (1992c) In utero and lactational exposure of male rats to 2,3,7,8-tetrachlorodibenzo-p-dioxin. 3. Effects on spermatogenesis and reproductive capability. Toxicol Appl Pharmacol 114: 118–126

Mason GGF, Witte A-M, Whitelaw ML, Antonsson C, McGuire J, Wilhelmsson A, Poellinger L, Gustafsson J-Å (1994) Purification of the DNA binding form of dioxin receptor. J Biol Chem 269: 4438–4449

Mastroiacovo P, Spagnolo A, Marin E, Meazza L, Bertollini R, Segni G (1988) Birth defects in the Seveso area after TCDD contamination. JAMA 259: 1668–1672

McIntyre WR, Samuels HH (1985) Triamcinolone acetonide regulates glucocorticoid-receptor levels by decreasing the half-life of the activated nuclear-receptor form. J Biol Chem 260: 418–427

Moore JA (1973) Characterization and interpretation of kidney anomalies associated with 2,3,7,8-tetrachlorodibenzo-p-dioxin (TCDD). Teratology 7: 24A

Muller M, Renkawitz R (1991) The glucocorticoid receptor. Biochem Biophys Acta 1088: 171–182

Nanda R, van der Linden RPGM, Jansen HWB (1970) Production of cleft palate with dexamethasone and hypervitaminosis A in rat embryos. Cleft Palate J 15: 176–181

Nebert DW, Puga A, Vasiliou V (1993) Role of the Ah receptor and the dioxin-inducible [Ah] gene battery in toxicity, cancer, and signal transduction. Ann N Y Acad Sci 685: 624–640

Neubert D, Dillman I (1972) Embryotoxic effects in mice treated with 2,4,5-trichlorophenoxyacetic acid and 2,3,7,8-tetrachlorodibenzo-p-dioxin. Arch Pharmacol 272: 243–264

Neubert D, Zens P, Rothenwallner A, Merker HJ (1973) A survey of the embryotoxic effects of TCDD in mammalian species. Environ Health Perspect 5: 67–79

Okey AB, Vella LM, Harper PA (1989) Detection and characterization of a low affinity form of cytosolic Ah receptor in livers of mice nonresponsive to induction of cytochrome P_1-450 by 3-methylcholanthrene. Mol Pharmacol 35: 823–830

Okey AB, Riddick DS, Harper PA (1994) Molecular biology of the aromatic hydrocarbon (dioxin) receptor. Trends Pharmacol Sci 15: 226–232

Okret S, Poellinger L, Dong Y, Gustaffson J-A (1986) Down-regulation of glucocorticoid receptor mRNA by glucocorticoid hormones and recognition by the receptor of a specific binding sequence within a receptor cDNA clone. Proc Natl Acad Sci USA 83: 5899–5903

Olson JR, Holscher MA, Neal RA (1980) Toxicity of 2,3,7,8-tetrachlorodibenzo-p-dioxin in the Golden Syrian hamster. Toxicol Appl Pharmacol 55: 67–78

Osborne R, Cook JC, Dold KM, Ross L, Gaido K, Greenlee WF (1988) TCDD receptor: Mechanisms of altered growth regulation in normal and transformed human ketatinocytes. In: Langenbach R (ed), Tumor promoters: biological approaches for mechanistic studies and assay systems. Raven New, York, pp 407–416

Perdew GH (1988) Association of the Ah receptor with the 90-kDa heat shock protein. J Biol Chem 263: 13802–13805

Perdew GH (1992) Chemical cross-linking of the cytosolic and nuclear forms of the ah receptor in hepatoma cell line 1c1c7. Biochem Biophys Res Commun 182: 55-62

Perdew GH, Poland A (1988) Purification of the Ah receptor from C57BL/6J mouse liver. J Biol Chem 263: 9848–9852

Peterson RE, Gray LE Jr (1995) Developmental male and female reproductive toxicity of 2,3,7,8-tetrachlorodibenzo-p-dioxin in rats and hamsters. Eur J Pharmacol (Environ Toxicol Pharmacol) 293: 1–40

Peterson RE, Theobald HM, Kimmel GL (1993) Developmental and reproductive toxicity of dioxins and related compounds: cross-species comparisons. Crit Rev Toxicol 23: 283–335

Pinsky L, DiGeorge AM (1965) Cleft palate in the mouse: a teratogenic index of glucocorticoid potency. Science 147: 402–403

Poland A, Glover E (1980) 2,3,7,8-Tetrachlorodibenzo-p-dioxin: segregation of toxicity with the Ah locus. Mol Pharmacol 17: 86–94
Poland A, Glover E (1990) Characterization and strain distribution pattern of the murine Ah receptor specified by the Ah^d and Ah^{b-3} alleles. Mol Pharmacol 38: 306–312
Poland A, Knutson JC (1982) 2,3,7,8-Tetrachlorodibenzo-p-dioxin and related halogenated aromatic hydrocarbons: examination of the mechanism of toxicity. Annu Rev Pharmacol Toxicol 22: 517–554
Poland A, Glover I, Kende AS (1976) Stereospecific, high affinity binding of 2,3,7,8-tetrachlorodibenzo-p-dioxin by hepatic cytosol: evidence that the binding species is receptor for induction of arylhydrocarbon hydrozylase. J Biol Chem 251: 4936–4940
Poland A, Glover E, Bradfield CA (1991) Characterization of polyclonal antibodies to the Ah receptor prepared by immunization with a synthetic peptide hapten. Mol Pharmacol 39: 20–26 (erratum in Mol Pharmacol 1991, 39: 435)
Pollenz RS, Sattler CA, Poland A (1994) The Aryl hydrocarbon receptor and aryl hydrocarbon receptor nuclear translocator protein show distinct subcellular localizations in Hepa lclc7 cells by immunofluorescence. Mol Pharmacol 45: 428–438
Pongratz I, Mason GGF, Poellinger L (1992) Dual roles of the 90-kDa heat shock protein hsp90 in modulating functional activities of the dioxin receptor. J Biol Chem 267: 13728–13734
Pratt RM (1985) Receptor-dependent mechanisms of glucocorticoid and dioxin-induced cleft palate. Environ Health Perspect 61: 35–40
Pratt RM, Hassell JR (1975) Appearance and distribution of carbohydrate-rich macromolecules on the epithelial surface of the developing rat palatal shelf. Dev Biol 45: 192–198
Pratt RM, Martin GM (1975) Epithelial cell death and cyclic AMP increase during palatal development. Proc Natl Acad Sci USA 72; 874–877
Pratt RM, Salomon DS (1981) Biochemical basis for the teratogenic effects of glucocorticoids. In: Juchau MR (ed) Biochemical basis of chemical teratogenesis. Elsevier/North Holland, New York, pp 179–193
Pratt RM, Dencker L, Diewert VM (1984) 2,3,7,8-Tetrachlorodibenzo-p-dioxin-induced cleft palate in the mouse: evidence for alterations in palatal shelf fusion. Teratogen Carcinogen Mutagen 4: 427–436
Probst MR, Reisz-Porszasz S, Agbunag RV, Ong MS, Hankinson O (1993) Role of the aryl hydrocarbon receptor nuclear translocator protein in aryl hydrocarbon (dioxin) receptor action. Mol Pharmacol 44: 511–518
Reggiani GM (1989) The Seveso accident: Medical survey of a TCDD exposure. In: Kinbrough R, Jensen A (eds) Halogenated biphenyls, terphenyls, naphthalenes, dibbenzodioxins and related products. Elsevier Science, New York, pp 445–470
Reisz-Porszasz S, Probst MR, Fukunaga BN, Hankinson O (1994) Identification of functional domains of the arylhydrocarbon receptor nuclear translocator protein (ARNT). Mol Cell Biol 14: 6075–6086
Reyes H, Reisz-Porszasz S, Hankinson O (1992) Identification of the Ah receptor nuclear translocator protein (Arnt) as a component of the DNA binding form of the Ah receptor. Science 256: 1193–1195
Rogan WJ, Gladen BC, Hung KL, Koong SL, Shih LY, Taylor JS, Wu YC, Yang D, Ragan NB, Hsu CC (1988) Congenital poisoning by polychlorinated biphenyls and their contaminants in Taiwan. Science 241: 334–336
Rosewicz S, McDonald AR, Maddux BA, Goldfine ID, Miesfeld RL, Logsdon CD (1988) Mechanism of glucocorticoid receptor down-regulation by glucocorticoids. J Biol Chem 263: 2581–2584
Salomon DS, Pratt RM (1976) Glucocorticoid receptors in murine embryonic facial mesenchyme cells. Nature 264: 174–177
Salomon DS, Pratt RM (1979) Involvement of glucocorticoids in the development of the secondary palate. Differentiation 13: 141–154

Salomon DS, Zubair Y, Thompson EB (1978) Ontogeny and biochemical properties of glucocorticoid receptors in mid-gestation mouse embryos. J Steroid Biochem 9: 95–107
Saxen I (1973) Effects of hydrocortisone on the development in vitro of the secondary palate in two inbred strains of mice. Arch Oral Biol 18: 1469–1479
Schantz SL, Bowman RE (1989) Learning in monkeys exposed perinatally to 2,3,7,8-tetrachlorodibenzo-p-dioxin (TCDD). Neurotox Teratol 11: 13–19
Schantz SL, Barsotti DA, Allen JR (1979) Toxcological effects produced in nonhuman primates chronically exposed to fifty parts per trillion 2,3,7,8-tetrachlorodibenzo-p-dioxin (TCDD). Toxicol Appl Pharmacol 48: 180
Schmidt JV, Carver LA, Bradfield CA (1993) Molecular characterization of the murine Ahr gene. Organization, promoter analysis, and chromosomal assignment. J Biol Chem 268: 22203–22209
Schwetz BA, Norris JM, Sparschu GL, Rowe VK, Gehring PJ, Emerson JL, Gerbig CG (1973) Toxicology of chlorinated dibenzo-p-dioxins. Environ Health Perspect 5: 87–99
Shah RM, Kilistoff A (1976) Cleft palate induction in hamster fetuses by glucocorticoid hormones and their synthetic analogues. J Embryol Exp Morphol 36: 101–108
Shuler CT, Halpern DE, Guo Y, Sank AC (1992) Medial edge epithelium fate traced by cell lineage analysis during epithelial-mesenchymal transformation in vivo. Dev Biol 154: 318–330
Silkworth JB, Cutler DS, Antrim L, Houston D, Tumasonis C, Kaminsky LS (1989) Teratology of 2,3,7,8-tetrachlorodibenzo-p-dioxin in a complex environmental mixture from the Love Canal. Fundam Appl Toxicol 13: 1–15
Sugimoto M, Kojima A, Endo H (1976) Role of glucocorticoids in terminal differentiation and tissue-specific function (a review). Dev Growth Differ 18: 319–327
Sutter TR, Greenlee WF (1992) Classification of members of the Ah gene battery. Chemosphere 25: 223–226
Sutter TR, Guzman K, Dold KM, Greenlee WF (1991) Targets for dioxin: genes for plasminogen activator inhibitor-2 and interleukin-1β. Science 254: 415–418
Swanson HI, Bradfield CA (1993) The AH- receptor: genetics, structure and function. Pharmacogenetics 3: 213–230
Swanson HI, Perdew GP (1991) Detection of the Ah receptor in rainbow trout: use of 2-azido-3-[^{125}I] iodo-7,8-dibromodibenzo-p-dioxin in cell culture. Toxicol Lett 58: 85–95
Thomas PT, Hinsdill RD (1979) The effect of perinatal exposure to tetrachlorodibenzo-p-dioxin on the immune response of young mice. Drug Chem Toxicol 2: 77–98
Vanden Heuvel JP, Clark GC, Kohn MC, Tritscher AM, Greenlee WF, Lucier GW, Bell DA (1994) Dioxin-responsive genes: examination of dose-response relationships using quantitative reverse transcriptase-polymerase chain reaction. Cancer Res 54: 62–68
Vedeckis WV, Ali M, Allen HR (1989) Regulation of glucocorticoid receptor protein and mRNA levels. Cancer Res [Suppl] 49: 2295s–2302s
Vekemans M (1982) The genetical difference between the DBA/2 and the C57BL/6 strains in cortisone-induced cleft palate. Can J Genet Cytol 24: 797-805
Vos JG, Moore JA (1974) Suppression of cellular immunity in rats and mice by maternal treatment with 2,3,7,8-tetrachlorodibenzo-p-dioxin. Int Arch Allergy 47: 777–794
Walker BE (1967) Induction of cleft palate in rabbits by several glucocorticoids. Proc Soc Exp Biol Med 125: 1281–1284
Whitelaw M, Pongratz I, Wilhelmsson A, Gustafsson J-Å, Poellinger L (1993a) Ligand-dependent recruitment of the ARNT coregulator determines DNA recognition by the dioxin receptor. Mol Cell Biol 13: 2504–2514
Whitelaw ML, Göttlicher M, Gustafsson J-Å, Poellinger L (1993b) Definition of a novel ligand binding domain of a nuclear bHLH receptor: co-localization of ligand and hsp90 binding activities within the regulable inactivation domain of the dioxin receptor. EMBO J 12: 4169–4179

Whitlock JP (1987) The regulation of gene expression by 2,3,7,8-tetrachlorodibenzo-p-dioxin. Pharmacol Rev 39: 147–161

Whitlock JP (1990) Genetic and molecular aspects of 2,3,7,8-tetrachlorodibenzao-p-dioxin action. Annu Rev Pharmacol Toxicol 30: 251–277

Whitlock JP (1993) Mechanistic aspects of dioxin action. Chem Res Toxicol 6: 754–763

CHAPTER 31

Endocrine Disruptors: Effects on Sex Steroid Hormone Receptors and Sex Development*

W.R. KELCE and L. EARL GRAY, Jr.

A. Introduction

Steroid hormone receptors control fundamental events in embryonic development and sex differentiation through their function as ligand-inducible transcription factors that either activate or repress transcription of target genes. The consequences of disrupting these processes can be especially profound during development due to the crucial role hormones play in controlling transient and irreversible developmental processes. For example, without androgen or the androgen receptor (AR; e.g., androgen insensitivity syndrome), the phenotype of the human fetus is female irrespective of genetic sex (FRENCH et al. 1990). In addition to genetic anomalies such as androgen insensitivity, drugs and environmental chemicals are capable of disrupting the action of functional steroid hormone receptors in fish, wildlife, and humans. Laboratory studies have confirmed many abnormalities of reproductive development observed in the field and have, in some cases, provided mechanisms to explain the effects in wildlife and humans. In males, exposure to environmental estrogens in utero may be responsible for the reported decline in human sperm counts (i.e., 50% over the last 50 years) and the apparent increase in the incidence of cryptorchid testes, testicular cancer, and hypospadias (CARLSEN et al. 1992; GIWERCMAN et al. 1993; SHARPE and SKAKKEBAEK 1993; AUGER et al. 1995). In females, exposure to estrogenic chemicals during development may contribute to earlier age at puberty and the increased incidence of endometriosis and breast cancer (DAVIS et al. 1993). Concerns regarding the effects of pesticides during development led the National Academy of Sciences in 1993 to release the report "Pesticides in the Diets of Infants and Children", suggesting that the young are a special concern with respect to pesticide exposures (NRC 1993). Scientists currently are grappling with research needs in this new area, and new test protocols to screen for endocrine effects are necessary.

---

*This manuscript has been reviewed in accordance with the policy of the National Health and Environmental Effects Research Laboratory, U.S. Environmental Protection Agency, and approved for publication. Approval does not signify that the contents necessarily reflect the views and policies of the Agency, nor does mention of trade names or commercial products constitute endorsement or recommendation for use.

While diethylstilbestrol (DES) exposure serves as a hallmark example of how perinatal exposure to a chemical can seriously alter human reproductive development, current discussion of "endocrine disruptors" continues to myopically focus on drugs and environmental chemicals exhibiting estrogenic activity; little consideration has been given to other endocrine mechanisms that may be of equal or even greater concern. A great deal of misinformation regarding endocrine disruptors has been communicated in the popular press and, unfortunately, in scientific journals as well. Examples of this misinformation include the following: (a) nonestrogenic chemicals, e.g., *p,p′*- dichlorodiphenyl-trichloroethane (*p,p′*-DDE) are repeatedly reported to be estrogenic; (b) there is little appreciation that many endocrine disruptors, i.e., 2,3,7,8-tetrachlorodibenzo-*p*-dioxin (TCDD), *p,p′*-DDE, vinclozolin metabolite M2) are extremely potent reproductive toxicants; (c) wildlife data is often dismissed as correlative, ignoring examples of clear-cut, cause and effect relationships between chemical exposure and reproductive alterations, e.g., DDT metabolite effects in birds, polychlorinated biphenyl (PCB) effects in fish and environmental estrogen effects in domestic animals; (d) there is little recognition that subtle, low-dose reproductive effects observed in the laboratory will be all but impossible to detect in typical epidemiological studies due to the fact that human reproductive function is highly variable (e.g., time to fertility, fecundity, and sperm measures) and that significant delays occur between developmental exposure and patient presentation with reproductive abnormalities/problems; and (e) there is little appreciation for the complexity of endocrine homeostasis, often leading to naive conclusions such as the recent suggestion that the presence of environmental estrogens will "cancel out" the effects environmental antiestrogens.

In this chapter, we attempt to debug some of this misinformation and provide a thorough, but by no means exhaustive, discussion of drug and environmental chemicals that alter reproductive development via agonist or antagonist effects on sex steroid hormone receptors, i.e., AR, estrogen receptors (ER), and progesterone receptors (PR). A brief discussion of normal sex development and sex steroid hormone receptor structure and the function and mechanisms of antagonist action is followed by a discussion of selected endocrine disruptors for the AR, ER, and PR, respectively. For each receptor, the biochemical and molecular mechanisms by which specific agonists and antagonists act will be discussed, along with the consequences of exposure to these chemicals during critical developmental stages of life. We include selected examples of steroid hormone-, drug-, and/or environmental chemical-induced alterations in reproductive development from human, laboratory, and wildlife studies and, where possible, compare the potency of these chemicals (Tables 1, 2). The reader is referred to a recent review (SCHARDEIN 1993) regarding the ability of many of the chemicals discussed in this chapter to induce teratogenicity in rodents and humans. Due to space limitations, review articles and selected manuscripts have been used as references wherever possible; no attempt has been made to exhaustively cite the original literature.

Table 1. Relative binding affinites of steroid and pharmaceutical ligands for the estrogen, androgen, and progesterone receptors

Estrogen receptor		Androgen receptor		Progesterone receptor	
Chemical name	RBA	Chemical name	RBA	Chemical name	RBA
17β-Estradiol[a]	100	Testosterone(T)[a]	100	Progesterone[a]	100
2-Hydroxy-estradiol[b]	25	17β-Estradiol[a]	25	17β-Estradiol[a]	0.1
2-Methyl-estradiol[a]	41	5α-Dihydrotestosterone[a]	120	19-Nortesosterone[k]	8.9
4-Hydroxy-estradiol[b]	42	Aldosterone	< 0.1	5α-Dihydrotestosterone[a]	3.3
4-Hydroxy-tamoxifen[c]	250	Casodex[f]	1.8	7α, 17α-Dimethyl--19-nor-T[c]	148
5α-Dihydrotestosterone[a]	< 0.1	Cortisol/corticosterone[a]	< 0.1	Aldosterone[a]	1.1
Aldosterone[a]	< 0.1	Cyproterone Acetate[g]	9.4	Chlormadinone Acetate[a]	321
CI-628[c]	5	Demegestone/RU 2453[a]	1.1	Corticosterone[a]	5.1
CI-680[c]	34	Dexamethasone[a]	< 0.1	Cortisol[a]	< 0.1
Cortisol / corticosterone[a]	< 0.1	Diethylstilbestrol[d]	< 0.1	Demegestone / RU 2453[a]	420
Dexamethasone[a]	< 0.1	Flutamide[g]	< 0.1	Dexamethasone[a]	0.4
Diethylstilbestrol[d]	250	Gestrinone / RU 2323[a]	83	Gestodene[l]	921
Estriol[a]	15	Hydroxyflutamide[g]	2.0	Gestrinone / RU 2323[a]	48
ICI 164, 384[c]	19	Megestrol acetate[a]	19	Levonorgestrel[l]	541
ICI 182, 780[e]	90	MPA[a]	52	Megestrol Acetate[a]	118
Keoxifen[e]	118	Nilutamide[h]	2.0	MPA[a]	306
LY 139481[e]	143	Norethidrone[a]	44	Norethidrone acetate[a]	64
LT 177018[e]	200	Norgestrel[a]	87	Norethidrone[a]	263
MER-25[e]	0.06	Progesterone[a]	6	Norgestimate[l]	124
Moxestrol[a]	12	Promegestone / RU 5020[a]	1.4	Norgestrel[a]	907
Nafoxidine[c]	15	R1881[a]	203	Promegestone / RU 5020[a]	533
Progesterone[a]	< 0.1	RU 2010[a]	71	R1881[a]	208
R1881[a]	< 0.1	RU 58841[i]	55	RU 2010[a]	47
RU 26988[a]	< 0.1	RU 59063[f]	300	RU 26988[a]	0.3
Tamoxifen[c]	5	RU 56187[f]	92	RU 38486[j]	530
Toremifene[e]	1	RU 57073[f]	163	Testosterone[e]	1.6
Triamcinolone acetonide[a]	< 0.1	RU 38486[j]	17	Triamcinolone acetonide[a]	15
Trioxifene[e]	20	Triamcinolone Acetonide[a]	< 0.1	–	–

RBA, relative binding affinity; MPA, medroxyprogesterone acetate.
[a]RAYNAUD et al. 1980.
[b]SCHUTZE et al. 1994.
[c]PASQUALINI et al. 1988.
[d]KELCE et al. 1995.
[e]CHANDER et al. 1993.
[f]TEUTSCH et al. 1994
[g]WAKELING et al. 1981.
[h]HARRIS et al. 1993.
[i]BATTMANN et al. 1994.
[j]VAN LOOK and BYGDEMAN 1989.
[k]JÄNNE and BARDIN 1984.
[l]PHILLIPS et al. 1992.

Table 2. Relative binding affinity (RBA) of select environmental endocrine disrupting chemicals to estrogen (ER), androgen (AR), and progesterone (PR) receptors

Chemical	RBA to ER	RBA to AR	RBA to PR
Atrazine[a,b]	0	ND	ND
Chlordecone[a,b]	++	+	++
Coumesterol[c]	+++	0 / +	ND
Endosulfan[a,b]	0	ND	ND
Genistein[c]	+++	0	ND
HPTE[b]	++	++	++
Lindane[a,b]	0	ND	ND
Methoxychlor[b]	0	0	0 / +
Nonylphenol[a,b]	++	++	ND
o,p′-DDT[a,b]	++	+	++
Octylphenol[a,b]	++	+	ND
p,p′-DDD[a,b]	0	+	0
p,p′-DDE[a,b]	0	++	0
p,p′-DDT[a,b]	0	+	0
V metabolite M1[a,b]	0	+	0
V metabolite M2[a,b]	0	++	+
Vinclozolin (V)[a,b]	0	0	0
Zearalenone[c]	+++	+	ND

DDT, dichlorodiphenyl-trichloroethane; DDD, dichlorodiphenyl-dichloroethane; DDE, 1,1-dichloro-2,2-dichlorophenylethylene; HPTE, *p*-hydroxyphenyl-trichloroethylene; +++, very active; ++, active; +, weakly active; 0, inactive; ND, no data.
[a]KELCE et al. 1994, 1995.
[b]LAWS et al. 1995.
[c]W.R. KELCE and S.C. LAWS, unpublished data.

Finally, as opinions regarding the best methods to screen for endocrine disruptor activity are varied, ranging from exclusively using in vitro systems to exclusively using in vivo screens, we have compiled, at the end of this chapter, what we feel are the best of each into a comprehensive investigational strategy intended to identify not only the chemical/metabolite of concern, but also to provide information as to the mechanism responsible for the phenotypic effects.

B. Sex Differentiation and Chemical Effects

As even the most severe alterations in sex development are not lethal, abundant information exists regarding the role of sex steroid hormones in mammalian reproductive development (for several excellent reviews, see WILSON 1978 and GEORGE and WILSON 1988). Genetic sex is determined at fertilization, where the presence of a Y chromosome serves to govern the differentiation of the indifferent gonads into testes. Prior to gonadal sex differentiation, the embryo has the potential to develop a male or female phenotype. Following gonadal sex differentiation, testicular secretions (i.e., testosterone and Müllerian inhibiting substance) induce the differentiation of the male internal duct system and external genitalia, resulting in the male phenotype. In the human embryo, the onset of testicular testosterone synthesis occurs approximately 65 days after fertilization and serves to induce the differentiation of the Wolffian duct system into the epididymis, vas deferens, and seminal vesicles; its metabolite, 5α-dihydrotestosterone (DHT), induces the development of the prostate and male external genitalia. It is generally held that, in the absence of these testicular secretions, the female phenotype is expressed independent of the presence of an ovary. However, the ovary does secrete estrogens during development, leading some to speculate that estrogens may play an active role in feminization of the female reproductive tract (see Sect. E.I).

Exposure to hormonally active chemicals during sex differentiation can produce morphological pseudohermaphrodism, meaning that XY males appear phenotypically female and XX females appear phenotypically male (SCHARDEIN 1993; GRAY 1992). Laboratory studies using rodents demonstrate that chemicals known to produce alterations of sex differentiation in humans induce similar alterations in rodents (GRAY 1992). For example, developmental exposure to the potent estrogen DES causes urogenital malformations and cancer in the reproductive tracts of both humans and rodents. In addition, mutations in genes regulating sex development result in profound and unique effects on the sexual phenotype of both humans and laboratory species (POLANI 1981). While the timing of events in sex differentiation certainly differs between humans and laboratory species, the basic mechanisms of sex differentiation are homologous in all mammals. It follows, then, that chemicals affecting reproductive development in rodents and other mammals should be considered potential human reproductive toxicants. Finally, we should not ignore reproductive alterations in avian, reptilian, and amphibian species, as considerable homology also exists with these different classes of vertebrates.

Chemical exposure during sex differentiation is of special concern for several reasons. First, the development of the reproductive system is sensitive to low-dose chemical effects. Second, while the chemical exposure may be transient, the effects of chemical exposure, including future reproductive behavior, are irreversible. Third, functional alterations often are not discovered until after puberty or even later in life, leading to underestimates of chemically induced effects on reproductive development. Lastly, developmental abnormalities cannot be predicted from similar chemical exposures in adult animals, as adult systems are fully differentiated and capable of compensating for chemical insult. Developmental reproductive toxicity data, then, are often critical in the assessment of noncancer health effects of endocrine disrupting chemicals.

While some classes of developmental reproductive toxicants act to inhibit steroid hormone biosynthesis, transport, and/or degradation, the remainder of this chapter will specifically focus on those chemicals whose developmental reproductive effects are mediated through sex steroid hormone receptors.

Under normal conditions, the sex steroid hormone receptor system functions astonishingly well. The naturally occurring sex steroids of the androgen, estrogen, and progestin classes exhibit short half-lives and reduced oral activity and are inactivated in the circulation, all of which safeguard against the possibility of prolonged or inappropriate receptor action. It is with the development of synthetic, orally active steroids with longer half-lives and the production of environmental contaminants exhibiting reduced receptor specificity that the ability to interact inappropriately with multiple steroid hormone receptors emerged (JÄNNE and BARDIN 1984).

C. Steroid Hormone Receptor Structure and Function

Steroid hormone receptors control fundamental events in embryonic development and sex differentiation through their function as ligand-inducible transcription factors that either activate or repress transcription of target genes. Steroid hormone receptors are composed of four functional domains (Fig. 1): (1) a hypervariable amino terminal region involved in the regulation of transcriptional activation and receptor stabilization via slowing the rate of ligand dissociation and receptor degradation; (2) a highly conserved, cysteine-rich DNA-binding domain involved in DNA binding, transcriptional activation, receptor dimerization, and nuclear transport; (3) a hinge region involved in nuclear translocation, dimerization, and transcriptional activation; and (4) a carboxyl-terminal ligand-binding domain involved in hormone binding, transcriptional activation, receptor dimerization, and heat shock protein (HSP) interactions (SIMENTAL et al. 1991; LUKE and COFFEY 1994; LANDERS and SPELSBERG 1991; ZHOU et al. 1995). AR, PR, and the glucocorticoid (GR) and mineralocorticoid (MR) receptors comprise a subclass of the steroid hormone receptor superfamily (Fig.1) based primarily on sequence homology in the DNA-binding domain and the ability to bind similar hormone response elements (HRE). Nuclear receptors in this class activate transcription of steroid hormone-dependent genes using HRE like that found in the mouse mammary tumor virus long terminal repeat (MMTV-LTR); the receptors form homodimers and recognize a half-site consensus TGTTCT sequence arranged as an inverted repeat or closely related sequence (KUPFER et al. 1993). The other major nuclear receptor subclass (Fig.1) consists of ER and the thyroid hormone (THR), retinoic acid (RAR), retinoid X (RXR), and vitamin D (VDR) receptors, based on their ability to bind another HRE consensus sequence; these receptors bind as homodimers or heterodimers (i.e, THR, RAR, RXR, and VDR) to a TGACCT or closely related sequence. The HRE half-site sequences for this receptor subclass are arranged in direct repeats, palindromes, or inverted palindromes (KUPFER et al. 1993; YEN et al. 1995; UMESONO et al. 1991); specificity of hormone action is determined by the spacing between the half-sites (YEN et al. 1995; UMESONO et al. 1991; NÄÄR et al. 1991).

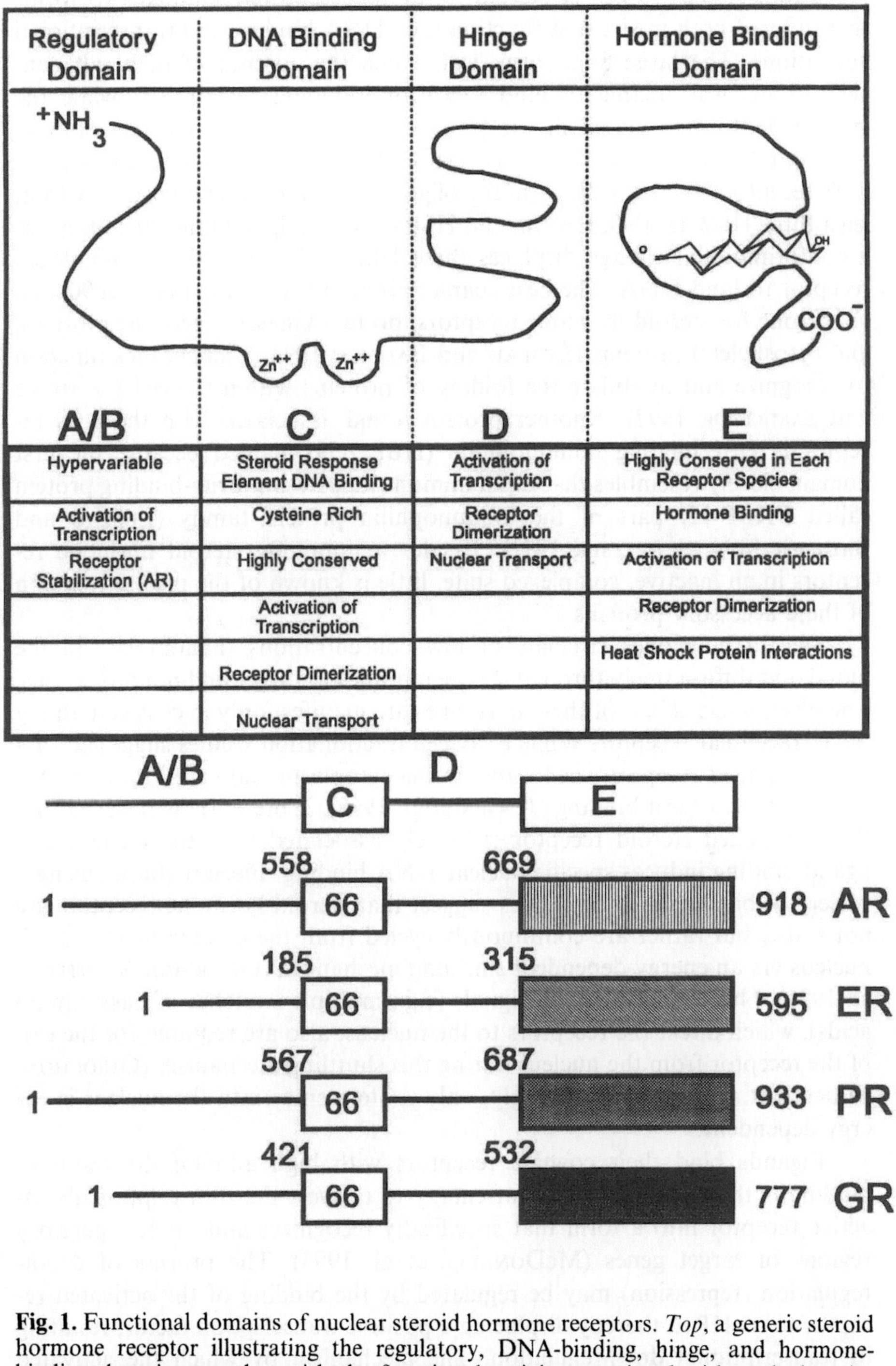

Fig. 1. Functional domains of nuclear steroid hormone receptors. *Top*, a generic steroid hormone receptor illustrating the regulatory, DNA-binding, hinge, and hormone-binding domains together with the functional activity associated with each domain. *Bottom*, comparative peptide and domain lengths of the superfamily of nuclear steroid hormone receptors. Note that the progesterone receptor (*PR*) illustrated is the longer type B splice variant. *AR*, androgen receptor; *ER*, estrogen receptor. (Adapted from LANDERS and SPELSBERG 1991)

Initial studies to purify steroid hormone receptors identified two forms that differed both in size and function (i.e., DNA binding and transcriptional activation). The large 8-9S form, isolated in the absence of high salt concentrations, was unable to bind DNA or initiate transcription, while the smaller 4S form, isolated in the presence of high salt, was active both in binding DNA and activating transcription. It now is known that unactivated 8-9S receptors exist in cells as bulky oligomeric complexes associated with at least three HSP, HSP90, HSP70, and HSP56. When ligand binds the receptor, a conformational change displaces these HSP and allows the uncomplexed receptor to bind DNA. The best characterized of these proteins, HSP90, is a chaperone for steroid hormone receptors, protein kinases, oncogene proteins, and cytoskeletal proteins (LEBEAU and BAULIEU 1994). Chaperones function to recognize and modulate the folding of proteins within the cell (GETHING and SAMBROOK 1992). Another protein found associated with the 8-9S receptor is HSP-binding immunophilin (HBI; p59), named because the first domain closely resembles that of an immunosuppressant drug-binding protein called FKBP-12, part of the immunophilin protein family (LEBEAU and BAULIEU 1994; SCHREIBER 1991). Besides maintaining steroid hormone receptors in an inactive, complexed state, little is known of the precise function of these accessory proteins.

Steroid hormones circulate at low concentrations (nanomolar) in the blood and diffuse freely through cell membranes to target and nontarget cells. The physiologic effects of these hormones are manifest only in cells containing the intracellular receptor. While early cell fractionation studies suggested that the unliganded receptor was located in the cytoplasm and translocated to the nucleus upon ligand binding (JENSEN et al. 1968), it presently is believed that the unliganded steroid receptor is loosely associated with the nucleus and ligand binding induces specific nuclear DNA binding. Elegant studies using a variety of biochemical techniques suggest that steroid hormone receptors are not static, but rather are continuously cycled from the cytoplasm to the cell nucleus via an energy-dependent shuttling mechanism (GUIOCHON-MANTEL et al. 1991). Nuclear localization signals (e.g., multiple stretches of basic amino acids), which direct the receptors to the nucleus, also are required for the exit of the receptor from the nucleus during this shuttling mechanism (GUIOCHON-MANTEL et al. 1994). Interestingly, only protein entry into the nucleus is energy dependent.

Ligands bind their cognate receptors with high affinity; the result of binding is the release of sufficient energy to convert the transcriptionally inactive receptor into a form that specifically recognizes and binds regulatory regions of target genes (McDONNELL et al. 1993). The process of downregulation (repression) may be regulated by the binding of the activated receptor to its HRE, thereby displacing a positive transcription factor, resulting in transcriptional downregulation. The mechanism by which the activated receptor promotes transcription is not understood in detail; however, it appears that the ligand-activated receptor promotes the assembly of the transcription factors on target DNA (KLEIN-HITPASS et al. 1990). These protein–protein interactions are thought to stabilize the transcription factors at the promoter regions of target genes (i.e., the TATA box or equivalent structures) to initiate transcription (O'MALLEY and TSAI 1992).

The mechanism by which steroid hormones specifically regulate gene expression is a paradox, given that the biological functions regulated by steroid receptors in the same class are distinct, yet the receptors recognize the same HRE. At least five mechanisms may explain the specificity of hormone action: (1) a given target cell may express only one type of receptor; (2) a given target cell may metabolically inactivate all but one receptor ligand; (3) regulatory sequences distinct from the HRE may exist and be differentially bound by the receptors; (4) the ligand-bound receptor may interact differently with other transcription factors specific to the target cell (PEARCE and YAMAMOTO 1993); and/or (5) HRE may be structurally or functionally different in tissues at different temporal stages such as during embryogenesis (i.e., tissue differentiation and development), enhancing the susceptibility of developing tissues to different or multiple receptors or to chemical insult.

Steroid hormone receptors are phosphoproteins in the absence of ligand. Upon ligand binding, multiple serine/threonine residues (and tyrosine residues in the ER) located primarily in the N-terminal domain (Fig. 1) become phosphorylated (i.e., the receptor is two to seven times more phosphorylated than in the basal state; BRINKMANN 1994). While it seems likely that phosphorylation has the potential to regulate steroid hormone receptor function, the effects of phosphorylation on receptor function are not yet clear. Receptor functions that have been linked to phosphorylation include HSP interactions, activation of hormone binding, nuclear import, nucleocytoplasmic shuttling, receptor dimerization and subsequent DNA binding, transcription factor interactions, and receptor half-life (BRINKMANN 1994). In fact, phosphorylation itself has been reported to stimulate transcriptional activation induced by steroid hormone receptors in the absence of ligand (PARKER 1993). While most studies of receptor phosphorylation have focused on "activation" functions, it also seems likely that phosphorylation (i.e., induced by environmental or pharmaceutical "antihormones") can induce steroid hormone receptor "inactivation." The role of phosphorylation in subsequent receptor function remains an active area of research and may prove fundamental to endocrine disruptor mechanisms.

D. Mechanisms of Antihormone Action

It has been suggested that steroid agoinsts bind productively to their cognate receptors using all of their binding energy to promote receptor activation. In contrast, steroid antagonists (i.e., antihormones) either bind nonproductively, using much of their binding energy either for the stabilization of the ligand receptor complex or conformational changes that fail to promote transcriptional activation, or, because of their lower affinity for the receptor, they release insufficient energy for the full receptor activation (WAKELING 1992). Once bound to the receptor, antihormones may interfere with transcriptional activation by one or more of the following mechanisms. Antihormones may: (a) competitively inhibit steroid hormone binding; (b) induce inappropriate conformational changes in the receptor protein; (c) alter receptor phosphorylation; (d) retard the dissociation of the heteromeric receptor complex (i.e., associated HSP and additional protein factors); (e) alter the ability of receptors to dimerize; (f) alter nuclear accumulation of the receptor; (g) fail to induce the association of the receptor with its HRE; (h) alter the ability of the receptor to interact with DNA, recruit adaptor proteins to or interact with the general transcription apparatus; or (i) alter the ability of the preinitiation complex to modulate RNA polymerase activity. Due to the complexity of this process, data from ligand-binding studies alone cannot be used to predict in vivo potency.

While all steroid receptor ligands bind to the same region of the receptor, agonists and antagonists appear to interact with different amino acid residues within the receptor ligand-binding domain, resulting in altered receptor conformation. Using limited protease digestion, it has been demonstrated that conformational changes occur in the AR (KUIL and MULDER 1994), ER (GRONEMEYER et al. 1992; MCDONNELL et al. 1994), and PR (ALLAN et al. 1992a,b). The nature of the conformational change depends on whether the receptor binds hormone or antihormone. A single amino acid substitution (threonine-877 to alanine; VELDSCHOLTE et al. 1992a) in the ligand-binding domain of the AR in LNCaP cells alters the specificity of ligand binding such that the presence of low doses of estrogens (HOROSZEWICZ et al. 1983; SCHULZ et al. 1985), progestins (SCHUURMANS et al. 1988), and antiandrogens such as hydoxyflutamide, cyproterone acetate, and nilutamide (KEMPPAINEN et al. 1992; VELDSCHOLTE et al. 1992b; WILDING et al. 1989; SCHUURMANS et al. 1990) promote LNCaP cell growth and secretion of androgen-dependent proteins; LNCaP cells do not contain ER or PR (BERNS et al. 1986; SCHUURMANS et al. 1988). A single point mutation in the ER ligand-binding domain reduces agonist binding without altering the binding affinity for the antihormone tamoxifen (DANIELIAN et al. 1993; WRENN and KATZENELLENBOGEN 1993). Finally, a single amino acid mutation in the ligand-binding domain of the PR prevents RU486 binding, but has no effect on progesterone binding (BENHAMOU et al. 1992). Taken together, these results indicate that hormones and anthihormones exhibit different structural requirements for binding; once bound, the differences in chemical position within the ligand-binding domain translate to altered receptor conformations and altered ability to activate gene transcription (MCDONNELL et al. 1994). As there appears to be a relationship between ER polymorphisms and susceptibility to breast cancer in humans (ANDERSEN et al. 1994), concern exists that natural mutations may increase susceptibility to various agonists and antagonists in the environment. In other words, the presence of naturally occurring receptor mutants may predispose or sensitize a cell, organ, or individual to chemical insult.

Prevalent models of antihormone action suggest two different mechanisms leading to transcriptional inhibition. Type I antagonists bind to the receptor and prevent binding of the receptor to DNA (TRUSS et al. 1994), while type II antagonists bind to the receptor, induce DNA binding, but fail to initiate transcription. A caveat to this working model is that recent experiments using dimethyl sulfate (DMS) footprinting of genomic DNA strongly suggest, at least for the PR, that both types of antihormones prevent DNA binding in vivo (TRUSS et al. 1994). The ability of some antihormones to induce receptor DNA binding, then, may be an in vitro artifact (see Sect. GI). In addition to direct competition with natural ligand for binding to steroid hormone receptors, antihormone action can result from receptor and/or physiological "cross-talk," i.e., transcriptional activation can be altered by factors other than the cognate receptor such as plasma membrane second messengers, other cellular components, or even other steroid hormone receptors.

Several steroid hormone receptors (e.g., avian PR, human ER, and the VDR) can be activated in a ligand-independent manner through the neurotransmitter dopamine or through growth factors (MANI et al. 1994). Regulation of gene expression via cross-talk between cell membrane receptors and intracellular steroid hormone receptors could provide a means for rapid response to environmental or physiological events. In fact, steroid hormone receptors also may be present at the cell membrane to mediate second messenger transduction responses (BRANN et al. 1995), such as that observed for the cell surface PR that mediates calcium influx in human sperm (BLACKMORE et al. 1994). This sperm surface receptor, however, appears to be more like the γ-aminobutyric acid ($GABA_A$) receptor than the intracellular PR (SHI and ROLDAN 1995).

Drug/chemical insults affecting cell physiology also may disrupt normal receptor function (i.e., physiological cross-talk). For example, unsaturated fatty acids, located primarily in cellular membrane phospholipids, inhibit binding of estrogens to the ER, progestins to the PR, and androgens to the AR; saturated fatty acids, in contrast, are without effect (MITSUHASHI et al. 1988). As fatty acids are released into the cytoplasm via the action of phospholipases, chemicals that alter phospholipase activity may secondarily alter steroid hormone action. Cellular proteins also can interact directly with nuclear receptors to inhibit transcriptional activity. Calreticulin, for example, can bind to the AR DNA-binding domain to inhibit the ability of the AR to bind DNA and activate transcription (DEDHAR et al. 1994). Expression of other or multiple steroid hormone receptors in target cells may result in receptor competition for transcriptional machinery (MEYER et al. 1989; KUMAR 1994) or the formation of nonproductive receptor heterodimers, as recently has been suggested for ER modulation of androgen-induced AR transcriptional activity (KUMAR et al. 1994). With a limited supply of intracellular enzymes and transcription factors, the most abundant receptor types will, in the presence of ligand, successfully compete for transcriptional machinery, preventing the action of less abundant ligand-bound receptors. In addition to antihormone action at the level of the receptor, direct competition at the level of the HRE or even at the level of transcription factor interactions is possible; pharmaceuticals currently are being designed based upon their ability to compete with ligand-activated receptors for binding to HRE (HENDRY 1993; BRANN et al. 1995), and several estrogenic chemicals have been reported to induce endogenous DNA adducts, potentially being involved in estrogen-induced cancers (LIEHR et al. 1986).

E. Androgen Receptor

I. Introduction

Sex steroid hormone antagonists originally were thought to function as simple competitive inhibitors of natural ligand binding. However, it now is clear that these antihormones are actively involved in the process of inhibiting androgen action. Conformational differences in the AR induced by hormone or antihormone binding are readily detectable using limited proteolytic digestion (KUIL and MULDER 1994). The altered structural conformation of the ligand-binding domain most likely explains the functional ability of some antiandrogens (e.g., hydroxyflutamide) to inhibit AR dimerization and transcriptional activation. The binding of antagonists with moderate affinity for the AR also fails to stabilize the receptor, effectively reducing its half-life in the target cells (ZHOU et al. 1994). In other words, natural ligands with high affinity for the AR slow receptor degradation by prolonging nuclear retention, thereby limiting recycling of the receptor from the nucleus to the cytosol (ZHOU et al. 1994). Rapid receptor recycling induced by antihormone exposes the AR to cytoplasmic degradative enzymes (ZHOU et al. 1994).

Tissue concentrations of steroid hormone receptors are regulated by ligand binding via alterations in the rate of transcription of new receptor mRNA, stability of receptor mRNA, and/or the turnover rate of receptor proteins. Castration results in a ten fold increase in AR mRNA and a two fold increase in AR protein within 72 h. Simultaneous administration of androgen prevents the compensatory rise in AR mRNA and protein (JÄNNE and SHAN 1991). The AR, GR, PR, and ER genes all possess promoters that are GC rich and lack TATA boxes (GROSSMAN et al. 1994). The GC box previously has been shown to be involved in transcriptional regulation of TATA-less promoters. Interestingly, the AR contains multiple promoters, as do the GR, PR, and ER genes, which presumably function in AR autoregulation as well as regulation by other mechanisms (GROSSMANN et al. 1994). It is interesting to note the diversity in the regulation of AR levels. AR mRNA and protein levels in Sertoli cells are regulated by follicle-stimulating hormone (FSH) a cyclic adenosine monophosphate (cAMP)-mediated hormone (BLOK et al. 1992; LINDZEY et al. 1993). In this context, the mouse AR gene contains in its 5′-

flanking sequence putative binding sites for CRE (cAMP response element), AP2 (activating transcription factor-2), and AP1, all of which can potentially respond to cAMP (LINDZEY et al. 1993). Clearly, our understanding of intracellular mechanisms responsible for maintaining steroid hormone receptor number and the subsequent ability of these receptors, either alone or in combination with other intracellular mechanims, to induce the appropriate biological response is in its infancy. It should be apparent, however, that multiple control points exist and are potential targets for disruption by drugs and/or environmental chemicals.

Using immunostaining and reverse transcriptase polymerase chain reaction (RT-PCR), AR is found in most rodent tissues, with the highest levels being located in the male reproductive tract (MIZOKAMI and CHANG 1994). Genetic evidence indicates that the same AR protein that mediates sex differentiation in the developing fetus is responsible for androgen action in post embryonic life (BENTVELSEN et al. 1994). In other words, mutations that impair AR function in humans or rodents prevent androgen action during development and in mature animals (BENTVELSEN et al. 1994). Immunoblots prepared from urogenital tract tissues of gestational day 17 male and female rats recognize a 110-kDa protein band characteristic of the adult AR (BENTVELSEN et al. 1994). The ligand-binding characteristics of AR isolated from the embryonic urogenital tract also are similar to those isolated from mature adult reproductive tract tissues (BENTVELSEN et al. 1994).

The most common birth defects seen in humans resulting from inhibition of AR action are alterations in the development of the external genitalia (SWEET et al. 1974). The first 12 weeks of gestation are considered the critical period for the sex differentiation of the human external genitalia (KALLOO ct al. 1993). During this time, high fetal androgen levels induce ventral folding and fusion of the urethral folds to form the penis and fusion of the labioscrotal folds to form the scrotum. In the second and third trimesters, androgen-dependent growth of these structures occurs. Human male external genitalia at 18–22 weeks of gestation exhibit intense positive immunohistochemical staining for AR, but not for ER (KALLOO et al. 1993); this questions whether maternal or environmental estrogens directly influence or alter the development of the human male external genitalia (KALLOO et al. 1993).

In contrast to androgen-dependent sex differentiation in males, the development of normal female external genitalia presumably occurs due to the absence of androgens (KALLOO et al. 1993). Interestingly, external genitalia from human female embryos were found to contain AR using immunohistochemical staining methods. The detection of AR in the external genitalia of the developing human female provides a mechanism by which these structures can be masculinized in the presence of androgen during the first trimester. Masculinization of female external genitalia accounts for the vast majority of all cases of ambiguous genitalia in humans (KALLOO et al. 1993).

II. Androgen Receptor: Mechanism and Effects of Select Chemicals

1. Drugs

First, we will mention several of the more popular nonsteroidal pharmaceuticals whose mechanism of action is to inhibit AR activity. These chemicals were developed primarily for the pharmaceutical treatment of androgen-dependent prostate carcinoma and benign prostatic hyperplasia/hypertrophy. Hydroxyflutamide acts as a pure AR antagonist: it binds the AR, is efficiently imported into the nucleus, but, like the active vinclozolin metabolites and p,p′-DDE (see below), fails to initiate transcription (KEMPPAINEN et al. 1992). ICI 176,334 (Casodex; Zeneca Pharmaceuticals, Wilmington, Delaware USA) binds to the AR and fails to induce receptor accessory protein dissociation, DNA binding, or transcriptional activation (WAKELING 1992; FREEMAN et al. 1989). Nilutamide is a nonsteroidal antiandrogen with weak affinity for the AR; it has a long biological half-life and exhibits potent antiandrogenic activity in vivo, likely reflecting its inhibitory action on androgen synthesis in addition to inhibition of AR binding (HARRIS et al. 1993). Developmental reproductive alterations induced by drugs specifically designed to inhibit androgen action have previously been described (NEUMAN 1977; IMPERATO-MCGINLEY et al. 1992) and resemble alterations produced by natural mutations in the AR gene (POLANI 1981; FRENCH et al. 1990) and those induced by environmental antiandrogens (GRAY et al. 1994). The relative ability of select drugs and environmental chemicals to compete with endogenous ligand for binding to the AR are listed in Tables 1 and 2, respectively.

2. Androgens

As discussed above, androgens are absolutely required for the masculinization of the reproductive tract of the fetal male. Paradoxically, exogenously administered testosterone adversely affects sex differentiation in male rats, as males treated neonatally with high doses of testosterone exhibit small testes with hypospermatogenesis and reduced prostate, seminal vesicle and epididymal weights (WILSON and WILSON 1943). The effects on the prostate and epididymis are thought to be mediated via reduced 5α-reductase activity (BARANAO et al. 1981). These results indicate that, while androgens are required for male reproductive development, their presence in excess is detrimental.

3. Estrogens

Antiandrogenic effects seen in male animals exposed in utero or neonatally to estrogens resemble those seen in human males exposed to DES during development. Male mice given DES perinatally develop epididymal cysts, hypospadias, phallic hypoplasia, inhibition of growth and descent of the testes, and underdevelopment or absence of the vas deferens, epididymis, and seminal vesicles (MCLACHLAN 1981). In vitro ligand-binding experiments suggest that the estrogenic chemicals estradiol, DES, chlordecone, and *o,p′*-DDT all compete with endogenous androgens for binding to the AR (KELCE et al. 1995). Once bound to the receptor, these estrogenic chemicals act as AR antagonists by inhibiting transcriptional activation. These results suggest that the demasculinizing action of estrogenic chemicals on male offspring after in utero exposure may result from an antagonistic interaction of the "estrogenic" compound with the androgen receptor, rather than acting exclusively through the ER. In clear support of this hypothesis is the report that human male external genitalia at 18–22 weeks of gestation exhibit intense positive immunohistochemical staining for the AR, but not the ER (KALLOO et al. 1993). The fact that ER was not detectable in human male external genitalia during development questions whether estrogenic chemicals act through the ER to directly influence the development of the male external genitalia.

Although human-wildlife exposure to high levels of suspect environmental "estrogens" (see Sect.E.II.6) in the United States is typically less than that observed 20–40 years ago, high levels of these chemicals still persist in some areas of the North American continent. Although it is generally assumed that problems with DDT and it metabolities are behind us, high environmental levels of DDT and its metabolities still persist. American robin eggs from orchard areas in British Columbia in 1991 contained extremely high levels of both *p,p′*-DDE and *p,p′*-DDT (total burden of 130 mg/kg; ELLIOT et al. 1994). These concentrations are high enough to conclude that considerable hazard exists to birds of prey that display egg shell thinning at 2 mg/kg (ELLIOT et al. 1994). Recent publications also suggest that the human population may be at risk. DDT metabolite levels, which often exceed WHO guidelines (50 ppb) by more than tenfold, are found in human breast milk in highly expose populations in India, Turkey, and Canada. High levels of these lipophylic chemicals in human breast milk presumably result from consumption of contaminated food including marine mammals (DEWAILLY et al. 1989), milk, and dairy products (BOUWMAN et al. 1990; BATTU et al. 1989). Daily intakes of *p,p′*-DDT and its metabolites through consumption of contaminated milk by 1- to 3-year-old children exceed their acceptable daily WHO intake three-to-fivefold, suggesting that a well-founded risk to infants exists (BATTU et al. 1989), particularly to firstborn infants, who receive a disproportionate amount of the maternal body burden of these chemicals.

4. Progestins

Depending on the dose, ligand structure, and individual tissue responsiveness, progestins interact with a functional AR to induce androgenic (mimic androgen action), synandrogenic (potentiate androgen action), and anti-androgenic (inhibit androgen action) effects (JÄNNE and BARDIN 1984). In the kidney, androgens act via the AR to stimulate β-glucuronidase and ornithine decarboxylase activity (BARDIN et al. 1973; PAJUNEN et al. 1982). Progestins such as medroxyprogesterone acetate (MPA), but not progesterone itself, also act via the AR to stimulate the activity of these enzymes (JÄNNE and BARDIN 1984). As these effects are not demonstrable in Tfm/y mice (BULLOCK et al. 1975; MOWSZOWICZ et al. 1974), androgen-insensitive rats (BARDIN et al. 1973), or humans with testicular feminization (PEREZ-PALACIOS et al. 1981,), it is clear that the androgenic action of these progestins requires a functional AR.

Cyproterone acetate is a synthetic derivative of hydroxyprogesterone that exhibits antiandrogenic as well as progestational activity (NEUMANN 1977; BRINKMANN et al. 1983; HUANG et al. 1985). Exposure to cyproterone acetate during development induces antiandrogenic effects in the male mouse, rat, guinea pig, hamster, sheep, pig, and dog consistent with the effects induced by testicular feminization (NEUMANN 1977). The antiandrogenic activity of cyproterone acetate results from competitive inhibition at the level of the AR. Cyproterone acetate is not considered a true antiandrogen, as it exhibits both agonist (at high concentrations) and antagonist (at lower concentrations) activity in vivo (KEMPPAINEN et al. 1992). At high concentrations, cyproterone acetate, progesterone, and the synthetic antiprogestin RU486 not only bind the AR, but they promote nuclear transport, androgen response element (ARE) DNA binding, and transcriptional activation (KEMPPAINEN et al. 1992). This correlates precisely with the in vivo observations that, at high concentrations, progestational steroids act as androgens by stimulating the growth of the male reproductive tract and virilization of the female fetus (KEMPPAINEN et al. 1992).

5. Vinclozolin

The fungicide vinclozolin alters sex differentiation in male rats through inhibition of AR action (KELCE et al. 1994). Perinatal exposure to vinclozolin induces hypospadias, ectopic testes, vaginal pouches, agenesis of the ventral prostate, and nipple retention in male rat offspring, while the female offspring are unaffected (GRAY et al. 1994). Exposure to 50 mg vinclozolin/kg per day during development (gestational day 14 to postnatal day 3) induces infertility and reduced ejaculated sperm counts in adult male offspring, primarily due to the presence of severe hypospadias. Concentrations as low as 6 mg/kg per day induce permanent reductions in ventral prostate weight following developmental exposure; however, even long-term, high-dose exposure (100 mg/kg per day for 25 weeks) does not produce infertility in adult male rats. These results suggest that the developing fetus is very sensitive to endocrine disruptors such as vinclozolin, which produces malformations at dosage levels that have little or no reproductive effect in adults.

The biochemical and molecular mechanism responsible for the antiandrogenic effects of vinclozolin has been elucidated (Fig. 2). While the parent chemical vinclozolin is not an effective inhibitor of AR binding, two primary metabolites, M1 and M2, do compete effectively with endogenous ligand for binding to the AR (KELCE et al. 1994). Once bound to the receptor, both active metabolites target the AR to the nucleus, but fail to initiate transcription

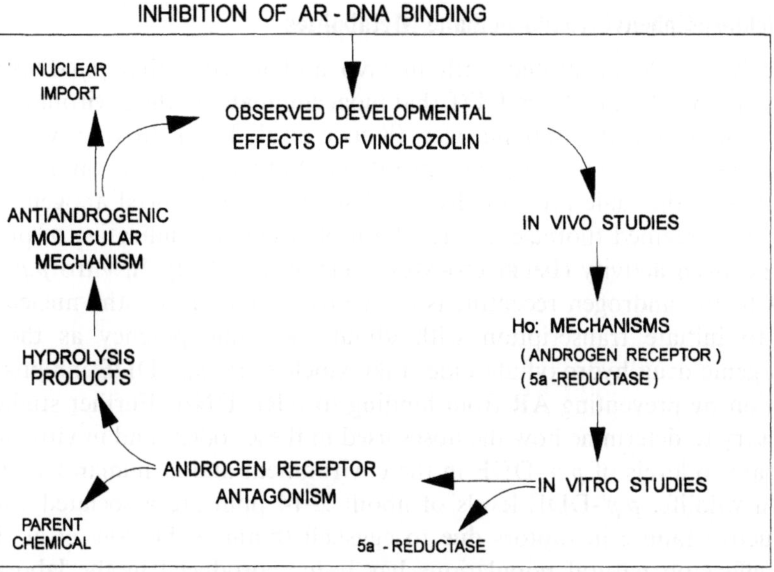

Fig. 2. Sequence of *in vivo* and *in vitro* studies used to screen vinclozolin for hormonal activity and identify the mechanism of action of this endocrine disruptor. *In -vitro* studies were completed to verify the developmental effects of vinclozolin observed by the registrant. The results of these *in -vivo* studies suggested that vinclozolin acted as an antiandrogen either by inhibiting androgen receptor (AR) activity or by inhibiting the enzymatic activity of 5α-reductase (i.e., the enzyme that converts testosterone to the more active androgen 5α-dihydrotestosterone). *In -vitro* studies to investigate this hypothesis (*Ho*) suggested that vinclozolin does not inhibit 5α-reductase activity, but does exhibit a weak ability to compete with endogenous ligand for binding to the AR. Subsequent *in -vitro* studies determined that two hydrolysis products of vinclozolin, M1 and M2, are far better inhibitors of AR binding than vinclozolin; this suggested that the parent chemical probably is not the active form *in -vivo*. Molecular studies with the transfected human AR confirmed the ability of M1 and M2 to bind AR and additionally determined that both chemicals inhibit transcriptional activation. The mechanism of transcriptional inhibition was not related to altered import of the anti - hormone-bound AR complex into the nucleus, but rather failure of the anti -hormone-bound AR to bind androgen response element DNA. The observed developmental effects of vinclozolin, then, are probably mediated by the vinclozolin hydrolysis products M1 and M2, both of which bind AR and act as antiandrogens by inhibiting AR transactivation

(Wong et al., in press). The molecular mechanism responsible for transcriptional inhibition was investigated using DNA mobility shift assays. M1 and M2 inhibit the ability of AR to bind ARE DNA, thereby preventing transcription (Wong et al., in press). Interestingly, at higher concentrations and in the absence of DHT (i.e., agonist ligand), M2 acts as an AR agonist, leading to the speculation that AR heterodimers (an androgen-bound AR monomer dimerized with an antiandrogen-bound AR monomer) are functionally antagonistic, while AR homodimers (either androgen or antiandrogen bound to both AR monomers) are agonists (Wong et al., in press).

6. Dichlorodiphenyl-trichloroethane Metabolites

Recently, we demonstrated both in vivo and in vitro that *p,p'*-DDE, the persistent metabolite of *p,p'*-DDT that bioaccumulates in the environment, is a potent environmental antiandrogen (KELCE et al. 1995). In vivo, when *p,p'*-DDE was administered to pregnant rats (100 mg/kg per day) on gestational days 14–18, the male progeny displayed significantly reduced anogenital distance and retained thoracic nipples, both of which are indicative of prenatal antiandrogen activity (IMPERATO-MCGINLEY et al. 1992). In vitro *p,p'*-DDE binds to the androgen receptor, is efficiently imported into the nucleus, but fails to initiate transcription with about the same potency as the antiandrogenic drug hydroxyflutamide. Like vinclozolin, *p,p'*-DDE inhibits transcription by preventing AR from binding to ARE DNA. Further studies are necessary to determine how the doses used in these rodent and in vitro studies compare to levels of *p,p'*-DDE in the environment and in human tissues.

In wildlife, *p,p'*-DDE levels of about 2–14 ppm are associated with reproductive failure in raptors due to eggshell thinning (SPITZER et al. 1978). This effect on natural populations has been reproduced in the laboratory, where DDT and *p,p'*-DDE exposures are found to inhibit calcium ATPase and carbonic anhydrase and reduce calcium levels in the oviduct of susceptible avian species (SPITZER et al. 1978). Ring doves given 10 ppm *p,p'*-DDT exhibit increased hepatic activity and decreased serum estradiol levels, leading to a delay in egg laying, a decrease in bone calcium deposition, and reduced eggshell weight. DDT exposure in Lake Apopka has led to developmental reproductive abnormalities in alligators (GUILETTE et al. 1994). Alligator eggs from Lake Apopka contained levels of *p,p'*-DDE (5.8 ppm) 20–100 fold higher than those required to inhibit AR transcriptional activity in vitro (KELCE et al. 1995) and to adversely affect avian sex differentiation. Among the multiple mechanisms by which *p,p'*-DDE acts as an endocrine disruptor, we suspect that the antiandrogenic activity of *p,p'*-DDE (starting at 0.064 ppm; KELCE et al. 1995) may play a role in the alterations in sex differentiation seen in both avian (FRY and TOONE 1981) and reptilian species (GUILLETTE et al. 1994).

The potency of *p,p'*-DDE as an AR antagonist is novel and troubling given that *p,p'*-DDE is the major DDT-derived residue in food (SZOKOLAY et al. 1977; SPINDLER 1983) and human body fat (WYLLIE et al. 1972; BARQUET et al. 1981) and comprises 50%–80% of the total DDT-derived residues in human breast milk mobilized to the infant (ADAMOVIC et al. 1978; O'LEARY et al. 1970; ROGAN et al. 1986). Although median levels of *p,p'*-DDE measured in serum (12.6 ppb) and placental tissues (6.8 ppb) from women in the United States within the last 10 years are below those required to inhibit androgen action in vitro, maximum levels in some women can exceed these concentrations (e.g., 180 ppb in serum and 74 ppb in placenta; ROGAN et al. 1986), presumably due to slow metabolic clearance. Even greater concern exists for developing wildlife species and humans in other countries, where DDT remains in use or is present in contaminated ecosystems. In South Africa, for example, BOUWMANN et al. (1991) found that serum from individuals living in DDT-treated dwellings (i.e., for malaria control) contained median DDT/DDE levels of 140.9 ppb; DDT/DDE levels in breast milk were even higher (475 ppb in whole milk, assuming 3% milkfat), leading those authors to postulate a potential risk to infants (BOUWMAN et al. 1990). In the mid-1960s, when DDT was still used in the United States, high concentrations of *p,p'*-DDE were found in tissues from stillborn infants in Atlanta, Georgia (650 ppb in brain, 850 ppb in lungs, 2740 ppb in heart, 980 ppb in liver, 3570 ppb in kidney, and 860 ppb in spleen; CURLEY et al. 1969). Although the concentration of *p,p'*-DDE at the level of the human AR in vivo is not known, the above reports suggest that human *p,p'*-DDE levels can exceed those that inhibit human AR transcriptional activation in vitro.

7. Plant Products

A natural plant product has been shown to possess antiandrogenic activity. Permixon is a liposterolic extract from the fruit of the American dwarf palm tree Serenoa Repens B, native to Florida (CARILLA et al. 1984). In clinical trials, this palm tree extract was found to lessen the signs and symptoms of benign prostatic hyperplasia including, dysuria, nocturia, and poor urinary flow (CHAMPAULT et al. 1984). The antiandrogenic effects of permixon are mediated through direct action at the level of the AR (IC_{50}, 367 μg/ml) and via inhibition of 5α-reductase activity (IC_{50}, 88 μg/ml) (CARILLA et al. 1984). To the best of our knowledge, effects of permixon on the development of the male or female reproductive systems have not been reported.

F. Estrogen Receptor

I. Introduction

The external genitalia of human female embryos stain intensely for ER, suggesting that estrogen may actively influence female sex differentiation (KALLOO et al. 1993). While it is known that the fetal mammalian ovary does not contribute to estrogen production or female sex differentiation, a role for maternal estrogens has not been excluded. Recent experiments with estrogen receptor knockout mice, however, suggest that homozygous mutant females (i.e., females lacking functional ER) contain normal female reproductive tract structures (KORACH 1994). These results suggest that the development of the rodent female reproductive system is not dependent on the presence of functional estrogen receptors. Interestingly, the development of the female reprodutive system does appear to be susceptible to the effects of too much estrogen, as discussed below in Sect F.II.2. The relative abilities of select drugs and environmental chemicals to compete with endogenous ligand for binding to the ER are listed in Tables 1 and 2, respectively.

The role of estrogens in the developing male reprodutive system is unclear. Immunohistochemical studies demonstrate ER expression in fetal mouse gonads on days 13 and 15, after which time the gonads loose ER expression (GRECO et al. 1992). A putative molecular target for estrogens in the developing gonad may be the Müllerian inhibiting substance (MIS) gene in Sertoli cells (HUTSON et al. 1982; NEWBOLD et al. 1984; GUERRIER et al. 1990), where estrogens presumably function to decrease MIS expression. Experiments with ER knockout mice, however, suggest that estrogens are not required for fetal male development, but they do appear to be important in maintaining semen quality at adulthood (LUBAHN et al. 1993). At the other extreme in transgenic mice that overexpress ER, sexual differentiation of the male reproductive system and subsequent fertility are normal (DAVIS et al. 1994). The role of environmental chemicals with estrogen agonist activity on male reproductive development and health is the subject of a recent report initiated by the Danish Environmental Protection Agency (DANISH EPA 1995), to which the reader is referred for additional and far more comprehensive information.

II. Estrogen Receptor: Mechanism and Effects of Select Chemicals

1. Drugs

With the intense interest in developing successful antiestrogenic drugs for use in breast cancer therapy, abundant information exists regarding the biochemical and molecular mechanisms of these antihormone ligands. Upon binding ligand, the ER binds as a dimer to estrogen response element (ERE) DNA and interacts with appropriate transcription factors to activate transcription of estrogen-responsive genes (KATZENELLENBOGEN et al. 1993). Estrogen antagonists compete with endogenous estrogens for binding to the ER, but fail to initiate transcription, thus preventing or blocking the effects of endogenous estrogens. Antiestrogens typically contain bulky polar or basic side chains in their structure, which is essential for antiestrogenic activity; removal of these side chains results in the chemical having agonist activity (KATZENELLENBOGEN et al. 1993). As an excellent example, when the basic side chain is removed from the pure ER antagonist ICI 164,384, the natural ligand 17β-estradiol is produced. Recently it has been determined that amino acids near cysteine-530 in the hormone-binding domain of the ER are involved in discriminating between ER agonists and ER antagonists (KATZENELLENBOGEN et al. 1993). These structure–activity relationships currently are being used to design more effective ER antagonists for use in endocrine-dependent breast cancer therapy.

In animals, ICI 164,384, unlike tamoxifen, has no agonist activity on the uterus, vagina, or mammary glands and competes with high affinity with endogenous ligand for binding to the ER (WAKELING and BOWLER 1991). More importantly, when tamoxifen and ICI 164,384 are administered simultaneously, the trophic action of tamoxifen on the uterus and the mammary gland tissue was eliminated (WAKELING and BOWLER 1991). These results suggest that pure antiestrogens, such as ICI 164384, can completely negate the effects of estrogenic chemicals in vivo and that endogenous estrogens, partial ER antagonists, and pure ER antagonists act through a common ER mechanism (WAKELING and BOWLER 1991). ICI 164,384 binds the ER hetero-ligomeric complex and induces the dissociation of HSP (MCDONNELL et al. 1994); however, it fails to initiate transcription, presumably by preventing ER dimerization and ERE DNA binding (JORDAN 1990). ICI 164,384 also alters the turnover of the ER as well as its intracellular localization (MCDONNELL et al. 1994). Another pure antiestrogen, ICI 182,780, binds to the ER five fold better than ICI 164, 384 and is approximately tenfold more potent in vivo (WAKELING and BOWLER 1991).

Tamoxifen, which is one of the most frequently prescribed anticancer drugs in the United States, is an ER antagonist, inhibiting 17β-estradiol stimulation of breast tumor growth (JORDAN et al. 1987). In breast cancer cells, tamoxifen appears to block not only the binding of natural estrogens to the ER, but also inhibits the mitogenic activity of some growth factors presumably by altering growth factor receptor levels or subsequent receptor function (FREISS et al. 1990). Although tamoxifen has a low affinity for the ER, it is a very efficient antiestrogen, because it is slowly inactivated, making it available in high concentrations at the level of the target (BAULIEU 1987). In contrast, 4-OH tamoxifen has a very high affinity for the ER and high antiestrogenic activity in vitro, but is a poor inhibitor in the intact animal, as it is rapidly metabolized and thus not available at sufficient concentrations (BAULIEU 1987). Other triphenylethylenes also interact with the ER and prevent the tertiary changes necessary for the appropriate activation of the ER; these complexes can bind the ERE, but are unable to initiate the full agonist activity (JORDAN 1990). In contrast, in other tissues, tamoxifen acts as an ER agonist, having beneficial effects on the preservation of trabecular bone mass (LOVE et al. 1992) and lipid metabolism (LOVE et al. 1990), but undesirable actions on endometrial cell proliferation (RAMKUMAR and ADLER 1995). It has been proposed that tamoxifen acts as an ER antagonist for gene activation and an ER agonist for gene repression (RAMKUMAR and ADLER 1995). The agonist/antagonist activity of the ER is thought to result from differences in the activity of the two transcriptional activation functions (TAF) in most nuclear hormone receptors. TAF-1, located in the N-terminal domain of the ER, is thought to mediate agonist responses of the ER, while antagonist responses appear to be derived from inhibition of TAF-2, located in the hormone-binding domain (BERRY et al. 1990).

Finally, progesterone may induce antiestrogenic effects. Progesterone acts as an antiestrogen via several different mechanisms including the following: (a) reducing estrogen secretion by suppressing gonadotropins and ovarian function; (b) reducing estrogen receptor levels presumably by reducing ER synthesis; (c) directly inhibiting the mitogenic effect of estrogen on the endometrium; and (d) increasing metabolism of the potent estrogen 17β-estradiol to the less active metabolite estrone (E1; via increased 17β-hydroxysteroid dehydrogenase activity) (MAUVAIS-JARVIS et al. 1987). These studies highlight the complex nature of ligand–receptor interactions and predict numerous direct and/or indirect mechanisms by which drugs and environmental chemicals may disrupt steroid hormone receptor function.

2. Diethylstibestrol

DES, a potent synthetic estrogen, provides a grim example of how in utero exposure to a potent endocrine disruptor with estrogenic activity can seriously alter reproductive development in humans. Although a few cases of masculinized females and demasculinized males were noted in the late 1950s, most of the effects of DES did not become apparent until after the children attained puberty. Transplacental exposure of the developing fetus to DES at critical periods leads to abnormalities of the urogenital tracts of both rodents and primates. The administration of DES to rodents during perinatal life reduces fertility, produces structural abnormalities of the oviduct, uterus, cervix, and vagina (MCLACHLAN et al. 1982; MCLACHLAN 1981; HENRY and MILLER 1986), produces squamous metaplasia in the uterus and part of the cervix (ENNIS and DAVIES 1982), and doubles the incidence of mammary carcinoma (HUSEBY and THURLOW 1982) in female offspring. Some of the effects of DES can be produced by exposure to other estrogenic substances, such as estradiol (BERN et al. 1983) and RU 2858 (VANNIER and RAYNAUD 1980). The synthetic estrogens may be more active than natural ligand in vivo, as they have less affinity for plasma-binding proteins such as alpha fetoprotein and testosterone–estradiol-binding globulin (VOM SAAL et al. 1992; MIZEJEWSKI and JACOBSON 1987), allowing more chemical to freely enter target cells.

Similar results are seen with antiestrogens such as LY117018, which paradoxically exhibits ER agonist activity in developing fetal rats (HENRY and MILLER 1986). Other antiestrogens such as tamoxifen, nafoxidine, and clomiphene also act as ER agonists when present at high concentrations neonatally, as evidenced by their ability to induce atrophic ovaries and uteri, oviducts exhibiting squamous metaplasia, accelerated vaginal opening, and complete cessation of estrous cycles at 4 months of age (CHAMNESS et al. 1979; CLARK and MCCORMACK 1977). Interestingly, these chemicals do not produce estrogen-like structural changes in the sexually dimorphic nucleus of the preoptic area (SDN-POA) (DOHLER et al. 1986), suggesting that the brain is somehow protected from estrogen overexposure.

Effects seen in male animals exposed in utero or neonatally to estrogens resemble those seen in DES-exposed men. When administered DES perinatally, male mice develop epididymal cysts, hypospadias, phallic hypoplasia, inhibition of growth and descent of the testes, and underdevelopment or absence of the vas deferens, epididymis, and seminal vesicles (MCLACHLAN 1981). Similar antiandrogenic effects are obtained in males using other potent estrogens such as estradiol and RU 2858 (VANNIER and RAYNAUD 1980).

These chemicals also produce estrogenic effects in males, as female structures derived from the Müllerian ducts persist in male mice after in utero administration of DES (NEWBOLD and McLACHLAN 1985; NEWBOLD et al. 1987). The fact that estradiol and DES both bind to the AR in vitro in high, but toxicologically relevant concentrations (KELCE et al. 1995) suggests that the "antiandrogenic" effects of estrogens may be mediated through inhibition of AR action.

Although DES and other potent synthetic ER agonists produce dramatic alterations in reproductive development, these chemicals are not found in the environment. However, plant estrogens, fungal toxins, pesticides, and toxic substances, including a few polychlorinated biphenols (PCB), have been shown to alter sex differentiation in an estrogen-like manner.

3. Plant Estrogens

The phytoestrogens and estrogenic fungal mycotoxins are widespread in nature and provide some of the most conclusive data demonstrating that environmental estrogens are toxic to mammalian reproductive function under natural conditions. FARNSWORTH et al. (1975) listed over 400 species of plants that contain potentially estrogenic isoflavonoids or coumestans suspected of being estrogenic based on biological grounds. In rats, neonatal exposure to the plant estrogen coumestrol (100 μg/day) accelerated the age at vaginal opening, increased the incidence of persistant vaginal cornification (PVC), and induced hemorrhagic follicles (100% at 40 days of age) (BURROUGHS et al. 1985). However, plants contain many other compounds in addition to estrogens that can affect reproductive performance. For example, it was also noted that, of the 525 species of plants that have been used as folkloric contraceptives or interceptives, more than half of these were abortifacients that stimulated uterine muscle contractions in vitro (SALUNKHE et al. 1989). In addition, many plant extracts possess antispermatogenic agents that reduce sperm counts, some of which appear to act directly on the testis or by altering hypothalamic–pituitary function (SALUNKHE et al. 1989). In recognition of the potency of these compounds and their potential to alter human reproduction, the World Health Organization of the UN has evolved common procedures for screening plant extracts for antifertility action.

Although most environmental estrogens are relatively inactive compared to steroidal estrogens or DES, they frequently occur in such high concentrations that they are able to induce reproductive effects in animals; the effects of these chemicals also can be additive (ADAMS 1989). For example, soybean products may contain up to 0.25% total isoflavones, which can induce signs of estrogenicity in swine. Pastures of clover containing formononetin cause fertility problems in sheep, having an estrogenic effect equal to 5–15 μg DES daily. Soybeans and alfalfa sprouts can contribute detectable amounts of plant estrogens to the human diet as well (ADAMS 1989). During World War II, people in Holland consumed large quantities of estrogenic tulip bulbs, resulting in women displaying signs of estrogenism, including uterine bleeding and other menstrual cycle abnormalities (LABOV 1977).

4. Zearalenone (A Fungal Estrogen)

Zearalenone is a mycotoxin. Despite its structural dissimilarity with estrogens, zearalenone binds the ER and activates transcription of estrogen-despendent genes (GENTRY 1986). The anabolic and uterotropic effects of zearalenone are consistent with exposure to excess estrogen; these effects include swollen vulva, hypertrophic myometrium, and vaginal cornification (OSWEILER et al. 1985). In domestic animals, feed contaminated by *Fusarium roseum* (i.e., zearalenone-producing fungi) induces adverse reproductive effects, including impaired fertility in cows and hyperestrogenism in swine and turkeys (ADAMS 1989). In the laboratory, neonatal exposure (1 mg, s.c. at birth) to zearalenone alters sex differentiation of the female rat and hamster reproductive system. In rats, zearalenone induces PVC (KUMAGAI and SHIMIZU 1982) and reduces ovarian size due to a lack of corpora lutea, while neonatally treated female hamsters display accelerated vaginal opening and abnormal male-like sex behavior as adults (GRAY et al. 1985).

5. Estrogenic Pesticides

It is known that perinatal administration of weakly estrogenic pesticides such as *o,p′*-DDT, kepone (chlordecone), or methoxychlor, induce estrogen-like alterations on central nervous system (CNS) sex differentiation, accelerate vaginal opening, induce PVC and delayed anovulatory syndrome (DAS), and masculinize sex-linked behaviors in female rats. Administration of *o,p′*-DDT (1 mg, s.c.) on the second, third, and fourth days of life to female rat pups induces PVC by 120 days of age; ovaries of the treated females contained follicular cysts and lacked corpora lutea (HEINRICHS et al. 1971). Kepone injection into newborn rat pups (0.2 mg per pup) advances vaginal opening by more than 10 days, accelerates the onset of PVC, and reduces ovarian weight by inhibiting corpora lutea formation (GELLERT 1978a). In addition, adverse reproductive development has been reported in avian species after in ovo exposure to environmental levels of *o,p′*-DDT or methoxychlor, such as feminized behavior, feminized gonads, and feminized reproductive tracts in male California gulls (FRY and TOONE 1981). The decline in the breeding success and local population levels of this species were attributed to estrogen-like reproductive alterations.

Administration of methoxychlor to rodent dams throughout gestation and lactation results in subtle alterations in the reproductive tract of male offspring, such as slightly smaller testes, epididymides, and lower sperm counts compared to controls (GRAY et al. 1989). The in vivo effects of methoxychlor are thought to be mediated by its *O*-demethylated metabolite *p*-hydroxyphenyl-trichloroethylene (HPTE), which competes for binding to the ER more effectively than the parent chemical (BULGER et al. 1978; JORDAN 1984).

Recently, however, we have demonstrated using comprehensive radioligand-binding assays that HPTE, but not methoxychlor, competes effectively for binding to the AR (K_i, 1 μM; LAWS et al. 1995). It remains to be definitively established, then, whether the in vivo effects of methoxychlor in the male are mediated via ER action or inhibition of AR action.

6. Estrogenic Polychlorinated Biphenyls and Phenols

A few of the 209 PCB congeners and PCB mixtures possess weak estrogenic activity, being able to induce uterotropic responses and estrogenic neuroendocrine alterations in female rats (GELLERT 1978b). Neonatal injection of Aroclor 1221 (10 mg, s.c.) on the second and third days of life accelerates the age vaginal opening and increases of persistent vaginal estrus and anovulatory cycles; in contrast, Aroclor 1224 is without effect (GELLERT 1978b). It has been proposed that these PCB may be rendered estrogenic through metabolic hydoxylation, yielding polychlorinated hydroxybiphenyls (KORACH et al. 1987). When a series of PCB congeners were compared for estrogenic activity, 4-hydroxy-2′,4′,6′- trichloro-biphenyl bound to the ER with the greatest affinity in vitro and increased uterine weight in vivo (KORACH et al. 1987).

Alkyphenol ethoxylates are nonionic surfactants used widely as detergents, emulsifiers, and wetting and dispersing agents in numerous household, agricultural, and industrial applications (AHEL et al. 1993). The nonyl-alkyl group is the most common; nonylphenol ethoxylates in addition to the above are used as spermicides in contraceptive foams, jellies, and creams. Recently, alkylphenol ethoxylates have been reported to be estrogenic in vivo and in vitro, presumably via the hydrolytic removal of the ethoxylate group generating nonyl- and octyl-alkylphenols, which are known to compete for binding to the ER (SOTO et al. 1991; MUELLER and KIM 1978; WHITE et al. 1994), to mimic the effects of estrogen in cell culture (SOTO et al. 1991) and to compete for binding to the AR (Table 2). More research is necessary, however, to understand the potential of these chemicals to induce adverse physiological effects.

G. Progesterone Receptor

I. Introduction

In the absence of ligand, the PR exists as a large multiprotein complex capable of responding to progestins via increased transcription of progestin-dependent genes. The inactive PR is complexed with HSP90, HSP70, two proteins that bind the immunosuppresive drug FK506 (FKBP52 and FKBP54), an immunophilin protein that binds the immunosuppresant cyclosporin A (cyclophilin-40), and p23, a precursor to the formation of the PR complex (JOHNSON and TOFT 1994). Association of the PR complex is energy dependent (JOHNSON and TOFT 1994). Like other steroid hormone receptors, the association of progesterone with its cognate receptor induces a dramatic conformational change, as determined using limited proteolytic digestion (ALLAN et al. 1992a,b). The change in receptor structure is prerequisite to HSP dissociation, receptor dimerization, DNA binding, and transcriptional activation (ALLAN et al. 1992a,b). The nature of this structural change is different when PR binds antihormones, suggesting that conformational changes again play an important role in the activation of steroid hormone receptors by agonists and in their neutralization by hormone antagonists (ALLAN et al. 1992b).

Two naturally occurring forms of the PR have been identified; PR-A is an n-terminally truncated version of the full-length PR-B isoform (HORWITZ 1992). Steroid hormone receptors (i.e., ER, GR, and PR) contain two transcription factor functions (TAF), located in the N-terminal domain (TAF-1) and/or the hormone-binding domain (TAF-2) (GRONEMEYER et al. 1992). The N-terminal region of the PR that is present in PR-B, but absent in PR-A, can participate in transcriptional activation depending both on the gene being transcribed and on the cell type being analyzed (HORWITZ 1992). As we learn more about the tissue-specific effects of these isoforms, we may find that the effects induced by PR agonists and antagonists are dependent upon the particular isoform expressed by the target cell.

Antiprogestins (e.g., RU 486) have important therapeutic potential in fertility control and in the treatment of hormone-dependent tumors, endometriosis, and breast cancer (MCDONNELL et al. 1993). Antiprogestins compete with natural ligands for binding to the PR, influence the conformation of the PR, and impair the dissociation of HSP from the PR, resulting in transcriptional inhibition (TRUSS et al. 1994; ALLAN et al. 1992b). Recent studies using DMS footprinting of genomic DNA strongly suggest that all types of antiprogestins prevent DNA binding of PR to the progesterone response element (PRE) in vivo (TRUSS et al. 1994). Antiprogestins induce a rapid disappearance of the agonist-induced footprints of PR as well as footprints of other transcription factors recruited by the receptor to the MMTV promotor (TRUSS et al. 1994). These results support earlier mechanistic studies suggesting that RU486 acts to stabilize the 8S untransformed heteromeric PR complex, thereby preventing its interaction with DNA (BAULIEU 1989). Additional evidence in support of this interpretation comes from studies with the GR, where RU486 (also a glucocorticoid receptor antagonist) fails to induce footprints in vivo on the HRE of the tyrosine aminotransferase promoter (BECKER et al. 1986). These results question current models suggesting that antihormone binding results in the formation of receptor dimers that occupy response elements and interact nonproductively with the transcription apparatus (MCDONNELL et al. 1994). The relative ability of select drugs and environmental chemicals to compete with endogenous ligand for binding to the PR are listed in Tables 1 and 2, respectively.

II. Progesterone Receptor: Mechanism and Effects of Select Chemicals

1. RU486 (Mifepristone)

Progesterone and synthetic agonists (e.g., R5020) bind the A and/or B forms of the PR, resulting in the formation of PR-A/PR-B homo- or heterodimers; the PR dimer, then, binds the PRE and initiates transcription of progesterone-dependent genes. The antiprogestin RU486 also binds to both forms of PR, however, binding of this antihormone results in transcriptional inhibition. The mechanism responsible for transcriptional inhibition induced by RU486 is thought to be the elimination of TAF-2 activity from the hormone-binding domain (GRONEMEYER et al. 1992). However, DMS footprinting experiments suggest that the RU486-bound PR is unable to activate transcription, because the antihormone receptor complex is unable to bind the PRE (TRUSS et al. 1994). Whatever the mechanism, RU486 acts to bind the PR and prevent PR action. Interestingly, RU486 does not bind to PR from some species, such as the chicken and hamster (BAULIEU 1989). Finally, both RU486 and another PR antagonist, ZK98734, bind to human plasma orosomucoid (LEBEAU and BAULIEU 1994), which presumably accounts for the long half-life of these chemicals (i.e., they are protected from metabolic inactivation).

Progesterone is essential for the formation of the secretory endometrium required for nidation and nourishment of the developing conceptus as well as the inhibition of myometrial contractility, ensuring that the uterus is maintained in a quiescent state during pregnancy (VAN LOOK and BYGDEMAN 1989). Withdrawal of progesterone or inhibition of progesterone action by antihormones prevents implantation and induces myometrial contractility, resulting in the expulsion of any previously implanted conceptus (VAN LOOK and BYGDEMAN 1989). The embryotoxicity of antiprogestins is particularly difficult to evaluate, as these chemicals act to terminate pregnancy. RU486 currently is used in France for this purpose. The protocol for pregnancy termination is oral administration RU486 (600 mg) followed by 400 μg of prostaglandin analogue 48 h later; expulsion usually occurs 4 h later (LEBEAU and BAULIEU 1994). The prostaglandins are far more efficient in inducing uterine contractions after RU486, because RU486 enhances the sensitivity of the myometrium to the prostaglandins in humans (LEBEAU and BAULIEU 1994).

In males, neonatal administration (1 mg every 2 days from day 1 to 15) of RU486 permanently alters testicular development in rats and delays the onset of puberty (VAN DER SCHOOT and BAUMGARTEN 1990). Sex behavior in these animals also was abnormal, as treated males rarely ejaculated and displayed abnormally high levels of female sex behavior following estradiol injections. Taken together, these results suggest that neonatal exposure to RU486 inhibits masculinization of the male reproductive and central nervous systems.

2. Androgens

Androgens exhibit the ability to compete with progestins for binding to the PR, with the 19 nor-androgens exhibiting the highest affinity (JÄNNE and BARDIN 1984). Once bound to the PR, androgens can act as PR agonists by inducing progesterone-dependent uteroglobin synthesis in the rabbit uterus. These effects appear to be mediated through the PR, as the potency to induce uteroglobin synthesis was related to the affinity of the androgen for the PR and because androgen-induced mRNA and protein synthesis was not inhibited by the simultaneous administration of the AR antagonist flutamide (JÄNNE and BARDIN 1984).

3. Estrogens

In many target tissues, estrogens induce PR expression (SAVOURET et al. 1994), which is an important role in the physiology of the estrous or menstrual cycles. In rabbits, the PR gene contains an ERE which confers the ability of estrogens to induce PR synthesis (SAVOURET et al. 1994). Interestingly, some environmental estrogens, such as chlordecone, *o,p'*-DDT, the nonylphenols, and the active methoxychlor metabolite HPTE, bind to the PR in vitro with about the same affinity that they display for the ER (LAWS et al. 1995). The fact that environmental endocrine disruptors can bind to multiple steroid hormone receptors should not be surprising, as similar effects have been reported both for steroidogenic drugs and for the steroid hormones themselves (Table 1, 2).

4. Progestins

Developmental alterations in the human male reproductive tract are induced by natural and synthetic progestins such as progesterone, hydroxyprogesterone, dimethisterone, norethidrone, and medroxyprogesterone. Developmental exposure to these chemicals results in hypospadias, ambiguous genitalia, and occasionally testicular atrophy in both rats and monkeys (SCHARDEIN 1993; PRAHALADA et al. 1985). While the mechanism responsible for these effects is not known, they most likely result from inhibition of AR action (see Sect E.II.3) and/or testicular androgen biosynthesis (POINTIS et al. 1984).

H. Hormone Disruption-Testing Strategies

Multigenerational reproductive toxicology tests are currently the only test protocols that encompass both developmental chemical exposure and continuous animal health monitoring (reviewed by PALMER 1981). These protocols, however, require neither endocrine data nor even "bioassay" measurements for hormonal activity. Multigenerational test guidelines which include measures of endocrine function (pubertal landmarks, estrous cyclicity, reproductive organ tweights, etc.) have been developed at the U.S. Environmental Protection

Agency (EPA; Makris 1995) and the U.S. Food and Drug Administration (FDA; Hubbard 1994). Our laboratory has developed an alternate reproduction test protocol (Gray et al. 1988; Zenick et al. 1994), which requires the measurement of endocrine end points and serum hormone levels; however, these tests are labor intensive, take more than 1 year to complete, and are expensive to conduct. Clearly, shorter-term, less expensive test protocols designed to identify potential endocrine disrupting chemicals and to elucidate mechanisms of endocrine toxicity are also required. The current version of the EPA's Safe Drinking Water Act (1995), if passed by Congress, mandates that such testing procedures be developed. It is our opinion that no single in vivo or in vitro test will adequately screen a chemical for endocrine disrupting activity. For example, screening chemicals in vitro for estrogenicity would fail to identify chemicals acting through other steroid hormone receptors, fail to identify chemicals which alter steroid hormone biosynthesis, transport, or degradation, and fail to identify chemicals that need metabolic activation. In vitro estrogenicity tests alone, then, would fail to predict the dramatic in vivo alterations induced by *p,p′*-DDE or vinclozolin. While screening procedures can be developed to detect chemicals that possess reproductive hormone or antihormone activity, no single test is likely to be effective; a battery of tests is far more likely to succeed.

We propose that a combined in vivo and in vitro test strategy be developed to screen chemicals for effects on the reproductive system. In vivo tests serve to identify chemicals with endocrine disrupting activity, while in vitro tests serve to delineate the chemical or metabolite responsible for the effects and provide information regarding the biochemical mechanism. Once the biochemical mechanism is understood, human susceptibility and risk assessment issues can be addressed in a more quantitative manner. Dosing animals with toxicants during puberty has proved successful in detecting estrogenicity and antiandrogenicity. In these studies, male and female weanling rats are dosed with a chemical from 22 to 50 days of age and subsequently necropsied. The age and weight at puberty, the weights of the reproductive organs, and serum hormone levels are measured. This simple protocol has been used successfully with chlordecone, methoxychlor, vinclozolin, and DDE. Using the above protocol, if vaginal opening was accelerated then one would determine the ability of the chemical to acutely stimulate uterine growth. A chemical that displayed uterotropic activity could be examined in vitro for ER binding, however, false negatives would result from this test if the chemical required metabolic activation or was insoluble. Chemicals that bound to the ER subsequently would be examined for their ability to initiate or inhibit transcription of estrogen-dependent target genes (i.e., to determine whether the chemical acts as a hormone or an antihormone).

The approaches employed by Kelce et al. (1994, 1995) and Gray et al. (1994) provide good examples of how to use in vivo studies to identify endocrine disrupting chemicals and in vitro studies to further characterize the biochemical and molecular mechanism of action (Fig 2). These studies identified the mechanism of action of the antiandrogenic chemical vinclozolin. With vinclozolin, studies by the registrant suggested that this chemical possessed antiandrogenic activity. Using the above in vivo protocol, vinclozolin, administered for 30 days starting at weaning, delayed puberty, reduced the weights of the sex accessory glands, and increased serum testosterone and luteinizing hormone (LH) levels, all of which are consistent with the endocrine profile of an antiandrogen. In a developmental study, vinclozolin produced antiandrogenic alterations in male rat sex differentiation, such as reduced anogenital distance, cleft phallus, hypospadias, ectopic testes, and retained thoracic nipples in male offspring (Gray et al. 1994). These in vivo experiments clearly identified vinclozolin as an antiandrogenic endocrine disruptor.

Subsequent in vitro studies demonstrated that, while the parent chemical vinclozolin exhibited little ability to bind to the AR, two primary hydrolysis products of vinclozolin were potent AR antagonists (Kelce et al. 1994). After determining that maternal serum concentrations of these metabolites were at levels around their respective K_i values for inhibition of androgen binding, it was suggested that vinclozolin developmental toxicity is likely mediated through its in vivo hydrolysis to the antiandrogenic metabolities M1 and M2 (Kelce et al. 1994). Molecular studies subsequently determined that the mechanism by which these active metabolites inhibit androgen-induced transcriptional activation is to bind to the AR and inhibit the ability of the AR to bind ARE DNA (Wong et al., in press).

I. Conclusions

The objective of this chapter was to show how in utero or perinatal exposure to drugs or environmental chemicals can mimic or block the action or steroid hormone receptors. Inclusion of the molecular mechanisms by which sex steroid hormone receptors interact with ligand to bring about alterations in transcription, protein synthesis, and subsequent cell function illustrates the numerous sequential steps in sex steroid hormone receptor action that can be mimicked or blocked by endocrine disrupting chemicals. It should be clear that either process (i.e., mimicking or blocking steroid hormone receptor action) in excess will induce adverse effects. These disruptions are especially devastating in developing animals, as even transient alterations in hormone action during embryogenesis produce permanent and irreversible anomalies. Finally, while the magnitude of the particular effect clearly will be related to target concentration, the timing of the chemical insult will dictate which particular tissue will be affected.

A review of the literature indicates that rodent reproductive system anomalies and reductions in sperm numbers can result from perinatal exposure to environmental chemicals. As the basic events in reproductive development are similar in all mammalian species, it is logical to conclude that similar effects could occur in humans. In fact, perinatal exposure to environmental chemicals may help explain the reported decline in human sperm counts and increased incidence of hypospadias, crytorchidism, and testicular cancers. The fact that so many chemicals alter reproductive development via such a large variety of mechanisms raises the question of how to best detect the hazard to the developing reproductive system posed by chemicals with unknown activities. Currently, the only test protocols that expose animals during development and monitor the reproductive function of the offspring are the multigenerational reproductive tests (reviewed by PALMER 1981). At present, these protocols do not require either endocrine data or even "bioassay" measurements. More recently, updated multogenerational guidelines have been developed (MAKRIS 1995), which include indirect (bioassay) indices of endocrine function (pubertal landmarks, estrous cyclicity, reproductive organ weights, etc). However, these long-term tests are expensive and take more than 1 year to complete; clearly, short-term tests for endocrine disruptor activity are warranted. As no single in vivo or in vitro test can adequately screen a compound for all mechanisms, we propose that a combined in vivo/in vitro test strategy by developed to screen chemicals for effects on the reproductive system. The approach that we employed (KELCE et al. 1994,1995; GRAY et al. 1994) is an example of how to detect and pursue the mechanism of action of endocrine disruptors in short-term laboratory studies.

While we have individually discussed numerous drugs and environmental chemicals that induce adverse developmental effects by interacting with sex steroid hormone receptors, it should be clear that we are continuously exposed to many chemicals everyday. Some of these chemicals will have endocrine disrupting activity and some of them will not. We may never understand exactly how all these chemicals interact to produce additive or synergistic effects; however, we can control our exposure to those endocrine disrupting chemicals that have been identified. As we learn more about these chemicals and their mechanisms of action through continued research efforts, we will become able to make informed decisions regarding our growing concern for wildlife as well as human health.

References

Adams RR (1989) Phytoestrogens. In: Cheeke PR (ed) Toxicants of plant origin, vol IV: phenolics. CRC Press, Boca Raton, pp 23–51

Adamovic VM, Sokic B, Smiljanski M-J (1978) Some observations concerning the ratio of the intake of organochlorine insecticides through the food and amounts excreted in the milk of breast-feeding mothers. Bull Environ Contam Toxicol 20: 280–285

Ahel M, McEvoy J, Giger W (1993) Bioaccumulation of the lipophylic metabolites of nonionic surfactants in freshwater organisms. Environ Pollut 79: 243–248

Allan GF, Tsai SY, Tsai M-J, O'Malley BW (1992a) Ligand-dependent conformational changes in the progesterone receptor are necessary for events that follow DNA binding. Proc Natl Acad Sci USA 89: 11750–11754

Allan GF, Leng X, Tsai SY, Weigel NL, Edwards DP, Tsai M-J, O'Malley BW (1992b) Hormone and antihormone induce distinct conformational changes which are central to steroid receptor activation. J Biol Chem 267: 19513–19520
Andersen TI, Heimdal KR, Skrede M, Tveit K, Berg K, Borresen AL (1994) Oestrogen receptor (ESR) polymorphisms and breast cancer susceptibility. Hum Genet 94: 665–670
Auger J, Kunstmann JM, Czyglik F, Jouannet P (1995) Decline in semen quality among fertile men in Paris during the past 20 years. N Engl J Med 332: 281–285
Baranao JLS, Chemes HE, Tesone M, Chiauzzi EH, Scacchi P, Calvo JC, Faigon MR, Moguilevsky JA, Charreau EH, Calandra RS (1981) Effects of androgen treatment of the neonate on the rat testis and the sex accessory organs. Biol Reprod 25: 851–858
Bardin CW, Bullock LP, Sherins R, Mowszowicz I, Blackburn WR (1973) Androgen metabolism and mechanism of action in male pseudohermaphroditism: a study of testicular feminization. Rec Prog Horm Res 29: 65–109
Barquet A, Morgade C, Pfaffenberger CD (1981) Determination of organochlorine pesticides and metabolites in drinking water, human blood serum, and adipose tissue. J Toxicol Environ Health 7: 469–479
Battmann T, Bonfils A, Branche C, Humbert J, Goubet F, Teutsch G, Philibert D (1994) RU 58841, a new specific topical antiandrogen: a candidate of choice for the treatment of acne, androgenetic alopecia and hirsutism. J Steroid Biochem Mol Biol 48: 55–60
Battu RS, Singh PP, Joia BS, Kalra RL (1989) Contamination of bovine milk from indoor use of DDT and HCH in malaria control programes. Sci Total Environ 86: 281–287
Baulieu EE (1987) Steroid hormone antagonists at the receptor level: a role for the heat-shock protein MW 90,000 (hsp 90). J Cell Biochem 35: 161–174
Baulieu EE (1989) Contragestion and other clinical applications of RU486, an anti-progesterone at the receptor. Science 245: 1351–1357
Becker PB, Gloss B, Schmid W, Strähle U, Schütz G (1986) In vivo protein-DNA interactions in a glucocorticoid response element require the presence of hormone. Nature 324: 686–691
Benhamou B, Garcia T, Lerouge T, Vergezac A, Gofflo D, Bigogne C, Chambon P, Gronemeyer H (1992) A single amino acid that determines the sensitivity of progesterone receptors to RU486. Science 255: 206–209
Bentvelsen FM, McPhaul MJ, Wilson JD, George FW (1994) The androgen receptor of the urogenital tract of the fetal rat is regulated by androgen. Mol Cell Endocrinol 105: 21–26
Bern HA, Mills KT, Jones LA (1983) Critical period for neonatal estrogen exposure in occurence of mammary gland abnormalities in adult mice. Proc Soc Exp Biol Med 172: 239–242
Berns EMJJ, de Boer W, Mulder E (1986) Androgen-dependent growth regulation and release of specific proteins by the androgen receptor containing human prostate tumor cell line LNCaP. Prostate 9: 247–259
Berry MD, Metzger D, Chambon P (1990) Role of the two activating domains of the oestrogen receptor in the cell-type and promoter-context dependent agonistic activity of the anti-oestrogen 4-hydroxytamoxifen. EMBO J 9: 2811–2818
Blakemore PF, Im WB, Bleasdale JE (1994) The cell surface progesterone receptor which stimulates calcium influx in human sperm is unlike the A ring reduced steroid site on the GABA (A) receptor chloride channel. Mol Cell Endocrinol 104: 237–243
Blok LJ, Themmen APN, Peters AHFM, Trapman J, Baarends WM, Hoogerbrugge JW, Grootegoed JA (1992) Transcriptional regulation of androgen receptor gene expression in Sertoli cells and other cell types. Mol Cell Endocrinol 88: 153–164
Bouwman H, Reinecke AJ, Cooppan RM, Becker PJ (1990) Factors affecting levels of DDT and metabolites in human breast milk from Kwazulu. J Toxicol Environ Health 31, 93–115

Bouwman H, Cooppan RM, Becker PJ, Ngxongo S (1991) Malaria control and levels of DDT in serum of two populations in Kwazulu. J Toxicol Environ Health 33: 141–155

Brinkmann AO (1994) Steriod hormone receptors: activators of gene transcription. J Pediatr Endocrinol 7: 1–8

Brinkmann AO, Lindh LM, Breedveld DI, Mulder E, van der Molen HJ (1983) Cyproterone acetate prevents translocation of the androgen receptor in the rat prostate. Mol Cell Endocrinol 32: 117–129

Brann DW, Hendry LB, Mahesh VB (1995) Emerging diversities in the mechanism of action of steroid hormones. J Steroid Biochem Mol Biol 52: 113–133

Bulger WH, Muccitelli RM, Kupfer D (1978) Interactions of methoxychlor, methoxychlor base-soluble contaminent and 2,2-bis(p-hydroxyphenol)-1,1,1-trichloroethane with rat uterine estrogen receptor. J Toxicol Environ Health 4: 881–893

Bullock LF, Barthe PL, Mowszowicz I, Orth DN, Bardin CW (1975) The effect of progestins on submaxillary gland epidermal growth factor: demonstration of androgenic, synandrogenic and antiandrogenic actions. Endocrinology 97: 189–195

Burroughs CD, Bern HA, Stokstad EL (1985) Prolonged vaginal cornification and other changes in mice treated neonatally with coumestrol, a plant estrogen. J Toxicol Environ Health 15: 51–61

Carilla E, Briley M, Fauran F, Sultan CH, Duvilliers C (1984) Binding of Permixon, a new treatment for prostatic benign hyperplasia, to the cytosolic androgen receptor in the rat prostate. J Steroid Biochem Mol Biol 20: 521–523

Carlsen E, Giwercman A, Keiding N, Skakkebaek NE (1992) Evidence for decreasing quality of semen during past 50 years. BMJ 305: 609–613

Chamness GC, Bannayan GA, Landry LA Jr, Sheridan PJ, McGuire WL (1979) Abnormal reproductive development in rats after neonatally administered antiestrogen (tamoxifen). Biol Reprod 21: 1087–1090

Champault G, Patel JC, Bonnard AM (1984) A double-blind trial of an extract of the plant Serenoa repens in benign prostatic hyperplasia. Br J Clin Pharmacol 18: 461–462

Chander SK, Sahota SS, Evans TRJ, Luqmani YA (1993) The biological evaluation of novel antioestrogens for the treatment of breast cancer. Crit Rev Oncol Hematol 15: 243–269

Clark JH, McCormack S (1977) Clomid or nafoxidine administered to neonatal rats causes reproductive tract abnormalities. Science 197: 164–165

Curley A, Copeland MF, Kimbrough RD (1969) Chlorinated hydrocarbon insecticides in organs of stillborn and blood of newborn babies. Arch Environ Health 19: 628–632

Danielian PSA, White R, Hoare SA, Fawell SE, Parker MG (1993) Identification of residues in the estrogen receptor that confer differential sensitivity to estrogen and hydroxytamoxifen. Mol Endocrinol 7: 232–240

Danish Environmental Protection Agency (1995) Male reproductive health and environmental chemicals with estrogenic potential (miljøprojekt no 290). Ministry of the Environment and Energy, Copenhagen

Davis DL, Bradlow HL, Wolff M, Woodruff T, Hoel G, Anton-Culver H (1993) Medical Hypothesis: xenoestrogens as preventable causes of breast cancer. Environ Health Perspect 101: 372–377

Davis VL, Couse JF, Goulding EH, Power SGA, Eddy EM, Korack KS (1994) Aberrant reproductive phenotypes evident in transgenic mice expressing the wild-type mouse estrogen receptor. Endocrinology 135: 379–386

Dedhar S, Rennie PS, Shago M, Magesteijn C-YL, Yang H, Filmus J, Hawley RG, Bruchovsky N, Cheng H, Matusik RJ, Giguere V (1994) Inhibition of nuclear hormone receptor activity by calreticulin. Nature 367: 480–483

Dewailly E, Nantel A, Weber J, Meyer R (1989) High levels of PCBs in breast milk of Inuit women from arctic Quebec. Bull Environ Contam Toxicol 43: 641

Dohler KD, Coquelin A, Davies F, Hines M, Shryne JE, Sickmoller PN, Jarzab B, Gorski RA (1986) Pre-and postnatal influence of an estrogen antagonist on differentiation of the sexually dimorphic nucleus of the preoptic area of male and female rats. Neuroendocrinology 42: 443–448

Eliot JE, Martin PA, Arnold TW, Sinclair PH (1994) Organochlorines and reproductive success of birds in orchard and non-orchard areas of central British Columbia, Canada, 1990–91. Arch Environ Contam Toxicol 26: 435–443

Ennis BW, Davies J (1982) Reproductive tract abnormalities in rats treated neonatally with DES. Am J Anat 164: 145–154

Farnsworth NR, Bingel AS, Cordell GA, Crane FA and Fong HHS (1975) Potential value of plants as sources of new antifertility agents II. J Pharm Sci 64: 717–740

Freeman SN, Mainwaring WIP and Furr BJA (1989) A possible explanation for the peripheral selectivity of a novel non-steroidal pure antiandrogen, Casodex (ICI 176 334). Br J Cancer 60: 664–668

Freiss G, Prebois C, Rochefort H, Vignon F (1990) Anti-steroidal and anti-growth factor activities of anti-estrogens. J Steroid Biochem Molec Biol 37: 777–781

French FS, Lubahn DB, Brown TR, Simental JA, Quigley CA, Yarbrough WG, Tan J, Sar M, Joseph DR, Evans BAJ, Hughes IA, Migeon CJ and Wilson EM (1990) Molecular Basis of Androgen Insensitivity. Rec Prog Horm Res 46: 1–42

Fry DM and Toone TK (1981) DDT-induced feminization of gull embryos. Science 213: 922–924

Gellert RJ (1978a) Kepone, mirex, dieldrin and aldrin: estrogenic activity and the induction of persistent vaginal estrus and anovulation in rats following neonatal treatment. Environ Res 16: 131–138

Gellert RJ (1978b) Uterotrophic activity of polychlorinated biphenyls and induction of precocious reproductive aging in neonatally treated female rats. Environ Res 16: 123–130

Gentry PA (1986) Comparative biochemical changes associated with mycotoxicosis other than aflatoxicosis and trichothecene toxicosis. In: Richard JR, Thurston JR (eds) Diagnosis of mycotoxicoses. Nijhoff, Dordrecht

George FW and Wilson JD (1988) Sex determination and differentiation. In: Knobil E, Neill J (eds) The physiology of reproduction. Raven, New York

Gething MJ, Sambrook J (1992) Protein folding in the cell. Nature 355: 33–45

Giwercman A, Carlsen E, Keiding S, Skakkebaek NE (1993) Evidence for increasing incidence of abnormalities of the human testis: a review. Environ Health Perspect [Suppl] 101: 65–71

Gray LE Jr (1992) Chemical-induced alterations of sexual differentiation: a review of effects in humans and rodents. In: Colborn T, Clement C (eds) Chemically induced alterations in sexual and functional development: the wildlife/human connection. Princeton Scientific, Princeton

Gray LE Jr, Ferrell JM, Ostby JS (1985) Alteration of behavioral sex differentiation by exposure to estrogenic compounds during a critical neonatal period: effects of zearalenone, methoxychlor, and estradiol in hamsters. Toxicol Appl Pharmacol 80: 127–136

Gray LE Jr, Ostby J, Sigmonn R, Ferrell J, Rehnberg G, Linder R, Cooper R, Goldman J, Laskey J (1988) The development of a protocol to assess reproductive effects of toxicants in the rat. Reprod Toxicol 2: 281–287

Gray LE Jr, Ostby J, Ferrell J, Rehnberg G, Linder R, Cooper R, Goldman J, Slott V, Laskey J (1989) A dose- response analysis of methoxychlor-induced alterations of reproductive development and function in the rat. Fund Appl Toxicol 12: 92–108

Gray LE Jr, Ostby JS, Kelce WR (1994) Developmental effects of an environmental antiandrogen: the fungicide vinclozolin alters sex differentiation of the male rat. Toxicol Appl Pharmacol 129: 46–52

Greco TL, Furlow JD, Duello TM, Gorski J (1992) Immunodetection of estrogen receptors in fetal and neonatal male mouse reproductive tracts. Endocrinology 130: 421–429

Gronemeyer H, Benhamou B, Berry M, Bocquel MT, Gofflo D, Garcia T, Lerouge T, Metzger D, Meyer ME, Tora L, Vergezac A, Chambon P (1992) Mechanisms of antihormone action. J Steroid Biochem Mol Biol 41: 217–221

Grossman ME, Lindzey J, Blok L, Perry JE, Kumar MV, Tindall DJ (1994) The mouse androgen receptor gene contains a second functional promoter which is regulated by dihydrotestosterone. Biochemistry 33: 14594–14600

Guerrier D, Boussin L, Mader S, Josso N, Kahn A, Picard J-Y (1990) Expression of the gene for anti-müllerian hormone. J Reprod Fertil 88: 695–706

Guillette LJ, Gross TS, Masson GR, Matter JM, Percival HF, Woodward AR (1994) Developmental abnormalities of the gonad and abnormal sex hormone concentrations in juvenile alligators from contaminated and control lakes in Florida. Environ Health Perspect 102: 680–688

Guiochon-Mantel A, Lescop P, Christinmaitre S, Loosfelt H, Perrotapplanat M, Milgrom E (1991) Nucleocytoplasmic shuttling of the progesterone receptor. EMBO J 10: 3851–3859

Guiochon-Mantel A, Delabre K, Lescop P, Milgrom E (1994) Nuclear localization signals also mediate the outward movement of proteins from the nucleus. Proc Natl Acad Sci USA 91: 7179–7183

Harris MG, Coleman SG, Faulds D, Chrisp P (1993) Nilutamide: a review of its pharmacodynamic and pharmacokinetic properties and therapeutic efficacy in prostate cancer. Drugs Aging 3: 9–25

Heinrichs WL, Gellert RJ, Bakke JL, Lawrence NL (1971) DDT administered to neonatal rats induces persistent estrus syndrome. Science 173: 642–643

Hendry LB (1993) Drug design with a new type of molecular modeling based on stereochemical complementarity to gene structure. J Clin Pharmacol 33: 1173–1187

Hendry EC, Miller RK (1986) The antiestrogen LY117018 is estrogenic in the fetal rat. Teratology 34: 59–63

Horoszewisc JS, Leong SS, Kawinski E, Karr JP, Rosenthal H, Ming Chu T, Mirand EA, Murphy GP (1983) LNCaP model of human prostate carcinoma. Cancer Res 43: 1809–1818

Horwitz KB (1992) The molecular biology of RU486. Is there a role for antiprogestins in the treatment of breast cancer. Endocr Rev 13: 146–163

Huang JK, Bartsch W, and Voigt KD (1985) Interactions of an antiandrogen (cyproterone acetate) with the androgen receptor system and its biological action in the rat ventral prostate. Acta Endocrinol (Copenh) 109: 569–576

Hubbard WK (1994) International conference on harmonization; guideline for detection of toxicity to reproduction for medicinal products. Fed Reg 59: 48746–48762

Huseby RA, Thurlow S (1982) Effects of prenatal exposure of mice to "low-dose" diethylstilbestrol and the development of adenomyosis associated with the evidence of hyperprolactinemia. Am J Obstet Gynecol 144: 939–949

Hutson JM, Ikawa H, Donahoe PK (1982) Estrogen inhibition of Mullerian inhibiting substance in the chick embryo. J Pediatr Surg 17: 953–959

Imperato-McGinley J, Sanchez RS, Spencer JR, Yee B, Vaugan ED (1992) Comparison of the effects of the 5α-reductase inhibitor finasteride and the antiandrogen flutamide on prostate and genital differentiation: dose response studies. Endocrinology 131: 1149–1156

Jänne OA, Bardin CW (1984) Steroid receptors and hormone action: physiological and synthetic androgens and progestins can mediate inappropriate biological effects. Pharmacol Rev 36: 35S–42S

Jänne OA, Shan L-X (1991) Structure and function of the androgen receptor. Ann N Y Acad Sci 626: 81–91

Jensen EV, Suzuki T, Kawashima T, Stumpf WE, Jungblut PW, DeSombre ER (1968) A two-step mechanism for the interaction of estradiol with rat uterus. Proc Nat Acad Sci USA 59: 632–636

Johnson JI, Toft DO (1994) A novel chaperone complex for steroid receptors involving heat shock proteins, immunophilins and p23. J Biol Chem 269: 24989–24993

Jordan VC (1984) Biochemical pharmacology of antiestrogen action. Pharmacol Rev 36: 245–276

Jordan VC (1990) "The only true estrogen is no estrogen". Mol Cell Endocrinol 74: C91–C95

Jordan VC, Fritz NF, Tormey DC (1987) Endocrine effects of adjuvant chemotherapy and long-term tamoxifen administration on node-positive patients with breast cancer. Cancer Res 47: 624–630

Kalloo NB, Gearhart JP, Barrack ER (1993) Sexually dimorphic expression of estrogen receptors, but not of androgen receptors in human fetal external genitalia. J Clin Endocrinol Metab 77: 692–698

Katzenellenbogen BS, Fang H, Ince BA, Pakdel F, Reese JC, Wooge CH, Wrenn CK (1993) Estrogen receptors: ligand discrimination and antiestrogen action. Breast Cancer Res Treat 27: 17–26

Kelce WR, Monosson E, Gamcsik MP, Laws SC, Gray LE Jr (1994) Environmental hormone disruptors: evidence that vinclozolin developmental toxicity is mediated by antiandrogenic metabolites. Toxicol Appl Pharmacol 126: 276–285

Kelce WR, Stone CS, Laws SC, Gray LE, Kemppainen JA, Wilson EM (1995) Persistent DDT metabolite, p,p′-DDE is a potent androgen receptor antagonist. Nature 375: 581–585

Kemppainen JA, Lane MV, Sar M, Wilson EM (1992) Androgen receptor phosphorylation, turnover, nuclear transport, and transcriptional activation. J Biol Chem 267: 968–974

Klein-Hitpass L, Tsai SY, Weigel NL, Allan GF, Riley D, Rodriguez R, Schrader WT, Tsai M-J, O'Malley BW (1990) The progesterone receptor stimulates cell-free transcription by enhancing the formation of a stable preinitiation complex. Cell 60: 247–257

Korach KS (1994) Insights from the study of animals lacking functional estrogen receptor. Science 266: 1524–1527

Korach KS, Sarver P, Chae K, McLachlan JA, McKinney JD (1987) Estrogen receptor-binding activity of polychlorinated hydroxybiphenyls: conformationally restricted structural probes. Mol Pharmacol 33: 120–126

Kuil CW, Mulder E (1994) Mechanism of antiandrogen action: conformational changes of the receptor. Mol Cell Endocrinol 102: R1–R5

Kumagai S, Shimizu S (1982) Neonatal exposure to zearalenone causes persistent anovulatory estrus in the rat. Arch Toxicol 50: 270–286

Kumar MV, Leo ME, Tindall DJ (1994) Modulation of androgen receptor transcriptional activity by the estrogen receptor. J Androl 15: 534–542

Kupfer SR, Marschke KB, Wilson EM, French FS (1993) Receptor accessory factor enhances specific DNA binding of androgen and glucocorticoid receptors. J Biol Chem 268: 17519–17527

Labov JB (1977) Phytoestrogens and mammalian reproduction. Comp Biochem Physiol 57: 3–9

Landers JP, Spelsberg TC (1993) Updates and new models for steroid hormone action. Ann N Y Acad Sci 637: 26–55

Laws SC, Carey SA, Kelce WR (1995) Differential effects of environmental toxicants on steroid receptor binding. Toxicologist 15: 294

Lebeau M-C, Baulieu EE (1994) Steroid antagonists and receptor associated proteins. Hum Reprod 9: 437–444

Liehr JG, Avitts TA, Randerath E, Randerath K (1986) Estrogen-induced endogenous DNA adduction: possible mechanism of hormonal cancer. Proc Natl Acad Sci USA 83: 5301–5305

Lindzey J, Grossman M, Kumar MV, Tindall DJ (1993) Regulation of the 5′ -flanking region of the mouse androgen receptor gene by cAMP and androgen. Mol Endocrinol 7: 1530–1540

Love RR, Newcomb PA, Weibe DA, Surawicz TS, Jordan VC, Carbone PP, DeMets DL (1990) Effects of tamoxifen therapy on lipid and lipoprotein levels in postmenopausal women with node-negative breast cancer. J Natl Cancer Inst 82: 1327–1332

Love RR, Mazess RB, Barden HS, Epstein SE, Newcomb PA, Jordan VC, Carbone PP, DeMets DL (1992) Effects of tamoxifen on bone mineral density in postmenopausal women with breast caner. N Engl J Med 326: 852–856

Lubahn DB, Moyer JS, Golding TS, Couse JF, Korach KS, Smithies O (1993) Alteration of reproductive function but not prenatal sexual development after insertional disruption of the mouse estrogen receptor gene. Proc Natl Acad Sci USA 90: 11162–11166

Luke MC, Coffey DS (1994) The male sex accessory tissues. In: Knobil E, Neill JD, (eds) The physiology of reproduction. Raven, New York, pp 1435–1487

Makris S (1995) Proposed revisions to the testing guidelines for developmental toxicity and two-generation reproductive toxicity studies conducted under FIFRA and TSCA – a progress update. Toxicologist 15: 287

Mani SK, Allen JMC, Clark JH, Blaustein JD, O'Malley BW (1994) Convergent pathways for steroid hormone-and neurotransmitter-induced rat sexual behavior. Science 265: 1246–1249

Mauvais-Jarvis P, Kuttenn F, Gompel A (1987) Antiestrogen action of progesterone in breast tissue. Horm Res 28: 212–218

McDonnell DP, Clevenger B, Dana S, Santiso-Mere D, Tzukerman MT, Gleeson MAG (1993) The mechanism of action of steroid hormones: a new twist to an old tale. J Clin Pharmacol 33: 1165–1172

McDonnell DP, Clemm DL, Imhof MO (1994) Definition of the cellular mechanisms which distinguish between hormone and antihormone activated steroid receptors. Cancer Biol 5: 327–336

McLachlan JA (1981) Rodent models for perinatal exposure to diethylstilbestrol and their relation to human disease in the male. In: Herbst AL, Bern HA (eds) Developmental effects of diethylstilbestrol in pregnancy. Thieme-Stratton, New York, pp 48–157

McLachlan JA, Newbold RR, Shah HC, Hogan MD, Dixon RL (1982) Reduced fertility in female mice exposed transplacentally to diethylstilbestrol (DES). Fertil Steril 38: 364–371

Meyer M-E, Gronemeyer H, Turcotte B, Bocquel M-T, Tasset D, Chambon P (1989) Steroid hormone receptors compete for factors that mediate their enhancer function. Cell 57: 433–442

Mitsuhashi N, Mizuno M, Miyagawa A, Kato J (1988) Inhibitory effect of fatty acids on the binding of androgen receptor and R1881. Endocrinol Jpn 35: 93–96

Mizejewski GJ, Jacobson HI (1987) Biological activities of alpha-fetoprotein. CRC Press, Boca Raton

Mizokami A, Chang C (1994) Induction of translation by the 5′ -untranslated region of human androgen receptor mRNA. J Biol Chem 269: 25655–25659

Mowszowicz I, Bieber DE, Chung KW, Bullock LP, Bardin CW (1974) Synandrogenic and antiandrogenic effect of progestins: comparison with nonprogestational antiandrogens. Endocrinology 95: 1589–1599

Mueller G, Kim UH (1978) Displacement of estradiol from estrogen receptors by simple alkylphenols. Endocrinology 102: 1429–1435

Näär AM, Boutin J-M, Lipkin SM, Yu VC, Holloway JM, Glass CK, Rosenfeld MG (1991) The orientation and spacing of core DNA-binding motifs dictate selective transcriptional responses to three nuclear receptors. Cell 65: 1267–1279

National Research Council (1993) Pesticides in the diets of infants and children. National Academy Press, Washington

Neumann F (1977) Pharmacology and potential use of cyproterone acetate. Horm Metab Res 9: 1–13

Newbold RR, McLachlan JA (1985) Diethystilbestrol-associated defects in murine genital tract development. In: McLachlan JA (ed) Estrogens in the environment II: influences on development. Elsevier North-Holland, New York, pp 288–318

Newbold RR, Suzuki Y, McLachlan JA (1984) Müllerian duct maintenance in hetertypic organ culture after in vivo exposure to diethylstilbestrol. Endocrinology 115: 1863–1868

Newbold RR, Bullock BC, McLachlan JA (1987) Mullerian remnants of male mice exposed prenatally to diethystilbestrol. Teratog Carcinogen Mutagen 7: 377–389

O'Leary JA, Davies JE, Edmundson WF, Reich GA (1970) Transplacental passage of pesticides. Am J Obstet Gynecol 107: 65–68

O'Malley BW, Tsai M-J (1992) Molecular pathways of steroid receptor action. Biol Reprod 46: 163–167

Osweiler GD, Carson TL, Buck WB, Van Gelder GA (1985) Clinical and diagnostic veterinary toxicology. Kendall-Hunt, Dubuque

Pajunen AE, Isomaa VV, Janne AO, Bardin CW (1982) Androgenic regulation of ornithine decarboxylase activity in mouse kidney and its relationship to changes in cytosol and nuclear androgen receptor concentrations. J Biol Chem 257: 8190–8198

Palmer T (1981) Regulatory requirements for reproductive toxicology: theory and practice. In: Kimmel C, Buelke-Sam J (eds) Developmental toxicology. Raven, New York, pp 259–288

Parker MG (1993) Steroid hormone action. Oxford University Press, New York, pp 56–57

Pasqualini JR, Sumida C, Giambiagi N (1988) Pharmacodynamic and biological effects of anti-estrogens in different models. J Steroid Biochem 31: 613–643

Pearce D, Yamamoto KR (1993) Mineralocorticoid and glucocorticoid receptor activities distinguished by nonreceptor factors at a composite response element. Science 259: 1161–1165

Perez-Palacios G, Chavez B, Escobar N, Vilchis F, Larrea F, Lince M, Perez AE (1981) Mechanism of action of contraceptive synthetic progestins. J Steroid Biochem 15: 125–130

Phillips A, Hahn DW, McGuire JL (1992) Preclinical evaluation of norgestimate, a progestin with minimal androgenic activity. Am J Obstet Gynecol 167: 1191–1196

Pointis G, Latreille MT, Richard MO, Athis PD, Cedard PD (1984) Effect of maternal progesterone exposure on fetal testosterone in mice. Biol Neonate 45: 203–208

Polani PE (1981) Abnormal sex development in man. 1. Anomalies of sex determining mechanisms. In: Austin R, Edwards RG (eds) Mechanisms of sex differentiation in animals and man. Academic, New York, pp 465–547

Prahalada S, Carroad E, Cukierski M, Hendrick AG (1985) Embryotoxicity of a single dose of medroxyprogesterone acetate (MPA) and maternal serum MPA concentrations in cynomolgus monkeys (Macaca fascicularis). Teratology 32: 421–432

Ramkumar T, Adler S (1995) Differential positive and negative transcriptional regulation by tamoxifen. Endocrinology 136: 536–542

Raynaud JP, Bouton MM, Moguilewsky M, Ojasoo T, Philibert D, Beck G, Labrie F, Mornon JP (1980) Steroid hormone receptors and pharmacology. J Steroid Biochem 12: 143–157

Rogan WJ, Gladen BC, McKinney JD, Carreras N, Hardy P, Thullen J, Tingelstad J, Tully M (1986) Polychlorinated biphenyls (PCB's) and dichlorodiphenyl dichloroethane (DDE) in human milk: effects of maternal factors and previous lactation. Am J Public Health 76: 172–177

Salunkhe DK, Adsule RN, Bhonsle KI (1989) Antifertility agents of plant origin. In: Cheeke PR (ed) Toxicants of plant origin. IV. phenolics. CRC Press, Boca Raton, pp 53–81

Savouret JF, Rauch M, Redeuilh G, Sar S, Chauchereau A, Woodruff K, Parker MG, Milgrom E (1994) Interplay between estrogens, progestins, retinoic acid and AP-1 on a single regulatory site in the progesterone gene. J Biol Chem 269: 28955–28962

Schardein JL (1993) Hormones and hormone antagonists. In: Schardein JL (ed) Chemically induced birth defects 2nd edn. Dekker, New York, pp 271–339
Schreiber SL (1991) Chemistry and biology of the immunophilins and their immunosuppressive ligands. Science 251: 283–287
Schulz P, Bauer HW, Fittler F (1985) Steroid hormone regulation of prostatic acid phosphatase expression in cultured human prostate carcinoma cells. Biol Chem Hoppe Seyler 366: 1033–1039
Schutze N, Vollmer G, Wunsche W, Grote A, Feit B, Knuppen R (1994) Binding of 2-hydroxyestradiol and 4-hydroxyestradiol to the estrogen receptor of MCF-7 cells in cytosolic extracts and in nuclei of intact cells. Exp Clin Endocrinol 102: 399–408
Schuurmans ALG, Bolt J, Voorhorst MM, Blankenstein RA, Mulder E (1988) Regulation of growth and epidermal growth factor receptor levels of prostate tumor cells by different steroids. Int J Cancer 42: 917–922
Schuurmans ALG, Bolt J, Veldscholte J, Mulder E (1990) Stimulatory effects of antiandrogens on LNCaP human prostate tumor cell growth, EGF-receptor level and acid phosphatase secretion. J Steroid Biochem Mol Biol 37: 848–853
Sharpe RM, Skakkebaek NE (1993) Are estrogens involved in falling sperm counts and disorders of the male reproductive tract? Lancet 341: 1392–1395
Shi Q-X, Roldan ERS (1995) Evidence that a $GABA_A$-like receptor is involved in progesterone-induced acrosomal exocytosis in mouse spermatozoa. Biol Reprod 52: 373–381
Simental JA, Sar M, Lane MV, French FS, Wilson EM (1991) Transcriptional activation and nuclear targeting signals of the human androgen receptor. J Biol Chem 266: 510–518
Soto AM, Justicia H, Wray JW, Sonnenschein C (1991) p-Nonylphenol: an estrogenic xenobiotic released from "modified" polystyrene. Environ Health Perspect 92: 167–173
Spindler M (1983) DDT: health aspects in relation to man and risk/benefit assessment based thereupon. Residue Rev 90: 1–34
Spitzer PR, Risebrough R, Walker W, Hernandez R, Poole A, Puleston D, Nisbet IC (1978) Productivity of ospreys in Connecticut-Long Island increases as DDE residues decline. Science 202: 333–335
Stanczyk FZ (1994) Structure-function relationships, potency, and pharmacokinetics of progestogens. In: Lobo RA (ed) Treatment of postmenopausal women: basic and clinical aspects. Raven, New York, pp 69–89
Sweet RA, Schrott HG, Kurland R, Culp OS (1974) Study of the incidence of hypospadias in Rochester, Minn, 1940–1970, and a case-control comparision of possible etiologic factors. Mayo Clin Proc 49: 52–58
Szokolay A, Rosival L, Uhnak J, Madaric A (1977) Dynamics of benzene hexachloride (BHC) isomers and others chlorinated pesticides in the food chain and in human fat. Ecotoxicol Environ Safety 1: 349–359
Teutsch G, Goubert F, Battmann T, Bonfils A, Bouchoux F, Cerede E, Gofflo D, Gaillard-Kelly M, Philibert D (1994) New nonsteroidal antiandrogens: synthesis and biological profile of high affinity ligands for the androgen receptor. J Steroid Biochem Mol Biol 48: 111–119
Truss M, Bartsch J, Beato M (1994) Antiprogestins prevent progesterone receptor binding to hormone responsive elements in vivo. Proc Natl Acad Sci USA 91: 11333–11337
Umesono K, Murakami KK, Thompson CC, Evans RM (1991) Direct repeats as selective response elements for the thyroid hormone, retinoic acid, and vitamin D_3 receptors. Cell 65: 1255–1266
Van der Schoot P, Baumgarten R (1990) Effects of treatment of male and female rats in infancy with mefepristone on reproductive function in adulthood. J Reprod Fertil 90: 255–266
Vannier B, Raynaud JP (1980) Long-term effects of prenatal oestrogen treatment on genital morphology and reproductive function in the rat. J Reprod Fertil 59: 43–49

Van Look PFA, Bygdeman M (1989) Antiprogestational steroids: a new dimension in human fertility regulation. Oxford Rev Reprod Biol 11: 2–60

Veldscholte J, Berrevoets CA, Ris-Stalpers C, Kuiper GGJM, Jenster G, Trapman J, Brinkman AO, Mulder E (1992a) The androgen receptor in LNCaP cells contains a mutation in the ligand binding domain which affects steroid binding characteristics and response to antiandrogens. J Steroid Biochem Mol Biol 41: 665–669

Veldscholte J, Berrevoets CA, Brinkman AO, Grootegoed JA, Mulder E (1992b) Antiandrogens and the mutated androgen receptor of LNCaP cells: differential effects on binding activity, heat-shock protein interaction, and transcription activation. Biochemistry 31: 2393–2399

vom Saal FS, Montano MM, Wang MH (1992) Sex differentiation in mammals. In: Colborn T, Clement C (eds) Chemically-induced alterations in sexual and functional development: the wildlife/human connection. Princeton Scientific, Princeton

Wakeling AE (1992) Steroid antagonists as nuclear receptor blockers. Cancer Surv 14: 71–85

Wakeling AE, Bowler J (1991) Development of novel oestrogen-receptor antagonists. Biochem Soc Trans 19: 899–901

Wakeling AE, Furr BJA, Glen AT, Hughes LR (1981) Receptor binding and biological activity of steroidal and nonsteroidal antiandrogens. J Steroid Biochem 15: 355–359

White R, Jobling S, Hoare SA, Sumpter JP, Parker MG (1994) Environmentally persistent alkylphenolic compounds are estrogenic. Endocrinology 135: 175–182

Wilding G, Chen M, Gelman EP (1989) Aberrant response in vitro of hormone-responsive prostate cancer cells to antiandrogens. Prostate 14: 103–115

Wilson JG, Wilson HC (1943) Reproductive capacity in adult rats treated prepubertally with androgenic hormone. Endocrinology 33: 353–360

Wilson JD (1978) Sexual differentiation. Annu Rev Physiol 40: 279–306

Wong C-I, Kelce WR, Sar M, Wilson EM (in press) Molecular mechanism of the environmental antiandrogen vinclozolin: antagonist versus agonist activities of antiandrogens. J Biol Chem

Wrenn CK, Katzenellenbogen BS (1993) Structure-function analysis of the hormone binding domain of the human estrogen receptor by region-specific mutagenesis and phenotypic screening in yeast. J Biol Chem 268: 24089–24098

Wyllie J, Gabica J, Benson WW (1972) Comparative organochlorine pesticide residues in serum and biopsied lipoid tissue: a survey of 200 persons in southern Idaho-1970. Pestic Monit J 6: 84–88

Yen PM, Wilcox EC, Chin WW (1995) Steroid hormone receptors selectively affect transcriptional activation but not basal repression by thyroid hormone receptors. Endocrinology 136: 440–445

Zenick H, Clegg ED, Perreault SD, Klinefelter GR, Gray LE (1994) Assessment of male reproductive toxicity: a risk assessment approach. In: Hayes AW, (ed) Principle and methods of toxicology, 3rd edn. Raven, New York, pp 937–988

Zhou ZX, Wong C-I, Sar M, Wilson EM (1994) The androgen receptor: an overview. Rec Prog Horm Res 49: 249–274

Zhou ZX, Lane MV, Kemppainen JA, French FS, Wilson EM (1995) Specificity of ligand-dependent androgen receptor stabilization: receptor domain interactions influence ligand dissociation and receptor stability. Mol Endocrinol 9: 208–218

Subject Index

MIX
Papier aus verantwortungsvollen Quellen
Paper from responsible sources
FSC® C105338
FSC
www.fsc.org

If you have any concerns about our products,
you can contact us on
ProductSafety@springernature.com

In case Publisher is established outside the EU,
the EU authorized representative is:
Springer Nature Customer Service Center GmbH
Europaplatz 3, 69115 Heidelberg, Germany

Printed by Libri Plureos GmbH
in Hamburg, Germany